Advances in Intelligent Systems and Computing

Volume 808

The series "Advances in Intelligent Systems and Computing" contains publications on theory, applications, and design methods of Intelligent Systems and Intelligent Computing. Virtually all disciplines such as engineering, natural sciences, computer and information science, ICT, economics, business, e-commerce, environment, healthcare, life science are covered. The list of topics spans all the areas of modern intelligent systems and computing such as: computational intelligence, soft computing including neural networks, fuzzy systems, evolutionary computing and the fusion of these paradigms, social intelligence, ambient intelligence, computational neuroscience, artificial life, virtual worlds and society, cognitive science and systems, Perception and Vision, DNA and immune based systems, self-organizing and adaptive systems, e-Learning and teaching, human-centered and human-centric computing, recommender systems, intelligent control, robotics and mechatronics including human-machine teaming, knowledge-based paradigms, learning paradigms, machine ethics, intelligent data analysis, knowledge management, intelligent agents, intelligent decision making and support, intelligent network security, trust management, interactive entertainment, Web intelligence and multimedia.

The publications within "Advances in Intelligent Systems and Computing" are primarily proceedings of important conferences, symposia and congresses. They cover significant recent developments in the field, both of a foundational and applicable character. An important characteristic feature of the series is the short publication time and world-wide distribution. This permits a rapid and broad dissemination of research results.

More information about this series at http://www.springer.com/series/11156

Valentina Emilia Balas · Neha Sharma
Amlan Chakrabarti
Editors

Data Management, Analytics and Innovation

Proceedings of ICDMAI 2018, Volume 1

Editors
Valentina Emilia Balas
Department of Automatics and Applied Software
Aurel Vlaicu University of Arad
Arad, Romania

Amlan Chakrabarti
Faculty of Engineering and Technology
A. K. Choudhury School of Information Technology
Kolkata, West Bengal, India

Neha Sharma
Audyogik Tantra Shikshan Sanstha's
IICMR
Pune, Maharashtra, India

ISSN 2194-5357 ISSN 2194-5365 (electronic)
Advances in Intelligent Systems and Computing
ISBN 978-981-13-1401-8 ISBN 978-981-13-1402-5 (eBook)
https://doi.org/10.1007/978-981-13-1402-5

Library of Congress Control Number: 2018947293

This Springer imprint is published by the registered company Springer Nature Singapore Pte Ltd.
The registered company address is: 152 Beach Road, #21-01/04 Gateway East, Singapore 189721, Singapore

Preface

These two volumes constitute the Proceedings of the International Conference on Data Management, Analytics and Innovation (ICDMAI 2018) held from January 19 to 21, 2018, which was jointly organized by Computer Society of India (Div II), Computer Society of India (Pune Section) and Institute of Industrial and Computer Management and Research (IICMR), Pune. The conference was supported by industry leaders like TCS, IBM, Ellicium Solutions Pvt. Ltd, Omneslaw Pvt. Ltd., and premier academic universities like Savitribai Phule Pune University, Pune; Lincoln University, Malaysia; Defence Institute of Advanced Technology, Pune. The conference witnessed participants from 14 industries and 10 international universities from 14 countries. Utmost care was taken in each and every facet of the conference, especially regarding the quality of the paper submissions. Out of 488 papers submitted to ICDMAI 2018 from 133 institutions, only 76 papers (15.5%) were selected for oral presentation. Besides quality paper presentation, the conference also showcased four workshops, four tutorials, eight keynote sessions, and five plenary talks by the experts of the respective fields.

The volumes cover a broad spectrum of fields such as computer science, information technology, computational engineering, electronics and telecommunication, electrical, computer application, and all the relevant disciplines. The conference papers included in this proceedings, published post-conference, are grouped into four areas of research such as data management and smart informatics, big data management, artificial intelligence and data analytics, and advances in network technologies. All four tracks of the conference were much relevant to the current technological advancements and had the Best Paper Award in each of their respective tracks. Very stringent selection process was adopted for paper selection, from plagiarism check to technical chairs' review to double-blind review; every step was religiously followed. We are thankful to all the authors who have submitted papers for keeping the quality of the ICDMAI 2018 conference at high levels. The editors would like to acknowledge all the authors for their contributions and also the efforts taken by reviewers and session chairs of the conference, without whom it would have been difficult to select these papers. We have received important help from the members of the International Program Committee.

We appreciate the role of special sessions organizers. It was really interesting to hear the participants of the conference highlighting the new areas and the resulting challenges as well as opportunities. This conference has served as a vehicle for a spirited debate and discussion on many challenges that the world faces today.

We especially thank our Chief Mentor, Dr. Vijay Bhatkar, Chancellor, Nalanda University; General Chair, Dr. P. K. Sinha, Vice Chancellor and Director, Dr. S. P. Mukherjee International Institute of Information Technology, Naya Raipur (IIIT-NR), Chhattisgarh; other eminent personalities like Dr. Rajat Moona, Director, IIT Bhilai, and Ex-Director General, CDAC; Dr. Juergen Seitz, Head of Business Information Systems Department, Baden-Wuerttemberg Cooperative State University, Heidenheim, Germany; Dr. Valentina Balas, Professor, Aurel Vlaicu University of Arad, Romania; Mr. Aninda Bose, Senior Publishing Editor, Springer India Pvt. Ltd; Dr. Vincenzo Piuri, IEEE Fellow, University of Milano, Italy; Mr. Birjodh Tiwana, Staff Software Engineer, Linkedin; Dr. Jan Martinovic, Head of Advanced Data Analysis and Simulations Lab, IT4 Innovations National Supercomputing Centre of the Czech Republic, VŠB Technical University of Ostrava; Mr. Makarand Gadre, CTO, Hexanika, Ex-Chief Architect, Microsoft; Col. Inderjit Singh Barara, Technology Evangelist, Solution Architect and Mentor; Dr. Rajesh Arora, President and CEO, TQMS; Dr. Deepak Shikarpur, Technology Evangelist and Member, IT Board, AICTE; Dr. Satish Chand, Chair, IT Board, AICTE, and Professor, CSE, JNU; and many more who were associated with ICDMAI 2018. Besides, there was CSI-Startup and Entrepreneurship Award to felicitate budding job creators.

Our special thanks go to Janus Kacprzyk (Editor-in-Chief, Springer, Advances in Intelligent Systems and Computing Series) for the opportunity to organize this guest-edited volume. We are grateful to Springer, especially to Mr. Aninda Bose (Senior Publishing Editor, Springer India Pvt. Ltd) for the excellent collaboration, patience, and help during the evolvement of this volume.

We are confident that the volumes will provide state-of-the-art information to professors, researchers, practitioners, and graduate students in the area of data management, analytics, and innovation, and all will find this collection of papers inspiring and useful.

Arad, Romania — Valentina Emilia Balas
Pune, India — Neha Sharma
Kolkata, India — Amlan Chakrabarti

Organizing Committee Details

Dr. Deepali Sawai is a computer engineer and is the alumnus of Janana Probodhini Prashala, a school formed for gifted students. She has obtained Master of Computer Management and doctorate (Ph.D.) in the field of RFID Technology from Savitribai Phule Pune University. She is a certified Microsoft Technology Associate in Database and Software Development. She has worked in various IT industries/ organizations for more than 12 years in various capacities.

At present, her profession spheres over directorship with several institutions under the parent trust Audyogik Tantra Shikshan Sanstha (ATSS), Pune. At present, she is Professor and Founder Director, ATSS's Institute of Industrial and Computer Management & Research (IICMR), Nigdi, a Postgraduate, NAAC-accredited institute affiliated to Savitribai Phule Pune University, recognized by DTE Maharashtra conducting MCA programme approved by AICTE, New Delhi; Founder Director, City Pride School, a NABET accredited school, affiliated to CBSE, New Delhi; Founder Director, ATSS College of Business Studies and Computer Applications, Chinchwad, Graduate Degree College affiliated to Savitribai Phule Pune University. Along with this, she is also carrying the responsibility as Technical Director of the parent trust ATSS and the CMF College of Physiotherapy conducting BPth affiliated to MUHS, Nasik, and MPth affiliated to Savitribai Phule Pune University.

She has so far authored six books for computer education for school children from grade 1 to 9 and three books for undergraduate and postgraduate IT students. She has conducted management development programs for organizations like nationalized banks, hospitals, and industries. She has chaired national and international conferences as an expert. Her areas of interest include databases, analysis and design, big data, artificial intelligence, robotics, and IoT.

She has been awarded and appreciated by various organizations for her tangible and significant work in the educational field.

Dr. Neha Sharma is serving as Secretary of Society for Data Science, India. Prior to this, she has worked as Director, Zeal Institute of Business Administration, Computer Application and Research, Pune, Maharashtra, India; as Dy. Director, Padmashree Dr. D. Y. Patil Institute of Master of Computer Applications, Akurdi, Pune; and as Professor, IICMR, Pune. She is an alumnus of a premier College of Engineering affiliated to Orissa University of Agriculture and Technology, Bhubaneshwar. She has completed her Ph.D. from the prestigious Indian Institute of Technology (ISM), Dhanbad. She is Website and Newsletter Chair of IEEE Pune Section and served as Student Activity Committee Chair for IEEE Pune Section as well. She is an astute academician and has organized several national and international conferences and seminars. She has published several papers in reputed indexed journals, both at national and international levels. She is a well-known figure among the IT circles of Pune and well sought over for her sound knowledge and professional skills. She has been instrumental in integrating teaching with the current needs of the industry and steering the college to the present stature. Not only loved by her students, who currently are employed in reputed firms; for her passion to mingle freely with every one, she enjoys the support of her colleagues as well. She is the recipient of "**Best Ph.D. Thesis Award**" and "**Best Paper Presenter at International Conference Award**" at the national level by Computer Society of India. Her areas of interest include data mining, database design, analysis and design, artificial intelligence, big data, and cloud computing.

Acknowledgements

We, the editors of the book, Dr. Valentina Balas, Dr. Neha Sharma, and Dr. Amlan Chakrabarti, take this opportunity to express our heartfelt gratitude toward all those who have contributed to this book and supported us in one way or the other. This book incorporates the work of many people all over the globe. We are indebted to all those people who helped us in the making of this high-quality book which deals with state-of-the-art topics in the areas of data management, analysis and innovation.

At the outset, we would like to extend our deepest gratitude and appreciation to our affiliations: Dr. Valentina Balas to the Department of Automatics and Applied Software, Faculty of Engineering, University of Arad, Romania; Dr. Neha Sharma to IICMR, Nigdi of Savitribai Phule Pune University, India; and Dr. Amlan Chakrabarti to A. K. Choudhury School of IT, University of Calcutta, India, for providing all the necessary support throughout the process of book publishing. We are grateful to all the officers and staff members of our affiliated institutions who have always been very supportive and have always been companions as well as contributed graciously to the making of this book.

Our sincere appreciation goes to our entire family for their undying prayers, love, encouragement, and moral support and for being with us throughout this period, constantly encouraging us to work hard. "Thank You" for being our backbone during this journey of compilation and editing of this book.

About the Book

This book is divided into two volumes. This volume constitutes the Proceedings of the 2nd International Conference on Data Management, Analytics and Innovation 2018, or ICDMAI 2018, which was held from January 19 to 21, 2018, in Pune, India.

The aim of this conference was to bring together researchers, practitioners, and students to discuss the numerous fields of computer science, information technology, computational engineering, electronics and telecommunication, electrical, computer application, and all the relevant disciplines.

The International Program Committee selected top 76 papers out of 488 submitted papers to be published in these two book volumes. These publications capture promising research ideas and outcomes in the areas of data management and smart informatics, big data management, artificial intelligence and data analytics, advances in network technologies. We are sure that these contributions made by the authors will create a great impact in the field of computer and information science.

Contents

About the Editors

Prof. Dr. Valentina Emilia Balas is currently Professor at the Aurel Vlaicu University of Arad, Romania. She is the author of more than 270 research papers in refereed journals and international conferences. Her research interests are in intelligent systems, fuzzy control, soft computing, smart sensors, information fusion, modeling and simulation.

She is Editor-in-Chief of International Journal of Advanced Intelligence Paradigms (IJAIP) and International Journal of Computational Systems Engineering (IJCSysE); is Member in editorial boards for national and international journals; and serves as Reviewer for many international journals.

She is General Co-chair to seven editions of International Workshop Soft Computing Applications (SOFA) starting from 2005. She was the editor for more than 25 books in Springer and Elsevier. She is Series Editor for the work entitled Elsevier Biomedical Engineering from October 2017.

She participated in many international conferences as General Chair, Organizer, Session Chair, and Member in International Program Committee. She was Vice President (Awards) of International Fuzzy Systems Association (IFSA) Council (2013–2015), responsible with recruiting to European Society for Fuzzy logic and Technology (EUSFLAT) (2011–2013), Senior Member of IEEE, Member in Technical Committees to Fuzzy Sets and Systems and Emergent Technologies to IEEE CIS, and Member in Technical Committee to Soft Computing to IEEE SMC.

Prof. Dr. Neha Sharma is serving as a Secretary of Society for Data Science, India. Prior to this, she has worked as Director, Zeal Institute of Business Administration, Computer Application and Research, Pune, Maharashtra, India; as Dy. Director, Padmashree Dr. D. Y. Patil Institute of Master of Computer Applications, Akurdi, Pune; and as Professor, IICMR, Pune. She is an alumnus of a premier College of Engineering affiliated to Orissa University of Agriculture and Technology, Bhubaneshwar. She has completed her Ph.D. from prestigious Indian Institute of Technology (ISM), Dhanbad. She is Website and Newsletter Chair of IEEE Pune Section and served as Student Activity Committee Chair for IEEE Pune Section as well. She is an astute academician and has organized several national and international conferences and seminars. She has published several papers in reputed indexed journals, both at national and international levels. She is a well-known figure among the IT circles of Pune and well sought over for her sound knowledge and professional skills. She has been instrumental in integrating teaching with the current needs of the industry and steering the college to the present stature. Not only loved by her students, who currently are employed in reputed firms; for her passion to mingle freely with every one, she enjoys the support of her colleagues as well. She is the recipient of "**Best Ph.D. Thesis Award**" and "**Best Paper Presenter at International Conference Award**" at the national level by Computer Society of India. Her areas of interest include data mining, database design, analysis and design, artificial intelligence, big data, and cloud computing.

Prof. Dr. Amlan Chakrabarti is *ACM Distinguished Speaker*, who is presently the *Dean Faculty of Engineering and Technology* and Director of the A. K. Choudhury School of Information Technology, *University of Calcutta*. He obtained M.Tech. from University of Calcutta and did his *doctoral research at the Indian Statistical Institute, Kolkata*. He was *Postdoctoral Fellow at the School of Engineering, Princeton University*, USA, during 2011–2012. He is the recipient of *DST BOYSCAST Fellowship Award* in the area of engineering science in 2011, *Indian National Science Academy Visiting Scientist Fellowship* in 2014, *JSPS Invitation Research Award* from Japan in 2016, *Erasmus Mundus Leaders Award* from European Union in 2017, and *Hamied Visiting Fellowship of the University of Cambridge* in 2018. He is Team Leader of European Center for Research in Nuclear Science (CERN, Geneva) ALICE-India project for University of Calcutta and also a key member of the CBM-FAIR project at Darmstadt, Germany. He is also Principal Investigator of the Center of Excellence in Systems Biology and Biomedical Engineering, University of Calcutta, funded by MHRD (TEQIP-II).

He has published around 120 research papers in refereed journals and conferences. He has been involved in research projects *funded by DRDO, DST, DAE, DeITy, UGC, Ministry of Social Empowerment, TCS, and TEQIP-II*. He is Senior Member of IEEE, Secretary of IEEE CEDA—India Chapter, and Senior Member of ACM. His research interests are quantum computing, VLSI design, embedded system design, computer vision and analytics.

Part I
Data Management and Smart Informatics

Improved Rotated Local Binary Pattern

Divya Khare, D. R. Gangodkar and Saurabh Dwivedi

Abstract Content-based image retrieval is the implementation of computer vision techniques to image retrieval problem, that is, issue of looking for images taking from high-end cameras in large image dataset. It aims to finding pictures of interest from an extensive picture database using the visual content of the pictures. "Content-based" implies that the hunt will break down the genuine content of the picture instead of the metadata, for example, tags, descriptions, and keywords linked with the picture. The term "content" in this context may point to shapes, hues, surfaces, or some other data that can be derived from the picture itself. Graphical processing unit is helpful in most picture handling applications because of multithread execution of algorithms, programmability and minimal effort. The substantial quantities of pictures have increased challenges to computer to store and oversee information adequately and productively. Features can be extracted parallel from the pictures with graphical processing unit utilizing different procedures. Utilizing Graphical processing unit based feature extraction algorithm can perform feature extraction easily and in fast and efficient way. The NVIDIA CUDA is fundamentally new computing architecture technology that enables the graphical processing unit to solve difficult, time taking, complex problems. This paper aims to find out the similarity between images that is query image and image present in dataset. The paper aims at the performance of various content-based image retrieval algorithms, which include, local binary pattern and rotated local binary pattern. An improvement has been made in Rotated local binary pattern where mapping function has been incorporated so as to give better results, with the help of mapping we are able to sort the higher and lower priority pixel values of query image and images present in our database.

D. Khare (✉) · D. R. Gangodkar · S. Dwivedi
Graphic Era University, Dehradun, India
e-mail: divya.khare216@gmail.com

D. R. Gangodkar
e-mail: dgangodkar@yahoo.com

S. Dwivedi
e-mail: saurabh.dwivedi11@gmail.com

V. E. Balas et al. (eds.), *Data Management, Analytics and Innovation*,
Advances in Intelligent Systems and Computing 808,
https://doi.org/10.1007/978-981-13-1402-5_1

Keywords CBIR · GPU · Improved RLBP · LBP · RLBP · SSIM

1 Introduction

Content-based image retrieval is a process of computer vision algorithm usage to solve image retrieval problem, that is, issue of looking for digital images in large image dataset [1]. An image have two types of features, one is high level feature another is low-level feature, high-level features are those features by which we classify object in real life for example emotions, background. The low-level features are those objective values based on pixel values for example color, contrast, and illumination. There are also some basic features of an image on the basis of those features CBIR is performed; those features are technical quality, blur, color, composition, face detection, and spatial envelope. Technical quality deals with things such as sharpness, contrast and noise. These are used to measure the objective quality of the image, and identify mistakes for example pictures which are out of focus, blur manages components to quantify fogginess in a picture, color incorporates features to analyze the hue and saturation of a picture, composition removes the auxiliary data about the shape in a image, including measures of picture symmetry and picture arrangement, face detection is as clear as crystal. Individuals may incline towards photographs having faces rather than the one without individuals in them. Spatial envelope is a gathering of a few features that attempt to group scene types, in view of these incorporate "instinctive nature", which measures the distribution of edge orientations: predominantly horizontal and vertical is less natural than an even mix, and roughness, which measures the general complexity of the picture, among others. CBIR of image is based on similarities in their contents, i.e., surfaces, hues, shapes and so on, viewed as the lower level elements of a picture [2]. These ordinary methodologies for image retrieval depend on the calculation of the similarity query and images. It aims to finding pictures of interest from an extensive picture database using the visual content of the pictures [3]. "Content-based" implies that the hunt will break down the genuine content of the picture instead of the metadata, like tags, keywords, etc., associated with the picture [4]. The term "content" in this context may refer to hues, shapes, surfaces, or some other data that can be derived from the picture itself. The objective of the work is to find similarity among images performing content-based image retrieval using graphical processing unit. The dataset comprises of large number of images belonging to various class, for example, class labels such as Indoor, Low Resolution, Inverted, Noisy, Outdoor, etc. The dataset used is WANG dataset [5], which is a subset of Corel dataset, and the work has been carried out using tool MATLAB. An improvement has been made in Rotated local binary pattern where mapping function has been incorporated so as to give better results, with the help of mapping we are able to sort the higher and lower priority pixel values of query image and images present in our database. Higher priority values are those, which are present in query image as well as in images present in our database.

Lower priority values are those, which not present in query image when we compare it with images present in our database. Further give higher priorities values to structural similarity index matching (SSIM) to perform CBIR.

2 Related Work

In CBIR many authors have worked in the field feature extraction and similarity matching of images. Zhang et al. [6] discussed about the performance of CBIR using saliency detection. The paper focuses on the CBIR system based on SIFT and region segmentation to identify CBIR. Zhang et al. [6] proposes a novel higher order local pattern descriptor, local derivative pattern. LDP is general framework to encode directional pattern features based on local derivative variations. The limitation of local binary pattern is, it does not capture the more detailed information as compared to Local derivative pattern. LBP only encode the relationship between the central point and its neighbors [6]. The LDP encodes various distinctive spatial relationships contained in a given local region by extracting higher order local information. Mehta et al. [7] talked about around two novel rotation invariant texture descriptors. They are based on LBP, which is best and every now and again utilized texture descriptors. Despite the fact that LBP effectively catches the local structure, it is not rotation invariant. The concept of dominant direction has been used, where circular neighborhood and descriptor is figured regarding it. The weights related with the pixels in its neighborhood are shifted circularly taking into account the dominant direction as well. The techniques proposed by author are checked for the class of texture classification and the execution is contrasted and the first LBP and its existed expansions. The various authors have worked on features of images. Features are important, for it helps us to extract information from an image. The authors have discussed about various features and how they impact the user perspective for a particular image. The work on content-based image retrieval and image descriptor like LBP, LDP, RLBP has been done by authors in [8–11]. There are limitations in the previous work done like, rotated local binary pattern (RLBP) works for rotated images as well as for normal images but it fails to remove if there is any noise present in the image. It reads that noise as a pixel value and fails to give accurate results.

3 Background

Local binary pattern just considers the signs of the differences to figure the final descriptor. The information related to the magnitude of the differences is totally disregarded [6]. The magnitude gives complimentary information that has been used to build the discriminative power of the operator. Local binary pattern (LBP) does not work for rotated images. It only works for normal images. If user

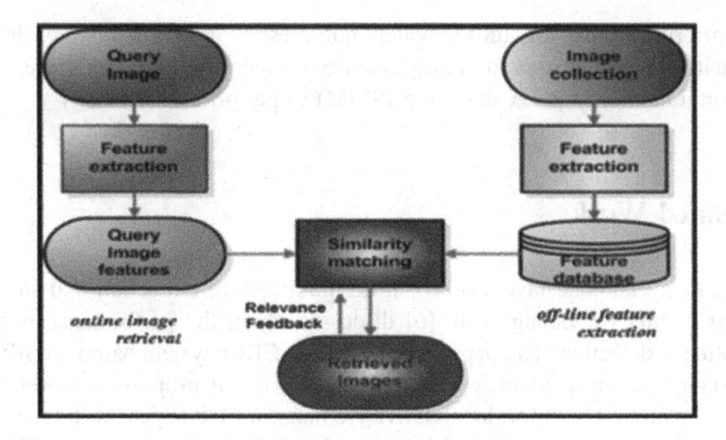

Fig. 1 Typical architecture of CBIR

provides a query, which is rotated, in that case it fails to find out the angle on which the image is rotated. It only works for the images whose angle is same for both query image and the image present in database. Rotated local binary pattern (RLBP) works for rotated images as well as for normal images but it fails to remove if there is any noise present in the image. It reads that noise as a pixel value and fails to give accurate results. It also fails to store the pixel values permanently it only do temporary buffering as long as process continues once pixels values are converted in the form of 0 and 1 it stores the weight associated with pixel value. Due to this limitation in RLBP there is scope of improvement in it. The improved version of RLBP is further discussed.

CBIR frameworks have turned into a dependable tool for many picture database applications and they were utilized as a part of different fields and spheres of human activity. There is a developing enthusiasm for CBIR frameworks due to the constraints inherent in metadata-based frameworks, and in addition the vast scope of conceivable uses for proficient image retrieval frameworks. The CBIR innovation has been utilized as a part of a plenty of uses, for example, unique fingerprint identification, advanced libraries, crime prevention, medicinal diagnosis, historical research, structural and building design, publishing and advertising, art, education, geographical information and remote sensing systems (Fig. 1).

Various applications of CBIR system

(a) **Medical Applications**—The utilization of content-based retrieval can bring about services that can profit biomedical systems. Three extensive areas can right away exploit CBIR strategies: educating, diagnostics, and research. Clinicians typically utilize comparative cases for case-based thinking in their decision-making process. In the medical field, a few diseases require the medicinal professional to pursuit and survey comparable X-rays or examined pictures of a patient before giving an answer.

(b) **Digital Libraries**—There are a few advanced libraries that give services based on image. One case is the computerized exhibition hall of butterflies, went for building an advanced accumulation of Taiwanese butterfly. This advanced library incorporates a module responsible for CBIR based on color, texture, and patterns.
(c) **Crime Prevention**—One of the main jobs of police is to distinguish and capture offenders. Nonetheless, to accomplish that, the bureau of security examination must distinguish the personality of crooks as fast as possible and with a high accuracy rate. For a long time, the crime rate is expanding so that the police must manage countless pictures that stored in a database. Once another picture is arrived, it must be contrasted with these pictures with group it accurately. Unmistakably, doing this job physically takes quite a while in this way; the requirement for criminal acknowledgment framework is firmly highlighted here.
(d) **Web Searching**—The most important application, in any case, is the Web, as a major part of it is dedicated to pictures, and looking for a particular picture in reality is an overwhelming task. Various business and exploratory CBIR frameworks are currently accessible, and many web searches are presently furnished with CBIR offices. Today it is evaluated that there are 30 billion pictures in Imageshack, Facebook holds 35 billion photographs.

4 Proposed Work

Rotated local binary pattern (RLBP) works for rotated images as well as for normal images, it select center of image surrounded by eight neighbors and find out dominant direction and shift pixel corresponding towards that direction. There is weight associated with each pixel value and to find out the total value the weights where pixel values are greater than center are added together. RLBP only considers the weight associated with the pixels and makes use of it during operation. It only stores the set of values, which comes out by using dominant direction [12]. It does not perform uniform bit shift, which is very helpful in removing any kind noise present in the image. Bit shift provide improvement in matching. If there is any noise present in an image, it reads that noise as a pixel value and fails to give more accurate results.

The main idea behind proposing improved RLBP is to perform bit shift so that any noise, which is present in image, should be removed and the results, which is obtained, should be more accurate. We also incorporated mapping values in our work. Color mapping is a function that maps the colors of one image to the colors of another (target) image. We indexed the pixel values of query images as well as images of our database and store those values. We maintain a array where we store the list of images. After storing those values we find out which pixel values are present in both query image and images in our dataset. We shuffle the image values, which have higher priority and low priority. After shuffling we pass the higher

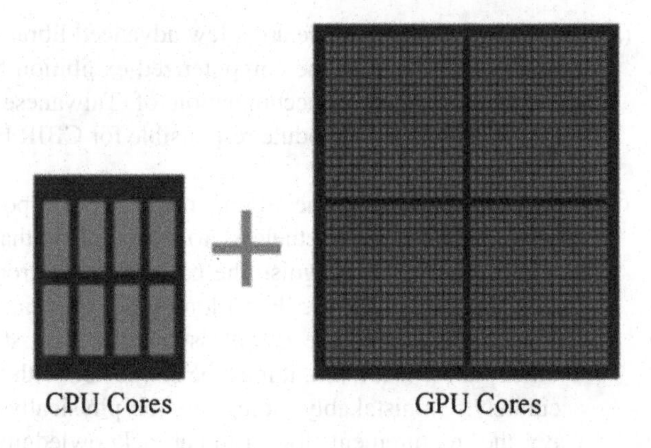

Fig. 2 Smaller number of CPU versus larger number of GPU cores

priority values to structural similarity index matching function to find out the best possible similarity in images. The idea is to implement RLBP with five parameters as

$$\text{IRLBP} = \text{RLBP}\ (I, R, N, M, H),$$

where, I is the input image, R is the radius, N is the neighbor pixels, M is Mapping Function and H is the Histogram. Incorporating mapping function gives better results as it works well for rotated images also as discussed. The SSIM value, i.e., the structural similarity index matching also comes out to be better (Fig. 2).

GPU array function is used to transfer an array from MATLAB to the GPU. In our work we are actually transferring the image from our code to GPU. Many MATLAB built in function support GPU array input argument, whenever any of these functions is called with at least one GPU array as an input argument, the function executes on the GPU and generates a GPU array as the result [1]. The function cat(3,imadjust) is used for increasing the contrast in the image for accuracy. "timeit()" function is used for CPU time, which measures the typical time(in seconds) required to run the function specified by the function handle. GPU "timeit()" is preferable to "timeit()" for function that use the GPU, because it ensure that all options on the GPU have finished before recording the time and compensates for the overhead. For operations that do not use a GPUtimeit() offers greater precision. After having all time intervals the total is computed and the results are processed.

Steps involved in proposed work

- First we input query image.
- Resize the resolution of our query image same as the resolution of images present in our database using MATLAB built in function.
- Find out the center of image and neighbors surrounding it (Fig. 3).
- In this matrix center and neighbors are calculated by dividing 5 rows by 2 and 5 columns by 2. We obtain the pixel 2.5 as our center.
- Find the angle of query image using

Fig. 3 Arrangement of center and pixel values

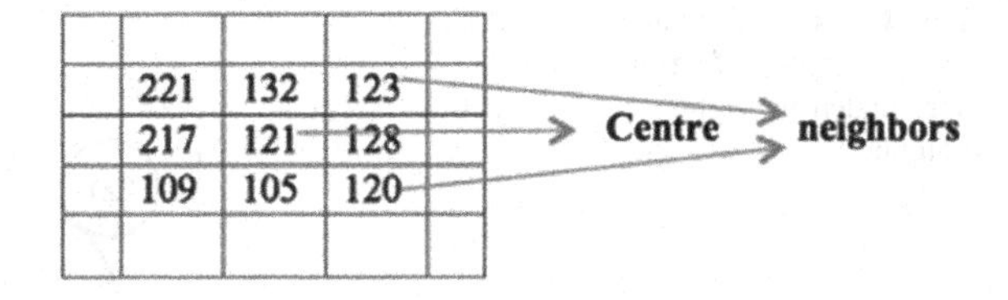

$$a = 2 * \pi/\text{neighbors},$$

where neighbors are the pixels surrounded by center pixel and spoint is an array.

$$\text{spoints}(i, 1) = -\text{radius} * \sin((i - 1) * a);$$
$$\text{spoints}(i, 2) = \text{radius} * \cos((i - 1) * a);$$

- Find the dominant direction (*D*) of an image. The neighbor whose difference is maximum with center pixel is marked as dominant direction [7].

$$D = \arg\max|\text{gp}-\text{gc}|$$

gp and gc are values of center pixel and neighbors respectively. Store all the dominant directions values of images present in an our database in an array *D*[*i*].

- Matching continued further by extracting features of both query image and image present in database using MATLAB function built in (Extractfeature), and matching the pixel values of query image and image present in database.
- To achieve more accurate results we included the concept of mapping, where we assign index number to each pixel value of our image and use function *imfilter* to separate the higher and lower priority values. Higher priority values are those, which are same in query image and image present in our database. Store those values in an array and further give those values to SSIM (Structural similarity index matching) to achieve more accurate results. This is how indexing performed as shown in figure store (Fig. 4).

5 Implementation and Results

The work has been carried out using the tool MATLAB. Utilizing MATLAB for GPU computing gives you a chance to accelerate your applications with GPUs more effectively than by utilizing C or Fortran. With the recognizable MATLAB language you can exploit the CUDA GPU computing technology without learning the complexities of GPU structures or low-level GPU computing libraries.

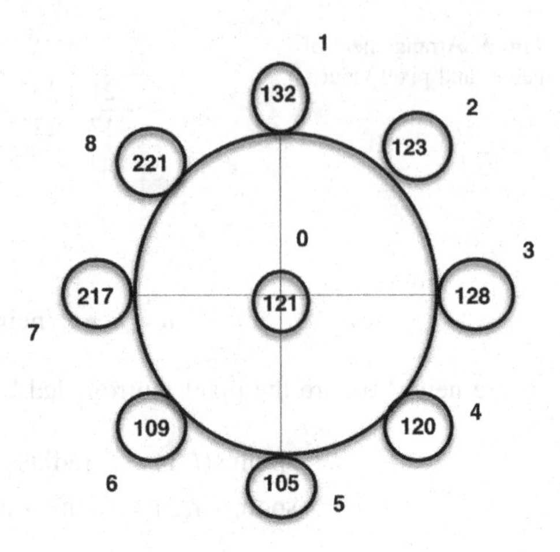

Fig. 4 Pixel values and index numbers of an image surrounded by eight neighbors

In the flow diagram, Fig. 5, comparison of rotated local binary pattern using three parameters and rotated local binary pattern using five parameters is shown. First we input our query image for RLBP using three parameters those are (*I*, *R*, *N*).

In case of five parameter, that is (*I*, *R*, *N*, *M*, *H*), *I*, *R*, *N* stands same as discussed, additional *M* and *H* stands for mapping parameter and histogram respectively. We performed this task on mixed dataset, which is made up of different classes of images. For three parameters, we obtain the structural similarity index matching less as compared with five parameters. For three parameters, SSIM value is 3.9279 and for five parameters, SSIM value is 4.2191. Further shown the improvement of 7.41% as compared to RLBP for three parameters.

In Fig. (6), comparison of matching results is shown for rotated local binary pattern for three parameters and five parameters for mixed classes of images. The blue bar shows SSIM value for RLBP using three parameters and red bar shows SSIM values for RLBP using five parameters. The values shown in vertical line of bar graph shows structural similarity index matching values. Figure represent that matching percentage for RLBP using five parameters is better as compared to RLBP using three parameter.

In Fig. (7), comparison of matching results is shown for rotated local binary pattern for three parameters and five parameters for same class of images. The blue bar shows SSIM value for RLBP using three parameters and red bar shows SSIM values for RLBP using five parameters. The values shown in vertical line of bar graph shows structural similarity index matching values. Figure represent that matching percentage for RLBP using five parameters is better as compared to RLBP using three parameter.

In Fig. (8), comparison of execution time of CPU and GPU is shown for local binary pattern. The blue bar shows the time taken by CPU to compute set of images in seconds and red bar shows time taken by GPU to compute set of images in seconds.

The values shown in vertical line of bar shows, the execution time in seconds.

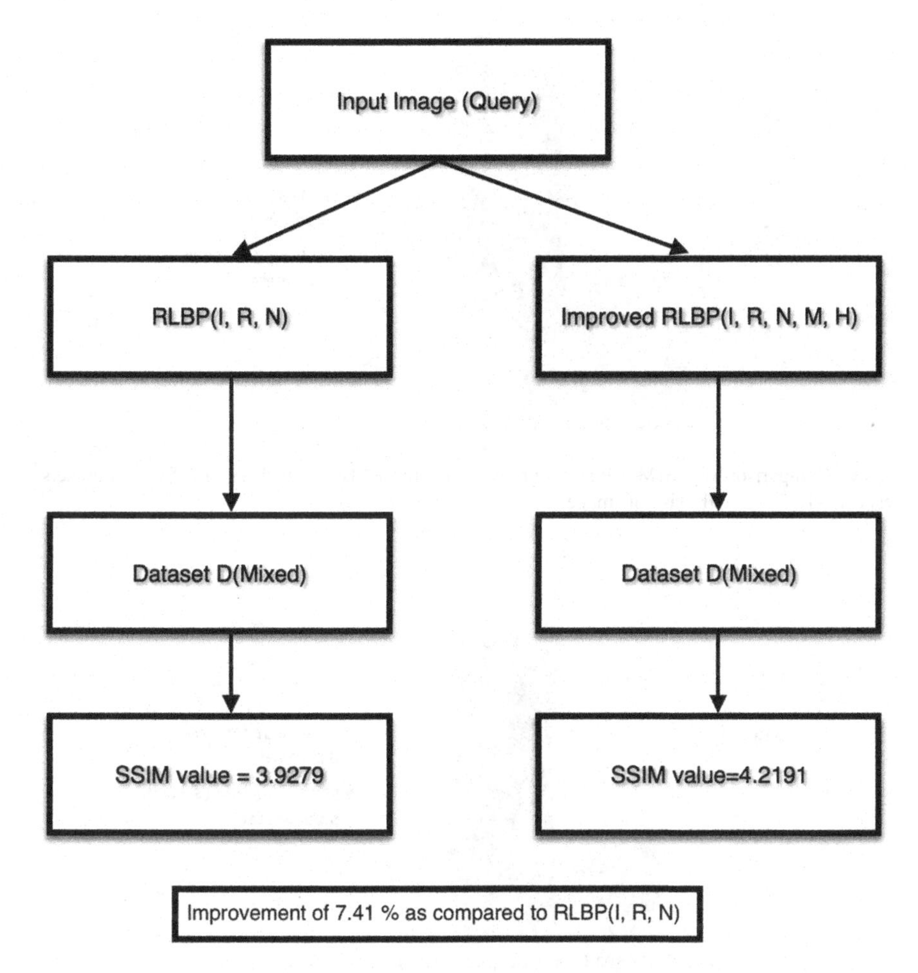

Fig. 5 Flow diagram of result obtained

In Fig. (9), comparison of execution time of CPU and GPU is shown for RLBP using three parameter, RLBP using five parameters and RLBP using five parameters with GPU. The blue bar shows the time taken by GPU to compute set of images in seconds for RLBP using five parameter, purple bar shows time taken by CPU to compute set of images in seconds for RLBP using five parameters and green bar shows the time taken by CPU to compute set of images in seconds for RLBP using three parameter.

The values shown in vertical line of bar shows, the execution time in seconds. Figure represents that GPU compute the set of images and gives results more frequently as compared to CPU.

In Fig. (10), the best five matches obtained based on local binary pattern is shown, we give a query image and obtain best five matches the drawback of this result is that we obtain our query image in result also but it is on second number.

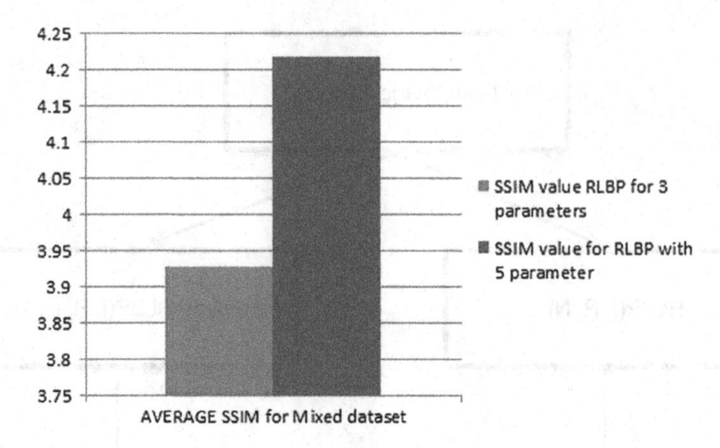

Fig. 6 Comparison of SSIM value results obtained for RLBP with three and five parameters respectively for mixed class of image

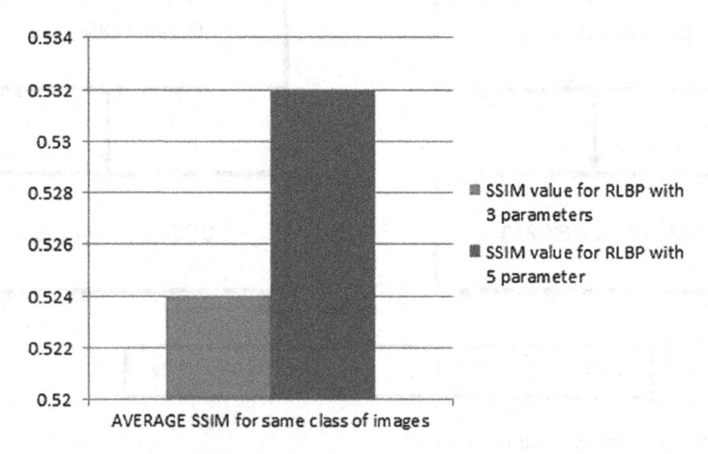

Fig. 7 Comparison of SSIM value results obtained for RLBP with three and five parameters respectively for same class of images

In Fig. (11), the best five matches obtained based on rotated local binary pattern using three parameters is shown, we give a query image and obtain best five matches the drawback of this result is that we are not able to get our query image and matches are also not way too similar with our query.

In Fig. (12), the best five matches obtained based on rotated local binary pattern using five parameters is shown, we give the same query image as we given in RLBP using three parameters and obtain best five matches. Here we obtain that the similarity index is more as compared with RLBP using three parameters and LBP. We obtain five best matches which are approximately 90% similar with our query image. So we concluded that rotated local binary pattern using five

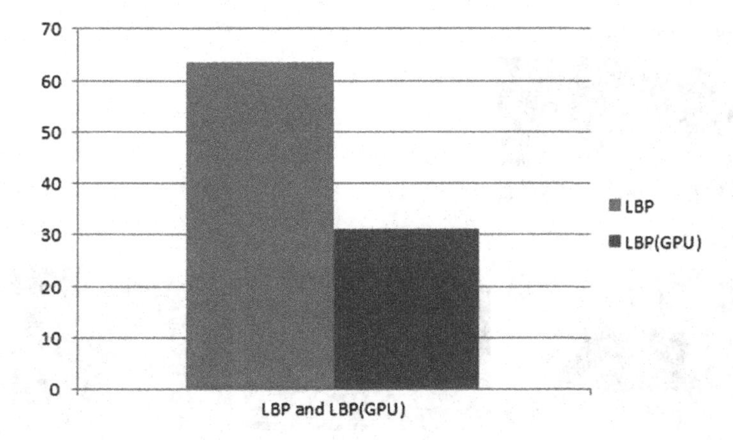

Fig. 8 Comparison of execution time for LBP using CPU and GPU respectively

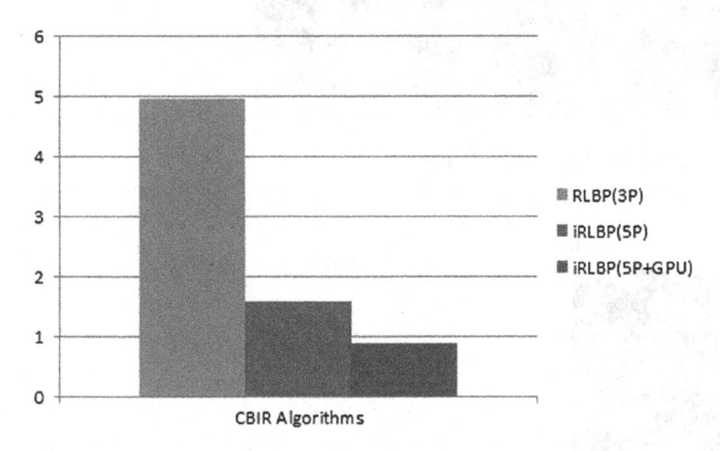

Fig. 9 Comparison of execution time by RLBP with three parameters, RLBP with five parameters using CPU and RLBP with five parameters using GPU

parameters give better matching results as compared with rotated local binary pattern using three parameters and with local binary pattern.

We have used two datasets to perform our work, the WANG dataset [5], the WANG database is a subset of 1,000 pictures of the Corel stock photograph database which have been physically chosen and which frame 10 classes of 100 pictures each. The WANG database can be considered similar to common stock photo retrieval tasks with a few pictures from every class and a potential user having a picture from a specific classification and searching for comparative pictures which not been utilized by other media.

The 10 different classes of images are utilized for importance estimation: given a question picture, it is accepted that the client is looking for pictures from a similar

Query image

Best five match

Fig. 10 Best five matches based on LBP

QUERY IMAGE:

Best five match:

Fig. 11 Best five matches based on RLBP

Query Image:

Best Five Matches:

Fig. 12 Best five matches based on improved RLBP

class, and in this manner, the rest of the 99 pictures from a similar class are viewed as applicable and the pictures from every single different class are viewed as immaterial. Stimuli dataset [5] has 17 classes, each class contain 40 images (resolution 384 × 256). We have mixed the images of all the 17 classes present and prepared a new dataset called as *mixed*. The work is performed on both dataset.

The Table 1 below, shows the result obtained on various classes, where image size is 384 × 256, the table shows time taken by LBP without using GPU, time taken by LBP with using GPU, time taken by RLBP without using GPU, time taken by improved RLBP without using GPU, time taken by improved RLBP with using GPU.

Class Line drawing shows best matching result for all the algorithms and class Fractal shows the worst matching results for all the algorithms. The work has been carried out using the tool MATLAB.

Utilizing MATLAB for GPU computing gives a chance to accelerate applications with GPUs more effectively than by utilizing C or Fortran. With the recognizable MATLAB language we can exploit the CUDA GPU computing technology without learning the complexities of GPU structures or low-level GPU computing libraries.

Table 1 Result obtained on various class of images

Name of classes	Size of image	Time taken by LBP without using GPU	Time taken by LBP with using GPU	Time taken by RLBP without using GPU	Time taken by IRLBP without using GPU	Time taken by IRLBP with using GPU
Action	384 × 256	56.52	28.16	5.68	1.12	0.68
Affective	384 × 256	66.81	31.41	5.94	1.08	0.74
Art	384 × 256	66.24	32.51	5.23	1.16	0.84
Black white	384 × 256	64.41	34.71	5.29	1.24	0.72
Cartoon	384 × 256	63.50	30.07	3.28	1.04	0.75
Fractal	384 × 256	96.87	47.79	6.79	2.83	1.24
Indoor	384 × 256	46.78	20.58	4.7	1.8	0.85
Inverted	384 × 256	66.32	34.42	5.11	2.19	1.47
Jumbled	384 × 256	57.50	27.57	3.41	1.32	0.74
Line drawing	384 × 256	67.88	31.27	2.92	1.02	0.62
Mixed dataset	384 × 256	46.28	24.91	6.21	2.73	1.17

6 Conclusion

When using RLBP with five parameters, mapping plays an important role in correctly identifying the similarity amongst various images. When we perform RLBP with three parameters on dataset with less number of images and RLBP with five parameters, the algorithm with five parameters gives a better result with similar class of images, even when the query image is rotated left by 90°, it is able to identify and map the corresponding pixels correctly. The dataset used in the work is "stimuli" and "wang" dataset which is a subset of "Corel 10K containing 10,000 images of low quality. When the same algorithm is executed on dataset with multiple images from different classes then RLBP with five parameters outperforms RLBP with three parameters and the result improves by approx. 8%. While doing content-based image retrieval using GPU, the execution time decreases a lot in case of GPU based RLBP. The work also shows the best matches while we perform CBIR using a query image.

References

1. Picard, D., Revel, A., & Cord, M. (2012). An application of swarm intelligence to distributed image retrieval. *Information Sciences: An International Journal, 192,* 71–81.
2. Ke, Y., Tang, X., & Jing, F. (2006). The design of high-level features for photo quality assessment. In *Proceedings of the 2006 IEEE Computer Society Conference on Computer Vision and Pattern Recognition*, Vol. 1, Washington, DC, USA, June 17–22, 2006, pp. 419–426.
3. Redi, J. A., Hobfeld, T., Korshunov, P., Mazza, F., Povoa I., & Keimel, C. (2013). Crowdsourcing-based multimedia subjective evaluations: A case study on image recognizability and aesthetic appeal. In *Proceedings of the 2nd ACM International Workshop on Crowdsourcing for Multimedia*, Barcelona, Spain, October 22–22, 2013, pp. 29–34.
4. Datta, R., Joshi, D., Li, J., & Wang, J. Z. (2006). Studying aesthetics in photographic images using a computational approach. In *Proceedings of the 9th European conference on Computer Vision*. Graz, Volume Part III, Austria, May 07–13, 2006, pp. 288–301.
5. Zhang, X., & Chen, X. (2016). Robust sketch-based image retrieval by saliency detection. In *Proceedings of the 22nd International Conference on MultiMedia Modeling*, Vol. 9516, Miami, FL, USA, January 04–06, 2016, pp. 515–526.
6. Zhang, B., Gao, Y., Zhao, S., & Liu, J. (2010). Local derivative pattern versus local binary pattern: Face recognition with high-order local pattern descriptor. *IEEE Transactions on Image Processing, 19,* 533–544.
7. Mehta, R., & Egiazarian, K. (2016). Dominant rotated local binary patterns (DRLBP) for texture classification. *Pattern Recognition Letters, 71,* 16–22.
8. Isola, P., Parikh, D., Torralba, A., & Oliva, A. (2011). Understanding the intrinsic memorability of images. In *Proceedings of the 24th International Conference on Neural Information Processing Systems*, Granada, Spain, December 12–15, 2011, pp. 2429–2437.
9. Mazza, F., Silva, M. P., Callet, P., & Heynderickx, I. (2015). What do you think of my picture? Investigating factors of influence in profile images context perception. In *Proceedings of Human Vision and Electronic Imaging XX*, San Francisco, United States, March 2015.
10. Isola, P. Xiao, J., Torralba, A., & Oliva, A. (2011). What makes an image memorable? In *Proceedings of the 2011 IEEE Conference on Computer Vision and Pattern Recognition*, Washington, DC, USA, June 20–25, 2011, pp. 145–152.
11. Mazza, F., Da Silva, M. P. & Le Callet, P. (2014). Would you hire me? Selfie portrait images perception in a recruitment context. In *Proceedings of Human Vision and Electronic Imaging XIX*, Vol. 9014, San Francisco, California, United States, February 02, 2014, pp. 90140X–90140X.
12. Mazza, F., Silva, M. P, & Callet, P. (2015). Think again about my picture: Different approaches investigating factors of influence in profile images context perception. In *Proceedings of Sino-French Workshop on Information and Communication Technologies*, Nantes, France, January, 2015.

An Introduction to Quantum Search Algorithm and Its Implementation

Jose P. Dumas, Kapil Soni and Akhtar Rasool

Abstract Quantum computing is new era of computing, used in modern world to solve complex problems which can be solved using a supercomputer and those problems can be solved in an efficient manner using quantum computer. Quantum computing uses properties of superposition and entanglement to solve complex problems like NP Hard, and by using them the quantum computer may find the solutions in faster manner compared to classical computers. Grover's search is introduced for searching in an unstructured database that is used to locate a particular item in the database. It provide a speedup of $\sqrt{N}$ over classical. This article describes Grover's search with an example, applications and limitations. Also explores the functionality of quantum circuit, oracle circuit that is particular to Grover's. This article concludes with the Grover's search advantage over classical search.

Keywords Diffusion operator · Oracle · Quantum grover's search
Qubits

1 Introduction

High-performance computers are used to solve many complex problems that cannot be solved using current generation of computers, such as in areas like medicine, science, defense, and so on. One of the promising types of high performance computer is based on the usage of Quantum effect for computation. The tasks that

J. P. Dumas (✉) · K. Soni · A. Rasool
Maulana Azad National Institute of Technology, Bhopal 462003,
Madhya Pradesh, India
e-mail: dumasjosep7@gmail.com

K. Soni
e-mail: prof.kapilsoni@gmail.com

A. Rasool
e-mail: akki262@gmail.com

V. E. Balas et al. (eds.), *Data Management, Analytics and Innovation*,
Advances in Intelligent Systems and Computing 808,
https://doi.org/10.1007/978-981-13-1402-5_2

are impossible to solve by current generation can now be solved using Quantum computing. It is derived from quantum mechanics, which is a core branch of physics [1].

Only one quantum computer has been available commercially, that was developed by D-wave systems in 2011. However, it does not have any sort of evidence, in order to prove that it operates with the real quantum effects. In order to create a quantum computer the essential equipment's are, fundamental storage, quantum bit, that holds simultaneous many states to be realized. The quantum computer is being created in research laboratories with few qubits [2].

The most famous algorithms having a greater boost in quantum computing were, the Shor's and Grover's. Grover's Search, uses superposition and entanglement. These are the main phenomena which are responsible for the exceptional computational power over classical computers or a supercomputer that is available in this time. That is the searching can be done within a limit of $O\ (\sqrt{N})$ [3–5].

To know about the working of quantum computer, various simulators such as Libquantum, jquantum, are used. These simulators simulate the working of quantum computer on classical computers. To study about quantum computer, write programs in various simulators that are executed by classical computers. C-language is the platform in which the simulator Libquantum operates. In order to simulate operations of quantum bits, the simulator must be used, calculate all states that are in entanglement. It is both time and space consuming [6, 7].

2 Quantum Qubits

Qubit or quantum bits are the basic block of quantum computing. Qubits are used to store the current state in quantum computers. This is made up of three components: a memory in order to hold data being processed, a CPU which is used to process the qubits and input/output used to give and extract data from quantum computers. The difference between a bit and qubit is bits are in either 0 or 1. But qubit is a combination of these.

Uses the Bra-ket notation to describe the state of qubit:

$$x = |0> \text{ and } y = |1> \tag{1}$$

The above equation says the qubit value of x is 0 while qubit y is 1 (it is pronounced as "ket 0 and ket 1"). This state corresponds to classical bits state 0 and 1. The |0> and |1> are column vectors represented below.

$$|0> = \begin{bmatrix} 0 \\ \overline{1} \end{bmatrix} \text{ and } |1> = \begin{bmatrix} 1 \\ \overline{0} \end{bmatrix} \tag{2}$$

2.1 Quantum Superposition

As per laws of quantum physics states, when a particle enters in superposition it simultaneously exit in both states. According to quantum computers it means these states can exit in superposition of 0s and 1s. In a classical computer there is only two states it is 1 or 0. But in this, qubit can be in complex superposition of states [8] (Fig. 1).

In the above figure, vector representation of single qubit is shown. Here qubit is visualized as unit vector on the plane. So according to this, depending on the complex numbers *a* and *b* it can exist either in |0> or |1> or in combination of | 0> or |1> . By using this property it can compute 2^n computations at a time (*n* is the number of qubits). Assume it make use of 500 qubits and by using this property, it can do 2^{500} calculations in a single step.

Superposition with qubit *x* is shown below

$$x = a|0> + b|1>, \tag{3}$$

where amplitude are *a* and *b*, which are complex numbers where $|a|^2$ and $|b|^2$ measures the probability of the qubit, and if any attempt of measurement result in state |0> with a probability $|a|^2$ and |1> with a probability of $|b|^2$. For example, let

$$x = a|0> + b|1>, \ |a|^2 = 0.2, \ |b|^2 = 0.8 \tag{4}$$

In the above example, the qubit *x* is measured |0> with a probability of $|a|^2$ while the other is measured |1> with a higher probability of $|b|^2$. That is the qubit *x* exist in a state known as |1> . Naturally

$$|a|^2 + |b|^2 = 1 \tag{5}$$

The main difference between a bits and qubits, in case of bit it is able to determine its value. But qubits value can be determined only by measuring its probability and when these measurements is done superposition property is destroyed.

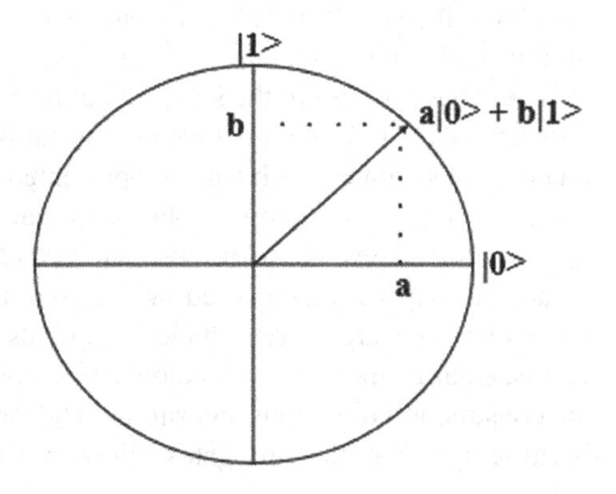

Fig. 1 Single qubit represented as unit vector

2.2 *Quantum Entanglement*

Entanglement is the major phenomenon exhibit by qubits. This phenomenon occurs when a pair of particles is generated or when these particles are separated by very far distance. By measuring first qubit it can tell something about the second qubit. It means states of entangled qubits cannot be described independent to each other.

Taking both these properties together to create an enormously enhanced computing power. For example, classical computer can store only one of four binary bits (00, 01, 10, or 11) at a time when a 2-bit register is used, but quantum computer can store all at a time. And when the qubits increases the capacity also increases exponentially. When entanglement is done no measurement need to be applied on second system. For example

$$x = \frac{1}{\sqrt{2}}(|00> + |11>) \tag{6}$$

Here according to above equation, the second qubit have equal probability to exist in either 0 or 1 if the first qubit is not measured. But if it has been measured, then second qubit will have a probability of 0 or 100%. The measurement of qubits is dependent to each other. In reality this can be thought as, there are two boxes and two balls one is blue and the other is red. The balls are assigned randomly to these boxes. Then by knowing what ball is inside the first box, the ball in second can be determined. This is a practical example for better understandability of entanglement. This property that is exhibited by qubits is known as entanglement.

3 Quantum Basic Gates

In a quantum computer, information is also processed using gates. Quantum gates are basics of all quantum algorithm and they are similar to classical logic gates. These gates provide the same functionality as classical logic gates. Various gates are available and each gate has a specific purpose. These gates are being used to transform the qubits value from one state to another according to properties of various gates. Each quantum gates can be expressed by unitary matrix [9, 10].

These basic gates are the smallest building block of several algorithms including Grover's search. Quantum Circuitry is used to graphically describe the various quantum algorithms. Each gate is represented by a block with lines which is used to represent input and output. Quantum gates are able to manipulate arbitrary multi-partite states including the superposition of states, which are frequently entangled. All these gates used in quantum are reversible. The advantage of using reversible gates are energy efficiency and also it maintain the property of quantum that is, entanglement. Shown below are major three gates which are very useful for the construction of various algorithms. They are also called universal quantum gates because by using this, any gates can be made.

Fig. 2 CNOT gate with 4×4 unitary matrix

$$\begin{pmatrix} 1&0&0&0 \\ 0&1&0&0 \\ 0&0&0&1 \\ 0&0&1&0 \end{pmatrix}$$

3.1 The Controlled-NOT Gate

The CNOT gate, 2-bit reversible gate is similar to classical NOT gate but a control bit is added to control the not operation. That is first bit is control bit while the other is the target. And if the first bit is 1 then only it performs the NOT operation. That is to change the bit value of second bit the control bit has to be 1 (Fig. 2).

In the figure given below it has two inputs (A, B) and two outputs (A′, B′). According to first bit (A) value, the second bit negated or not negated.

3.2 The Toffoli Gate (CCNOT)

This reversible gate has two control bits. There will be change of third input bit if, the first two input bits are both 1. In other words, values of the third input bit is flipped the first two control bits are true. It is named after Tommaso Toffoli. Toffoli gates can be used to implement any classical circuit (Fig. 3).

Here 2-bit gate, where there are three inputs (A, B, C). Out of them two are control bits (A, B). If control bits is set to 1 then the results in third bit output (C′) get flipped. By using this gate, there is two control line which can be used to control the operation.

3.3 The Hadamard Gate

Single qubit gate which is useful out of all gates. It is named, after French mathematician who is Jacques Hadamard. Hadamard gates are used to create superposition of states. Many quantum computer algorithms make use Hadamard gate to get superposition of states [11] (Fig. 4).

Fig. 3 CCNOT gate with 8×8 unitary matrix

$$\begin{pmatrix} 1&0&0&0&0&0&0&0 \\ 0&1&0&0&0&0&0&0 \\ 0&0&1&0&0&0&0&0 \\ 0&0&0&1&0&0&0&0 \\ 0&0&0&0&1&0&0&0 \\ 0&0&0&0&0&1&0&0 \\ 0&0&0&0&0&0&0&1 \\ 0&0&0&0&0&0&1&0 \end{pmatrix}$$

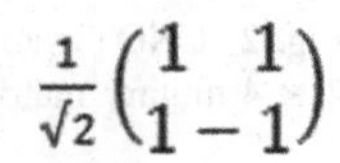

Fig. 4 Hadamard gate with propagator

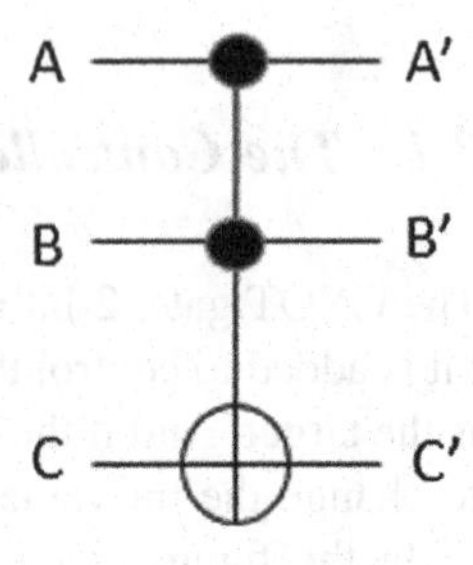

Fig. 5 CCNOT gate

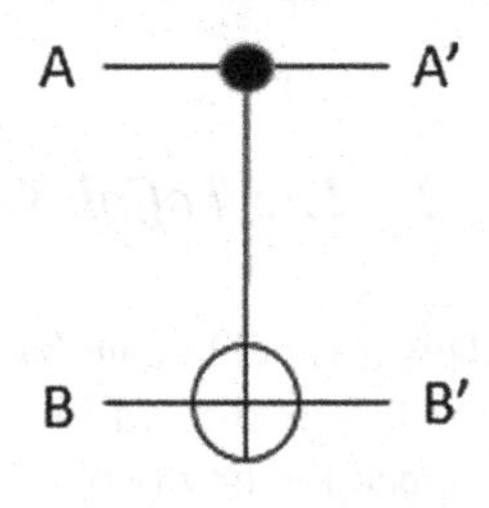

Fig. 6 CNOT gate

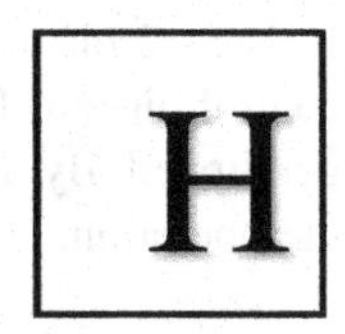

Fig. 7 Hadamard gate

By using these Universal gate, all other gates (X-OR, AND, OR, etc.) can be made. The graphical representation of the gates is shown below (Figs. 5, 6 and 7).

4 Grover's Search

In 1996 Lov Grover found this to find an element from an unsorted database. It provides a speedup of $\sqrt{N}$ in magnitude, (where size of the database is N) over classic algorithm. It was with great practical use. Grover Search is a probabilistic model, since it runs several times to give the result with highest probability [12]. It is proven that it is one of the quantum algorithms which provides optimal and also results in the best possible database search over classical computers. Grover's Search performs a search over a set of $N = 2^n$ items to find unique element within $O(\sqrt{N})$ (Fig. 8).

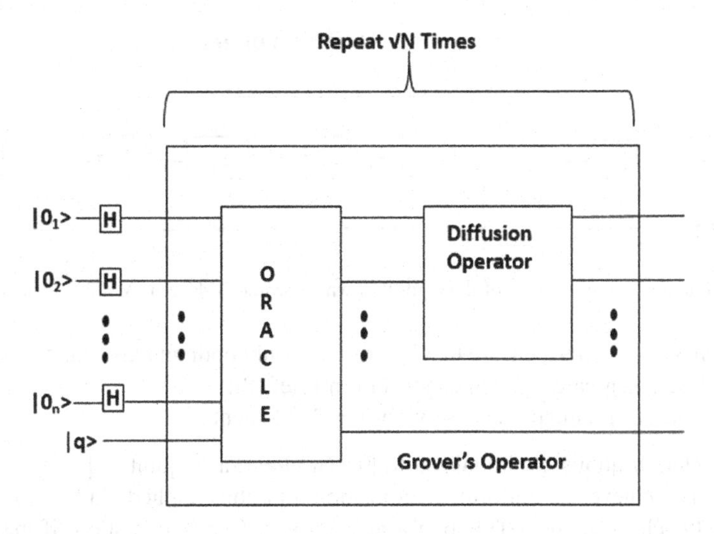

Fig. 8 Circuit for Grover's search

The number to be searched is pre-entered. The classical information and the measured quantum register are given to oracle function. The oracle depends on the specific instance of the search problem. The diffusion operator block is also known as inversion about the average operator and it in order to increase the probability of the marked state's amplitude is amplified. From the above figure, first all states are initialized to |0> except the state to be found. After that it is allowed to pass through Hadamard gate so that equal superposition is obtained. After that it is allowed to pass through oracle function and the value to be found is negated and other values remains same. After that it pass through diffusion operator so that it amplify the state to be found.

4.1 Grover's Algorithm

Given an unstructured database of N elements, search for a single key matching a given search criterion, using classical computation resources, takes $O(N)$ time to search at least half of the elements. Grover's algorithm can locate the key using $\sqrt{N}$ only queries to an oracle that judges a match with the search criterion. To do this we need specific Grover's quantum circuit for a given quantum algorithm. Its circuit is shown below (Fig. 9).

U_w is Grover's Oracle function (varies according to search application) which used to map the search element (element required to find). It returns 1 if match found else returns 0. In order to obtain more security the oracle function is repeated $\sqrt{N}$ times. The main thing about Grover's search is, since it is a probabilistic model there is a chance that result diverge from the correct answer if it is allowed to run so

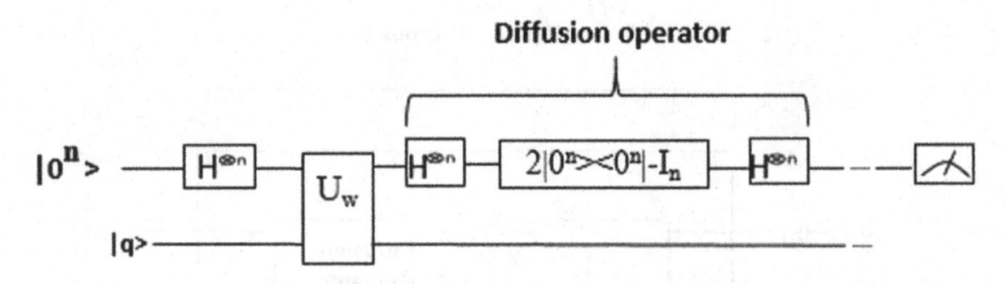

Fig. 9 Graphical representation of the algorithm where search is done in $N = 2^n$ elements

many times. So we must ensure that it is running only optimum amount of time and it should be taken care off. The algorithm in brief [8]:

Given an unstructured database with $N = 2^n$ elements

Step 1 Here n qubits are initialized to |0> and the oracle qubit to |1>.

Step 2 To achieve the uniform superposition all qubits applied Hadamard gate.

Step 3 Oracle is applied. This performs a form of CNOT operation. If the input satisfies the search criteria, then oracle bit is flipped. Where x is the correct state if $f(x) = 1$, else returns 0.

$$x > \dot{\rightarrow} (-1)^{f(x)} |x > \tag{7}$$

Step 4 Apply Diffusion Operator that is Conditional Phase-Shift gate on all qubits except the oracle qubit. This gate negates the probability amplitude of the | 000 … 0> basis state, leaving that of the others unaffected. It can be realized using a combination of X, H and C^{n-1}—NOT gates' as shown. Conditional Phase-Shift is given below [13].

$$2|0^n > < 0^n| - I_n \tag{8}$$

Step 5 Repeat steps 3–5 $\frac{\pi}{4}\sqrt{\frac{N}{M}}$ times, the number of Keys (M) matching the search criterion [14].

The above steps are, how the algorithm finds the desired element. Algorithm has $n + 1$ qubits. First initialization step, here the database size is taken as N and if the size is N there should be log N qubits are required. The element to be found is initialized to |1> and others are initialized to |0>. The algorithm has 3 major steps, first is applying Hadamard gate as a result all states are equally superpositioned. That is, splitting the qubits into every possible inputs. These mix qubits into uniform amount of amplitude. Suppose x^* is the search element (Fig. 10).

Then comes the Oracle function, its function is simple it output 1 if desired element is found else 0. The search element (x^*) is given externally to oracle. If the input, not found then the Eigen value $(-1)^0$ results in 1. Here Oracle function is invoked twice since $\sqrt{N}$ value is 2. And the eigen value is -1 for the element to be found, it result in flipping of amplitude to $\frac{-1}{\sqrt{N}}$ (Fig. 11).

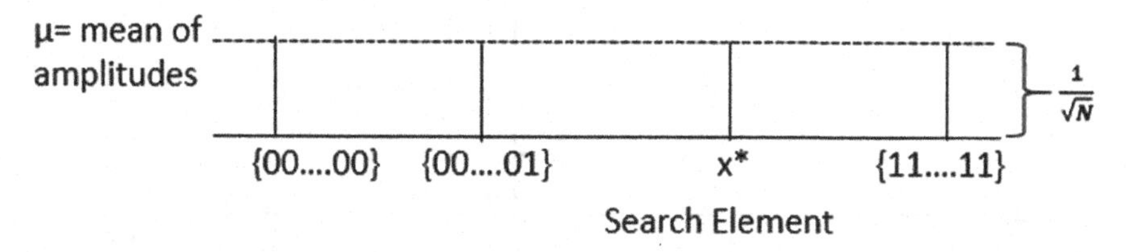

Fig. 10 Equal superposition of all states

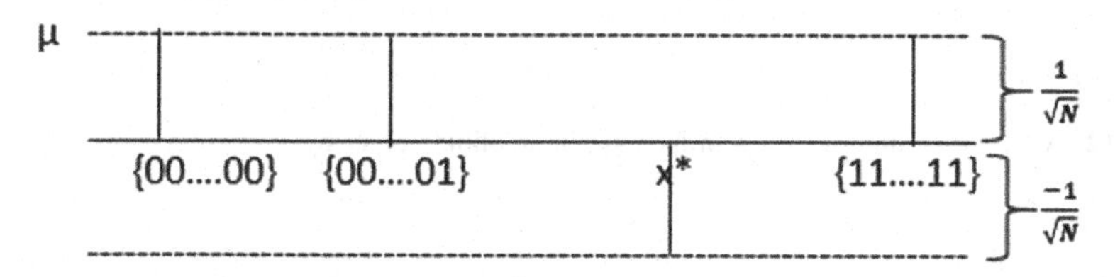

Fig. 11 The Oracle negate the value of search element

Diffusion operator comes next, it increase the amplitude by $\frac{3}{\sqrt{N}}$ roughly. After oracle again Hadamard gate is applied it result in state change, which once again reflects the input looking for to x-axis. It results in the increase of getting the output, the probability of expected value goes up.

In unitary matrix form, $Z_0 = 2|0^n\rangle\langle 0^n| - I_n$, where I is identity matrix. Suppose the input is $|x\rangle$ having no 0^n component.

$$Z_0|x> \; = 2|0^n> <0^n|x> - |x> \; = -|x> \tag{9}$$

In the above equation, $<0^n|x>$ equals to zero, so that desired output $-|x>$ is obtained. That is if the input is not present, it doesn't do anything else it results in increase of amplitude.

Phase inversion followed inversion by mean increase the amplitude by $\frac{2}{\sqrt{N}}$. Diffusion operator is used to increase the amplitude of search element absolutely (Fig. 12).

After 2nd iteration the amplitude increases by $\frac{5}{\sqrt{N}}$, after third iteration it becomes $\frac{7}{\sqrt{N}}$ and so on. And after $\sqrt{N}$ steps the amplitude is increases and after such iteration the desired result is obtained (Fig. 13).

4.2 *Grover's Search Example*

Consider a system having $N = 4 = 2^2$, where N is total number of elements and it requires $n = 2$ Qubits in order to simulate the search. A state to be searched is represented by '$|x_0>$' and the corresponding bit string to be searched is |10 (Fig. 14).

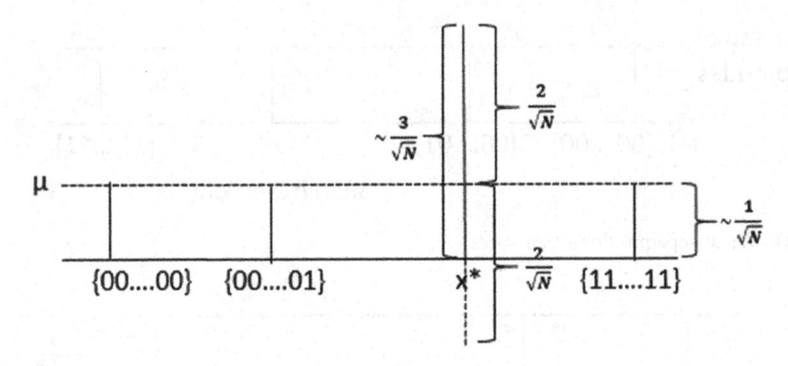

Fig. 12 After 1st iteration there is an increase in amplitude of x^* by $\frac{3}{\sqrt{N}}$

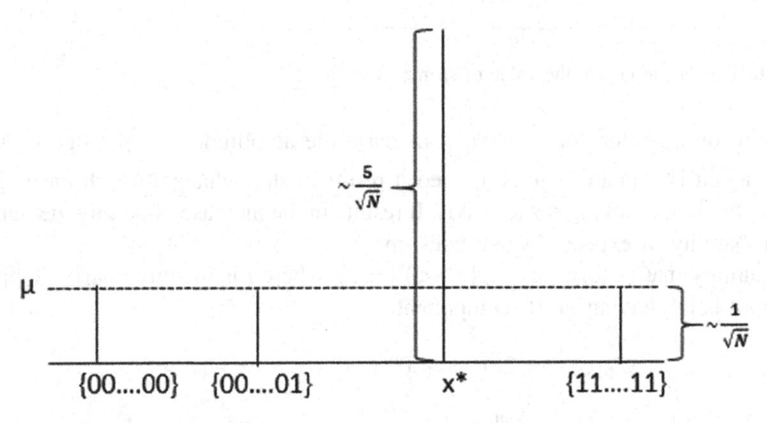

Fig. 13 After 2nd iteration there is an increase in amplitude of x^* by $\frac{5}{\sqrt{N}}$

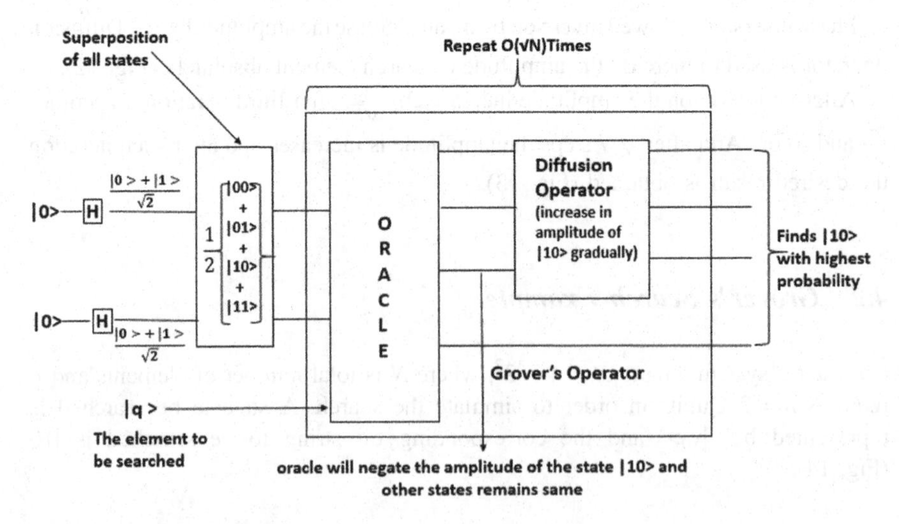

Fig. 14 Block diagram of Grover's example

The ket notation with two qubits is given below

$$|x> \; = a_0|00> + a_1|01> + a_2|10> + a_3|11> \tag{10}$$

According to algorithm, all n qubits is initialized to "|0>" and a_0, a_1, a_2, a_3 the amplitude of the corresponding states.

The next step is to apply Hadamard gate, in order to obtain uniform amplitudes associated with each and every state. That is for four states available and it is $\frac{1}{\sqrt{N}} = \frac{1}{\sqrt{4}} = \frac{1}{2}$

$$H^2|00> \; = \frac{1}{2}|00> + \frac{1}{2}|01> + \frac{1}{2}|10> + \frac{1}{2}|11> \; = \frac{1}{2}\sum_{0}^{3}|x> \tag{11}$$

In order to visualize how the algorithm work, geometric interpretation of the amplitudes are necessary (Fig. 15).

According to algorithm we have to perform two iterations. It is obtained by

$$\frac{\pi}{4}\sqrt{4} = \frac{\pi}{4}2 = \frac{\pi}{2} \approx 1.6 = 2 \tag{12}$$

In each iteration the oracle is called first and then inversion about its mean is done. The oracle negates amplitude of qubit state |10> and thus results in (Fig. 16)

$$|x> \; = \frac{1}{2}|00> + \frac{1}{2}|01> - \frac{1}{2}|10> + \frac{1}{2}|11> \tag{13}$$

After this perform step 5 that is diffusion operator (conditional phase shift), which will increase the amplitudes of the corresponding state by their difference from their mean, decrease if the difference is negative (Fig. 17).

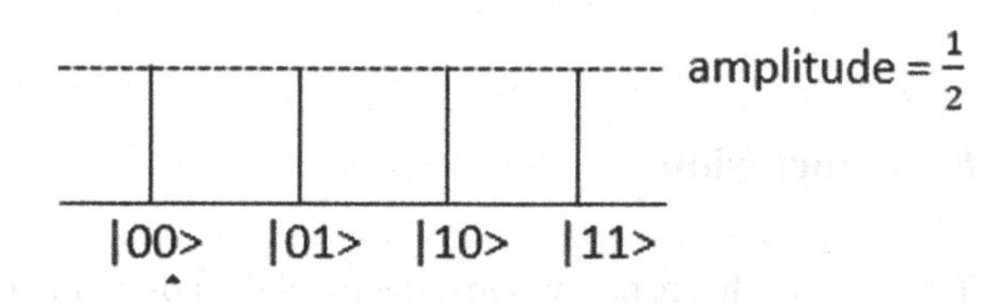

Fig. 15 Equal superposition of all states

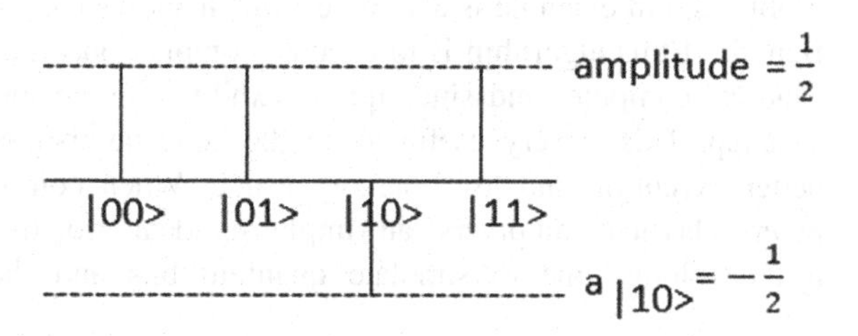

Fig. 16 The Oracle negate the value of |10>

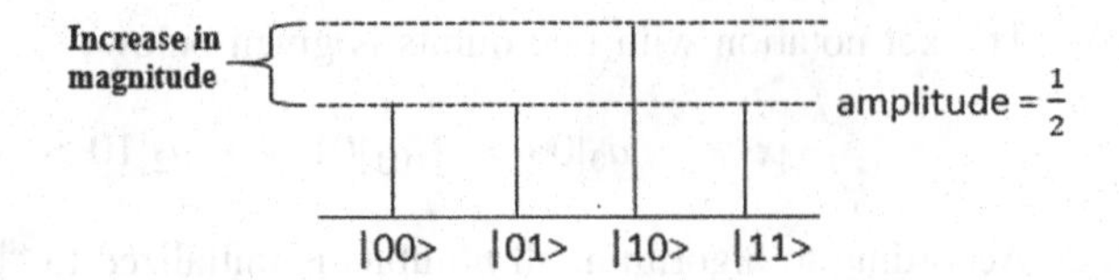

Fig. 17 Increase in the order of magnitude of |10> state

Then the second iteration performed same as above, until the correct result is obtained. Now when the system is observed, the correct solution, |10> is obtained with high probability. After second iteration again the magnitude of |10> increases. Grover's algorithm is driven by probability.

5 Applications

Grover's search is mainly used in Graph coloring, Boolean satisfiability, Triangle finding, TSP (Traveling salesman problem), Pattern Matching, N-queen Problem, etc. The pattern matching problem, the best known algorithm will take $O\ (MN)$ where M is the pattern size and N is entire string from where the matching is done. This can be easily done using Grover's Search. Basic application of Grover's Search is database search, which is used nowadays for searching a webpage using several search engines. By using Grover's Search the search can be done much faster compared to searching mechanism that is currently available.

Transcendental logarithm problem [15] is a problem where Grover's search is made used build signature schemes. This algorithm not only can speed up solving some NP-complete problems, but also is very useful for attacking some symmetric key cryptosystems which have a low length key. Some enlightening on how to improve the security of existing cryptosystems. Grover's Search can be also used to crack passwords. Grover's Search can be also used to solve almost all NP-Hard and NP-Complete problems in a faster and efficient manner and it takes low memory.

6 Conclusion

The focus of this paper was to explore all the possibility of Quantum Grover Search which could be useful to tackle various problems like NP-Hard, NP-complete problems. An example is also given which shows the working of Grover's search. Benefit of this algorithm is to execute complex operations in a faster manner than classical computer and since qubits exhibit superposition it can gain exponential speedup. This is very useful to handle large number of calculations. It provides better execution and low memory usage. When competing with quantum computers, classical computers can simply run ideal error-free quantum algorithms. But it takes long time to simulate quantum bits and the technology required to implement quantum computer might be available now. Only through simulation over classical computer, the research can be done using Grover's search. Since it is a probabilistic model there is a chance that result diverge from the correct answer if it is allowed to run so many times. So we must ensure that it is running only optimum amount of time and it should be taken care off. Quantum Grover's search is next generation search technique which can be used to solve complex problems.

References

1. Zalka, C. (1999). Grovers quantum searching algorithm is optimal. *Physical Review A, 60,* 2746–2751.
2. Nielsen, M. A., & Chuang, I. L. (2011). *Quantum computation and quantum information.* Cambridge: Cambridge University Press.
3. Grover, L. K. (1997). Quantum mechanics helps in searching for a needle in a haystack. *Physical Review Letters, 78,* 325–328.
4. Farhi, E., & Gutmann, S. (1998). Analog analogue of a digital quantum computation. *Physical Review A, 57,* 2403–2405.
5. Lloyd, S. (2000). Quantum search without entanglement. *Physical Review A, 61,* 010301(R).
6. Glassner, A. (2001). Quantum computing. *IEEE Computing Graphics Applications, 21*(5), 86–95.
7. Shor, P. W. (1997). Polynomial-time algorithms for prime factorization and discrete logarithms on a quantum computer. *SIAM Journal on Computing, 26*(5), 1484.
8. Viamontes, G. F., Markov, I. L., & Hayes, J. P. (2003). Gate-level simulation of quantum circuits. In *IEEE Asia South Pacific Design Automation Conference.*
9. Saha, A., Chongder, A., Mandal, S.B., & Chakrabarti, A. (2015). Synthesis of vertex coloring problem using grover's algorithm. In *IEEE International Symposium on Nanoelectronic and Information Systems.*
10. Grover, L. K. (1996). A fast quantum mechanical algorithm for database search. In *Proceedings, 28th Annual ACM Symposium on the Theory of Computing (STOC)*, pp. 212–219.
11. Kai, K. (2015). Two improvements in Grover's algorithm. In *IEEE Control and Decision Conference (CCDC).*
12. ElGamal, T. (1985). A public-key cryptosystem and a signature scheme based on discrete logarithms. *IEEE Transactions on Information Theory, 31*(4), 469–472.
13. Karafyllidis, G. (2005). Quantum computer simulator based on the circuit model of quantum computation. *IEEE Transactions on Circuits and Systems I, 52,* 1590–1596.
14. Li, X., Song, K., Sun, N., & Zhao, C. (2013). Phase matching in grover's algorithm. In *IEEE Control Conference (CCC),* 2013 32nd Chinese.
15. Tang, Y., & Shenghui, S. (2014). Application of grover's quantum search algorithm to solve the transcendental logarithm problem. In *IEEE International Conference on Computational Intelligence and Security.*

Smart Waste Management for Segregating Different Types of Wastes

Rashi Kansara, Pritee Bhojani and Jigar Chauhan

Abstract The trend of IoT is increasing day by day, making each and everything related to daily life a smart thing. Making things smart also reduces the need for manpower; in addition, the work is done appropriately as it decreases human intervention. The garbage management is one of the major problems as it causes many diseases if the garbage is not collected on time. Smart bins are used to collect all the garbage from a particular area until the bins are filled up to a threshold level. When the bin is 80% filled, the location of the bin is displayed to the garbage collector on his mobile application using the GPS module attached to the bin. With respect to the above theory, we are proposing an idea in which there is a prototype that the bin when gets filled up to a threshold level, i.e. when 80% of the bin is filled, the GPS module attached to the bin will show an alert to the garbage collector truck of that area that the bin in his vicinity needs to be emptied. A shortest path to empty all the filled bins or the bins with the threshold level is also displayed on the mobile application that saves fuel and time and more work can be done. When the bin is filled or senses hazardous gas using a gas sensor, it will close the lid until the garbage truck arrives to empty it. Once the lid is closed, only the garbage collecting truck assigned to that area can open the lid. The garbage is then dumped to the garbage collecting truck, which in turn is composed of a unit that comprises individual machines that have the ability to segregate different types of waste using robotic arm by detecting the properties of that waste.

Keywords Dry waste · GPS · GSM · HC-SR04 Ultrasonic Sensor
Infrared transmitter and receiver · IoT · Magnet · MQ4 Gas Sensor
RFID tag · Robotic arm · Smart bin · Threshold level · Wet waste

---

R. Kansara (✉) · P. Bhojani · J. Chauhan
Information Technology, Universal College of Engineering, Vasai, Maharashtra, India
e-mail: rashikansara@gmail.com

P. Bhojani
e-mail: pritee.bhojani32@gmail.com

J. Chauhan
e-mail: jigar.chauhan@universal.edu.in

V. E. Balas et al. (eds.), *Data Management, Analytics and Innovation*,
Advances in Intelligent Systems and Computing 808,
https://doi.org/10.1007/978-981-13-1402-5_3

1 Introduction

As we are noticing urbanization is the major cause of the increased and unfriendly environment waste generated from households, hospitals and factories are increasing day by day. In the absence of bins, people scatter the waste on the road, making the roads unpleasant and sometimes this waste gets piled up and animal's squatter on the waste and consume it, and in addition it also attracts flies, rats and other creatures that spread diseases. If waste management is not properly handled, it may affect the nature as well as the humans residing in the nearby vicinity. Plastic waste is generated in huge quantities but the method used to dispose it is inappropriate and also it does not get decomposed in the soil, and in addition burning of plastic is harmful to the nature. Decomposable waste is used to make fertilizers or biogas, and hence they are eco-friendly causing no harm to nature. Metallic wastes easily get attracted to the magnet and they can be melted and formed into a required shape to reuse it. Glass wastes can be easily smelted and reused by recycling it, which in turn reduces the cost of the new products made. Biomedical wastes are very harmful as they can spread the diseases very easily. Toxic metals in e-waste are very harmful to the environment.

The garbage collecting trucks do not collect the waste from roads on time. These trucks collecting the waste from roads deliver the waste to the dumping ground making a pile of waste, which then results in a huge mountain of waste. This waste consists of metals, biodegradables, plastics, e-waste, etc. These wastes are not segregated according to the properties, as there is no hi-tech technology available. Smart waste management for segregating different types of wastes is needed, as there is a lack of knowledge and awareness among the people to dispose the waste differently according to their properties.

In this paper, we propose a smart bin that when gets filled closes the lid of the bin so that no more waste can be added that will fall on the ground and cause harmful diseases. Also, if harmful gas is present in the bin it will close the lid until the bin is emptied into the garbage collecting trucks. The smart garbage collecting truck consists of a small mechanism that when the bin is emptied into the truck all the waste gets segregated into three different large bins. The bins in the truck after sorting of waste consist of dry, wet and metallic waste in three different bins.

2 Drawbacks of Existing System

The existing system notifies the fullness of the bin but there is no system that does not allow intake of more garbage when the bin is full, and if the intake is not restricted people will empty their bins and all the litter will be on the floor causing many diseases and a bad odour as well as the city becomes dirty. Also, it is necessary to close the lid if the bin consists of harmful gasses that can harm the nature. This feature is more useful near the industrial areas. The garbage collected from these bins is then emptied into the dumping ground, i.e. there is no proper waste management. The waste garbage is not decomposed or reused efficiently. There is no garbage truck that segregates the waste inside it using small machinery, hence reducing the manpower and providing an efficient way to manage the waste. There is no system that segregates the waste according to parameters such as dry waste, wet waste and metallic waste. The following are the drawbacks of some of the existing devices:

- This surveyed literature [1] has the following drawbacks:

1. It only considers the bins that are filled equal to if its level exceeds 80% of the bin or contains harmful gas.
2. If the trucks are overloaded, it discards its further route of collection.
3. Inadequate data about garbage collecting time and area of the bin.
4. Inadequate facility for monitoring the system, trail the truck and bins. There is no facility for quick response to critical cases like accident of the truck and breakdown.
5. No way for clients to complaint about bins being full if the bin system breaks down.

Solution:

1. Each area consists of minimum two bins to avoid people facing problem of bin being closed due to bin being fully filled or containing harmful gas until the truck arrives at the desired location.
2. If the trucks are overloaded, then the trucks assigned to the nearby area will be notified about it to collect the garbage and one of the trucks will accept the request to avoid people facing problems of waste, as the bin lid will be closed once it is fully filled or contains harmful gasses.
3. Every time the waste in the bin reaches the threshold level, it shows a task to the truck assigned to that area and when the truck reaches that area using the shortest path suggested it will respond to the task as completed, and time and area of that bin will be notified for further information.
4. The path is suggested considering the shortest path to save fuel and time and, in addition, with the requests of the bins that are to be emptied immediately.
5. The alternative truck or the truck of the nearby area will collect the garbage by accepting the request until the truck of that area gets back to normal or get repaired; it will shut down its system indicating that it will not accept any requests and it is the duty of the nearby truck to accept the request or in worst case the alternative truck will accept the request. There will be an alternative truck for every 25 km radius in case of emergency, and if truck is overloaded or if the truck of any particular area has met any accident, then there are many emergency requests from that area.

- This surveyed literature [2] has a disadvantage that some bin gets filled up much faster than the next scheduled time for collection which causes an overflow of waste.

Solution:

It is due to the static time for waste collection like the traditional system, whereas we are having a dynamic waste collection system which is smart. For the areas with more waste generation like near the factories or hospitals and nursing homes, we will be provided with extra bins so that people do not face any problems of truck arriving late as these are the places which have a high demand of waste collection.

- This surveyed literature [3] has the disadvantage that it uses RFID card to indicate when it empties the bin. Every time the bin is emptied, the status of the bin is updated and to empty the bin, RFID card is used.

Solution:

RFID is the unique identity of every bin and if it breaks down, the status of the bin cannot be updated when the garbage truck has emptied the bin. The time at which it is emptied is also not updated. The proposed system uses a battery for the system inside the bin to work.

- In this surveyed literature [4], smart bin segregates the waste inside the bin itself. It segregates into decomposable wastes and plastic wastes. It uses decomposable waste to make biogas out of it. It uses three ultrasonic levels to check the level of waste in bin each at 120° angle and four load cells to check the waste level inside the bin.

Solution:

Unnecessary extra costs of more ultrasonic sensors and load cell do the same work. Also, if the waste consists of metals and glasses, it will not be separated and mixed with the decomposable wastes.

- In this surveyed literature [5], it has the basic functionality of finding the fill level of the bin and updating the server about the amount of waste collected to predict the amount of waste generated in future.
- In this surveyed literature [5], the smart bin checks the amount of the waste in the bin using ultrasonic sensor and information about the waste is sent to the server to make future plans to estimate the quantity of waste that will be produced. The volume of the waste is checked twice a day.

Solution:

It takes real-time information about the waste. The amount of waste emptied is updated to the server whenever the garbage collecting truck empties the bin. The ultrasonic sensor measures the level of waste at regular intervals.

- In this surveyed literature [6], the people have to segregate the waste and throw it into their respective bins. It gives some points according to the weight of the waste thrown on their respective cards. This makes the human to remove more and more waste in the greed of money.

  Solution:

  People having less knowledge of the products might not put the waste in its respective bin but in other bins. This might be problematic and also the humans have to segregate the waste, whereas here the machines and sensors will perform this work without human intervention.
- In this surveyed literature [7], it is equipped with an LCD screen to show the fill percentage of the bin and the humans have to segregate the waste before throwing them into the bin. It is the duty of the users to maintain different wastes in different bins at your home and accordingly dump them in their respective bins before throwing them into the bin.

3 Comparison with Existing Literature

Presently, all the waste collected from houses is gathered near the corner of the street. These wastes are not collected by the garbage collecting trucks daily. Sometimes, it might take few days. If the waste is not emptied daily, then the bins get overflowed and all the waste on the roads makes the city unpleasant. Also, it becomes home for many diseases. When the waste is collected from these places, it is dumped in one place. These wastes are not decomposed or reused based on the type of the waste.

In this surveyed literature [1], its purpose is to schedule the trucks by suggesting the nearest route to empty the bins that are almost filled with wastes or contain some harmful gasses. These smart bins provide data about the level of the bin filled and the level of the dangerous gas contained in it. The system also shows approximate time to collect the waste, status of the bin, the approximate time taken to fill the bin and nearest route to collect the waste. The data are summarized and based on those data the reports are generated.

In this surveyed literature [2], the smart bin identifies the fullness of litter bin and reports the readings and sensor status. The bin contains a GPS sensor to obtain the current location of the deployed bin. It consists of a back-end server to analyse whether the predefined rules and tags are fulfilled to generate the event accordingly.

This surveyed literature [3] consists of a smart alert system for waste collection from the bins by offering an urgent emptying of bins alert on the web server of the municipality. The cleaning of the bins is done based on verification. The garbage level is measured using ultrasonic waves interfaced with Arduino UNO. After cleaning, the driver of waste collecting truck confirms the task of emptying the bin. The task is performed using RFID tag. Remote monitoring is done to reduce the manual work process of monitoring and verification.

In this surveyed literature [4], a smart bin will separate plastic material from the waste inside the bin itself so that the decomposable waste can be used as fertilizers or to produce biogas. Ultrasonic sensor is used to find the level of the bin filled. It uses GSM module to communicate with the server, while the GPS module is for the identification of the bin.

In this surveyed literature [5], a smart waste sorter machine was used to sort out different types of wastes. The wastes can be sorted out according to metal waste, paper waste, plastic waste and glass waste. It consists of a weight sensor to measure the amount of particular waste collected.

In this surveyed literature [5], the smart bin consists of sensors to identify the fullness of the bin. It also provides shortest routes, and from past feedback it also predicts the future state with respect to factors such as traffic congestion in an area where the bins are placed, cost efficiency balance, rate at which the bins get filled, etc. to improve collection efficiency and it determines the bins to be emptied in early stages.

In this surveyed literature [6], a smart recycle bin was used that uses a plastic card which is smart to measure the weight of the waste. The bins contain RFID technology to track the waste in the bins. Here, the smart bin knows which type of waste is present in what quantity. The waste is disposed according to glass, paper, aluminium can and plastic products. When waste is thrown, it consists of reward points on the user's card using RFID. The points depend on the amount of wastes disposed and what type of waste is disposed. The accumulated points can be encashed in the bank account or rebate for a product. This is done to create awareness among people to dispose the garbage according to different types.

In this surveyed literature [7], a cloud-based waste management system was used. Here, the bins are equipped with sensors and the waste level status is notified and uploaded on the cloud. All appointed people can access the data from cloud. It also helps in suggesting shortest path according to the status of the bin to save fuel and time.

4 System Overview

The following are the components used in the smart bin:

1. MQ-4 Gas Sensor

MQ-4 Gas Sensor is a simple and easy to use sensor. It is a compressed natural gas sensor which is appropriate to sense the components of the natural gas present in the air (Fig. 1).

2. Ultrasonic Sensor

HC-SR04 Ultrasonic Sensor measurement functionality is used without having any contact. Its range is between 2 and 400 cm. Its accuracy can reach up to 3 mm. Each module of the sensor is contained with a control circuit, transmitter and receiver (Fig. 2).

Fig. 1 MQ-4 Gas Sensor

Fig. 2 HC-SR04 Ultrasonic Sensor

3. GPS Module

GPS module is used to receive the data from GPS satellites. It sends the geographical location (Fig. 3).

Fig. 3 GPS module

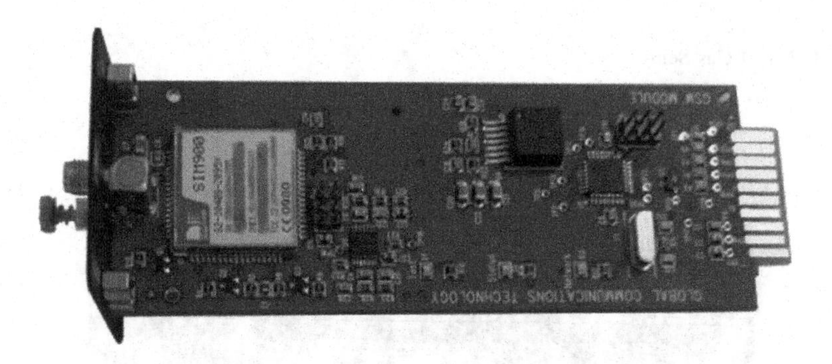

Fig. 4 GSM module

4. GSM Module

GSM module is used to send real-time location of bin received from the GPS module to the GPS-based system present in the smart truck. The digital system has an ability to carry 64 kbps to 120 Mbps of data rates (Fig. 4).

5. Microcontroller

Arduino Mega 2560 is the microcontroller used to connect all the sensors and battery for power supply (Fig. 5).

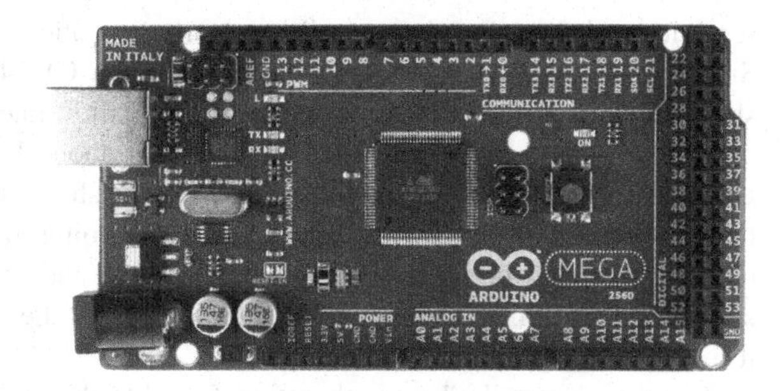

Fig. 5 Microcontroller Arduino Mega 2560

Fig. 6 RFID module

6. RFID Module

RFID sends the converted radio waves. These waves are sent from RFID tag to the controllers. The RFID reader is placed on the device. The RFID reader and the RFID tag should have the same frequency (Fig. 6).

5 Proposed System

In this paper, we propose a smart waste garbage truck that will collect the waste from a smart garbage bin. This bin with the help of some sensors will analyse the level of garbage and accordingly alarm the preventive measures. The ultrasonic sensor is used to measure the amount of the waste in the bin. When the garbage in the bin reaches the threshold level, i.e. when the bin is 80% filled, it will show an alert to the garbage collector truck on its mobile application to empty the bin as soon as possible. If the bin is fully filled, the lid of the bin closes and does not open until the garbage collecting truck empties the bin. The ultrasonic sensor—HC-SR04—uses ultrasonic waves to measure the level of the garbage in the bin. The gas sensor—MQ4 is used to detect whether the hazardous natural gas like methane is present in the bin. If the gas sensor detects the harmful gasses, it closes the lid of the bin until the bin is emptied by the garbage collecting trucks. The person in the garbage collecting truck will have a card, which consists of RFID tag and using this RFID tag the lid of the bin is opened. The bin consists of a battery for keeping all the sensors working and to keep the track of the bin. It also consists of a GSM module to communicate with the server. The GPS module attached to the bin will send its real-time location to all the garbage collecting trucks in its vicinity. Similarly, waste collector truck would possess a GPS-based system, which will show all the garbage bins within the truck's vicinity. Each truck will have a radius of 5 km to empty all the bins that show an alert about the bins present in that area. Using analytics, the mobile application shows the shortest path to empty all the bins that show the alert, by saving fuel and time and doing more of work. This helps in reducing the pile of wastes present on the road and also decreases the diseases spreading through this waste, and in addition to this the animals do not squatter on it to eat the waste.

The waste from the bins is then transferred to the garbage collecting truck. The garbage collecting truck on its way segregates the waste, thus utilizing the time and no heap of waste is created. The segregated waste is then transferred to their respective places, where they are reused or decomposed appropriately. There is a moving belt inside the truck in a circular shape on which the waste from the bins is emptied. The belt consists of a mechanism to segregate waste. The infrared transmitter and receiver are used to detect whether objects are present on the belt or not. If the waste is detected on the belt, the conveyor belt starts moving. As soon as the conveyor belt starts moving the blower starts, the blower is used to remove all the dry waste from the conveyor belt and put them into one of the bins. The magnet present at one side of the belt will attract all the metals and transfer it to one of the bins. Rest of the wet wastes is put into the bin using the robotic arm present on the belt to one of the bins. All the dry, wet and metallic wastes are put into bin number 1, 2 and 3, respectively. It also includes some other contaminants. It does not fully segregate the waste.

The graph is plotted with parameters for daily, weekly, monthly, quarterly, semi-annually and annually available data. It also consists of a feature that shows a graph for the amount of each waste collected from respective area to analyse the different types of wastes generated. All the data collected are stored in the database server. All the data are calculated using analytics (Figs. 7, 8, 9 and 10).

6 Conclusion and Future Work

In most of the places, waste management is done manually. This increases more time to perform some task as well as manpower. Also, the work done is not perfect and the waste is not segregated to either reuse it or decompose it appropriately. When waste is not managed properly, it directly affects the human health as well as the nature. The waste when not managed properly creates environment pollution and makes people prone to diseases. IoT is a major technology evolving to make such things simpler and automatic with less human intervention. This paper makes an attempt to dispose the waste properly into the smart dustbin by humans, to collect the waste by garbage collecting trucks from the bins on time and it closes the lid of the bin when the bin is full or contains harmful gasses. The GPS gives an alert to the garbage collector to empty the bin, and when the lid is closed, the garbage

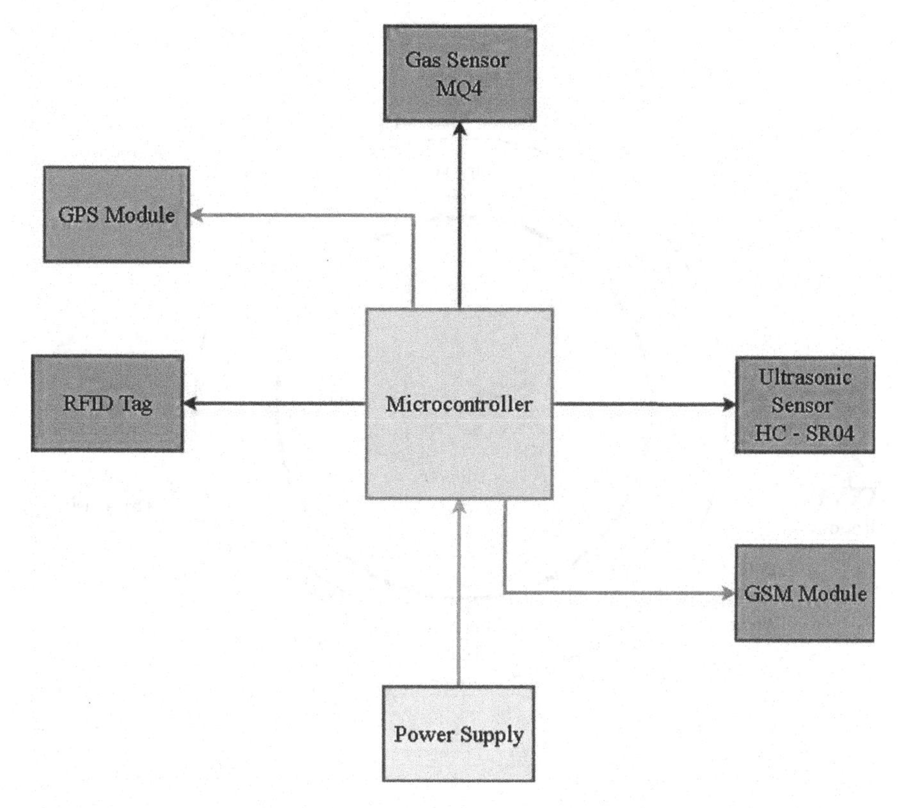

Fig. 7 Smart garbage bin architecture

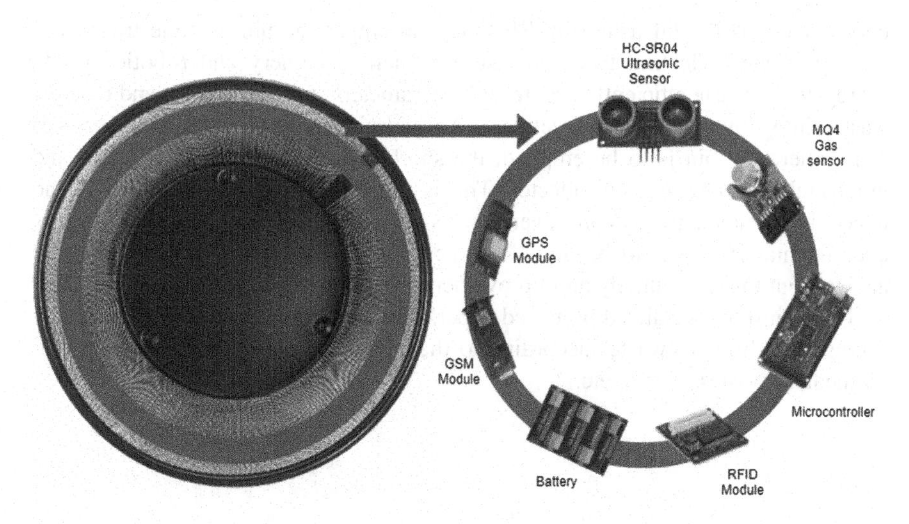

Fig. 8 System diagram for smart garbage bin

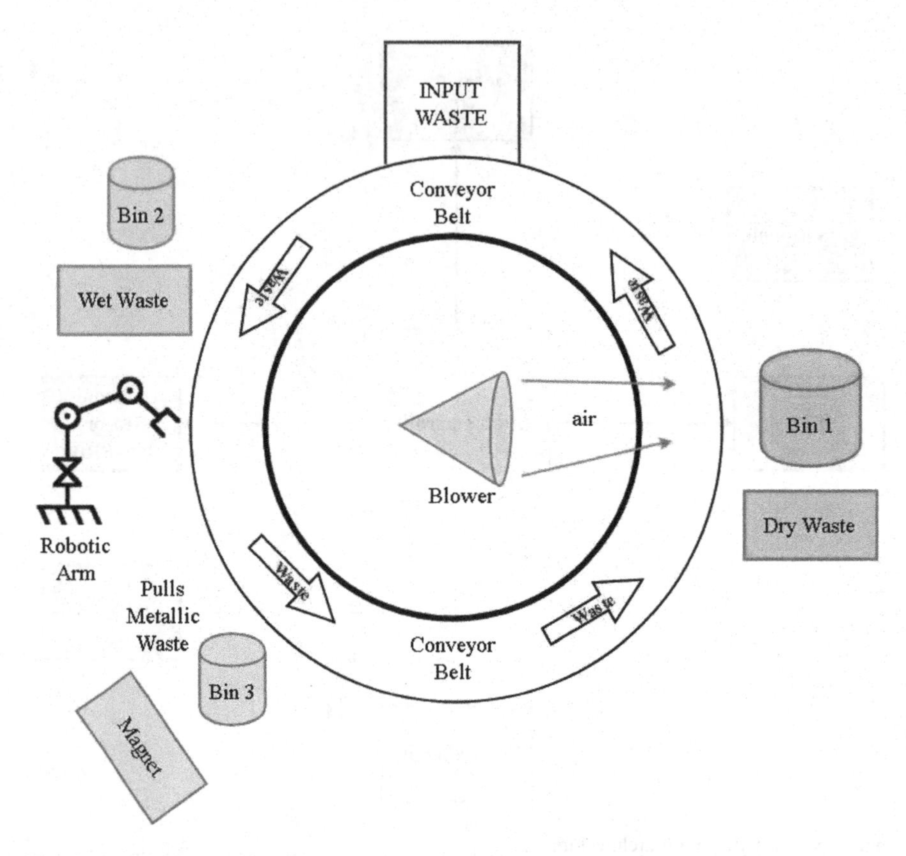

Fig. 9 System diagram fir waste segregation in garbage truck

collector opens the lid using the RFID tag, to empty the bin into the truck. The smart garbage collecting truck consists of small machinery and robotic arm to segregate the waste efficiently according to parameters such as dry, wet and metallic wastes into different bins, respectively. An android application shows the places from where the bin is to be emptied, the shortest path to empty all the bins and graphs related to the wastes collected. This is a research-oriented survey paper, and hence implementation system overview is provided. As this is the prototype, experimentation is yet to be done, and hence no results are compared in result analysis but this is definitely an improvement in existing systems available till date.

This is just a generalized proposed paper. This can be done in detail in future by segregating different wastes according to their types such as plastic, glass, paper, biomedical wastes, waste, etc.

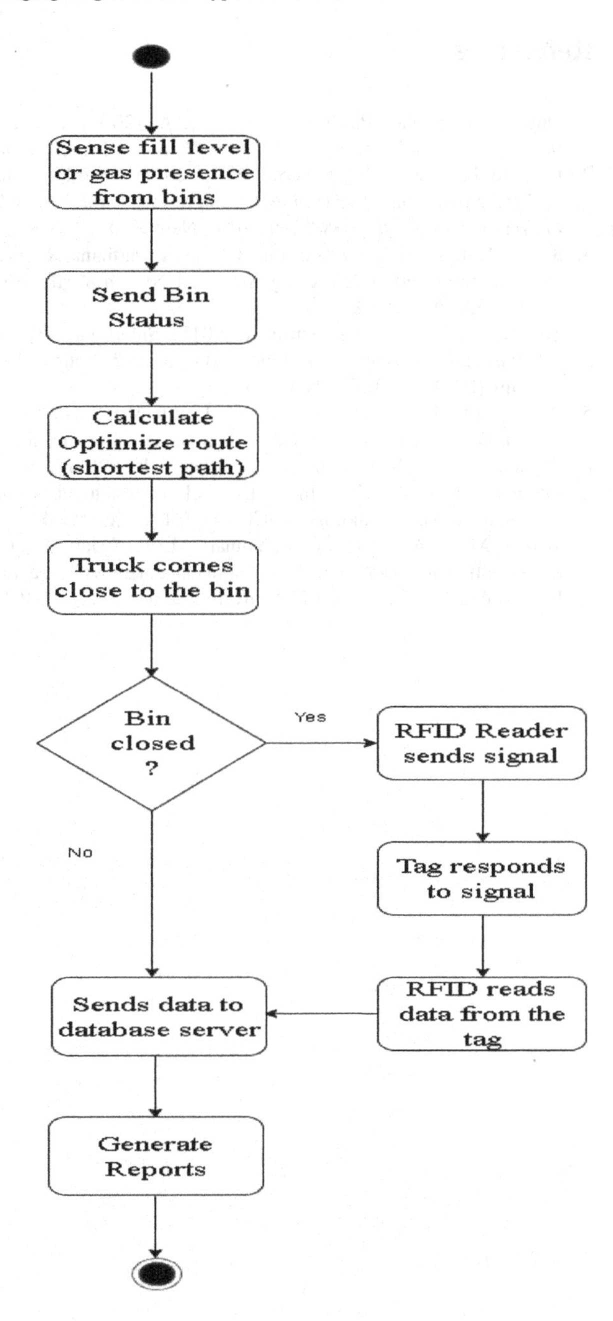

Fig. 10 Anatomy of system structure for mobile application

References

1. Dugdhe, S., Shelar, P., Jire, S., & Apte, A. (2016). Efficient waste collection system. In *International Conference on Internet of Things and Applications (IOTA)*. IEEE, New York.
2. Folianto, F., Low, Y. S., & Yeow, W. L. (2015). Smartbin: Smart waste management system. In *2015 IEEE Tenth International Conference on Intelligent Sensors, Sensor Networks and Information Processing (ISSNIP)*. IEEE, New York.
3. Sathish Kumar, N., Vuayalakshmi, B., Jenifer Prarthana, R., & Shankar, A. (2017). IOT based smart garbage alert system using Arduino UNO. In *Region 10 Conference (TENCON), 2016 IEEE*. IEEE, New York.
4. Thakker, S., & Narayanamoorthi, R. (2015). Smart and wireless waste management. In *2015 International Conference on* Innovations in Information, Embedded and Communication Systems (ICIIECS). IEEE, New York.
5. Russel, Md. M. H., Chowdhury, M. H., Uddin, Md. S. N., Newaz, A., & Talukder, Md. M. M. (2013). Development of automatic smart waste sorter machine. Research Gate.
6. Shyam, G. K., Manvi, S. S., & Bharti, P. (2017). Smart waste management using Internet-of-Things (IoT). In 2017 2nd International Conference on Computing and Communications Technologies (ICCCT). IEEE, New York.
7. Wahab, M. H. A., Kadir, A. A., Tomari, M. R., & Jabbar, M. H. (2015). Smart recycle bin: A conceptual approach of smart waste management with integrated web based system. In 2014 International Conference on IT Convergence and Security (ICITCS). IEEE, New York.

Performance Evaluation and Analysis of Feature Selection Algorithms

Tanuja Pattanshetti and Vahida Attar

Abstract Exorbitant data of huge dimensionality is generated because of wide application of technologies nowadays. Intent of using this data for decision-making is greatly affected because of the curse of dimensionality as selection of all features will lead to over-fitting and ignoring the relevant ones can lead to information loss. Feature selection algorithms help to overcome this problem by identifying the subset of original features by retaining relevant features and by removing the redundant ones. This paper aims to evaluate and analyze some of the most popular feature selection algorithms using different benchmarked datasets K-means Clustering, Relief, Relief-F, Random Forest (RF) algorithms are evaluated and analyzed in the form of combinations of different rankers and classifiers. It is observed empirically that the accuracy of the ranker and classifier varies from dataset to dataset. Novel concept of applying Multivariate co-relation analysis (MCA) for feature selection is made and results show improved performance over legacy based feature selection algorithms.

Keywords Classification · Clustering · Feature selection algorithms
Relief · Relief-F · Random forest

1 Introduction

In machine learning and data mining applications pertaining to real-world problems often there are issues with the data with *n*-dimensions where '*n*' can vary between several hundreds [1]. Processing time and memory requirements of the algorithms is hugely affected by the data with high dimensionality. Algorithm's performance gets degraded because of presence of irrelevant, redundant and noisy dimensions

T. Pattanshetti (✉) · V. Attar
College of Engineering Pune, Savitribai Phule Pune University, Pune, Maharashtra, India
e-mail: trp.comp@coep.ac.in

V. Attar
e-mail: vahida.comp@coep.ac.in

V. E. Balas et al. (eds.), *Data Management, Analytics and Innovation*,
Advances in Intelligent Systems and Computing 808,
https://doi.org/10.1007/978-981-13-1402-5_4

which is often referred to as due to the curse of dimensionality. In order to make sense out of a huge dataset (having hundreds of features and instances), it is cumbersome to analyze each and every feature. Feature Selection is a technique through which the irrelevant or redundant features in a dataset can be removed, making it easier to analyze the dataset with small subset of features [2]. Feature ranking is a similar approach wherein the attributes are ranked in accordance with their order of significance [3]. Here the real challenge lies in selection of optimal number of features such that the resulting subset of features is most relevant and provide maximum accuracy. Feature selection techniques help to curtail the issues related to high dimensions of data by identifying most relevant features in an effective and efficient manner.

In some cases, feature ranking algorithms are used along with classifiers for determining accuracy and this approach is coined as "Wrapper" method [3, 4]. Usually in this technique the classifier performance on validation set is used to predict the efficiency of identified set of features and with iterations the predictor accuracy can be optimized. It is implemented as a cross-validation approach in which the algorithm is trained to estimate the prediction accuracy of the identified feature subsets. This technique promises good results but iteratively training the algorithm impacts its speed. For this very reason wrapper method is seldom used on large datasets containing many features. Filter method on the other hand is much faster than wrapper and hence can be applied to large data sets containing many features and hence is its wide applicability.

Depending upon existence and applicability of labels feature selection methods based on data can be broadly classified into supervised and unsupervised methods. In unsupervised learning the training data set includes samples where labels are missing. Typical approaches used in this method are clustering, probabilistic generative models and identifying meaningful data transformations. The added advantage of having a class label in supervised methods improves the effectiveness to discriminate and distinguish features into mutually exclusive classes. Supervised feature selection algorithms make use of "Sparse learning" as it is one of the effective approaches. In case of unsupervised feature selection algorithms as the labels are missing, by making use of clustering technique the cluster labels are generated and based on these multi-cluster feature selection (MCFS) labels sparse learning similar to supervised method is adopted [5] thus making it applicable to classification or regression problem. As most real-world data is unlabeled and looking at the expense of transforming it to labeled one, unsupervised feature selection has thrown new challenges to researchers. The validation of feature selection algorithm by making use of classifier works well with labeled dataset but in case of unlabeled dataset validation of feature selection algorithm remains a challenge [6].

To overcome this limitation new approach of applying multivariate correlation analysis (MCA) for feature selection is proposed. MCA can be used on both labeled and unlabeled datasets as it removes the dependency on classifying the features on labeled attributes. In this approach correlation amongst features is calculated and the features with higher correlation score are chosen to be candidates of optimal features set. Depending upon the co-relation score the features are ranked. The classifier is trained iteratively by making use of these ranked features and respective classifier accuracy with particular feature subset is measured. For example, the four features ranked in descending order are $\{a, b, c, d\}$, then accuracy is evaluated with the combination of highest ranked features viz; $\{a, b\}$ with 70% accuracy $\{a, b, c\}$ with 80% accuracy and $\{a, b, c, d\}$ with 70% accuracy. Thus, the combination which has the highest accuracy percentage (in this case $\{a, b, c\}$) is selected and represents the subset with optimal number of features.

The rest of this paper is organized as follows. In Sect. 2, existing feature selections techniques are covered. In Sect. 3, proposed technique which makes use of multivariate correlation score for feature ranking is discussed. In Sect. 4, the evaluation of proposed technique is carried out and in Sect. 5 the paper is concluded with a glimpse of future work.

2 Literature Survey

Assume a feature set is given as, $F = \{f_1, f_2, \ldots f_n\}$, then machine learning tries to associate functional relationship $Y = f(X)$ between an input set X and output set Y, where $X = F$. However, obtaining output Y may requires only subset of input features $X \in F$. Feature selection helps us to extract subset of complete set which can predict the output Y with same or greater accuracy as compared to accuracy when using complete input set X, and while keeping computational cost minimum. Feature selection methods can be broadly classified into three categories namely filter, *wrapper* and *embedded* [2, 7]. Filter approach uses threshold measure to generate feature subset whereas wrapper approach uses prediction model by training a classifier to produce feature subsets. Wrapper methods are computationally expensive as they tend to call induction algorithm for every feature set generated. This has compelled researchers in finding alternative method that requires less computations. Embedded method is a collective technique wherein feature selection is done as a part of model construction process [5].

As most real-world problems include multi-class scenarios it is desirable that an algorithm operates in multi-class domains (as opposed to two-class domains) [6]. Thus, it becomes essential to choose algorithms that can operate on multi-class domain problems. While finding out relevant features it is noteworthy to consider combined effect of several features on result. Most of the earlier algorithms do not calculate relevance as a combination of more than one feature. One of the important requirements of feature selection algorithms is that it should work on nominal as well as categorical values so as to work with real-world datasets [7]. The multivariate correlation analysis technique is used to find correlation between multiple variables and to evaluate how many features of the subset are relevant. It selects optimal set of attributes which are relevant. The notion of relevancy is introduced by Kohavi and John [7]. Here the authors categorize three levels of feature relevancy as strong relevance, weak relevance and irrelevance. A feature Xi is relevant if removing that feature affects the classification accuracy whereas it is weakly relevant if, output is not fully dependent on input feature.

One of the basic methods of feature selection is brute-force that exhaustively evaluates all possible combinations of feature sets thus finding best feature subset. However, it is computational cost is very high and there exists danger of over-fitting. Hence many researchers use greedy methods, such as forward selection for assisting feature subset finding. Greedy Forward Search [8] method, starts with ranking the features and then including only those features that contribute to the accuracy. It assumes that there exists dependence between features which are highly ranked. Similar approach is adapted in FOCUS [9, 10] in which researchers start with empty feature set and uses BFS to find minimal combination of features which can accurately predict classes.

Relief [11] operates on two-class domains and fails to discriminate between features exhibiting redundancy. It employs learning approach which is instance based and relevance weight is given to each feature which is used as a metric to classify the feature and assign it to a particular class [11]. The strategy used for distinguishing makes this technique immune towards noise-tolerant as well as robust to the interactions among features. The features being ranked by weight help in selecting only those features to be part of final set whose weight is above the user-specified threshold level. It uses Euclidean distance method to find the proximity of feature to a particular class in terms of nearest hit and nearest miss. The training data is randomly sampled by the algorithm and for every single sampled instance the distance is calculated. The instance belongs to the same class if there is a hit whereas if it is a farthest instance it is assigned to the opposite class contributing to nearest miss [2]. The feature weight is updated based on the values of nearest hit and nearest miss. The feature will receive a low weight if it fails to distinguish between instances of same class and vice versa. It fails to effectively identify relevant features in multi-class domain and to discriminate between redundant features. It's more effective when applied for large numbers of training instances. Relief-F is an updated version of Relief that operates on multi-class problem [2]. Relief-F generalizes the behavior of Relief to classification. Relief-F [12] uses Manhattan distance to find nearest hit and nearest miss. Manhattan distance uses absolute value distance as opposed to squared error (Euclidean) distance in Relief. The techniques used for calculating distance can also impact feature ranking. Absolute value distance gives more robust results [11]. I-Relief [12] algorithm is used to solve multi-class problems as it makes use of statistical search approach. R-Relief-F is used to solve regression based problems similar to I-Relief using statistical approach.

Relief-F [12] used to derive the most relevant subset of features. It belongs to Relief family algorithm, works on the same basic principle that of Relief [11]. It calculates the feature weight for each attribute and then ranks them. The weight of the attribute in Relief-F is a measure which is obtained from difference in the values of instances for that particular attribute depending on how well it distinguishes these instances from same class or different class [11]. For this purpose, we use nearest hit H, for the same class and nearest miss M, nearest neighbor from different class for the same instance. This adds up quality estimation measure '*W*', weight of that attribute. For example, for an instance *I* and *H*, if they have separate values for an attribute *A*, if the attribute separates these two instances within same class the quality estimation measure *W* for this attribute is decreased. Whereas for instances *I* and *M*, if they have different values for attribute *A* and the attribute separates them into two instances in different classes then the quality estimation measure for this attribute is increased. This process is repeated '*m*' times, which is user-defined parameter and here it is equal to the number of instances in the dataset. The K-value basically distinguishes Relief-F from Relief [13]. Selecting optimal value of '*K*' is crucial as this implies the number of iterations to be carried to identify the number of hits and misses. *K*-value is related to complexity of the problem and depends on number of instances. If '*K*' chosen is very small then essence of robustness is lost and if chosen very large then it may unnecessarily influence the ranking by exhibiting biased behavior. This problem is dealt by iteratively calculating the weights of parameters till there is convergence. Thus, Relief-F is more robust compared to Relief. Relief-F algorithm performs better in case of dependent attributes when there is split in decision. I-Relief [14] focuses on the comparison of differences on sample dataset. If samples are not in close proximity then their contribution to the relevance weight will be zero [13]. Random Forest (RF) works similar to decision tree algorithm.

In Fast Correlation-Based Filter approach [15] relation between feature-feature and feature-class is determined. In first step ranking of correlated features using Symmetrical Uncertainty Coefficient is done and a threshold over coefficient is applied to limit number of features. In the second step redundant features are removed by ranking the features based on their correlation. Kolmogorov Smirnov-Correlation-Based Filter test [16] is used to determine the correlation between various features and their class labels. Unsupervised feature selection approach can work with large datasets. It uses feature similarity between pairs of features calculated with custom indexing mechanism. The features are clustered and features representing most of the information are chosen from these clusters. This approach is redundancy tolerant.

In Heuristic Search Algorithm [17] user selects the value of '*d*' (*d* number of features from set *F* with total number of features '*p*' ($d \leq p$)) making it unsuitable for real-world problems. Determining optimal value of '*d*' is a challenge as if *d* is small, it may reject relevant features and even single excluded feature may lead to poor concept description. On the other hand, if *d* is set to higher value then it may lead to inclusion of irrelevant features. These types of algorithms further exhibit poor performance when features interact with each other. Ensemble feature selection repetitively select features using data re-sampling and final aggregates various runs to select commonly ranked features. Such approaches help to reduce biasing in feature ranking.

Rankers are used for preprocessing and efficient classifiers are needed to match with them to give appropriate subset of features. The weight estimators like Gain Ratio and Gini do not identify the conditional dependency between the attributes optimally [18].

K-means clustering works better with large dataset, it partitions data based on the criterion function of square-error and then identifies the high-density region and assigns the clusters. The working is as mentioned below in following manner.

a. Select randomly K clusters and calculate their centroids ('M' is the centroid with 'K' clusters in 'n' dimensional space)

$$M_K = \left(\frac{1}{n}\right) \sum_{i=i}^{n_m} x_i \mid (1 \leq m \leq k) \tag{1}$$

b. Assign each instance to a particular cluster.
c. Calculate Euclidean distance and square-error for each instance and based on this assign the instances to new centroids.

$$e_k = \sum_{i=0}^{n_m} dist\ (x_i,\ M_k) \quad where \quad 1 \leq m \leq k \tag{2}$$

This indicates within cluster variation.

$$J = \sum_{i=1}^{k} e_i^2 \quad where\ J\ \ is\ squared\ error\ function \tag{3}$$

d. Obtain these results and repeat till convergence.

K-means is chosen here because of its simplicity. Time complexity and space complexity $O(n * k)$ and $O(k + n)$ respectively are relatively better. As this algorithm does not depend on the order of the instances this helps to get similar results each time clustering is performed.

Vapnik [19] introduced Support vector machine (SVM) for feature selection, it's a supervised data classification method. SVM works by generating a hyper-plane which could distinguish data points into different classes. In most of the datasets there are multiple hyper-planes separating classes, in such case SVM choose hyper-plane that is farthest from its nearest data points. Newly arrived data points are derived based on its location in the half space. SVM [20] performs better with large features set. When data is categorical and data distribution is linear then linear SVM is applied, however when data is nonlinear kernel SVM is applied. When SVM is used for feature selection the data is represented into very sparse representation allowing very simple classification rule and controlling over-fitting by maximizing the margin. RF algorithm [4] exhibits good performance in problems involving both classification and regression. However, its accuracy gets affected by high-dimensional data. Multi valued features are favored by RF but it is tolerant to noise. While constructing tree from raw samples the features are randomly sampled so the generated node might get biased by irrelevant features. In another approach RF score is used for ranking informative features.

Koller and Shamai [21] uses Markov blanket as a method for feature selection. Their approach uses backward elimination in which the feature is removed if the class label is not dependent on feature. Similar approach taken by Langley and Sage's [22] called OBLIVION which is a combination of wrapper approach with K-NN approach. Their system initially contains all the features, those features were removed that does not have effect accuracy when removed. Finally, it halts when accuracy starts declining. Table 1 shows application of few feature selection algorithms from previous work summarized below.

Table 1 Application of feature selection algorithms

Algorithm name	Type	Approach used	Benefit	Drawback/ purpose/results
Relief	Filter	Relevance evaluation	Scalable to high dimensionality data set	Can't eliminate redundant features/ preprocessing data
Correlation-based feature selection	Filter (Multivariate)	Symmetric uncertainty	Handles irrelevant, redundant features	Can't handle numeric class problems/ evaluation function is biased toward highly correlated features class
Fast correlation-based filter	Filter	Predominant correlation as a goodness measure	Reduces dimensionality	Can't handle feature redundancy/ efficient for high dimensional data
Fast clustering-based feature subset selection	Filter	Graph-theoretic clustering method	Dimensionality is hugely reduced	Works well only for Microarray Data/proposed and evaluated
Condition dynamic mutual information feature sel.	Filter	Conditional dynamic Mutual Information	Better performance	Sensitive to noise/ proposed and evaluated
Affinity propagation—sequential feature selection	Wrapper	Applied SFS on clusters to get best subset	Faster than sequential feature selection	Accuracy is not better than SFS/ APSFS is proposed based on affinity propagation clustering

(continued)

Table 1 (continued)

Algorithm name	Type	Approach used	Benefit	Drawback/ purpose/results
Evolutionary local selection algorithm	Wrapper	K-Means Algorithm used for clustering	Covers a large space of possible feature combinations	Cluster quality decreases with increased number of features
Wrapper based feature selection using SVM	Wrapper	Feature selection —forward sequential used, SVM-evaluation.	Better accuracy, faster Computation.	Selection algorithm of iterative nature
Hybrid feature selection	Hybrid	Filter-mutual information. Wrapper-feature sel.	Improves accuracy	High computation cost for high dimensional data set

3 Proposed System

As applications are becoming more complex and data stored by these applications can be very large; gaining information from such unprocessed data could misguide machine learning algorithms [4]. Using feature selection algorithms could greatly reduce number of computations [2, 3] and also help machine learning algorithms to work on unbiased datasets. Over the labeled dataset supervised feature selection algorithm works well, however selecting threshold level is a real challenge which can be solved by using classifiers. For unlabeled datasets feature selection algorithms cannot be trusted and thus we lack a good measure of accuracy. Therefore, problem of feature selection over unlabeled dataset has attracted more attention. As discussed earlier most of the feature selection/ranking approaches consider each attribute as standalone feature but in real world more than one attributes contribute to final output. Final target variable should be extracted from these multi-feature interactions. Based on the findings of state of art literature, the technique called multi-feature correlation score is proposed and can be used for determining the relevant features.

The architecture of proposed system is shown in Fig. 1. The input to this system is raw dataset consisting of multiple features which is preprocessed and normalized and then the matrix representation of each sample is created. Each samples matrix representation is then used for calculating the correlation score between every feature pair. This calculated correlation value for every pair is then passed to ranker algorithm which then ranks those pairs in descending order. The evaluation is carried out by comparing various feature selection algorithms with proposed technique and the results gathered over labeled and unlabeled datasets.

Consider a unlabeled dataset $X = [x_1, x_2, \ldots, x_n]$ which consists of n number of samples and each sample consists of m number of features which can be represented as, $x_i = \{f_1, f_2, \ldots, f_m\}$. Extracting relevance of each feature using multi-variable

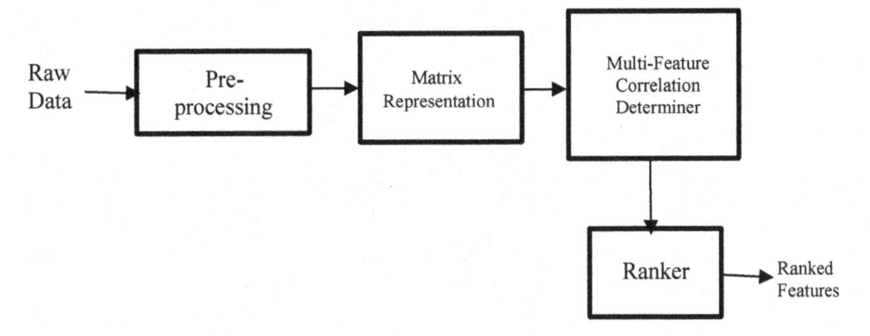

Fig. 1 System architecture

correlation technique is a three-step approach. Initially each sample is represented in two-dimensional matrix form represented as,

$$x_i = \begin{bmatrix} f_1 & 0 & 0 & 0 & 0 & 0 \\ 0 & f_2 & 0 & 0 & 0 & 0 \\ 0 & 0 & f_3 & 0 & 0 & 0 \\ .. & 0 & 0 & 0 & f_{m-1} & 0 \\ 0 & 0 & 0 & 0 & 0 & f_m \end{bmatrix} \qquad (4)$$

$x_i = v_{j,k}$ where, $1 \leq i \leq m, 1 \leq j \leq m, 1 \leq k \leq m, j! = k.$

Each diagonal element of matrix shown above represents a single feature from the sample. In second step, Euclidian distance between any two features is calculated. This information is represented in matrix form as shown below,

$$d_i = \begin{bmatrix} 0 & e(f_1 - f_2) & .. & e(f_1 - f_{m-1}) & e(f_1 - f_m) \\ 0 & 0 & .. & e(f_1 - f_{m-1}) & e(f_2 - fm) \\ 0 & 0 & .. & .. & .. \\ 0 & 0 & 0 & 0 & e(f_{m-1} - f_m) \\ 0 & 0 & 0 & 0 & 0 \end{bmatrix}, \qquad (5)$$

where, e is distance function and $e(fi - fj) = 0$ iff, $i == j$, also $|e(fi - fj)| = |e(fj - fi)|$.

Thus, two triangular area maps obtained are mirror images and are exactly similar in nature. Thus, only upper/lower triangle values can be calculated which serves the purpose. These two steps are repeated for every sample in the dataset X. The mean μ and standard deviation σ is computed using the equation given below.

$$\mu_{i,j} = \sum_{j=1}^{n} x_{i,j} \tag{6}$$

$$\mu = \sum_{i=1}^{n} di \tag{7}$$

$$\sigma = \frac{1}{n}\sum_{i=1}^{n} d_i - \mu \tag{8}$$

For every sample the covariance matrix is calculated as follows:

$$c_i = \frac{1}{n-1}\begin{bmatrix} \sigma(f_1,\mu_1) & \cdots & \sigma(f_1,\mu_m) \\ \vdots & \ddots & \vdots \\ \sigma(f_m,\mu_1) & \cdots & \sigma(f_m,\mu_m) \end{bmatrix} \tag{9}$$

Finally, the mean of all covariance matrices is calculated for the given dataset to rank features based on their relevance, as given in the equation.

$$c_{\text{mean}} = \sum_{i=1}^{n} c_i \tag{10}$$

From the mean covariance matrix C_{mean}, it can be inferred by how much amount feature f_i is correlated with feature $f_j, f_{j+1}, \ldots, f_m$. The values of covariance matrix are sorted in ascending order to rank the features according to the score. The threshold τ is applied to remove the features which are having low ranking score.

4 Experimental Results

To verify the applicability of proposed feature selection algorithm, the initial experiments are carried out on publicly available, preprocessed and normalized datasets such as weather and breast cancer datasets. Initially, the results are carried out by extracting relevant features using well known FSA like Relief, Relief-F. Table 2 gives the statistics of each of the dataset used for carrying out evaluation. Relevant features are extracted at different intervals by gradually decreasing the threshold level. Using the training dataset only relevant features are used for training a classifier. Accuracy of classifier gets impacted by the order and relevance of the features; this helps in identifying the relevant features. RF is selected as a classifier for experimentation and the results collected are given below. By giving the training dataset only relevant features are used for training a classifier and accuracy is measured. This helps in offering a clear view of how accuracy gets

Table 2 Benchmark dataset statistics

S. No.	Dataset	Total number of samples	Number of features	Number of training samples	Number of test samples
1	Weather forecast	1648	14	1483	165
2	Breast cancer Wisconsin (diagnostic)	569	33	480	89

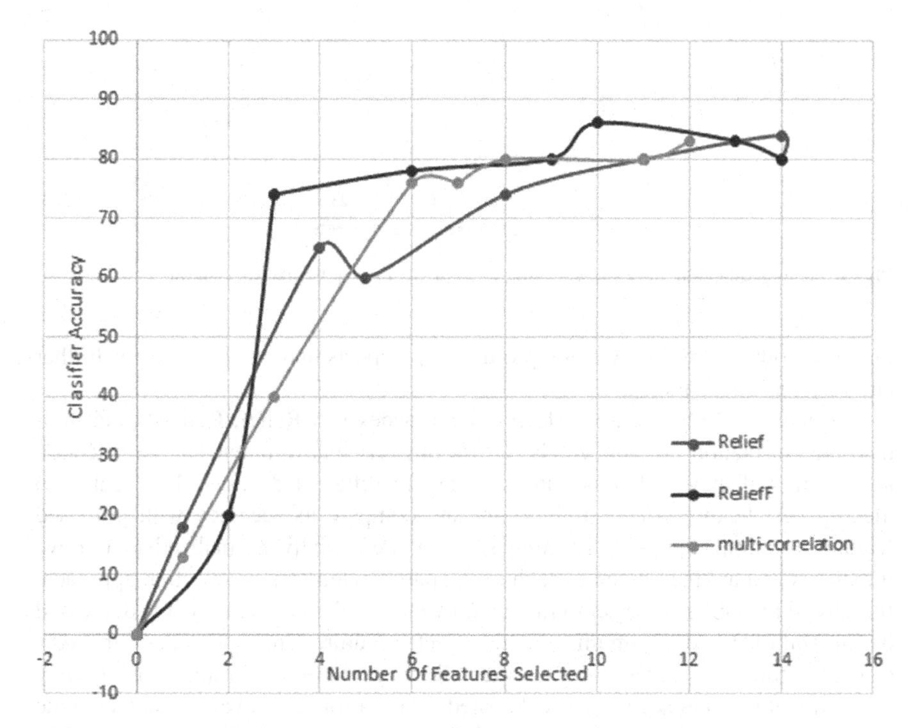

Fig. 2 Feature selection algorithms accuracy for weather forecast dataset

impacted while features are included into relevancy list and in understanding which features are relevant.

For unsupervised approach before using benchmark dataset [2] for determining the accuracy of proposed feature selection algorithm the labels are removed from the datasets and the ranked features are extracted at different threshold levels. These features are used for training the classifiers and accuracy is determined. Figures 2 and 3 show the impact of feature selection algorithm on classifier accuracy. To determine the classifier accuracy the classifier training is carried with increasing training feature set (features were added one by one). The results suggest that

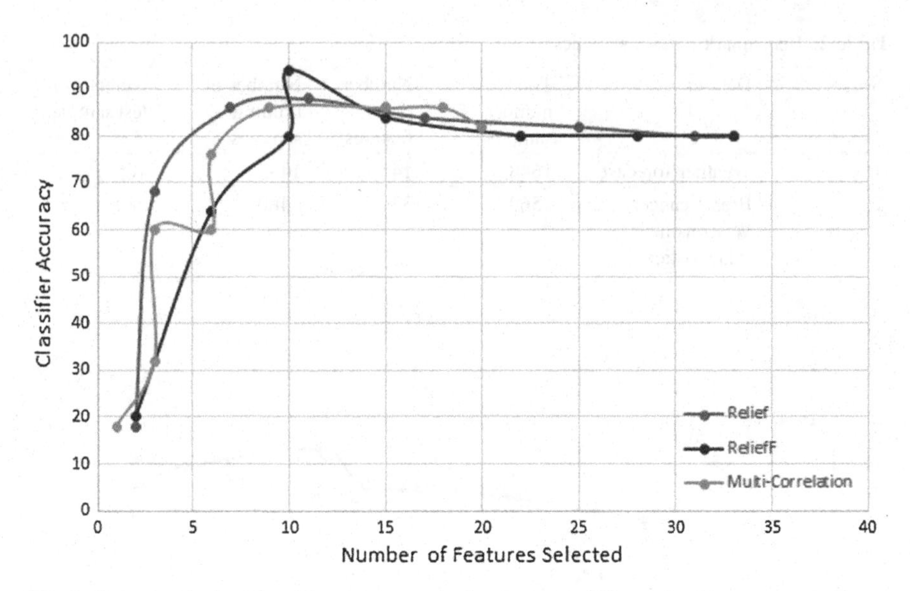

Fig. 3 Feature selection algorithms accuracy for breast cancer Wisconsin dataset

ranking produced by MCA holds promising prospects and can be used for further analysis and applications.

For analysis three features selection approaches viz; Relief, Relief-F and multivariate correlation are applied. It is very obvious that different feature selection algorithms will have difference in accuracy for different datasets. The goal is to identify best feature selection algorithm and compare its accuracy with proposed feature selection approach. Training samples are initially normalized to remove impurities and are then given to each of the participated feature selection approach. Initially, threshold level is set to maximum value '0.9' and gradually it is decreased by interval of 0.1 to minimum '0' and extracted features are observed. With every changing threshold level, the classifier (Random Forest) is trained using only selected features and accuracy is calculated. The 10-fold cross validation technique is used to verify the results obtained from different feature selection algorithms. The results collected are given below. Figure 2 shows the classifier accuracy on weather dataset.

After carrying out experiments over different feature selection algorithms, it is observed that model performance decreases as number of features gets increased. The order in which features are getting introduced plays crucial role in determining the accuracy of classifier.

To verify how feature selection helps in reducing the number of features and its effect on accuracy of classifiers, feature selection algorithms over breast cancer dataset is applied and the results are shown in Fig. 3. The results provided clearly shows that, feature ranking approaches removed irrelevant features from dataset.

The chosen breast cancer dataset has 33 features. Only 10 features from total 33 features are highly relevant. As features relevancy decreases adding these irrelevant features has negative impact on classifier. For breast cancer dataset, the maximum performance accuracy observed is 94% over Relief-F, whereas Relief and multivariate correlation approach attained nearly 86 and 90% of the accuracy respectively. The graph shows that as more features get included for classification the classifier performance starts steadily decreasing. It is also observed that multivariate correlation technique is useful and can be used for feature selection.

5 Conclusion

Many real-world problems currently existing have vast number of features for example data in bioinformatics. It becomes difficult to process such data and comprehend as many more features are often redundant and not relevant. Feature selection can be employed to identify the features which really contribute towards decision making and only such features should be considered as candidates for decision making process. Computation speed of algorithm can further be improved effectively by only focusing on such relevant and non-redundant features. This paper proposes multivariate correlation-based approach for feature selection. The evaluation results collected over benchmark dataset shows that this approach can be used for feature selection. Multivariate correlation score for feature ranking is found to be promising approach. In MCA, highly correlated features are represented as +1 and if features are not correlated then it is represented as −1. This helps users to remove those features from dataset whose Co-relation is less than 0. If classifier attribute is known then we can measure Co-relation value against classification index and for unsupervised datasets we can rely on correlation score of feature for its ranking. By applying feature selection, we are able to improve prediction quality.

References

1. Pattanshetti, T., & Attar, V. (2017). Survey of Performance Modeling of big data applications. In *7th IEEE International Conference on Cloud Computing, Data Science and Engineering, Confluence-2017*.
2. Guyon, I., & Elisseeff, A. (2003). An introduction to variable and feature selection. *Journal of Machine Learning*, pp. 1157–1182.
3. Chandrashekar, G., & Sahin, F. (2013). *A survey on feature selection methods* (pp 16–28). Amsterdam: Elsevier.
4. Genuer, R., Poggi, J., & Tuleau-Malot, C. (2010). Variable selection using random forests. *Pattern Recognition, 31* (14), 2225–2236.
5. Wang, S., Tang, J., & Liu, H. (2015). Embedded unsupervised feature selection. In *Proceedings of the Twenty-Ninth AAAI Conference on Artificial Intelligence*.
6. Mitra, P., Murthy, C., & Pal, S. K. (2002). Unsupervised feature selection using feature similarity. *IEEE Transaction Pattern Analysis Machine Intelligence, 24*(4).

7. Kohavi, R., & John, G. H. (1997). Wrappers for feature subset selection. *Artificial Intelligence*, pp 273–324.
8. Caruana, R. A., & Freitag, D. (1994) Greedy attribute selection. In *Proceedings of the Eleventh International Conference on Machine Learning* (pp. 28–36).
9. Kira, K., & Rendell, L. (1992). The feature selection problem: Traditional methods and a new algorithm. In *AAAI Proceedings*.
10. Almuallim, H., & Dietterich, T. G. (1991). Learning with many irrelevant features", Proceedings of the Ninth National Conference on Artificial Intelligence, San Jose, CA: AAAI Press, pp. 547–552,1991.
11. Kira, K., & Rendell, L. A. (1999). A practical approach to Feature Selection. In *9th International Conference on Machine Learning* (pp. 249–256).
12. Robnik-Šikonja, M., & Kononenko, I. (2003). Theoretical and empirical analysis of Relief-F and R-Relief-F. *Machine Learning, 53,* 23–69.
13. Kononenko, I. (1994). Estimating attributes: Analysis and extensions of RELIEF. In *European Conference on Machine Learning* (pp. 171–182).
14. Sun, Y. (2007). Iterative RELIEF for feature weighting: Algorithms, theories, and applications. *IEEE Transactions on Pattern Analysis and Machine Intelligence, 29*(6).
15. Yu, L., & Liu, H. (2003). Feature selection for high-dimensional data: A fast co-relation-based filter solution. In *Proceedings of the Twentieth International Conference on Machine Learning*.
16. Duch, W., & Biesiada, J. (2005). Feature selection for high-dimensional data: A Kolmogorov-Smirnov co-relation-based filter solution. In *Advances in soft computing* (pp. 95–104). Berlin: Springer.
17. Moore, A. W., & Lee, M. S. (1994). Efficient algorithms for minimizing cross validation error. In *Proceedings of the Eleventh International Conference on Machine Learning* (pp. 190–198).
18. Statnikov, A., Aliferis, C., Tsamardinos, I., Hardin, D., & Levy, S. (2005). A comprehensive evaluation of multi-category classification methods for microarray gene expression cancer diagnosis. *Bioinformatics*, pp 631–643.
19. Vapnik, V. (1998). *The nature of statistical learning* (2nd ed.). New York: Springer.
20. Gilad-Bachrach, R., Navot, A., & Tishby, N. (2004). Margin based feature selection—Theory and algorithms. In *21st International Conference on Machine Learning*.
21. Koller, D., Sahami, M. (1996). Toward optimal feature selection. In *International Conference on Machine Learning* (pp 284–292).
22. Langley, P., & Iba, W. (1993). Average-case analysis of a nearest neighbor algorithm. In *Proceedings of the Thirteenth International Joint Conference on Artificial Intelligence Chambery, France* (pp. 889–894).

Vertex Importance Extension of Betweenness Centrality Algorithm

Jiří Hanzelka, Michal Běloch, Jan Martinovič and Kateřina Slaninová

Abstract Variety of real-life structures can be simplified by a graph. Such simplification emphasizes the structure represented by vertices connected via edges. A common method for the analysis of the vertices importance in a network is betweenness centrality. The centrality is computed using the information about the shortest paths that exist in a graph. This approach puts the importance on the edges that connect the vertices. However, not all vertices are equal. Some of them might be more important than others or have more significant influence on the behavior of the network. Therefore, we introduce the modification of the betweenness centrality algorithm that takes into account the vertex importance. This approach allows the further refinement of the betweenness centrality score to fulfill the needs of the network better. We show this idea on an example of the real traffic network. We test the performance of the algorithm on the traffic network data from the city of Bratislava, Slovakia to prove that the inclusion of the modification does not hinder the original algorithm much. We also provide a visualization of the traffic network of the city of Ostrava, the Czech Republic to show the effect of the vertex importance adjustment. The algorithm was parallelized by MPI (http://www.mpi-forum.org/) and was tested on the supercomputer Salomon (https://docs.it4i.cz/) at IT4Innovations National Supercomputing Center, the Czech Republic.

J. Hanzelka (✉) · M. Běloch · J. Martinovič · K. Slaninová
IT4Innovations, VŠB—Technical University of Ostrava, 17. Listopadu 15/2172, 708 33 Ostrava, Poruba, Czech Republic
e-mail: jiri.hanzelka@vsb.cz

M. Běloch
e-mail: michal.beloch@vsb.cz

J. Martinovič
e-mail: jan.martinovic@vsb.cz

K. Slaninová
e-mail: katerina.slaninova@vsb.cz

J. Hanzelka
Department of Computer Science, FEECS, VŠB—Technical University of Ostrava, 17. Listopadu 15/2172, 708 33 Ostrava, Poruba, Czech Republic

V. E. Balas et al. (eds.), *Data Management, Analytics and Innovation*,
Advances in Intelligent Systems and Computing 808,
https://doi.org/10.1007/978-981-13-1402-5_5

Keywords Betweenness centrality · High performance computing
MPI · Traffic network

1 Introduction

The graph theory is proving to be a useful tool in the variety of research fields. A graph can describe objects and their communication and/or relationship. This has become very advantageous in areas like biology and chemistry [1–3], sociology [4–7], linguistics [8, 9], physics and informatics [10–14]. The structure and topology of the graph or network thus become the focus of the research. Although the meaning of vertices and edges is different across the scientific fields, the underlying idea remains the same. For example, the interaction of proteins in biology is not different to an interaction of two people in a social network. Proteins and people are represented by vertices and their interaction by edges. It is the weight of an edge that allows the quantification of the interaction.

The shared interest is finding the important or influential vertex in a given graph. For this purpose, the various centrality scores are capable of an answer. There is a plethora of them available, like closeness, degree, eigenvector or Ketz centrality, but the most common and widely adopted is the betweenness centrality score [15]. The role of betweenness centrality score is to identify the busiest vertices in the network. This score is computed using the knowledge of the shortest paths in the network. Therefore, it adopts the information from the edges. But the vertices also contain valuable information that should not be neglected.

Particularly, interesting and growing area where the graph theory has its place is the urban traffic monitoring. The latest development of the society sees the increase of the vehicles in the cities as more people are able to afford a car. Therefore, the advanced planning of the infrastructure and the traffic control become more important. One approach is to make a connection with the known physical problem. Daganzo [16] is treating traffic flow as a hydrodynamic model and solves it using differential equations. More general approach to traffic flow is based on the application of extended Kalman filter [17, 18]. Other authors noted the trend pattern in traffic flow and analyzed it as time series [19, 20]. Since the traffic flow is happening on the roads, it is natural to consider the road network as well. The graph theory helps in visualization of the transportation network, where the vertices represent crossroads, turnings, exits from the parking lot, etc., and edges represent roads. The oriented weighted graph makes the faithful description of the real network. Additionally, the weights of the edges can be used to represent various information about the conditions of a given section like speed, weather condition, time of the day, etc.

The graph of the city transportation network is complicated even for smaller cities. Therefore, it is desirable to extract as much information as possible from it. The common way is to use the betweenness centrality score that gives relevant answers about the vertex importance. The role of betweenness centrality score is to identify the busiest vertices in the network. This approach is examined by authors in

[21–24]. However, some authors argue that human rationality is bounded [25] and their choice of the shortest path is not equivalent to the topological choice. Another problem is a high dynamic of the network. Thus, the authors raised the question whether betweenness alone is enough to describe the traffic flow. While the concern is certainly valid, there is still a lot of research to be done on the topic.

For our purpose of traffic monitoring, the betweenness centrality is satisfactory. The problem we discovered is that the betweenness centrality uses the information about the shortest paths in the graph and it does not take into account the vertices which are passed when going from the origin to the destination. People do not use the network just for the sake of using it. They enter it at some origin point because they care for the destination. Let us consider a major event is happening in the city like a concert or political meeting. We can expect increased traffic towards this vertex when the people want to reach this event and when the people are leaving it. We could change the weights of all edges lying on the shortest paths to the destination to account the increase in traffic, but it would require the change of a significant amount of values. Moreover, when the people will leave the event, it would be difficult to assess which values to increase, since the destination points are unknown. Thus, we propose the assignment of the weight to the vertex. It can also help solve the dynamic problem mentioned above.

While the motivation behind the modification was its use in the traffic network, we believe its usefulness goes beyond it. While analyzing the social behavior of a group, the ability to increase the importance of a given vertex (individual) can be used to describe a sudden change in the group dynamics. When studying the spread of information in a network, credible and dependable sources can have increased importance. These are just few examples where the concept of the vertex importance can be considered.

This paper is organized as follows. Section 2 briefly reminds of the betweenness centrality. Section 3 describes the proposed modification. Section 4 shows the results of the performance and Sect. 5 shows the visualization and examples, both carried out on real traffic data. Section 6 concludes the paper.

2 Betweenness Centrality

Betweenness centrality or in full form Shortest path betweenness centrality first appeared in sociology [26]. Its main purpose was to quantify the individual's influence over the information flow in the social network. It has been generalized and adopted since as one of the centrality measures in the network. The higher value of betweenness centrality means the greater importance within the network. Specifically, within the context of the traffic network it usually represents a problematic section of the traffic flow.

Betweenness centrality of a vertex is defined as the ratio of the number of the shortest paths between an origin and a destination that pass through the vertex and the number of all the shortest paths between an origin and a destination.

Let us describe the traffic network with the graph $G = (V, E)$, where V is the set of vertices pair-wise connected by edges forming the set E. We define a path from $s \in V$ to $t \in V$ as an alternating sequence of vertices and edges, beginning with s and ending with t, such that each edge connects its preceding vertex with its succeeding vertex. The symbol $\sigma_{st} = \sigma_{ts}$ denotes the number of the shortest paths between vertex $s \in V$ and $t \in V$, and $\sigma_{st}(v)$ is the number of the shortest paths between s and t that goes through $v \in V$. Betweenness centrality BC for the vertex $v \in V$ is then defined as

$$BC(v) = \sum_{s \neq v \neq t \in V} \frac{\sigma_{st}(v)}{\sigma_{st}} \tag{1}$$

To get the shortest paths we use standard Dijkstra algorithm [27]. Betweenness centrality is computed using Brandes algorithm [28]. The pseudo code is mentioned below as Algorithm 1.

Algorithm 1. Original betweenness centrality for directed graph with edge lengths.

input : directed graph $G = (V, E)$ with edge lengths $\lambda : E \to \mathbb{R} > 0$
data : priority queue Q with keys $dist[\cdot]$
stack S (both initially empty) and for all $v \in V$
$dist[v]$: distance from source
$Pred[v]$: list of predecessors on shortest paths from source
$\sigma[v]$: number of shortest paths from source to $v \in V$
$\delta[v]$: dependency of source on $v \in V$
output: betweenness $c_B[v]$ for all $v \in V$ initialized to 0

1 **foreach** $s \in V$ **do**
2 **foreach** $w \in V$ **do** $Pred[w] \leftarrow$ empty list ;
3 **foreach** $t \in V$ **do** $dist[t] \leftarrow \infty$; $\sigma[t] \leftarrow 0$;
4 $dist[s] \leftarrow 0$; $\sigma[s] \leftarrow 1$ enqueue $s \to Q$
5 **while** *Q not empty* **do**
6 extract $v \leftarrow Q$ with minimum $dist[v]$; push $v \to S$
7 **foreach** *vertex w such that* $(v, w) \in E$ **do**
8 **if** $dist[w] = dist[v] + \lambda(v, w)$ **then**
9 $dist[w] \leftarrow dist[v] + \lambda(v, w)$
10 insert/update $w \to Q$ with new key
11 $\sigma[w] \leftarrow 0$
12 $Pred[w]$ *gets* empty list
13 **end**
14 **if** $dist[w] = dist[v] + \lambda(v, w)$ **then**
15 $\sigma[w] \leftarrow \sigma[w] + \sigma[v]$
16 append $v \to Pred[w]$
17 **end**
18 **end**
19 **foreach** $v \in V$ **do** $\delta[v] \leftarrow 0$;
20 **while** *S not empty* **do**
21 pop $w \leftarrow S$
22 **foreach** $v \in Pred[w]$ **do** $\delta[v] \leftarrow \delta[v] + \frac{\sigma[v]}{\sigma[w]} \cdot (1 + \delta[w])$;
23 **if** $w \neq s$ **then** $c_B[w] \leftarrow c_B[w] + \delta[w]$;
24 **end**

3 Modification

As mentioned in the introduction, the purpose of this modification is to describe the situation better, when the importance will change for one vertex or some number of vertices.

We identified that there are two natural types of events that can occur in experiments or real situations. The first event arises when the vertex is the source of the shortest paths. It means every shortest path leading from this point is influenced by the increase. To simulate this event, we selected the parameter *alpha* to describe the rise of the vertex importance.

The second event arises when the vertex is more important as a destination point for the betweenness calculation. This kind of behavior influences all the shortest paths that lead to the destination vertex. For this phenomenon, we designated the parameter *beta*.

Therefore, the importance parameter of the vertex in a modified version of the betweenness centrality algorithm consists of two positive real numbers *alpha* and *beta*.

- *alpha* $\in \mathbb{R} \geq 0$ (source vertex importance)—increases betweenness on the shortest paths directing from this vertex
- *beta* $\in \mathbb{R} \geq 0$ (destination vertex importance)—increases betweenness on the shortest paths directing to this vertex

The betweenness gain on the shortest paths is proportional to the chosen values. For example, if *alpha* is equal to 2 for vertex A ($alpha_A = 2$), then betweenness centrality gain on vertices that are on the shortest paths from vertex A to all other vertices is twice as big. This situation is illustrated in Fig. 1. The betweenness centrality is calculated just from the vertex A to all other vertices and the red edge means that it is lying on the shortest path from vertex A to all other vertices.

Change in *beta* value can be seen in Fig. 2, where betweenness centrality is also calculated just from vertex A to all other vertices. We set *beta* equal to 2 for vertex D ($beta_D = 2$). In this case, the betweenness is doubled for every surrounding

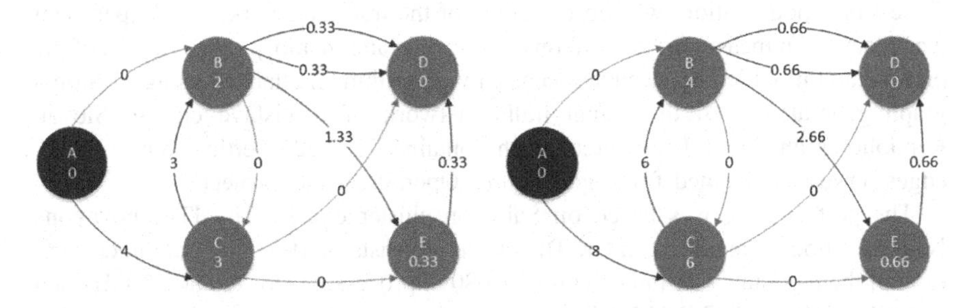

Fig. 1 Comparison of betweenness gain for vertex A, when $alpha_A = 1$ on the left and $alpha_A = 2$ on the right

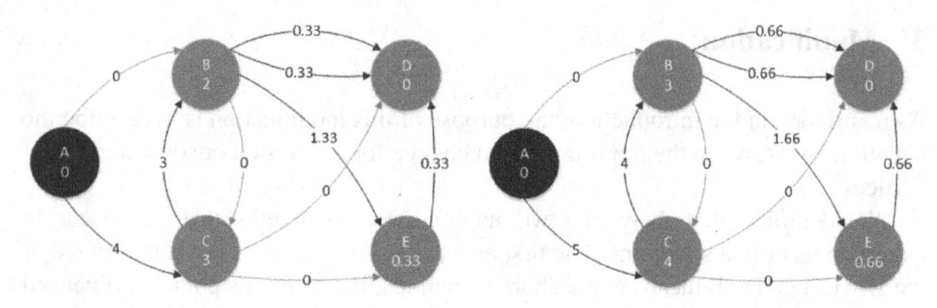

Fig. 2 Comparison of betweenness gain for vertex A, when $beta_D = 1$ on the left and $beta_D = 2$ on the right

vertex. The amount of betweenness gain for vertices that are further from the destination vertex slowly dissipates.

The pseudo code of this modification can be seen in Algorithm 2. The modification is on line 23 where we added the parameters *alpha* and *beta*. If *alpha* and *beta* values are set to value 1, we get original unmodified algorithm from Ulrik Brandes as seen in Algorithm 1.

Algorithm 2. Modification of original betweenness centrality for directed graph with edge lengths.

19 **foreach** $v \in V$ **do** $\delta[v] \leftarrow 0$;
20 **while** S *not empty* **do**
21 pop $w \leftarrow S$
22 **foreach** $v \in Pred[w]$ **do**
23 $\delta[v] \leftarrow \delta[v] + \frac{\sigma[v]}{\sigma[w]} \cdot (alpha[s] \cdot beta[w] + \delta[w])$
24 **end**
25 **if** $w \neq s$ **then** $c_B[w] \leftarrow c_B[w] + \delta[w]$;

4 Performance Test on a Traffic Network

To test our modification, we chose a graph of the traffic network. This experiment was focused on measuring the difference in speeds and memory requirements of the original and modified betweenness centrality algorithm parallelized using MPI on a graph generated from the actual traffic network of Bratislava city in Slovak Republic as an input. The source graph contained 225,526 vertices and 529,767 edges. Data are obtained from open source OpenStreetMap project[1].

The performance was tested on Salomon cluster operated by IT4Innovations National Supercomputing Center. The cluster consists of 1008 compute nodes and each of them contains 2x Intel Xeon E5-2680v3 processors clocked at 2.5 GHz and 128 GB DDR4@2133 RAM. When the experiments were performed, the operating

[1]https://www.openstreetmap.org

Table 1 Comparison of the original and modified version of the betweenness centrality algorithm on the graph of Bratislava city traffic network

Computing nodes	MPI processes	Speed(s)		Memory requirements per process (MB)		Speed decrease (%)	Memory increase (%)
		Orig. BW	Mod. BW	Orig. BW	Mod. BW		
1	24	812.21	883.35	86.0	89.5	8.76	4.07
2	48	414.74	452.91	95.4	98.9	9.20	3.67
4	96	208.45	226.52	88.5	92.0	8.67	3.95
8	192	104.88	113.68	89.0	93.0	8.39	4.49

system was CentOS 6.7 and the code was compiled using Intel C++ Compiler 17.0. To increase the speed for a given architecture, *-xHost* flag was used in a compilation. This flag allows to use the highest instruction set available on our Haswell architecture.

Table 1 shows the speed and memory requirements of the original and modified version of the betweenness centrality algorithm. The test was performed for a different number of computing nodes from one to eight. On each computing node, a process per processor core was created. So, one node could have 24 processes, because each node had 2 × 12 cores. On average, each process consumed 89.73 MB of RAM for the original version of the algorithm and 93.35 MB for the modified version. The modified version of algorithm took on average about 8.76% longer when compared to the original version. Speed decrease (SD) was calculated according to formula $SD = (t_2 - t_1)/t_1 * 100$, where t_1 is time of original version of betweenness and t_2 is time of modified version of betweenness. We also tested the memory requirements of both versions of the algorithm. From the results, we can compute that the modified version of algorithm required around 4.05% more memory. To summarize, the overhead of modified version is acceptable for us.

5 Experiments

To better describe and provide visual example of our modification, we chose a graph of a traffic network again because it is easy to interpret and visualize. For this particular case, we can notice certain ambiguity between parameters *alpha* and *beta*. Consider the example of people going to work from certain suburb area in the morning. Do we increase the parameter *alpha* to show the increase in traffic coming from the source or parameter *beta* to show increase in traffic going to the destination? We usually want to increase both values, but we should base this on our knowledge of the local network and people's behavior. Therefore, this change should be used in the context of macro-simulation when *alpha* and *beta* values are

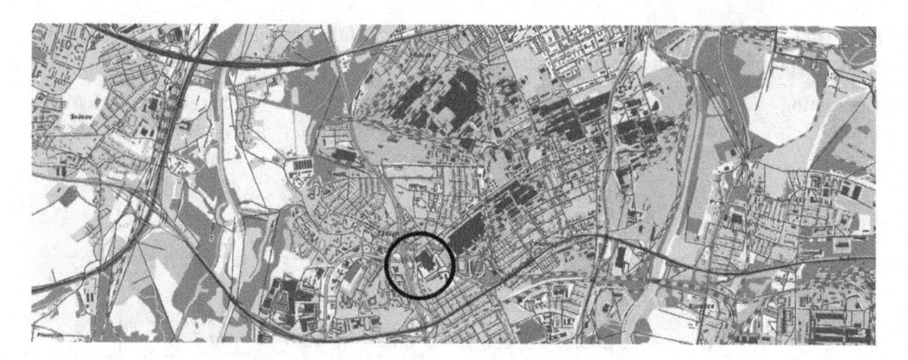

Fig. 3 Visualized betweenness centrality output for a part of Ostrava city. In black circle, there is a selected vertex with value *beta = 1*

selected in agreement with the analysis of mobility of the population. If we get the origin-destination matrix, we can use it to adjust these values for each vertex and the resulting betweenness can be more precise. We can also simulate differences of betweenness for different times on the same graph of the traffic network. Therefore, we can identify bottlenecks during weekdays in the morning, when a lot of people are going to work, or in the evening when they are going back home.

Our modified version of the betweenness centrality was integrated into the Floreon+ system[2] operated by IT4Innovations National Supercomputing Centre[3]. The objectives of the Floreon + system are monitoring, modeling and prediction of crisis situations. With the help of this system, we created examples of betweenness centrality algorithm results when we modified *alpha* and *beta* values. We used different values to illustrate the contrast between them and the default values. The subset of a real traffic network of Ostrava city, the Czech Republic was used as an input graph. The dimension of the observed area was 45 × 60 km. The visualization of the betweenness centrality is therefore confined to this area.

You can see the selected area of traffic network of Ostrava city in Fig. 3. Betweenness centrality score was obtained via the unmodified algorithm (we used default values *alpha = 1*, *beta = 1* in this example). The width of the red line denotes the value of betweenness score; the wider the line is, the bigger the betweenness centrality values are. At the bottom of this picture, there is a black circle with the selected vertex for which *beta* value will be changed.

Figure 4 shows the value of *beta* increased to 5000 for the selected vertex in the black circle. This change makes the vertex more important destination for the traffic. We can see a visible change of the line width directing to this vertex.

[2]https://floreon.it4i.cz

[3]http://www.it4i.eu

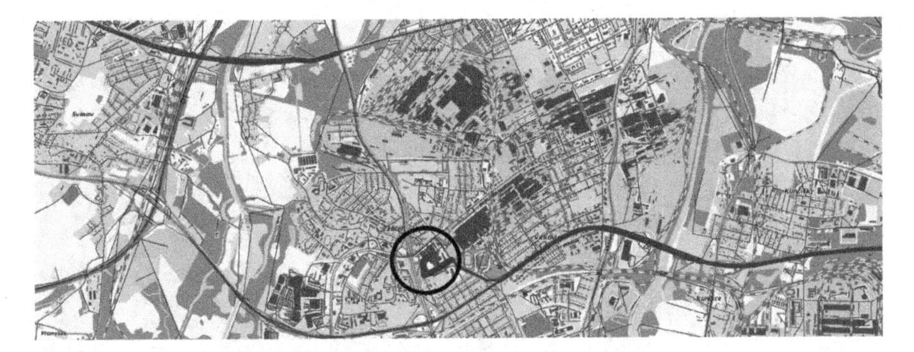

Fig. 4 Visualized betweenness centrality output for a part of Ostrava city. In black circle, there is a selected vertex with value *beta = 5000*

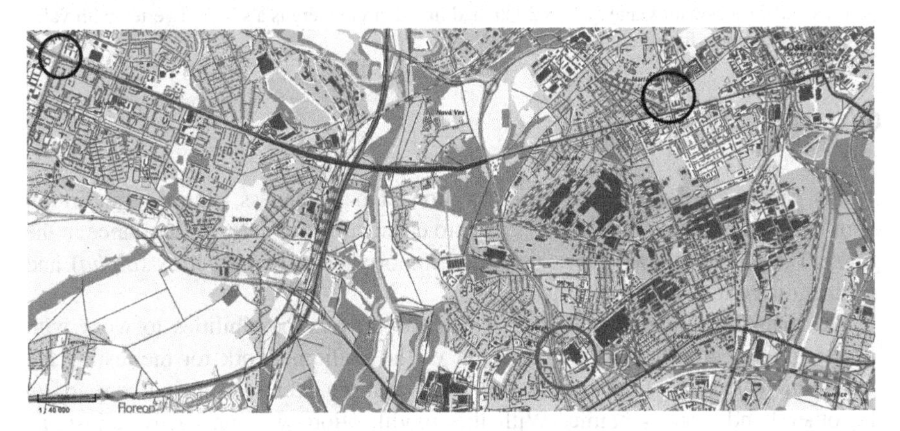

Fig. 5 Visualized betweenness centrality output for a part of Ostrava city. In black circles, there are selected vertices with value *beta = 1*, and in red circle there is a selected vertex with value *alpha = 1* (color figure online)

Figure 5 shows another example. The selected part of Ostrava city remains the same, but this time we consider three points of interest. Here we use the original betweenness centrality, i.e. *alpha = 1* and *beta = 1* for all the circles.

Then Fig. 6 shows how the betweenness changes in comparison to Fig. 5, when we set the value of *beta* to 2000 for the black circles and the value of *alpha* to 5000 for the red circle. This change is clearly visible and simulates higher traffic around the red circle where the selected vertex is more important as a source point for the shortest paths. Also, higher traffic is visible on main road between black circles, where the selected vertices are more important as a destination points.

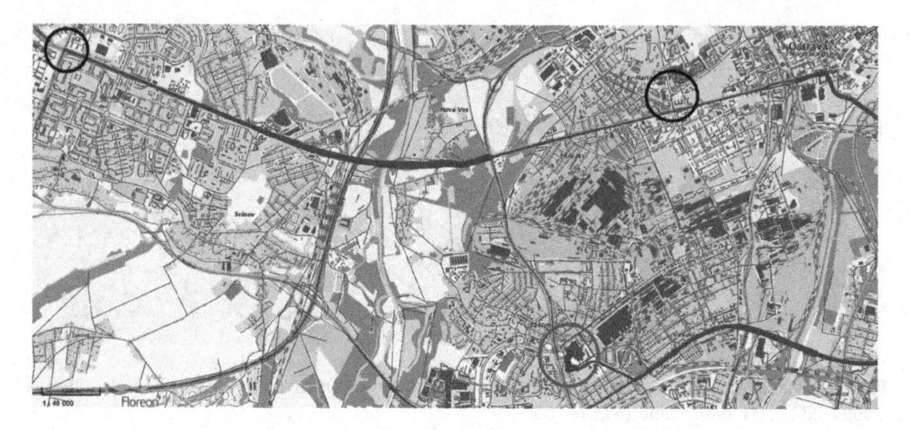

Fig. 6 Visualized betweenness centrality output for a part of Ostrava city. In black circles, there are selected vertices with value *beta = 2000*, and in red circle there is a selected vertex with value *alpha = 5000* (color figure online)

6 Conclusion

In this paper, we described the modification of betweenness centrality algorithm from Ulrik Brandes, so it is able to take into calculation the vertex importance in the graph. We identified two types of vertex importance—source vertex (*alpha*) and destination vertex (*beta*) importance.

The main objective of this change was to add new possibilities to work with betweenness centrality. We chose graphs of the traffic network for the testing the algorithm. This type of network is very static, but the events occurring here are on the other hand very dynamic. With this modification, we can easily adjust the betweenness score during the day cycle, and we can also quickly react to new events. It also allows us to anticipate problematic parts and bottlenecks in the traffic flow better.

The comparison of the original betweenness algorithm and our modified version shows favorable results for our modification. When they were both tested on the same traffic network depicting the city of Bratislava, the additional computation slowed the algorithm on average by 8.76%. The memory consumption increase was only 4.05%. The performance of the modification is therefore very good.

The attached visualization of the Ostrava city traffic network shows the interpretation of these two values and how they differ from each other.

The only thing we have not tackled in this paper is how to choose the values themselves. There is no quick and easy recommendation. It is imperative to know the structure of your network. It is important to perform the research on the mobility of the population for the particular traffic network. The origin-destination matrix can be used for more precise adjustment of the parameters. Monitoring of the network is also very useful to get information on movements within the network during the day cycle.

The future research will be focused on better understanding of the *alpha* and *beta* values. We also will focus on better analysis of our traffic network so we can better refine these coefficients.

Developing self-sustaining system capable of auto adjustment depending on the current situation is one of the goals of the project ANTAREX[4] and its use case focused on self-adapted server side/client side navigation system. The proposed extension of the betweenness centrality algorithm will be tested within this project on the server-side navigation system part to obtain a global view of traffic network and to be able to influence the traffic flow within the Smart City context.

Acknowledgements This work was supported by The Ministry of Education, Youth and Sports from the National Programme of Sustainability (NPU II) project 'IT4Innovations excellence in science—LQ1602', by the IT4Innovations infrastructure which is supported from the Large Infrastructures for Research, Experimental Development and Innovations project "IT4Innovations National Supercomputing Center—LM2015070", and partially by ANTAREX, a project supported by the EU H2020 FET-HPC program under grant 671,623, and by grant of SGS No. SP2017/182 "Solving graph problems on spatio-temporal graphs with uncertainty using HPC", VŠB—Technical University of Ostrava, Czech Republic.

References

1. Li, M., Wang, J., Chen, X., Wang, H., & Pan, Y. (2011). A local average connectivity-based method for identifying essential proteins from the network level. *Computational Biology and Chemistry, 35,* 143–150.
2. Xia, J., Sun, J., Jia, P., & Zhao, Z. (2011). Do cancer proteins really interact strongly in the human protein–protein interaction network? *Computational Biology and Chemistry, 35,* 121–125.
3. Hagmann, P., Cammoun, L., Gigandet, X., Meuli, R., Honey, C. J., Wedeen, V. J., et al. (2008). Mapping the structural core of human cerebral cortex. *PLoS Biology, 6,* 1479–1493.
4. Everett, M. G., & Borgatti, S. P. (1999). The centrality of groups and classes. *Journal of Mathematical Sociology, 23,* 181–201.
5. Szell, M., & Thurner, S. (2010). Measuring social dynamics in a massive multiplayer online game. *Social Networks, 32,* 313–329.
6. Wasserman, S., & Faust, K. (1994). *Social network analysis: Methods and applications.* Cambridge, England: Cambridge University Press.
7. Clifton, A., Turkheimer, E., & Oltmanns, T. F. (2009). Personality disorder in social networks: Network position as a marker of interpersonal dysfunction. *Social Networks, 31,* 26–32.
8. Vandenberghe, R., Wang, Y., Nelissen, N., Vandenbulcke, M., Dhollander, T., Sunaert, S., et al. (2013). The associative-semantic network for words and pictures: Effective connectivity and graph analysis. *Brain and Language, 127,* 264–272.
9. He, T., Zhao, J., Li, J. (2006). Discovering relations among named entities by detecting community structure. In *PACLIC20* (pp. 42–48).
10. Donges, J. F., Zou, Y., Marwan, N., & Kurths, J. (2009). The backbone of the climate network. *EPL (Europhysics Letters), 87.*

[4]http://www.antarex-project.eu/

11. Zhang, G. Q., Wang, D., & Li, G. J. (2007). Enhancing the transmission efficiency by edge deletion in scale-free networks. *Physical Review E, 76.*
12. Zhou, S., & Mondragón, R. J. (2004). Accurately modeling the internet topology. *Physical Review E, 70.*
13. Comin, C. H., & Da Fontoura Costa, L. (2011). Identifying the starting point of a spreading process in complex networks. *Physical Review E, 84.*
14. Kawamoto, H., & Igarashi, A. (2012). Efficient packet routing strategy in complex networks. *Physica A: Statistical Mechanics and its Applications, 391,* 895–904.
15. Klein, D. J. (2010). Centrality measure in graphs. *Journal of Mathematical Chemistry, 47,* 1209–1223.
16. Daganzo, C. F. (1994). The cell transmission model: A dynamic representation of highway traffic consistent with the hydrodynamic theory. *Transportation Research Part B: Methodological, 28,* 269–287.
17. Wang, Y., & Papageorgiou, M. (2005). Real-time freeway traffic state estimation based on extended Kalman filter: A general approach. *Transportation Research Part B: Methodological, 39,* 141–167.
18. Ngoduy, D. (2007). Real time multiclass traffic flow modelling-English M25 freeway case study. In *12th Conference of the Hong-Kong Society for Transportation Studies* (pp. 143–152).
19. Shang, P., Li, X., & Kamae, S. (2005). Chaotic analysis of traffic time series. *Chaos, Solitons & Fractals, 25,* 121–128.
20. Hong, W. C., Dong, Y., Zheng, F., & Lai, C. Y. (2011). Forecasting urban traffic flow by SVR with continuous ACO. *Applied Mathematical Modelling, 35,* 1282–1291.
21. Galafassi, C., & Bazzan, A. L. C. (2013). Analysis of traffic behavior in regular grid and real world networks.
22. Kazerani, A., & Winter, S. (2009). Can betweenness centrality explain traffic flow? In *12th AGILE International Conference on Geographic Information Science* (pp. 1–9).
23. Gao, S., Wang, Y., Gao, Y., & Liu, Y. (2012). Understanding urban traffic flow characteristics: A rethinking of betweenness centrality. *Environment and Planning B: Planning and Design.*
24. Zhao, P. X., & Zhao, S. M. (2016). Understanding urban traffic flow characteristics from the network centrality perspective at different granularities. *International Archives of the Photogrammetry Remote Sensing and Spatial Information Sciences, 41,* 263–268.
25. Piggins, A. (2012). Rationality for mortals: how people cope with uncertainty, by gerd gigerenzer. *The Journal of Positive Psychology, 7,* 75–76.
26. Freeman, L. C. (1977). A set of measures of centrality based on betweenness. *Sociometry, 40,* 35–41.
27. Dijkstra, E. W. (1959). A note on two problems in connexion with graphs. *Numerische Mathematik, 1,* 269–271.
28. Brandes, U. (2001). A faster algorithm for betweenness centrality. *Journal of Mathematical Sociology.*

Privacy Protection Data Analytics in Smart Home Environments with Secure Computation

N. Naveen and K. Thippeswamy

Abstract These days in smart home environments, one could find the absence of mechanisms to empower occupants so as to see and manage the data created by smart gadgets at peoples living places. By means of the expanding adoption physical devices such as remote systems, intelligent gadgets, and sensors homes have been getting to be smart home environments. Intelligent gadgets could obtain an unfathomable measure of delicate individual data. Nevertheless the incredible way of smart home data investigation has been building up defensive consideration. The gathering and processing of data of this data raises privacy concerns about how the people existing in a kind of a smart home environments could guarantee where this data would be shared just pertaining to their own particular great, as opposed to be shared, collected, used, or maliciously disclosed so as to meet the requirements which would damage their independence and security. Hence handling a sort of data ought to be exclusive to specific clients in charge of straight concern. This study proposes a framework displayed to keep up safety and saving protection so as to examine the data regarding from homes that are brilliant, in the absence of bargaining on utility data. This study deals with the implantation of a security protecting method of pertaining to the art of solving coding called cryptography as well as randomization has been utilized for keeping up the protection of touchy data pertaining to a person. Randomization is strategy; which adjusts unique data by the addition of a few noises arbitrarily to unique data which is independent of different reports. At this time cryptography strategy has been utilized to provide safety and security of sensitive attributes. Prior to the process of Randomization Data partition are performed in vertical and horizontal. At long last, giving right of entry of

N. Naveen (✉)
Department of Information Science & Engineering, Kalpataru Institute of Technology, Tiptur, Karnataka, India
e-mail: naveennmj2007@gmail.com

K. Thippeswamy
Department of Computer Science & Engineering, VTU PG Center, Mysore, Karnataka, India
e-mail: thippeswamy_k@hotmail.com

V. E. Balas et al. (eds.), *Data Management, Analytics and Innovation*, Advances in Intelligent Systems and Computing 808,
https://doi.org/10.1007/978-981-13-1402-5_6

sharing data is data is ensured against third party and valuable data is imparted to approved data per users for security counseling, and investigative reason.

Keywords Cryptography · Data collector · Data provider · Dataset Partition · Randomization

1 Introduction

Nowadays a number of senior citizens in developed countries is increasing rapidly and related on the research by UN this would increase double in 2050. Normally if elder people are living in home they need healthcare services to monitor and keep their health safely. If elder people population growing rapidly that much of healthcare services needs to be provided to them. But in developing countries parents and senior citizens people are living in individual home. In this situation these people needs preventive health care and keep their living situation securely in all situation. For providing health care and security of elder people, smart home system is introduced to monitor their heath report keep their living data safely.

In smart home system sensor are used to collect data from home about living peoples. This data should be transformed to data receiver for further services (healthcare service, research service) with secure manner to maintain privacy of each people in smart homes. In order to achieve the effective services by data analytics technology sensor data has to collected centrally and distributed with privacy mode. Distribution of preserving privacy of small scale data is concentrated broadly as of late. Miniaturized scale data encloses data where all of it consists of data around a single element, for example, a man, a family unit, or an association [1].

In the present days, tidy home surroundings have been expanding quickly. Progressively, material articles have begun to pick up the capacity to transmit data concerning their surroundings. By means of the expanding reception of sensors, savvy gadgets, and remote systems, houses are getting to be brilliant home situations. Situations of these kinds are exceptionally particular, savvy gadgets work together, course of action, distribute, as well as create derivations from the data caught concerning the condition of the home and the exercises of its inhabitants (and guests) [2].

According to Cook et al. [3, 4] by existing in a keen home setting its occupants could build the nature of ones life. Shrewd home look to improve a man's situation and lifestyle. Helped existing, a type of smart house, tries to screen a home to guarantee the inhabitant's wellbeing, empowering the matured and invalid populace to stay in their abodes for more time. In any case, reconnaissance functions in situations of this kind need protection procedures if the innovations are to be acknowledged by the inhabitants. It is because of the confidential way related to the home as well as intrusive way of observation.

In the present days keen home situations there is an absence of instruments to empower tenants to see and manage the data created by brilliant gadgets in their home [5]. Insightful gadgets might get a boundless measure of delicate individual data. The gathering and handling of this data elevates security worries concerning how the people existing in a shrewd situation of this kind, home situation can guarantee that this data would be mutual very soon for their personal great, instead of be gathered, distributed, utilized, or malignantly unveiled for reasons which may disregard their independence and protection.

To give aiding administrations from side to side data expository innovations, and as a rule sensor data should be gathered halfway and adequately and execute knowledge discovery algorithms. Knowledge discovery from records, procedures such as regression, clustering, association rules, decision trees, classification, genetic algorithm and so forth are utilized as a part of these days.

In any case, the gathered sensor data from smart homes symbolizes to individual and touchy data and could frequently uncover the total existing conduct of a person. In the meantime, it is not possible to execute investigation on data that might have changed mostly owing to the way of the arrangement in which it is critical to have the capacity to distinguish character, to which defensive mind should be outfitted. To beat this subject protection saving is the most excellent arrangement in this sort of dispersed surroundings.

Protection safeguarding data distributed have built up thought lately as talented methodologies have been utilized for sharing data where saving character security. Preferably investigation on encrypted data will be a flawless answer for saving security anyway, it is not a simple or a without cost errand.

Homomorphic encryption [6], attempts to deal with data investigation on scrambled data. Fontaine et al. [7] assesses the progressions in homomorphic encryption however, present study in encrypted data investigation stay wasteful to be utilized realistic applications. It gets to be important to formulate a plan which would permit implementation of data analytic/mining algorithms whilst safe-guarding security of observed people. This study tries to suggest security safe-guarding system related to cryptography as well as randomization ideas to give improved uprightness and secrecy related to facts as well as gives improved components to allocation of data. At this time prior to accomplishing randomization procedure on data, data partition technique has been utilized for partition data present as the form of a vertical and horizontal. In horizontal partition of data, every piece comprises of a set of which all the elements are contained in another set. Proceedings of a connection R wherever as perpendicular dividing of data, every fragment comprises of a set of which all the elements are contained in another set qualities of a connection R. In distributed environment partition techniques provide more data integrity.

A strategy of security safeguarding is randomization and this methodology attempts to protect data privacy through including random noise, while ensuring that the arbitrary commotion at a standstill jams the “signal” emitted from the data as a result the examples could be even now be precisely assessed. At this time cryptography strategy additionally utilized for gives option privacy of allocation of data. Data Randomization consolidated with cryptographic procedure additionally gives productive consequences. In the events that data change and encryption techniques have been applied in mix after that the data privacy is saved emphatically [8].

2 Related Work

The important issue of implementing geometric data [9] bother in various people allocated mining has been to safely bring together numerous geometric irritations which have been favored by various gatherings, individually. Three conventions have been created for perturbation unification.

Sharma [10] gives a broad study of various protection safeguarding data mining algorithms. The researcher have examined about benefits and negative marks of algorithm Randomization. Researchers proposed techniques are just estimated to the objective of protection conservation; it is required to advance immaculate those methodologies or build up several effective strategies.

As per Devi [11], an audit of the cutting edge strategies for security and examination of the agent strategy for protection safeguarding data mining and brings up their benefits and bad marks. Learning is matchless quality as well as the extra proficient one is about data soften up, one is less inclined to become victim to the detestable programmer double-dealers of data innovation.

Tipawan [12] as well as the group talked about on the discoveries that might be divided as four aspects

(i) Data mining;
(ii) Knowledge sorts and/or knowledge datasets;
(iii) Data mining tasks; and
(iv) Data mining strategies and applications utilized as a part of knowledge management.

This portrays the meaning of data mining by means of its usefulness. At that point it clarifies learning administration reason and different administration apparatuses coordinated in data administration. Finally, the utilizations of data mining procedures have been condensed and talked about. Sharma, Gupta and Jain's proposition [13] primarily manages data recovery framework. The territory wherever clients may have the capacity to search for records, data in the middle of report or a set of data that describes and gives data about other data. Archives on the network have been called as the Data recovery.

Adhvaryu [14] the progression in data mining methods assumes a vital part in several applications. In setting of protection and safety concerns, the issues created by affiliation principle mining procedure have been explored through numerous exploration researchers. It has been demonstrated so as to the abuse of this strategy might uncover the record proprietor's delicate and confidential data to unknown people. Numerous scientists are coming out with their push to protect security present within Association Rule Mining. This study tries to display the overview about the methods and algorithms utilized so as to safeguard protection as a part of association rule to partition the database horizontally.

Rajkumar [15] so as to safely provide individual definite delicate data from two data providers, whereby the shared data keeps up the vital data for later data mining tasks. Protected data conveyed areas the issue of disclosure delicate data while mining for helpful data. Researchers in this study deal with the issue of confidential data distributed, wherever individual exceptionality for the comparative plan of individuals has been confined through people of two kinds. Differential protection has a detailed security demonstration which creates undoubtedly approximately a foe's knowledge groundwork learning. A differentially private module guarantees that the probability of output (discharged data) has been promptly similarly from about the identical data sets and in this manner guarantees all outputs are merciless to any particular's data. However, a particular's protection has been not at danger considering enthusiasm for the data set. In particular, the study demonstrates an organized set up for differentially confidential data release pertaining to straightened-circulated data between gatherings in two in the semi-honest to goodness enemy replica. Initially a two-party tradition has been provided for the proponent system. The tradition could be utilized regarding the sub-tradition through a number of other algorithms which involves the segment related to exponents in distributed surroundings. In like manner, the research study tries to present a two-party algorithm which releases differentially confidential data securely as per the importance of protected varied party algorithm.

Shrivastva [16] data mining can be deemed as a procedure where data would be gathered from various bases as well as recommence it in valuable data. Data mining has been otherwise called knowledge discovery in database (KDD). Privacy and accuracy have been the vital concerns in data mining when data is distributed. A productive course for prospect data mining exploration would be the advancement of systems which consolidate security issues. The greater part of the strategies uses arbitrary stage methods to cover the data, for saving the protection of delicate data. Randomize reaction strategies have been created with the end goal of securing studies protection and keeping away from answers bias mainly. In RR system will add the certain level of randomness to the response to keep the data. The target of this proposition is to improve the security step in RR procedure utilizing four gathering plans. To begin with as indicated by the algorithm irregular attributes a, b, c, d had been viewed as, later the randomization are performed on each dataset as indicated by the estimations of theta. At that point CART and ID3 algorithm were connected on to the data randomized. The outcome demonstrates in kind of way through expanding the gathering, the security stage would increment.

The goal of privacy safeguarding system to ensuring private data has been defends from process of perceptive data. Smart houses have been raised in different literatures from privacy concerns about data. The design of technical solution is able to protect privacy and analyze complete data life cycle for smart homes. The data analytic technologies to provide a service through sensor data has been collected normally and efficiently performs the knowledge based on discovery algorithm. After collecting the sensor data through smart homes that represents the sensitive and personal data is to complete the trade of individual behaviors. At the same moment, it is impossible to perform data analytics and is transformed to a nature of solution.

Moncrieff et al. [17] propose innovative solutions to dynamically modify the confidentiality level in smart homes, while maintaining the functionality based on data masking techniques using environmental context to decrease the invasive nature of technology.

Meyer and Rakotonirainy [18] demonstrate the selected data to disclose a confidentiality administrator module through context-aware systems based on interact with each user.

Bagüés et al. [19] propose a framework to control the distribution of data surrounded by context-aware service through chain interaction, based on a set of user distinct privacy policies.

Drosatos and Efraimidis [20] introduces a privacy preserving based on cryptographic approaches to distribute the statistical investigation of data begins with sensors. The above solution to discuss about the address privacy concern is very precise which are required the proposed solution. The different process of data has been incorporates an innovative smart home blueprints. Fontaine et al. describes the advancement of encryption but, present study to encrypt the homomorphic data analytic remains efficient that is used in convenient applications. The smart homes system ensures the execution of mining algorithms and data analytic while preserve privacy of monitor individuals. The schema is reversible because that authorized personal details can be provided with personnel individual need of assistance. To finish the storage overhead and computation of the method has to be evaluated carefully. The role of privacy differs from cultures, jurisdiction and countries. In generally, the confidentiality is related with storage, assortment, use, sharing and processing of personally identify the data.

Chen and Zhao [21] propose a survey the privacy issues and data precautions about the entire lifecycle of data provide from cloud computing. The structure is based on their four areas to make a certain security. In our proposed system, the cloud identifies the authority user exclusive of expressive the authorized users are able to identify the stored data. In our system identify the features level of access control with authorized users are able to encrypt and decrypt the storing data. The system is used to protect the replay attack and support their modification, creation, and reading the data store from cloud system. The user access control and authentication scheme is robust and decentralized, but different access control systems of cloud are centralized. Therefore, the smart home appliance is capable to protect data such as all type of products and it contains their ingredients or other product attributes. The benefits of system can be securely realized by access the user data of home appliances through Smartphone or a browser.

3 Proposed Work

In huge applications, the entire data might be in single space known as multiple sites or centralized known as distributed database. Approaches have been proposed by several authors for both distributed database as well as centralized to secure confidential data. This study manages privacy protecting in distributed database environment whilst distributing discovered knowledge/hidden data to numerous legitimate people. Privacy safeguarding in data mining through utilizing cryptographic part foundation admittance organization is displayed in. The researchers have suggested another arrangement by coordinating the benefits of the main methodology that protects the protection of the data via utilizing a developed part which is stressed on the entire managed method besides the secondary method that utilizes strategies based on cryptographic with the perspective of reducing reduction of data as well as security.

3.1 Data Collector and Receiver

The data collector happens to be a task that is present in every smart home. It is in charge of gathering sensor data and exchanging all of those to the data cluster at standard interims. It can be configured from end to end a design document controlling each part of its usefulness. In the middle of others the principle angles it designs, are connection with the sensor data sources, the recurrence where it verifies for novel data, the location at where the data has to be sent, the protocol utilizing that builds up a connection and the arrangement where the data will be transferred.

The data receiver module gathers the data from the data collector. Data receiver plays out an algorithmic capacity for division between the diverse properties of the record, taking into account the current composition meaning of the record. The attributes have been grouped in view of regulations, empirical observations and linkage of public sources. In data handling requirements, a normal procedure for grouping has been established to be built up as well as it might call for a different exploration center. The change in algorithmic capacity yields have been put away independently so that to accomplish disengagement amongst delicate and de-sharpened data. The attributes which have been deemed primary/quasi-identifiers have been utilizing crypto strategies, before encrypting and putting away them, as well as their real worth into the dictionary storage identifier, in the event that they don't as of now subsist. The non-identifiers alongside the primary/quasi identifiers have been in de-distinguished capacity which they would put away (Fig. 1).

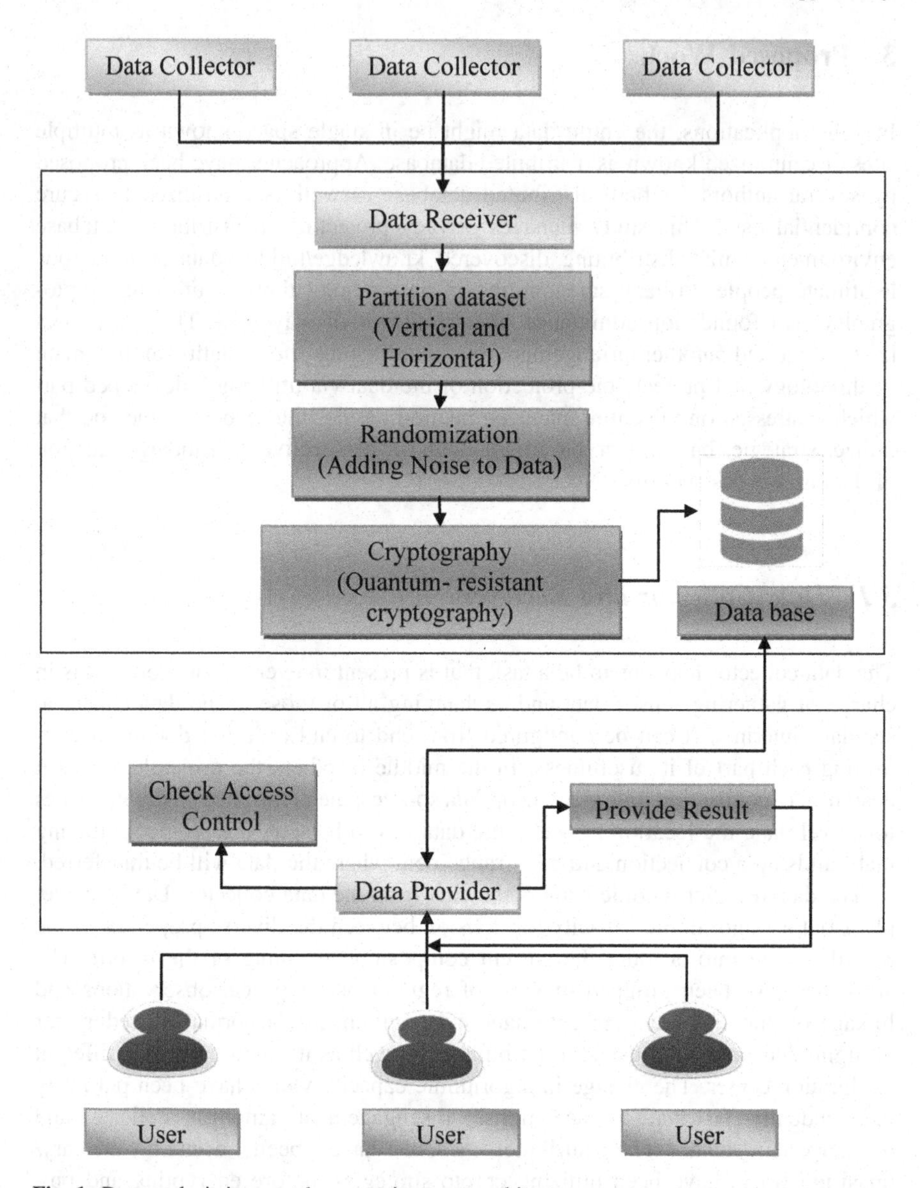

Fig. 1 Data analysis in smart home environment architecture

3.2 *Partition Method*

In distributed environment, the database has been divided into put out of joint parts and every spot comprises of only fragment. Data could be partition in various courses, for example, mixed, vertical and horizontal. In horizontal partition of data, a set of fragment containing within a set pertaining to the proceeding is present in every piece of a relation R wherever as vertical apportioning of data, a set containing within a set of properties is present in every section of a relation R. The partition method has been blended with the discontinuity anywhere data is being apportioned on a level plane and after that each divided section has been supplementary divided into pieces that are vertical and the other way around.

(1) ***Vertically Process***: In this work k-attribute record has been divided vertically into k single-attribute records. For this situation, the test would be data mining: taking in the insights of Boolean functions of the attributes, utilizing the single-characteristic inquiry and reaction instruments as outdated. A third possibility is a mix of a k-attributes record which is in a vertical position apportioned more than two or more records along with k_1 and k_2 (potentially covering) qualities, separately, where $k_1 + k_2$ k. data set i, $i = 1; 2$, could deal with k_i-ary useful inquiries, and the objective is to study connections flanked by the sensible yields, e.g.,\If $f_1(1;1;:;\ 1;k_1)$ holds, “do play a role in the improvement the probability where $f_2(2;1:;\ 2;k_2)$ holds?”, let us say f_i has been identified as the capacity on the property estimations in the ith for the files.
(2) ***Horizontal Process***: In horizontal partitioned appropriated record, distinctive arrangements of account with a familiar set of properties of entire record have been set at various destinations. The relationship among things or thing sets could be discovered right just if guidelines are resolved from consequences of aggregate arrangement of files across all the locales. In any case, either of the individual locale proprietors desires to give no single record to some place and that creates the concern a testing one which is separating fundamental data from all locales data exclusive of getting to personal account so as to produce association regulations.

3.3 Randomization Method

In the present security saving data mining innovation, the randomization technique has been considered in general the best. These strategies additionally provide knowledge discovery and balance among privacy preservation. At this time for utilizing this strategy some commotion has been inserted to the data to veil the areas of documentation [7] that is adequately vast as a result the character estimations of the documentation could by no means again be recouped. For the execution of randomization strategies, it is required to actualize two stages.

The given features are as per the following:

(1) data providers randomize their data and broadcast randomized data to data recipient;
(2) data receiver evaluations unique distribution of data utilizing distribution recreation algorithm.

This study tries to investigate the noise addition perturbation methods which change secret credits via the addition of clamor to give confidentiality. Noise addition functions through the inclusion or duplication of a random variable process or number that is randomized to private quantitative attributes. The stochastic quality is looked over an ordinary conveyance mean that is zero as well as a minor standard deviation. Deal with added substance commotion with the universal term that = 5 K + 5 (1) at which Z is termed to be the changed data spot, X is the first data point and is the arbitrary variable (clamor) with a distribution $5R \sim 5A$ (0,2). This has been later being made to add it to X. This X is later supplanted alongside Z owing to the data set to be distributed.

3.4 Cryptography Method

The method relying on cryptography ensures high level of data protection. Encryption strategy determines the [18] issues where individuals together direct mining assignments taking into account the confidential contributions they give. These mining assignments might occur among two contenders or even among untrusted parties. Hence it is the reason protection saving strategy requires actualizing to protect the data. There are different PPDM strategies, for example, the strategy on vertically and horizontally partition the data. Encryption strategy guarantees the exchange of data is safe and secure, yet this technique has been very little effective.

3.5 Data Provider

The past sub-segments the areas that have been safely gathering, processing and storing delicate data were tended to. With a specific end goal to understand the advantages of a framework the outcomes from data processing should be prepared that could be accessible to proper users. Here data provider accomplishes validation process for each client. At whatever point clients goes into this system, the data provider verifies the client part and whether this client is validated or not and gives access clients.

4 Result and Discussion

An advanced approach to ensure protection is to transfer detecting data on individual data provisions that are being claim and manage by the clients, empowering them to manage as well as confine individual data statement and activity entire management to their data. At this time, clarify test comes about so as to show the proficiency of our methodology as far as flexibility, safety, and execution.

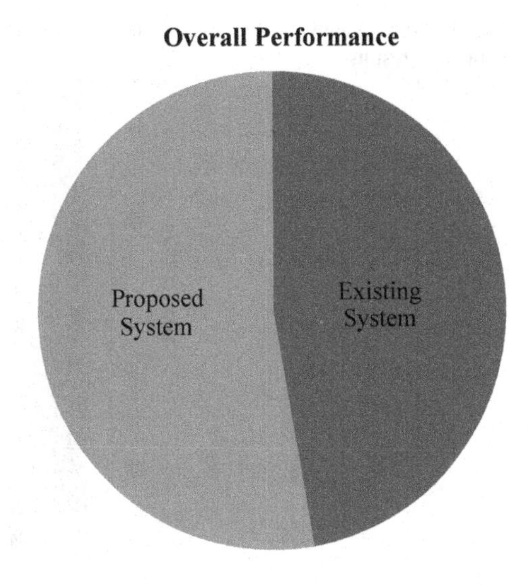

Fig. 2 Overall performance of proposed system

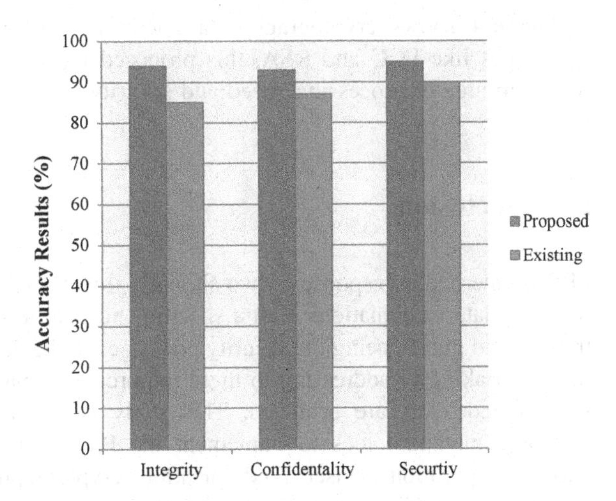

Fig. 3 Show methods accuracy results

Figure 2 indicates contrastive procedure amongst obtainable and planned arrangement of general framework. In disseminated surroundings distribution facts to the clients who are third has been a basic procedure. Since veiling touchy properties and in the meantime distributing related data to client is a subject of a honesty. In this suggested framework respectability of delicate properties has been accomplished through cryptography as well as randomization system. It gives improved execution when contrasted with the existing framework.

Figure 3 shows comparative results between existing and proposed system methods based on accuracy. Compare to existing system this proposed system gives more integerity and confidentiality when sharing the data in distributed environment.

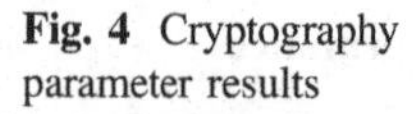

Fig. 4 Cryptography parameter results

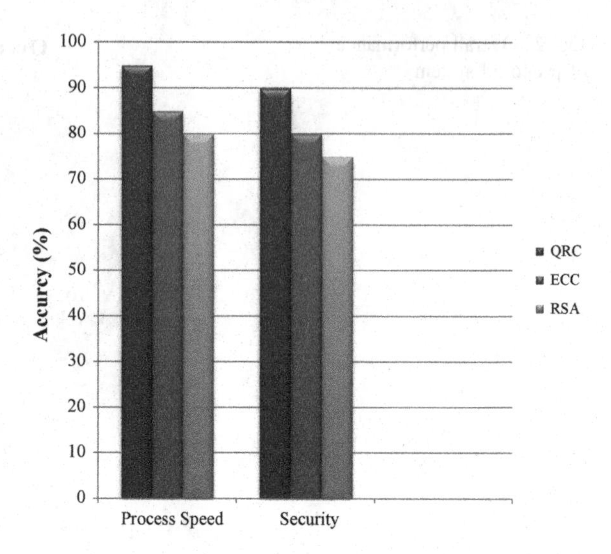

Figure 4 shows cryptography Parameter results, when compare to existing techniques like ECC and RSA, this proposed technique of QRC provides better results in area of processing speed and security.

5 Conclusion

PPDM emerged in response of two similarly imperative (and apparently dissimilar) required data examinations with a specific end goal to convey improved administrations and guaranteeing the security privileges of the data owners. Troublesome as the undertaking of addressing to these requirements may appear to be, a few substantial endeavors are achieved. This study is deemed to be a survey of the prevalent methodologies to implement the Privacy Preserving Data Mining, in particular: partition dataset, randomization, cryptography as well as synopsis. In Privacy Preserving Association standard mining over on horizontally partition database and vertical partition database to finds the global association rule from the neighborhood itemset that is incessant. In horizontally and vertically segment database utilize an alternate method to give protection. Randomization-based systems have been prone to assume a vital part in this domain. It demonstrated where underneath specific stipulations it is moderately simple to rupture the security insurance given by the irregular based clamor including procedure. The study accompanies a strategy, which accomplishes a sort of safe and secure sealed framework. Here modern cryptographic algorithms are used to achieve a secure way of data.

References

1. Wong, R. C.-W., Fu, A. W.-C., Wang, K, & Pei, J. (2007). Minimality attack in privacy preserving data publishing. Proceedings of the International Conference on Very Large Data Bases (VLDB), pp. 543–554.
2. Essa, A. (2000). Ubiquitous sensing for smart and aware environments. *IEEE Personal Communications, 7*(5), 47–49. https://doi.org/10.1109/98.878538.
3. Cook, D. J., Youngblood, M., Heierman, E. O., Gopalratnam, K., Rao, S., Litvin, A., & Khawaja, F. (2003). Mavhome: An agent-based smart home. In Proceedings of the First IEEE International Conference on Pervasive Computing and Communications 2003, pp. 521–524. https://doi.org/10.1109/percom.2003.1192783.
4. Cook, D. J., Youngblood, G. M., & Jain, G. (2008). Algorithms for smart spaces. In Technology for Aging, Disability and Independence: Computer and Engineering for Design and Applications. Wiley.
5. Langheinrich, M. (2001). Privacy by design—Principles of privacy-aware ubiquitous systems. In Ubicomp: Ubiquitous Computing, 2001, pp. 273–291. Available at http://Link.Springer. Com/Chapter/10.1007/3-540-45427-6_23.
6. Rivest, R., Adleman, L., & Dertouzos, M. (1978). On data banks and privacy homomorphisms. *Foundations of Secure Computation*, 169–177. Academic Press.
7. Fontaine, C., & Galand, F. (2007). A survey of homomorphic encryption for nonspecialists. *Journal on Data Security, 2004*(15).
8. Dhivakar, K., & Mohana. (2014). A survey on privacy preservation approaches and techniques. *International Journal of Innovative Research in Computer & Communication Engineering, 2*(11), November 2014.
9. Chen, K., & Liu, L. (2009). Privacy-preserving multiparty collaborative mining with geometric data perturbation, Dec 2009.
10. Sharma, D. (2012). A survey on maintaining privacy. In Data Mining. *IJERT* April
11. Devi, S. (2011). A survey on privacy preserving data mining: Approaches and techniques. *IJEST* March.
12. Silwattananusarn, T., & Tuamsuk, K. (2012). Data mining and its applications for knowledge management: A literature review from 2007 to 2012. *IJDKP* September 2012.
13. Shrma, S., Gupta, H., & Jain, P. (2012). A study survey of privacy preserving data mining. *IJRICE* April 2012.
14. Adhvaryu, R., & Domadiya, N. (2014). On horizontally partitioned database. *International Journal Of Engineering Development And Research IJEDR, 2*(1). ISSN 2321–9939 Ijedr1401035 (www.ijedr.org) 200 Research Trends In Privacy Preserving In Association Rule Mining (Pparm).
15. Rajkumar, M. N., Vijayalakshmi, A., & Shiny, M. U. (2016). Survey of secure two parties confidential data release in vertical partitioned data. *WSN, 45*(2), 126–136 Eissn 2392–2192.
16. Shrivastva, V. (2013). Randomized response technique in data mining. *International Journal on Recent and Innovation Trends in Computing and Communication, 1*(6), 569–574. ISSN 2321–8169. June 2013, Monika Soni Arya College of Engineering And It, Jaipur (Raj.)
17. Moncrieff, S., Venkatesh, S. et al. (2007). Dynamic privacy in a smart house environment. *IEEE International Conference on Multimedia and Expo*, pp. 2034–2037, July 2007.
18. Meyer, S., & Rakotonirainy, A. (2013). A survey of research on context-aware homes. *Australian Computer Society, 21,* 159–168.

19. Bagüés, S., Zeidler, A. et al. (2007). Sentry home—leveraging the smart home for privacy in pervasive computing. *International Journal of Smart Home, 1*(2), July 2007.
20. Drosatos, G., & Efraimidis, P. (2011). Privacy-preserving statistical analysis on ubiquitous health data. In 8th International Conference on Trust, Privacy and Security in Digital Business, pp. 24–36, Springer Verlag.
21. Chen, D., & Zhao, H. (2012). Data security and privacy protection issues in cloud computing. *International Conference on Computer Science and Electronics Engineering (ICCSEE), 1,* 647–651.

Part II
Big Data Management

A Review of Software Defect Prediction Models

Harshita Tanwar and Misha Kakkar

Abstract This paper analyzes the performance of various software defects prediction techniques. Different datasets have been analyzed for finding defects in various researches. The main aim of this paper is to study many techniques used for predicting defects in software.

Keywords Defect prediction models · Redundant metrics · Attribute selection process · Software quality

1 Introduction

As use of software is increasing in various fields such as hospital, IT companies, banking, etc. So, having defects free software is very important. A high quality of software can be obtained by using SDP model. SDP models identify the bugs in the particular software at the early stage that is at the stage of software development. This SDP model is trained with the help of software metrics or attributes. Effectiveness of SDP is based on the characteristics of various metrics of a particular software. These metrics are used to find whether a software contains the defective modules or not. Researches are done regarding selection of attributes in order to develop as much as effective SDP model.

To construct effective software defect prediction model first data is collected and then, analyzed. Many techniques can be used for preprocessing of data which includes data cleaning, feature selection, variable clustering, VIF, Spearman, redundant analyses etc. Datasets from these preprocessing techniques are then used for training SDP models. For constructing SDP models, many algorithms such as

H. Tanwar (✉) · M. Kakkar
Department of CSE, Amity University, Sec-125, Noida, Uttar Pradesh, India
e-mail: tanwar.harshita92@gmail.com

M. Kakkar
e-mail: mkakkar@amity.edu

V. E. Balas et al. (eds.), *Data Management, Analytics and Innovation*,
Advances in Intelligent Systems and Computing 808,
https://doi.org/10.1007/978-981-13-1402-5_7

KNN, NN, SVM, Naïve Bayes and random forest can be used. Prediction output then determines whether the dataset contains defect metrics or not.

The performances of these SDP models can be evaluated using performance indicator that is CA (Classifier Accuracy), AUC (area under curve), Precision and Recall etc. Also many SDP models such as random forest, fuzzy logic system, SAL, regression analyses etc. are introduced by researchers.

This review paper is organized as follows: Sect. 2 consists of review procedure part, Sect. 3 contains literature review part, Sect. 4 contains the conclusion part and last section contains the references.

2 Review Procedure

In order to analyze the performance of various SDP models, we have reviewed 20 relevant research papers out of 100 research paper. We find the relevant paper for review based on the following steps:

(i) Downloaded the research paper using the search keywords: Software Defect Prediction.
(ii) Read the title, Abstract and conclusion of research papers.
(iii) Selected the 20 relevant paper after reading the content of 100 research paper.
(iv) Results and conclusion of 20 paper is then analyzed thoroughly.

Figure 1 describes the flowchart used for defect prediction.

To analyze SDP, we formulate the following research questions to keep review focused

RQ1: what are the different techniques of software defects prediction?
RQ2: what are the measures that effect the performance of SDP models?
RQ3: How irrelevant data can introduce defects in software?
RQ4: what methods can be used for improving software defects prediction models?

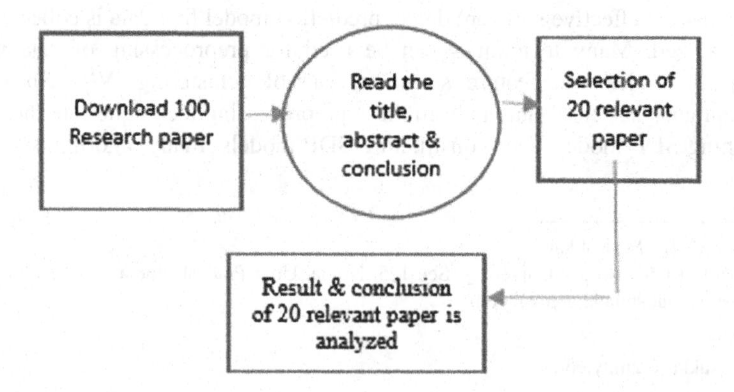

Fig. 1 Flowchart for review process

3 Literature Review

It has been analyzed from review of 20 research papers that mainly three techniques are used for implementing the SDP models that is classification, regression and clustering. Many researches on SDP model done by researcher are discussed below:

In [1], Ai-jamimi and Hamid proposed a fuzzy logic-based SDP model. The performance of this logic-based prediction model has been checked by real software projects data. They find this model as the most effective way to obtain dominant set of metrics. This in turn make fuzzy logic-based model more valid and satisfactory as compared to other models. Result showed that using all software metrics gives the lowest accuracy and less satisfaction as compared with the other set of metrics. The relevant set of metrics gives better result that is metrics obtained after removal of redundant metrics.

In [2], Koroglu et al. used seven old versions of software and their additional feature to find the defects of current versions. They compared several SDP process that is Naïve Bayes, decision tree, and random forest and finds the random forest has the highest predictive power as compared to other models. All these models are compared with the AUC value that is area under curve. They find that random forest has the highest AUC value.

In [3], Sharmin proposed a novel technique of attribute selection that is selection of attribute with log filtering (SAL). They used the log filtering to preprocess the data. Finally, comes to the conclusion that this method gives the more accuracy of SDP as compared to other techniques. This method is applied on several widely used publicly available datasets

In [4], Sethi and Gagandeep find that the artificial neural network (ANN) gives the better result as compared to fuzzy based logic model. ANN gives the more accurate value. It can be used in hybrid approach to a large dataset. These model is analyzed with the mean magnitude of relative error (MMRE) and balanced mean magnitude of relative error (BMMRE).

In [5], Suffian used the metrics in order to find the performance of different models that is regression model with other models. They find that regression analysis is most accurate as compared to other models. They used the p-value of 0.05 as the threshold for the selection of attributes of software.

In [6], Ami et al. proposed a novel approach of attribute selection method for construction of effective defect prediction model. This approach finds the attributes with high accuracy by calculating the total weight of each attribute and sorting each attribute based on total weight. They used the one classifier that is Naïve Bayes in their study in order to construct the SDP model.

In [7], Can et al. introduced a novel approach for software defect prediction PSO and SVM called as P-SVM model and observed that P-SVM has more accuracy than BP neural network, SVM Model and GA-SVM model. They found this model as most robust. The dataset used is only JM1 for proposing the novel approach of P-SVM.

In [8], Jiarpakdee finds after studying 101 available datasets that 10–67% of metrics of these datasets are redundant. Also, it has been observed that elimination of redundant metrics before constructing the SDP model is very important. It improves the performance of SDP model.

In [9], Wang et al. observed that multivariant Gauss Naïve Bayes has best performance as compared to all kind of classifiers. It is most effective defect prediction model. They also experiment with J48 in order to find the performance of multivariant Gauss Naïve Bayes. They found that MVGNB is most effective in predicting the defects at an early stage of software development.

In [10], Liu et al. proposed a SDP model for that service oriented software. They find the SDP model based on the present model, QDPSOMO. It provides better management of quality for software that depends on EXPERT COCOMO. It is formed by the combination of defect prediction, measurement and management.

In [11], Kakkar and Sarika Jain concluded from their research work that hybrid model of classifier or the combination of one or more classifier always gives the better result than any single classifier. The hybrid approach of selection of attribute gives more accuracy. It also helps us to analyze the impact of attribute selection and preprocessing of data on different SDP models. Performance of five classifiers has been compared, i.e., IBk, KStar, LWL, Random forest, and Random tree. It has been observed that LWL gave the accuracy of 92.23% and has best performance.

In [12], Verma and Kumar analyzed the multiple regression in their research work. They find the impact of clustering on defect prediction. Three clusters are formed. Result has shown that prediction model formed after clustering showed better result rather than applying prediction model on whole software project.

In [13], Yang et al. proposed a novel approach that is learning-to-rank (LTR) approach for the construction of SDP model. This approach helps to find the test resources more effectively by finding which module of software have more defects. They found that learning to rank approach gives better prediction accuracy as compared to linear model using LS. However, LTR in some cases is not giving as better result as given by Random Forest. LTR is not performing better in all cases.

In [14], Sawadpong and Allen use a exceptional handling for implementation of SDP model. They proposed exception-based software metrics. It is based on the structural attributes of exception handling call graphs. They came to the conclusion that if SDP model that is depends on exceptional based metrics gives more result as compared to conventional prediction model. They used the software repositories that have mined data and defect reports for their research.

In [15], Shuai et al. implemented Genetic algorithm with SVM (GA-CSSVM) on NASA datasets. They concluded that GA-CSSVM performed better as compared to increases normal SVM.

In [16], Gabriel Kofi Armah et al. performed Multilevel preprocessing by selecting the attributes twice and filtering instance thrice. Four K-NN classifier's preprocessing that is KNN-LWL, KStar, IBK, and IB1 results were analyzed and compared with random tree, random forest, and non-nested generalized classifier. Four performance parameter that is accuracy, recall, Area under curve (AUC) and

Table 1 Summary of studied research papers

S. No.	Title	Authors	Year and publication	Basic techniques	Dataset used	Methods for improving
1	Toward comprehensible SDP models using fuzzy logic	Ai-jamimi and Hamid	2016, IEEE	Fuzzy logic-based software prediction model	PROMISE data repository	Attribute selection using trapezium and triangular membership's functions of metrics
2	Defect prediction on a legacy industrial software : a case study on software with few defects	Y. Koroglu et al.	2016, ACM	Random forest	Data collected JIRA entries of previous seven older version of software	Ranking metrics using Information gain
3	Improved approach for SDP using artificial neural networks	T. Sethi et al.	2016, IEEE	ANN based techniques	20 genuine software venture datasets	Fuzzy logic-based approach for metrics evaluation
4	A study of redundant metrics in defect prediction datasets,	Jiarpakdee et al.	2016, IEEE	Analyze how redundant metrics effects the performance of SDP model	NASA defect datasets	1. Redundancy analysis 2. Spearman method 3. Variable clustering
5	Feature selection in SDP: a comparative study	Kakkar et al.	2016, IEEE	IBk, LWL, k-star, Random forest and random tree	NASA MDP datasets named CM1, JM1, KC1, KC3 and PC1	Hybrid attribute selection approach
6	SDP using exception handling call graphs : a case study	P. Sawadpong et al.	2016, IEEE	Exception-based software metrics SDP model	Defect data from Hadoop0.19.0 and hadoop0.20.20	Exceptional handling call graphs
7	Emperical study of defects dependency on software metrics using clustering approach	D. K. Verma et al.	2015, IEEE	Regression technique	NASA PC1 DATASETS	Clustering based metrics selection method
8	A learning-to-rank approach to software defect prediction	X. Yang et al.	2015, IEEE	Proposed a novel approach that is learning to rank (LTR) approach for the construction of SDP model	Eclipse datasets	Info gain metrics selection method

(continued)

Table 1 (continued)

S. No.	Title	Authors	Year and publication	Basic techniques	Dataset used	Methods for improving
9	An effective method for software defect prediction	S. Sharmin et al.	2015, IEEE	Proposed a novel technique of attribute selection	NASA datasets	SAL (Selection of attribute with log filtering)
10	Selecting best attributes for software defect prediction	Ami et al.	2015, IEEE	Finds the way of attribute selection such that performance of SDP increases	NASA datasets	Naïve Bayes classifier
11	A new model for software defect prediction using PSO and SVM	He Can et al.	2013, IEEE	Introduced a novel approach for SDP model using PSO and SVM called as P-SVM model	NASA dataset named JM1	Optimization theory
12	SDP using dynamic support vector machine	B. Shuai	2013, IEEE	Naïve Bayes, MLP, J48, Random Forest	NASA datasets CM1, KC1, PC1, JM1	Accuracy increases with feature selection but decreases at further stage as important features are lost
13	Multilevel data preprocessing for software defect prediction	Gabriel Kofi Armah et al.	2013, IEEE	KNN (LWL, KStar, IBk, IB1), non-nested generalized exemplars (NNGE), random tree and random forest	NASA Datasets CM1, PC1 JM1, KC2 KC1	Double selection of attributes and triple filtering of instances gives better accuracy than classifying training set directly
14	Assuring software quality using data mining methodology: a literature study	Arun Singh et al.	2013, IEEE	LR, Random forest, SVM, fuzzy programming association rule mining, NB, ANN and genetic algorithm	NA	Mining techniques help eliminate vestigial defects
15	A prediction model for system testing defects using regression analysis	M. Suffian	2012, International Journal of Soft Computing and Software Engineering	Analyze the performance of different model using metrics	Data is based on the software development using V-shape development model	Statistical approach

(continued)

Table 1 (continued)

S. No.	Title	Authors	Year and publication	Basic techniques	Dataset used	Methods for improving
16	A data-driven model for software reliability prediction	Jung-Hua Lo et al.	2012, IEEE	ARIMA and SVM	Dataset1: project from Rome air development Centre Dataset2: Project given by Hu et al.	Hybrid model performs better in SDP and decreases error rate
17	Naïve Bayes software defect prediction model	T. Wang and W. Li	2010, IEEE	Introduced machine learning algorithm for implementing SDP	NASA datasets named CM1, JM1, KC1, KC2 and PC1	Multivariants Gauss Naïve Bayes
18	A defect prediction model for software based on service oriented architecture using EXPERT COCOMO	Jun Liu et al.	2009, IEEE	Gives a SDP model for service oriented software	Genuine software	Used software that depends on Expert COCOMO
19	Defect prediction for embedded software	Ataç Deniz Oral et al.	2007, IEEE	MLP, Naïve Bayes, classification by voting feature intervals (VFI)	NASA datasets named CM1, PC1, PC3, PC4	Ensemble proves to be the best performer with 73% balance
20	Empirical assessment of machine learning based SDP techniques	Venkata et al.	2005, IEEE	Linear Regression, support vector logistic regression, support vector regression, pace regression, 1R, neural network for continuous and discrete gold field, NB, instance based learning, J48	NASA datasets named CM1, PC1, JM1, KC1	NB, IBL and NN perform better than other prediction models, also NB alone was best performer for 3 datasets

precision are used to compare them. Results showed that performance of Random Forest increased by performing double preprocessing.

In [17], Lo et al. combined SVM and Auto Regression Integrated Moving Average (ARIMA) for SDP. They analyzed that performance of hybrid model is better as compared to conventional prediction model and decreases error rate.

In [18], Oral et al. performed SDP by combining three classification techniques that is NB, voting feature interval and MLP using five datasets. He concluded that combination of these classifiers gives better performance to SDP models especially for embedded system.

In [19], Singh et al. analyzed the performance of different mining techniques that is Logistic Regression, random forest, C4.5, Association Rule Mining, Naïve Bayes, ANN, SVM, genetic algorithm and Fuzzy Programming. They concluded that Data Mining techniques are very helpful for removing minor defects.

In [20], Challagulla et al. compared 13 machine learning methods. They find that NB, neural network, and Instance-based learning performed better than other as compared to all other methods.

As seen from Table 1, there are many techniques use for the implementation of SDP models. Some of these techniques are fuzzy logics based, ANN based model, P-SVM model, Multivariant Gauss Naïve Bayes model, random forest method, regression analysis and many more.

NASA datasets are the most commonly used dataset for analyses of defects in software.

4 Conclusion

There are many techniques for constructing SDP models such as fuzzy logic-based software prediction, Naïve Bayes, neural network, random forest, SVM, P-SVM, etc. Different researcher performs preprocessing with different techniques and comes out with different conclusions. It has been observed that selection of attributes effects the performance of SDP model. There are many measures that effect the performance of SDP models that is AUC (area under curve), precision, recall, classifier accuracy, etc. However, introduction of irrelevant data decreases the performance of SDP model. Many methods are there for improving the performance of SDP that is multiple regression, multivariant Naïve Gauss Bayes, Info gain metrics selection method, SAL (Selection of attribute using log filtering), statistical approach, optimization theory, Exceptional handling call graphs etc. Based on the analysis, further new techniques can be introduced for constructing the better SDP models.

References

1. Ai-jamimi, H. A. (2016). Toward comprehensible software defect prediction models using fuzzy logic (pp. 127–130).
2. Koroglu, Y., Sen, A., Kutluay, D., Bayraktar, A., Tosun, Y., Cinar, M., & et al. (2016). Defect prediction on a legacy industrial software : A case study on software with few defects. In *2016 IEEE/ACM 4th International Workshop on Conducting Empirical Studies in Industry (CESI)* (pp. 14–20).
3. Sharmin, S. (2015). SAL: An effective method for software defect prediction (pp. 184–189).
4. Sethi, T., & Gagandeep. (2016). Improved approach for software defect prediction using artificial neural networks. In *2016 5th International Conference on Reliability, Infocom Technologies and Optimization (Trends and Future Directions)* (pp. 480–485).
5. Suffian, M. D. M., Ibrahim, S., Dhiauddin, M., Suffian, M. D. M., & Ibrahim, S. (2012). A prediction model for system testing defects using regression analysis. *International Journal of Soft Computing and Software Engineering, 2*(7), 69–78.
6. Mandal, P., & Ami, A. S. (2015). Selecting best attributes for software defect prediction. In *2015 IEEE International WIE Conference on Electrical and Computer Engineering* (pp. 110–113).
7. Can, H., Jianchun, X., Ruide, Z., Juelong, L., Qiliang, Y., & Liqiang, X. (2013). A new model for software defect prediction using Particle Swarm Optimization and support vector machine. In *2013 25th Chinese Control and Decision Conference* (pp. 4106–4110).
8. Jiarpakdee, J., Tantithamthavorn, C., Ihara, A., & Matsumoto, K. (2011). A study of redundant metrics in defect prediction datasets (pp. 37–38).
9. Wang, T., & Li, W. (2010). Naïve Bayes software defect prediction model. *IEEE*, no. 2006 (pp. 0–3).
10. Liu, J., Xu, Z., Qiao, J., & Lin, S. (2009). A defect prediction model for software based on service oriented architecture using EXPERT COCOMO. In *2009 Chinese Control and Decision Conference* (pp. 2591–2594).
11. Kakkar, M., & Jain, S. (2016, January). Feature selection in software defect prediction: A comparative study. In *2016 6th International Conference on Cloud System and Big Data Engineering (Confluence),* (pp. 658–663).
12. Verma, D. K., & Kumar, S. (2015). Emperical study of defects dependency on software metrics using clustering approach (pp. 0–4).
13. Yang, X., Tang, K., & Yao, X. (2015). A learning-to-rank approach to software defect prediction. *IEEE Transactions on Reliability, 64*(1), 234–246.
14. Sawadpong, P., & Allen, E. B. (2016). Software defect prediction using exception handling call graphs : A case study.
15. Shuai, B., Li, H., Li, M., Zhang, Q., & Tang, C. (2013). Software defect prediction using dynamic support vector machine. In *2013 9th International Conference on Computational Intelligence and Security (CIS)* (pp. 260–263).
16. Armah, G. K., Luo, G., & Qin, K. (2013). Multi_level data pre_processing for software defect prediction. In *2013 6th International Conference on Information Management, Innovation Management and Industrial Engineering (ICIII)* (pp. 170–174).
17. Lo, J.-H. (2012). A data-driven model for software reliability prediction. In *IEEE International Conference on Granular Computing*.
18. Oral, A. D., & Bener, A. B. (2007, November). Defect prediction for embedded software. In *22nd International Symposium on Computer and Information Sciences, 2007. ISCIS 2007* (pp. 1–6). New York: IEEE.
19. Singh, A., & Singh, R. (2013, March). Assuring Software Quality using data mining methodology: A literature study. In *2013 International Conference on Information Systems and Computer Networks (ISCON)* (pp. 108–113). New York: IEEE.
20. Challagulla, V. U. B., Bastani, F. B., Yen, I. L., & Paul, R. A. (2008). Empirical assessment of machine learning based software defect prediction techniques. *International Journal on Artificial Intelligence Tools, 17*(02), 389–400.

A Study on Privacy-Preserving Approaches in Online Social Network for Data Publishing

S. Sathiya Devi and R. Indhumathi

Abstract Online Social Networks (OSNs) have become major platform for social interactions, sharing personal experiences and providing other services. OSN providers provide significant services to its user for free of cost. Various privacy control mechanisms for users have been provided by OSNs to decide who can view their personal information. User's sensitive information could be leaked even when privacy rules are properly set by the service providers. Various users' data are collaborated for different analysing purposes. Many threats arise to user data in OSN. This paper discusses various types of threats that arise to user data and the technique which overcomes the attacks made on the user data.

Keywords Anonymization · Differential privacy · K-anonymity
OSN threats

1 Introduction

Online Social Networks (OSNs) have become a necessary part in modern life for person to stay connected to each other. In the current era, social networking has become a major trend for many Internet users. The Internet has over 3.17 billion users and there are around 2.3 billion active social media users. Around 85% online user use at least one of the OSN such as Facebook, Twitter and LinkedIn, etc. Which is used for building relationship, sharing personal experiences and providing other services. As Social Network sites are growing tremendously, the owners accumulate huge amount of information about OSN users. Various business organizations collect data from different OSN which helps to find hidden growth

S. Sathiya Devi (✉)
Department of CSE, BIT, Trichy, Tamil Nadu, India
e-mail: sathyadevi.2008@gmail.com

R. Indhumathi
Research Scholar, BIT, Trichy, Tamil Nadu, India
e-mail: induhari83@gmail.com

V. E. Balas et al. (eds.), *Data Management, Analytics and Innovation*,
Advances in Intelligent Systems and Computing 808,
https://doi.org/10.1007/978-981-13-1402-5_8

opportunities for an organization. Government also gathers various data that could help them to improve policy design and service delivery. Generally, the data collected by OSN operators is rich in content and the relationships that are very helpful to many third party consumers. The data generated through social network services is called as social networking data. It includes Profiles, login credentials, user message, tags, multimedia, their preferences, rating, interest and the data related to their browsing interest. Some forms of data are shared by various organizations for business purpose and public interest. The shared data may also contain sensitive information while sharing may lead to privacy breach. In order to overcome the privacy issue, many tools have been developed. The individual's personal information known as sensitive data has to be protected while publishing the data. So privacy plays a vital role in OSN. In an OSN there are three basic requirements as (i) Confidentiality, (ii) Integrity and (iii) Availability [1]. Privacy-Preserving Data Mining (PPDM) and Privacy-Preserving Data Publishing (PPDP) are techniques used for maintaining privacy. In PPDM, mining algorithms are developed to modify the user data to maintain privacy such as Trust Third Party (TTP), Data Perturbation technique, Secure Multiparty Computation and Game Theoretic approach and PPDP is used where the whole raw data are sanitized and then published to the third party for various analysing purposes. While publishing data, various threats occurs.

Though Social Network has lots of advantages, one of the big challenges is to overcome the threats that occur to user privacy. This paper provides a broad view of the recent studies on social network data threats, privacy attacks and privacy-preserving techniques. The rest of this paper is organized as follows: Sect. 2 analyses the various threats for the user and threats through service provider to user in Social Network. Section 3 describes the different type of Anonymization techniques used to sanitize the data. The Differential privacy which overcomes the drawback of Anonymization techniques is discussed in Sect. 4.

2 Threats in Social Network

Most of the users are unaware of various security risks in the Social Network. Fire et al. [2] listed out various threats in SN and they are classified into (i) Classic threats, (ii) Modern threats, (iii) Combinational threats and (iv) Targeting children and it's shown in Fig. 1. Generally Classical threats are performed in user account and their credential information. The various types of Classical threats are Malware, Phishing attack, Spammer, Cross-site scripting and Internet fraud. By using external software, the user system and all credential information is accessed in Malware. The attacker steals user identities and posts the messages on behalf of the user in Phishing attack. The attacker sends the spam message to all users in Social Network is called Spammer. Malicious scripts are injected into trusted web sites through the process of Cross-site scripting. The unauthorized person who hijacks the genuine user's account is called Internet fraud.

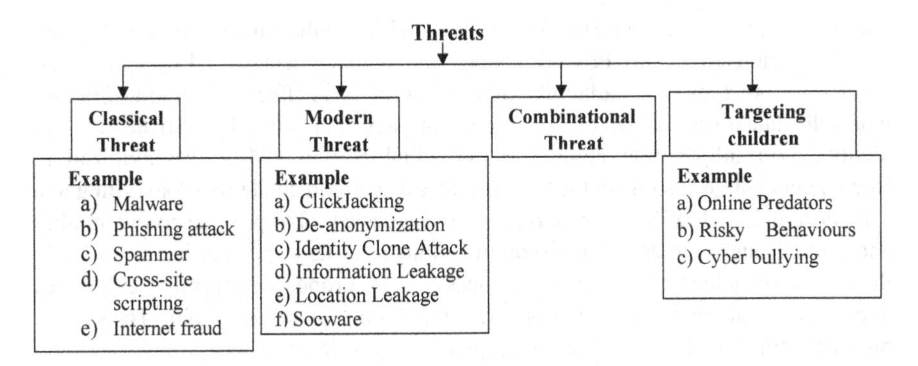

Fig. 1 Threats in social network

In modern threats, the user information is hacked by following ways such as Click Jacking, De-anonymization, Identity Clone Attack, Information Leakage, Location Leakage and Socware. The users are asked to click a link which leads to a fake page, through which the user information can be accessed is known as ClickJacking. By De-anonymization attack [3] the user details are gathered from cookies and the user real identity is uncovered from various OSN group. In the Identity Clone Attack, attacker deceives the user's friend to make healthy relationship with them by replicating the user's identity either in the same network or in the other network. The sensitive data about the user is leaked due to application weakness in Information Leakage. The location about the user is leaked from their account access and various posting of the user in the OSN sites is known as Location Leakage. In Socware the adversary takes the advantage of the social context of posts. The combinational threats are created by combining Classical threat and Modern threat. In the same way in OSN, many young children are targeted, where they communicate with the strangers and they share the personal information about them in OSN, which causes various attacks.

Beye et al. [1] described User-related privacy concern and Provider-related privacy concern in a Social Network. The privacy is breached by OSN user or unauthorized visitors in User-related privacy concern. Palen et al. [4] identified three privacy boundaries in User-related privacy and they are (i) Identity boundary, (ii) Disclosure boundary and (iii) Temporal boundary. Identity disclosure occurs when an identity of a person in a record is revealed. In Disclosure boundary information about the users are made more public than intended and Temporal disclosure, which manages the user's present and future action such as the information once posted in OSN, then it becomes impossible or difficult to remove that information in future. In Provider-related privacy concern, the SN user keeps full trust on the service provider. The data retention issue occurs when data posted by user in OSN cannot be removed permanently. The data are not deleted for various reasons, OSN employee or service providers have full access to the user browsing private information and this information are sold. In order to provide the privacy for the user, the information are anonymized and then sold to the third party. To protect privacy of all user data in OSN various privacy protecting technologies are used [5] and they are: (i) Anonymization, (ii) Decentralization, (iii) Privacy setting and management, (iv) Encryption and (v) Awareness, Law & Regulation. Anonymization is an effective method of PPDP, where the data is sanitized in order to protect privacy of an individual. So the personal identification information about the individual is removed in Anonymization. Buchegger et al. [6] listed out the possible advantages and challenges for shifting to a fully decentralized OSN. Encryption can be used as a tool to provide confidentiality and as the basis for integrity. Encryption is done in order to protect data from unauthorized users or the Service Provider. Lucas et al. [7] proposed encryption techniques to encrypt certain parts of a user's profile using public key cryptography. The service provider established various laws to maintain privacy for the user.

Apart from various threats, certain attacks are also possible in SN which is listed out in Table 1 [8]. Nagel [9] described a high level of classification of threats such as Privacy breach, Passive attack and Active attack. The attacks discussed in Table 1 come under these three classifications of threats.

Table 1 Various attack in social network

Plain Impersonation	Fake profiles are created for the user and the adversary participates in OSN activities on behalf of genuine user
Profile cloning	Adversary creates a fake profile for a user who already has a profile
Profile hijacking	Adversary gets control of the user's existing profile
Profile porting	Profile in one OSN is cloned to another OSN
ID theft	Adversary is able to convince everyone about the ownership of some others profile and they misuse the reputation of the real profile
Fake request	The adversary sends a fake request and through which, details about different users are obtained
Profiling	The user always shares their personal information in OSN, so various attacks are possible with the user profile
Secondary data collection	Adversary tries to collect data about user from secondary source
Crawling and harvesting	The whole OSN is targeted so that the information about user can be used for various marketing and business purpose
Image retrieval and analysis	Attacks are carried on multimedia data available on OSN
Communication Tracking	Adversary uses the communication information between the users and the posted comments by the user in OSN
Fake profile and Sybil attacks	Fake profile is generated by creating fake email address and Sybil account is created on behalf of a group
Group Metamorphosis	Group is formed by various user of same interest, but the owner of the group deviate the users according to his own interest and the group member will be unaware of it
Ballot stuffing and Defamation	Attacker tries to decrease the public interest of a target user
Censorship	OSN providers have control on overall data in a network, so that the adversary can use all the data and information of the user

2.1 Anonymization Operation

Social Network providers use a variety of techniques to provide security for user. In Sect. 2 various threats were discussed and one of the best methods to overcome the threat is Anonymization. Data analysis is done in two ways such as Interactive and Non-Interactive data analysis. In Non-Interactive data analysis the whole data is anonymized before publishing. While publishing the data to government and various organizations certain information about the individual is sanitized in order to preserve privacy, this process is called Anonymization. Table 2 represents the details of patients in a hospital. The attributes in Table 1 are classified as (i) Key attributes, (ii) Quasi-Identifiers (QID) and (iii) Sensitive attributes (SA). The name or the unique ID by which the individual can be directly identified are said to be key attributes and the sensitive attributes are the attributes that the individual does not want to disclose about them such as disease, salary, etc. When the combinations of different attributes are linked with external data, the user identity can be easily found then these types of attributes are called Quasi-Identifiers. Always the key attributes are removed during publishing the data for analysis purpose. For example in Table 2 the name attribute is removed. It is important to prevent disclosure of sensitive information of an individual during release of micro data. There are two types of information disclosure represented [10] as (i) Identity disclosure and (ii) Attribute disclosure. Identity disclosure arises when an individual is associated to a particular record in the released table. Attribute disclosure arises when new information about any individuals is exposed while associating the individual with the sensitive attribute. These two types of information disclosures are achieved by anonymizing the records before releasing. The Anonymization process is done by the following operation [11, 12]: (i) Generalization, (ii) Suppression, (iii) Anatomization, (iv) Permutation and (v) Perturbation. In Generalization, the QID and SA are replaced with some value, in order to hide QID and the reverse operation is said to be Specialization. Various type of Generalization schemes are Full-domain, Sub-tree, Sibling, Cell and Multidimensional. For example if the location of the individual is given as Tamil Nadu or Kerala then it can be replaced by India and the age 23 can be represented as >20. By performing Generalization, more records will have the same set of Quasi-Identifier values which is defined as Equivalence class. In Suppression, the QID and SA are replaced with special values

Table 2 Patient details

Name	Age	Gender	Pincode	Disease	Physician
Ram	23	Male	632,123	HIV	John
Raju	45	Male	632,189	Brain tumour	Sam
Siva	35	Male	632,023	HIV	John
Sita	60	Female	632,003	Cancer	Prem
Raja	65	Male	632,003	Heart Disease	Raj
Rani	33	Female	632,123	HIV	John

Table 3 The result of the operation of Generalization & suppression of Patient data

Age	Gender	Pincode	Disease	Physician
20–30	Male	632***	HIV	John
40–70	Male	632***	Brain tumour	Sam
20–40	Male	632***	HIV	John
50–70	Female	632***	Cancer	Prem
60–70	Male	632***	Heart Disease	Raj
20–40	Female	632***	HIV	John

Table 4 Quasi-Identifier table

QID	Age	Gender	Pincode
1	23	Male	632,123
2	35	Male	632,023
3	45	Male	630,189
4	60	Female	630,003
5	65	Male	632,003
6	33	Female	632,123

Table 5 Sensitive attributes table

QID	Disease	Physician
1	HIV	John
2	Brain tumour	Sam
3	HIV	John
4	Cancer	Prem
5	Heart Disease	Raj
6	HIV	John

(Example: "*"). Three types of suppression are Value, Record, and Cell. For example 620,023 can be replaced by 6*****. After applying Generalization and Suppression to Table 2 the result appears as Table 3.

Anatomization is a process in which the values of QID is not modified but disassociates the relationship between QID and SA. Here both QID and SA are represented in separate table and they are connected using common tuple id. The Table 2 is Anatomized and tuple id is provided for each tuple and then separated into two different tables such as Tables 4 and 5 respectively.

The next type of Anonymization method is Permutation where the attributes are disassociated by partitioning and shuffling within the group. For example Table 2 is portioned vertically or horizontally and different groups are formed. The result of vertical and horizontal partition is shown in Table 6. In Perturbation the original data value is replaced with synthetic data values. Various Perturbation methods are Additive noise, Data swapping and Synthetic data aggregation. Even though the attributes are sanitized through various operations discussed above, still privacy breach is possible. Various techniques used to maintain privacy during Anonymization are discussed in the next Section.

Table 6 Vertical and horizontal partition

Name	Age	Gender	Pincode	Disease	Physician
*	23	Male	632,123	HIV	John
*	45	Male	632,023	Brain tumour	Sam
*	35	Male	630,189	HIV	John
*	60	Female	630,003	Cancer	Prem
*	65	Male	632,003	Heart Disease	Raj
*	33	Female	632,123	HIV	John

3 Anonymization Techniques

As an incredible amount of information is being exchanged synchronously between various organizations for business transactions so individual's privacy is subject to risk. In order to maintain privacy of the individual, the data is anonymized by removing the personal information.

The processes carried out in Anonymization are represented in Fig. 2, where the record such as patient details are collected first and then the key attributes are removed. The Quasi-Identifiers are identified using different methods from the record. Finally the Anonymization techniques are applied in order to prevent any information disclosure.

3.1 *Identifying QI*

Identifying QI in a given record is a major part in Anonymization process. There are different methods available in literature to identify the QI. Mothwani et al. [13] proposed a method called (i) Distinct Ratio (ε) and (ii) Separation Ratio (δ). And also proposed greedy algorithms for the (ε, δ)-separation and distinct minimum key problems, which identifies small Quasi-Identifiers with provable size. According to Divanis et al. [14] Quasi-Identifier are detected by combining different attributes and they found the unique combination of record in dataset and Gkoulalas et al. [15] proposed two multi-threaded algorithms for discovering privacy Vulnerabilities in datasets. They introduced the Multi-Thread Unique Identification (MTUI) Algorithm that identifies the set of minimal attributes combinations that contain unique attribute

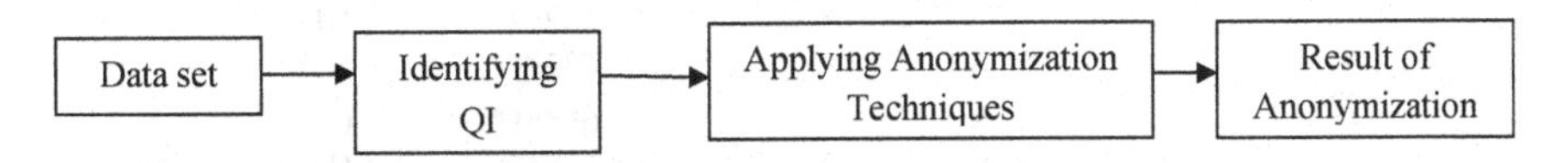

Fig. 2 Anonymization process

in the given record. They presented the Multi-Thread Risk Algorithm (MTRA), which is designed to calculate the vulnerability index for each combination of attributes in a dataset. After identifying the QI, the next step is to apply the appropriate Anonymization techniques, which is discussed in the next Section.

3.2 Privacy-Preserving Techniques

Various privacy-preserving techniques are applied after identifying the Quasi-Identifiers. Several techniques are available to protect from various attacks and they are: (i) ***K*-anonymity**, (ii) **ℓ-diversity**, (iii) ***t*-closeness**, (iv) **Slicing and** (v) **Differential privacy**.

The linkage attack is possible while combining the Quasi-Identifiers, Samarati and Sweeney [16] introduced K-anonymity which overcomes the linkage attack, where attributes are suppressed or generalized until each row is identical with at least $k - 1$ other rows. According to [16] a table satisfies k-anonymity if every record in the table is indistinguishable from at least $k - 1$ other record with respect to every set of quasi-identifier attributes, such a table is called a k-anonymous table. Table 7 is an anonymized output of Table 2 satisfying 3-anonymous since it has at least three indistinguishable tuple. Two types of attacks occur in K-anonymity such as (i) Background knowledge and (ii) Homogeneity attack. Background knowledge attack is possible when Table 7 is linked with external record. For an example in Background knowledge attack, suppose by knowing Raja's age and Pincode, adversary can judge that Raja corresponds to a record in the last equivalence class in Table 7 and if the adversary knows that Raja is suffering from chest pain. This background knowledge enables adversary to conclude that Raja has Heart disease. Homogeneity attack is possible in Table 7, since all age group within 36 have HIV disease. To overcome these types of attacks Machanavajjhala et al. [17] proposed a technique known as ℓ-diversity.

According to [17] a q^*-block is ℓ-diverse if contains at least ℓ "well-represented" (It guarantees that there are at least ℓ distinct values for the sensitive attribute in each equivalence class) values for the sensitive attribute. Each equivalence class of the table has ℓ-diverse values and there are three types of ℓ-diversity such as (i) Distinct ℓ-diversity, (ii) Entropy ℓ-diversity and (iii) Recursive (c, ℓ)-diversity. Table 8 shows an anonymized version of Table 2 satisfying 2-diversity which overcomes the attacks in k-anonymity. For example, a person cannot infer from

Table 7 3-anonymous Patient details

Age	Gender	Pincode	Disease
<36	*	632***	HIV
<36	*	632***	HIV
<36	*	632***	HIV
>39	*	6*****	Cancer
>39	*	6*****	Heart Disease
>39	*	6*****	Brain tumour

Table 8 2-diverse Patient detail

Age	Gender	Pincode	Disease
20–70	*	6321**	HIV
20–70	*	6321**	Brain tumour
20–70	*	6321**	HIV
20–70	*	6320**	Cancer
20–70	*	6320**	Heart Disease
20–70	*	6320**	HIV

2-diverse Table that Ram (23 year old and Pincode 632,123) has HIV disease, but it can be easily identified from 3-annoymous Table. Since it overcomes the attacks in *K*-anonymity, it also suffers some types of attack such as (i) Skewness attack, (ii) Similarity attack and (iii) Probabilistic inference attack. Skewness attack occurs when ℓ-diversity does not prevent Attribute disclosure. Similarity attack occurs when an adversary can learn important information from the anonymized table even though the equivalence classes are distinct. Probabilistic inference attack occurs when the adversary has a large variation between the prior and posterior beliefs. Li et al. [18] proposed a method known as *t*-closeness which overcomes the drawback of ℓ-diversity. According to [18] an equivalence class is said to have *t*-closeness if the distance between the distribution of a sensitive attribute in the class and the distribution of the attribute in the whole table is no more than a threshold *t*. Earth movers distance is used to find out the distance between the sensitive attributes. It overcomes the Homogeneity attack and Background knowledge attack in K-anonymity and protects against Attribute disclosure.

Bucketization [19] is an Anonymization technique which was introduced after *t*-closeness. According to [20] a tuple partition consists of several subsets of micro Table 2, such that each tuple belongs to exactly one subset. Each subset of tuples is called a bucket. The process of Bucketization is carried out by the following way. The first tuple is selected from Table 5 and collected in a bucket, next the second tuple is selected from Table 5 and it is compared with the first tuple in the bucket, if there is diverse from the first one, then it is selected for the same bucket, else the next tuple is selected from the Table 5 and compared in the same way with the tuples in the bucket. According to bucket size, the tuples are separated in the bucket. The main objective is to group sensitive attributes into small buckets and to make it difficult to understand the association among the sensitive information and identity information.

Table 9 represent the Bucketized output for sensitive attributes in Table 2, which has two buckets named as 1 and 2. Attributes in each bucket are selected as a distinct one so that ℓ-diversity is achieved. For example the different SA {HIV, Brain tumour, Cancer} is in bucket 1 and SA {HIV, Heart Disease, HIV} is in bucket 2. But Bucketization does not prevent Membership disclosure. Membership disclosure occurs when an attacker can infer with high probability that an individual's record is present in the published data. So a novel method called Slicing was proposed [20]. Where a given patient detail in Table 2 is sliced by attribute partition and tuple partition which also involves column generalization. Column generalization ensures that each column satisfies the *k*-anonymity requirement.

Table 9 Bucketized result of Patient detail

QID	Disease	Physician	Group
1	HIV	John	1
2	Brain tumour	Sam	
4	Cancer	Prem	
3	HIV	John	2
5	Heart Disease	Raj	
6	HIV	John	

Table 2 is sliced and represented as Table 10. Within each bucket of sliced table, the values in each column are randomly permutated to break the linking between different columns. For example in the first bucket of sliced Table 10 the values {(23, M), (35, M), (33, F)} and {(632,123, HIV), (620,023, Brain tumour), (620,023, HIV)} are randomly permutated so that the linking between the two columns within one bucket is hidden. Here each portioned data maintains ℓ-diversity. It also overcomes the problem of identification of Quasi-Identifiers in a huge dataset and it handles multiple dimensional data.

So far different Anonymization techniques were discussed for single SA, non-numerical SA and fixed bucket size. But more number of techniques is proposed in literature for multidimensional attributes, multiple numerical SA and having different bucket size. Liu et al. [21] proposed a new algorithm to overcome the problem of multiple numerical sensitive attributes by using the methods such as (i) Clustering and (ii) Multi-Sensitive Bucketization. Wang et al. [22] presented a flexible Bucketization scheme to address the issue in maintaining uniform or fixed bucket size which does not attain privacy. Susan et al. [23] proposed an algorithm to overcome the problem of high dimensional data and multiple sensitive data. Zhang et al. [24] proposed a better algorithm of K-anonymity for multiple sensitive attributes where more balanced distribution of sensitive attributes are made, which satisfies the demand of diversity. Aristodimou et al. [25] created algorithm which is based on K-anonymity through pattern-based multidimensional suppression and used feature selection for reducing the data dimensionality and to obtain K-anonymity they combined the attribute and record suppression.

Wong et al. [26] proposed k_i-anonymity model, where level of protection is decided first, instead of trusting an agency. Ma et al. [27] proposed k_i-degree anonymity with vertex and edge modification algorithm where a graph is modified by adding edges and vertices to preserve privacy. Liu et al. [28] proposed (α, k)-anonymity model based on clustering techniques which provides stronger privacy protection. Sei et al. [29] Considered all attributes in a dataset as sensitive QID and they proposed two algorithm as (i) Anonymization and (ii) Reconstruction Algorithm. Comas et al. [30] used micro aggregation to generate k-anonymous t-close data sets. The methods discussed above are done to provide better privacy in Non-Interactive analysis process. Privacy-preserving techniques for Interactive process are discussed in the next section.

Table 10 Anonymized result after slicing

(Age, Gender)	(Pincode, Disease)
(23,M)	(632,123, HIV)
(35,M)	(620,023, Brain tumour)
(33,F)	(620,023, HIV)
(45,M)	(600,003, Heart Disease)
(60,F)	(600,003, Cancer)
(65,M)	(632,123, HIV)

4 Differential Privacy

The Interactive data analysis is carried out by Differential Privacy. Dwork et al. [31] proposed Differential Privacy to prevent the attackers from detecting the presence or absence of a given person in a database. The Differential privacy [31] is defined as *Definition:* A mechanism M: $D \rightarrow T$ satisfies ε-differential privacy if for any possible output $O \in \text{Range}(M)$, and every pair D, $D^1 \in D$ distinct in only one record $\Pr[M(D) = O] \leq e^{\varepsilon} \ldots \Pr[M(D) = O]$. Where D and D^1 are two databases, e is the error added to output of the query and O is the output of the query. The process is diagrammatically represented in Fig. 3.

In Fig. 3, query is submitted to the data holder and according to the result of the query; certain amount of noise is added to the result and published to the analyser. Using Laplace or Gaussian method the noise is added to the query result.

The query is generated during real time in an Interactive data analysis. DP is a method used for releasing the results of statistical queries of sensitive data. DP overcomes the drawback of identifying whether an individual has participated in the analysis or not, by adding noise to the result in such a way that the result will not change and the privacy of the user is not violated. For Example, for the query "how many patients are having HIV below age 34" in Table 2, the answer is one and if the adversary knows the name below 34, then it is easily identifiable that Ram is having HIV. So here the noise is added such a way that the particular individual is not identified.

4.1 Addition of Noise (ε)

There are four mechanisms by which noise is added to the query and they are (i) Laplace mechanism, (ii) Gaussian mechanism, (iii) Exponential mechanism and (iv) Geometric mechanism [32]. The value of ε is selected as neither too large nor

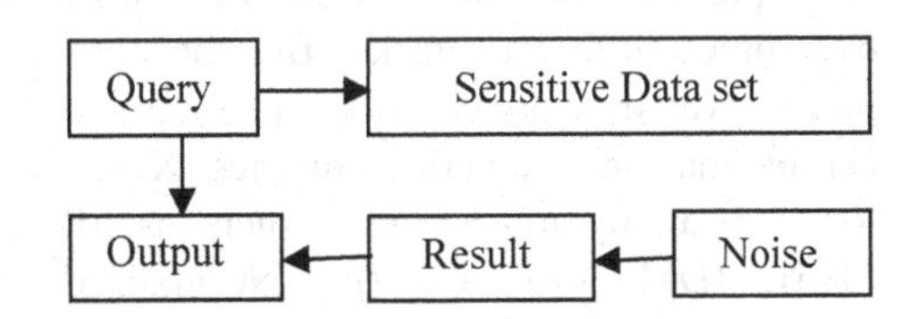

Fig. 3 Differential privacy process

too small, so that it will not compromise the privacy. Geng et al. [33] measured noise magnitude and noise power and then the noise is added using staircase distribution. Ji et al. [34] did not explicitly considered post-processing, since post-processing does not help in terms of improving the data utility so they added independent Laplacian noise to each component of the query output using Uniform Noise Mechanism and Discrete Laplacian Mechanism. Xu et al. [35] generated histograms where each dataset is transformed into statistical counts on each possible value. And they proposed two differential privacy mechanisms such as (i) NoiseFirst and (ii) StructureFirst, which provides accuracy in output for short range and long range queries respectively. These are the various ways by which noise is added to the query output in order to provide privacy. But different attacks are possible in DP, which is discussed in next Section.

4.2 Attacks in Differential Privacy

Even though DP overcomes the drawback of Anonymization, various types of attacks are possible in DP. The covert channel attack and Side channel attacks are the two types of attacks which occur in DP [36]. Lists outs the three types of attacks in Side channel as State attack, Privacy budget and Timing attack. In a DP, when an adversary quotes a query there are three possible outputs are seen and they are, (i) the result for the query, (ii) the time taken to execute the query and (iii) adversary may get a output for the query or not. A query consists of one or more mapping operation that process individual record in the data base and combines the result of mapping function using reducer code. The mapping operation is called as micro queries and the rest of the code is called the macro query. The State attack exploits a global variable to open a channel between micro queries. Here the adversary programmes in such a way to modify some internal state. An adversary can then look at the state to figure out whether the record is present in the dataset or not. Privacy budget is maintained for all databases and it refuses to answer any queries once the budget is exhausted. In DP lower privacy budget implies better privacy. In Timing attack, the user identity is identified by estimating the time taken to compute the query, for example a system takes one minute to answer a query if Ram has cancer and one micro-second otherwise, then based on query time the adversary may know whether Ram has cancer or not. To overcome these types of attacks, different methods are developed such as Privacy integrated query (PING) [37], Gupt [38] and Fuzz [36]. Ji et al. [34] proposed differentially private data publishing algorithm by building a noisy wavelet-coefficient synopsis of the multidimensional data.

In previous Section various Anonymization techniques were discussed which overcomes different attacks. But the common trade-off in PPDP and PPDM is privacy versus Quality, so the efficiency and effectiveness of Anatomized record can be analysed by various metrics. Xuezhen et al. [39] used three measurements for the anonymous model such as (i) Distortion ratio, (ii) Discernibility Metric (DM) [40] and (iii) Normalized Average EC size metric (CAVG).

The above-mentioned metrics are explained with help of a taxonomy tree. Each column is represented as tree and the leaf node are the all possible values in the column. The higher nodes represent collective terms for their child nodes. The distortion of a value is defined in terms of the height of the node that is generalized in a tree. Especially, for a node which has not been generalized, then the value is equal to 0. That means no distortion. The distortion (*d*) is represented in Eq. (1) as:

$$d = \sum_{i,j} h_{i,j} \ldots, \tag{1}$$

where $h_{i,j}$ is the height of the generalized value of QI q_i of the record r_j. Distortion ratio is equal to the distortion of the generalized Table divided by the distortion of the fully generalized table, where the fully generalized table is one with all QI values are generalized to the root of the taxonomy tree. Minimal Distortion is a penalty-based system where, whenever a value (for one record) is generalized, the distortion count is incremented.

Discernibility Metric (DM) [41] assigns a penalty to each tuple based on how many tuples in the transformed dataset are indistinguishable from it. The utility measure for *k*-anonymity is given as a tuple that inhabits a generalized equivalence class E of size$|E| = j$, $j > k$ incurs a "cost" of j. A tuple that is suppressed entirely incurs a cost of D, where D is the size of the entire database. Thus, the cost incurred by an Anonymization is given in Eq. (2).

$$C_{DM} = \sum_{|E| \geq k} |E|^2 + \sum_{|E| \leq k} |D||E| \ldots, \tag{2}$$

where the sets E refer to the equivalence classes of tuples in D induced by the Anonymization. The first sum computes penalties for each non-suppressed tuple and the second for suppressed tuples.

Normalized Average EC size metric (C_{AVG}) was proposed by LeFevre et al. [42] where the quality of *k*-anonymization is measured by the average size of QI-groups produced. The objective is to reduce the normalized average QI-group size and it is represented in Eq. (3).

$$C_{\text{AVG}} = \frac{|T|}{\text{numE} * k} \ldots, \tag{3}$$

where k is the records and numE is the number of ECs. The utility of anonymous data is measured by DM and C_{AVG}.

Iyengar [43] defined Loss Metric (LM) as the information loss caused by generalization. LM is defined as the number of nodes in a record's value has been made indistinguishable from (via generalization) compared to the total number of original leaf nodes in the taxonomy tree. Information loss (ILoss) [44] takes the same approach as LM where the information lost by generalizations is measured. It measures the fraction of domain values lost for an attribute by each generalization.

Li [20] used Rooted Mean Square Error (RMSE) to measure the predication accuracy. The RMSE is defined as the square root of the mean squared error in Eq. (4).

$$\mathrm{RMSE} = \sqrt{\frac{\sum_{i=1}^{n}\left(X_{\mathrm{obs},i} - X_{\mathrm{model},i}\right)^2}{n}} \cdots, \tag{4}$$

where X_{obs} is observed values and X_{model} is modelled values at time/place i.

For differential privacy [31] Mean Square Error (MSE) metric is used to evaluate the utility of sanitized histograms in different approach. Information-Gain-to-Privacy-Loss ratio (IGPL) [45, 46] is a trade-off measure for privacy and utility. It gives the trade-off between Information Gain and Privacy Loss (*PL/IG*). Here IG and PL are defined by information measure and privacy measure.

In this survey, we reviewed the recent developments in privacy maintenance. The main objective is to transform the original data into anonymous form to prevent from inferring sensitive information of individual record. The implementation of privacy-preserving techniques varies according to real lifetime applications. The process of various privacy-preserving methods leads to loss of information. There is trade-off between data utility and privacy. Data utility is compromised during privacy preserving and so the objective is to preserve privacy in such a way that the utility of data is increased.

5 Conclusion

This paper addressed various threats and attacks in PPDM and PPDP. Sharing information has become part of the regular activity of many individuals, companies, organizations and government agencies. So maintaining privacy during sharing information is a major task. There are various threats and attacks which occur during publishing user data for various analysis processes. The individual's data are not secured in an Online Social Network. The data are used for various processes such as public health centre, marketing and various analysis processes. Various techniques are developed to protect the privacy of the user. But attacks are still possible in the privacy protecting techniques. And to overcome that new techniques are followed such as k-anonymity, ℓ-diversity, t-closeness, Slicing, Bucketization and Differential Privacy. Despite of all these techniques, various attacks are possible from adversaries and it is difficult to maintain privacy of an individual. Future work mainly focuses on maintaining privacy and high data utility so that user data can be utilized for various data analysis process and Research work.

References

1. Beye, M., Jeckmans, A., Erkin, Z., Hartel, P. H., Lagendijk, R., & Tang, Q. (2012). Privacy in online social networks. In *Computational Social Networks: Security and privacy* (pp. 87–113). Springer Verlag, August 2012.
2. Fire, M., Goldschmidt, R., & Elovici, Y. (2014). Online social networks: Threats and solutions. *IEEE Communication Survey & Tutorials, 16*(4), 2019–2034 Fourth Quater 2014.
3. Narayanan, A., & Shmatikov, V. (2009). De-anonymizing social networks. In *30th IEEE Symposium on Security and Privacy* (pp. 173–187).
4. Palen, L., & Dourish, P. (2003). Unpacking 'privacy' for a networked world. In *CHI 2003* (pp. 129–136). ACM.
5. Mehmood, A., Natgunanathan, I., Xiang, Y., Hua, G., & Guo, S. (2016). Protection of big data privacy. *IEEE Access, 4,* 1821–1834.
6. Buchegger, S., & Datta, A. (2009). A case for p 2p infrastructure for social networks—opportunities and challenges. In *6th International Conference on Wireless On-demand Network Systems and Services* (pp. 161–168) February 2009.
7. Lucas, M. M., & Borisov, N. (2008). Flybynight: Mitigating the privacy risks of social networking. In *7th ACM workshop on Privacy in the electronic society (WPES)* (pp. 1–8). ACM.
8. Cutillo, L. A., Manulis, M., & Strufe, T. (2010). Security and privacy in online social networks. In *Chapter 23: Handbook of social network technologies and applications* (pp. 497–522), 15 October 2010.
9. Nagle, F. (2013). Privacy breach analysis in social networks. In *Lecture notes in social networks: Mining social networks and security informatics* (pp. 63–77). Berlin: Springer.
10. Gkoulalas Divanis, A., Loukides, G., & Sun, J. (2014). Publishing data from electronic health records while preserving privacy: A survey of algorithms. *Journal of Biomedical Informatics, 50,* 4–19, August 2014.
11. Panackal, J. J., & Pillai, A. S. (2013). Privacy preserving data mining: An extensive survey. In *International conference on multimedia processing, communication and information technology, MPCIT* (pp. 297–304).
12. Mehmood, A., Natgunanathan, I., Xiang, Y., Hua, G., & Guo, S. (2016). Protection of big data privacy. *IEEE Access, 4,* 1821–1834, May 9, 2016.
13. Motwani, R., & Xu, Y. (2007). Efficient algorithms for masking and finding quasi-identifiers. In *The conference on very large data bases (VLDB)* (pp. 83–93).
14. Gkoulalas Divanis, A., & Braghin, S. (2016). *Detecting Quasi Identifier in dataset.* United States Patent Application Publication, Pub No: US 2016/0342637 A1, Pub Date Nov 24, 2016.
15. Gkoulalas Divanis, A., & Braghin, S. (2015). Efficient algorithms for identifying privacy vulnerabilities. In *IEEE First International Smart Cities Conference (ISC2)* (pp. 1–8).
16. Sweeney, L. (2002). K-anonymity: A model for protecting privacy. *International Journal of Uncertainty, Fuzziness and Knowledge-Based Systems, 10*(5), 557–570.
17. Machanavajjhala, A., Gehrke, J., Kifer, D., & Venkitasubramaniam, M. (2006). ℓ-diversity: Privacy beyond k-anonymity. In *22nd international conference on data engineering (ICDE)* (pp. 24).
18. Li, N., Li, T., & Venkatasubramanian, S. (2007). t-Closeness: Privacy beyond k-anonymity and ℓ-diversity. In *IEEE 23rd International conference on data engineering (ICDE)* (pp. 106–115).
19. Yang, X, Wang, Y. Z., Wang, B., & Yu, G. (2009). Privacy preserving approaches for multiple sensitive attributes in data publishing. *Chinese Journal of Computers*, pp 574–587 Sept 2009.

20. Li, T., Li, N., Zhang, J., & Molloy, I. (2012). Slicing: A new approach for privacy preserving data publishing. *IEEE Transactions on Knowledge and Data Engineering, 24*(3), 561–574.
21. Liu, Q., Shen, H., & Sang, Y. (2015). Privacy-preserving data publishing for multiple numerical sensitive attributes. *Tsinghua Science And Technology, 20*(3), 246–254.
22. Wang, K., & Wang, P. (2016). Generalized bucketization scheme for flexible privacy settings. In *Information Sciences* (pp. 377–393). Elsevier Inc.
23. Susan, V. S., & Christopher, T. (2007). Anatomisation with slicing: A new privacy preservation approach for multiple sensitive attributes. SpringerPlus.
24. Zhang, L., Xuan, J., Si, R., & Wang, R. (2016). An improved algorithm of individuation K-anonymity for multiple sensitive attributes. *Wireless Personal Communication*. Springer Science Business Media New York, 2016.
25. Aristodimou, A., Antoniades, A., & Pattichis, C. S. (2016). Privacy preserving data publishing of categorical data through k-anonymity and feature selection. *Healthcare Technology Letters, 3*(1), 16–21.
26. Wong, K. S., & Kim, M. H. (2015). Towards a respondent-preferred k_i-anonymity model. *Front Information Technology Electronic Engineering, 16*(9), 720–731.
27. Ma, T., Zhang, Y., Cao, J., Shen, J., Tang, M., Tian, Y., Al-Dhelaan, A., & Al-Rodhaan, M. (2015). KDVEM: A k_i-degree anonymity with vertex and edge modification algorithm. *Computing, 97*(12), 1165–1184, December 2015. Springer Vienna.
28. Liu, X., Xie, Q., & Wang, L. (2015). A personalized extended (a, k)-anonymity model. In *Third international conference on advanced cloud and big data*. IEEE.
29. Sei, Y., Okumura, H., Takenouchi, T., & Ohsuga, A. (2017). Anonymization of Sensitive Quasi-Identifiers for l-diversity and t-closeness. *IEEE Transactions on Dependable and Secure Computing*, (99), 1–1.
30. Comas, J. S., Ferrer, J. D., Sánchez, D., & Martínez, S. (2015). T-Closeness through microaggregation: Strict privacy with enhanced utility preservation. *IEEE Transactions on Knowledge and Data Engineering, 27*(11), 3098–3110.
31. Dwork, C. (2006). Differential privacy. In *Proceedings of International Colloquium Automata, Languages and Programming (ICALP)* (pp. 1–12).
32. Nguyen, H. H., & Kim, J. (2013). Differential privacy in practice. *Journal of Computing Science and Engineering, 7*(3), 177–186, September 2013.
33. Geng, Q., & Viswanath, P. (2016). Optimal noise adding mechanisms for approximate differential privacy. *IEEE Transactions on Information Theory, 62*(2), 925–951.
34. Ji, Z., Xin, D., Jiadi, Y., Yuan, L., Minglu, L., & Bin, W. (2014). Differentially private multidimensional data publication. *Information Security*, Communications Supplement No. 1, 79–85.
35. Xu, J., & Zhang, Z. (2013). Differentially private histogram. *The VLDB Journal*, 797–822.
36. Haeberlen, A., Pierce, B. C., & Narayan, A. (2011). Differential privacy under fire. In *USENIX Security*.
37. McSherry, F. D. (2009). Privacy integrated queries: An extensible platform for privacy-preserving data analysis. In *35th SIGMOD international conference on management of data* (pp. 19–30), Providence, RI.
38. Mohan, P., Thakurta, A., Shi, E., Song, D., & Culle, D. E. (2012). GUPT: 'Privacy preserving data analysis made easy SIGMOD' 12 (pp. 20–24).
39. Xuezhen, H., Jiqiang, L., Zhen, H., & Jun, Y. (2014). A new anonymity model for privacy-preserving data publishing. Communications System Design. *China Communications*, pp 47–59, Sept 2014.
40. Fletcher, S., & Islam, M. Z. (2015). Measuring information quality for privacy preserving data mining. *International Journal of Computer Theory and Engineering, 7*(1), 21–28.
41. Bayardo, R. J., & Agrawal, R. (2005). *Data privacy through optimal k-anonymization* (pp. 217–228). IEEE Computer Society: In ICDE.

42. LeFevre, K., DeWitt, D. J., & Ramakrishnan, R. (2006). Mondrian multidimensional k-anonymity. In *ICDE'06* (p. 25). IEEE Computer Society.
43. Iyengar, V. (2002). Transforming data to satisfy privacy constraints. In *The Eighth ACM SIGKDD International Conference on Knowledge Discovery and Data Mining* (pp. 279–288). ACM.
44. Xiao, X., & Tao, Y. (2006). Anatomy: Simple and effective privacy preservation. In The *32nd International Conference on Very Large Data Bases* (pp. 139–150).
45. Fung, B., Wang, K., & Yu, P. (2007). Anonymizing classification data for privacy preservation. *IEEE Transactions on Knowledge and Data Engineering, 19*(5), 711–725.
46. Fletcher, S., & Islam, M. Z. (2015). Measuring information quality for privacy preserving data mining. *International Journal of Computer Theory and Engineering, 7*(1), pp. 21–28, February 2015.

Parallel Clustering for Data Mining in CRM

E. Manigandan, V. Shanthi and Magesh Kasthuri

Abstract In modern business conditions that are characterized by a stronger process of globalization, uncertainty, risk and competition, companies have to struggle every day to maintain market share and achieving better business results. In order to achieve this, the company must always be a step ahead of the competition. This means anybody must anticipate the needs of its clients and each client must access individual. This work is based on addressing this goal. Due to the fact that it is a large amount of data, it is simply impossible to do manual data analysis. Analyses are left to specially developed programs; a new kind of technology whose goal is precisely the solution of the problems that has been faced in Business Intelligence. Business Intelligence (BI) refers to be a broad set of applications and technologies for data collection, access to data and expert analysis of data, and in order to provide adequate support to the decision making process. BI represents a family of products that includes Data mining Algorithms, Data mining products for creating reports. Improving efficiency in this process is discussed in this work. The M-Clustering algorithm which is conceived in this work provides solution to data mining using clusters in twofolds—setting boundary limits during filtering and historical data processing. Define a set of data to be used for training which can be taken from filtering various attributes and the fields from the classifications set given. The data processing activity will be done using this training datasets to get expected result. This is evaluated for processing actual dataset or further execution for provisional trained dataset preparation. This work covers high-level view of the

E. Manigandan (✉)
Sri Sankara Arts & Science College, SCSVMV University, Enathur,
Kanchipuram 631561, Tamil Nadu, India
e-mail: ibmmani78@yahoo.com

V. Shanthi
Department of MCA, St. Joseph's College of Engineering,
Chennai 600119, Tamil Nadu, India
e-mail: drvshanthi@yahoo.co.in

M. Kasthuri
Wipro Technologies, Bengaluru 560100, Karnataka, India
e-mail: magesh.kasthuri@wipro.com

V. E. Balas et al. (eds.), *Data Management, Analytics and Innovation*,
Advances in Intelligent Systems and Computing 808,
https://doi.org/10.1007/978-981-13-1402-5_9

proposed system along with the processing steps used in the system. It also covers experimental evaluation carried out with customized algorithm implementation in WEKA tool and compared the processing efficiency of experimental data with k-means evaluation.

Keywords Big data · Clustering algorithm · Data classification
Data mining · K-Means · Parallel clustering · Spectral · Spatial algorithm
WEKA · Classification

1 Introduction

Lead generation is conceived as an idea to get better Return of Investment (ROI) through easy and meaningful way of marketing. The objective of this work is to define an effective way of lead generation using customized clustering algorithm. The customization is required for applying in the field of CRM to get better Return of Investment with less cost and time spent in data mining.

There are different customization done in clustering algorithm in data mining [1] like Spectral clustering, special clustering, binary tree clustering to name a few. But still clustering algorithm is not efficient enough in Big data analytics as the volume increases the process rate also proportionally increases [2]. Hence a customization is required in Clustering algorithm when applied for Big data analytics like in the field of data mining in CRM. This customization implemented in this work is done in such a way that the processing rate will have consistent performance when volume increases by addressing data volume increase with delta (use historical processing results with new data evaluation) volume processing which improves performance by implementing parallel clustering and hence higher Return of Investment (ROI) in marketing trend analysis.

As shown in Fig. 1, there are six stages in clustering algorithm where the data mining activity starts with input data collection which can be used for attribute selection and using the filtering criteria as fed by the user depending on the type of data mining to be done and taken further to store them as data set which can be used to create cluster by transforming the input data set into nodes of relational data in terms of cluster tree [3]. The next stage is to mine the node to do pattern evaluation which can be used interpret and evaluate the data and its relation which can be converted into evaluation results as needed by the user in terms of visualized reports.

2 Parallel Clustering

Clustering algorithm in data mining is the technique of grouping data elements that allows establishing/relating objects of data that are similar. Groupings are actually sorting elements in sets, in which achieve the greatest similarity data (for example,

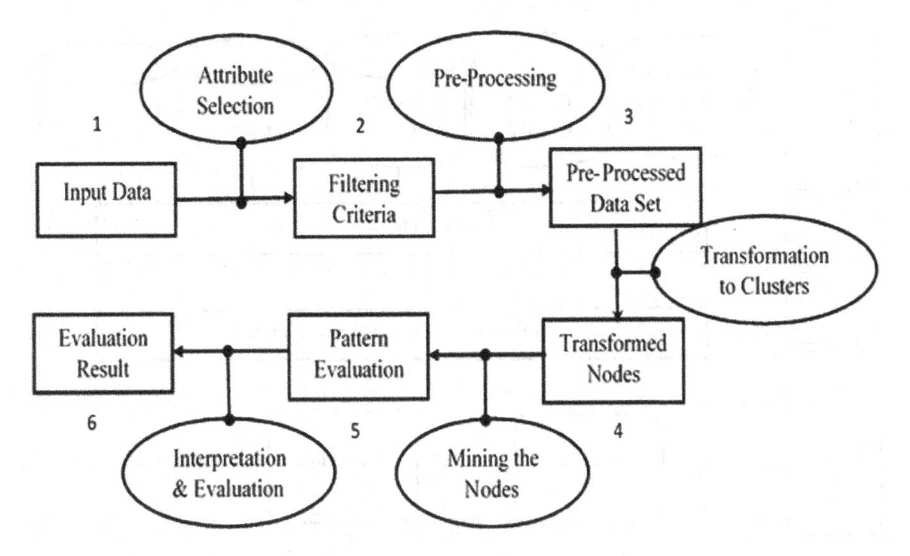

Fig. 1 Stages in clustering algorithm

customer segmentation—by age, occupation, income, consumption) [4]. This work is to customize the clustering algorithm by eliminating the pitfalls that traditional Clustering algorithm has and do an experimental evaluation of the Customized M-Clustering algorithm for real-time customer relationship data collection and do a comparison with other clustering algorithm with respect to cost effectiveness, efficiency, improvisation, usability and adoptability.

Nowadays, customers have multiple sources/channels to communicate with a company. Apart from calling them, they can use chat, email, Web and even SMS to reach your organization when they have a problem or need assistance with products and services. Moreover, customers can also use social media and blogging/review sites on the Web [5] to share their experiences about any product/service. With this drastic expansion in the communication channels, it becomes all the more important for companies to understand customer needs/perception and strategize to enhance customer experience and expand business.

Since M-Clustering is designed (as shown in Fig. 2) to handle data collected from multiple data sources, user can collect data from all possible sources suitable for the business evaluation and feed to the input feeder system to take it for processing further.

One cannot deny the fact of the growing popularity of social networking sites like Facebook and Twitter [6] and sites which provides high degree of user personalization and user intercommunication like LinkedIn and Research Gate. While yearly growth in the largest sites may have started to slow down, there is evidence that growth is accelerating in communities that have previously not had a high degree uses in a social networking site. Social networking sites are being used by people to understand each other in a better way and to explore themselves.

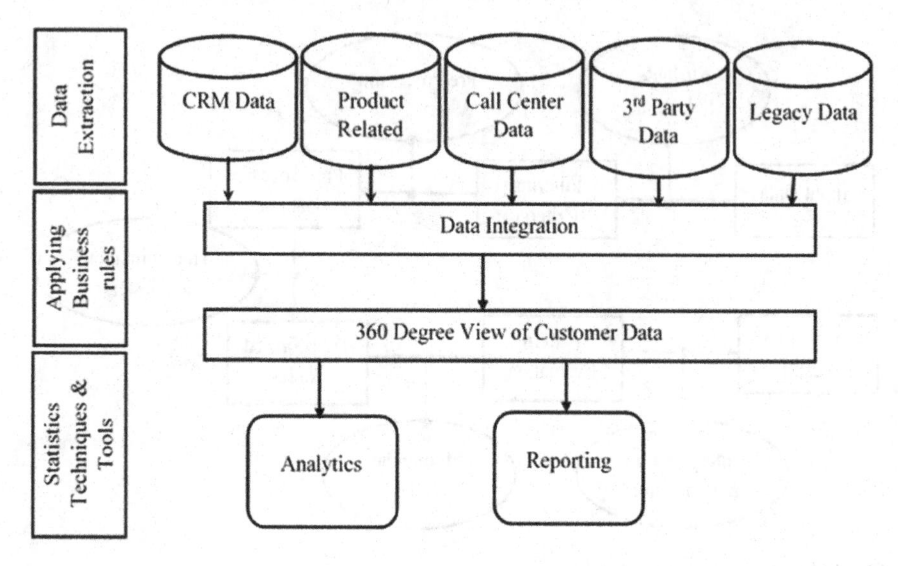

Fig. 2 High-level view of M-Clustering system

One of the interesting area of applying extensive data mining is social networking sites as they have huge volume of data flowing which can be used for relate information and get data analytics information. While the usage rates, public availability and media scrutiny all point to increased interest in the sites, there are number of impediments to capitalizing on data mining strategies for this area.

3 Implementing in CRM Data Analysis

In Big data analysis for customer relationship data (ex: Banking, Insurance) there is a huge time consuming process [7, 8] in reprocessing the data when new data gets added. If there is any manual tuning done, then the reprocessing time is still more and increases for every process tuning. When the frequency increases (ex: Banking KYC, Insurance customer preference of products [9]) then the impact of such Big Data solution is high.

One can understand that this customized mining has two goals: effective prediction and better turn-around time for result. This uses variables existing in the database from historical store which helps to evaluate unknown and future values of interest and concentrates on getting patterns including data and helps in user interpretation. As a result, this process differs in underlying application and also technical process involved in it which is explained in Fig. 3.

The key processing logic of M-Clustering algorithm starts with defining the data model used to store the data processing information in the Data service component. This also means, storing the data model in the own clustering representation in a

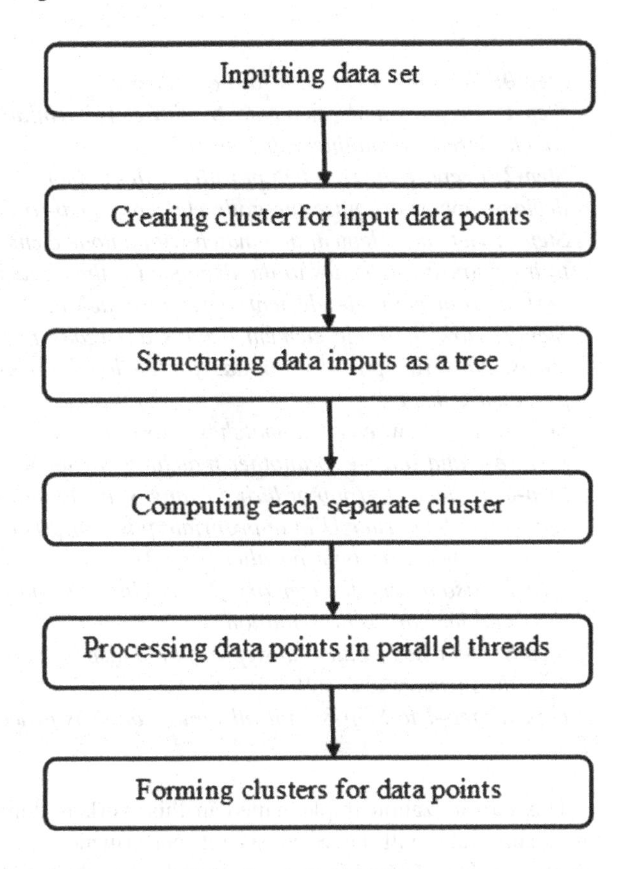

Fig. 3 Process steps of M-Clustering algorithm

matrix format associating each unique field in an indexed searchable fashion. This is an unsupervised solution which stores indexed model of fields based on each set of results it evaluates and uses them for field selection in subsequent evaluation. During Clustered evaluation, preprocessing steps like attribute filtering and attribute range setting can be done multiple times based on user requirement [10]. A typical example for range setting is Age limit (or age group) to be considered for a data mining execution of Internet usage statistics in a given city.

4 Purpose for Customizing Clustering Algorithm

M-Clustering algorithms are suitable for identification of class based data in spatial databases. But it is not limited to the large spatial databases alone but also can handle minimal requirements of domain knowledge to determine parameters given as input, clusters received for discovery which has higher efficiency on large databases. The high-level pseudo code for this customized parallel clustering algorithm is listed below

Step-0: *Feed the input data to the system*
Step-1: *Prepare node elements U with c as 1 (initially) and add elements which increases running value of c*
Step-2: *Prepare the tree T in parallel with all the node element groups which decides how to organize the node element group (ordering)*
Step-3: *Get each element and match with a node element if it can be attached to it or not. If it matches to the distance (c) then attach the element else find next appropriate node element group to attach it.*
Step-4: *Now the node element groups are ready to handle rest of the elements. Now run parallel threads for each node element group and start processing them.*
Step-5: *If a new element matches closer to an existing element (v) then remove v and -merge to another matching cluster w*
Step-6: *If element group ordering changes due to new element addition, then rearrange the cluster U in appropriate position. Also rearrange the same in the tree T which runs in parallel (step-2)*
Step-7: *Also update the memory Q with cluster groups x (tree representation in U and memory representation is x)*
Step-8: *Detached element in step-5 should also be re-organized as per cluster w in the memory also (Q)*
Repeat Step-4 to Step-8 until all input from S is processed.

This customization implemented in this work is done in such a way that the processing rate will have consistent performance when volume increases by addressing data volume increase with delta (use historical processing results with new data evaluation) volume processing which improves performance by implementing parallel clustering and hence higher Return of Investment (ROI) in marketing trend analysis [11].

5 Attribute Evaluation Using Weka

This work uses WEKA as the tool for implementing Clustered evaluation algorithm. There are two stages of algorithm implemented in WEKA. M-Clustering Attribute evaluation (classification) and data visualization with attribute relation as stored in clustered group of data nodes. The steps involved in using M-Clustering system using WEKA for data mining is explained in a simple flowchart of steps as given in Fig. 4.

The customization of M-Clustering algorithm is done some simple solution design by using Historical data evaluation to minimize the re-evaluation process and handling delta change of data from historical information. But there are more

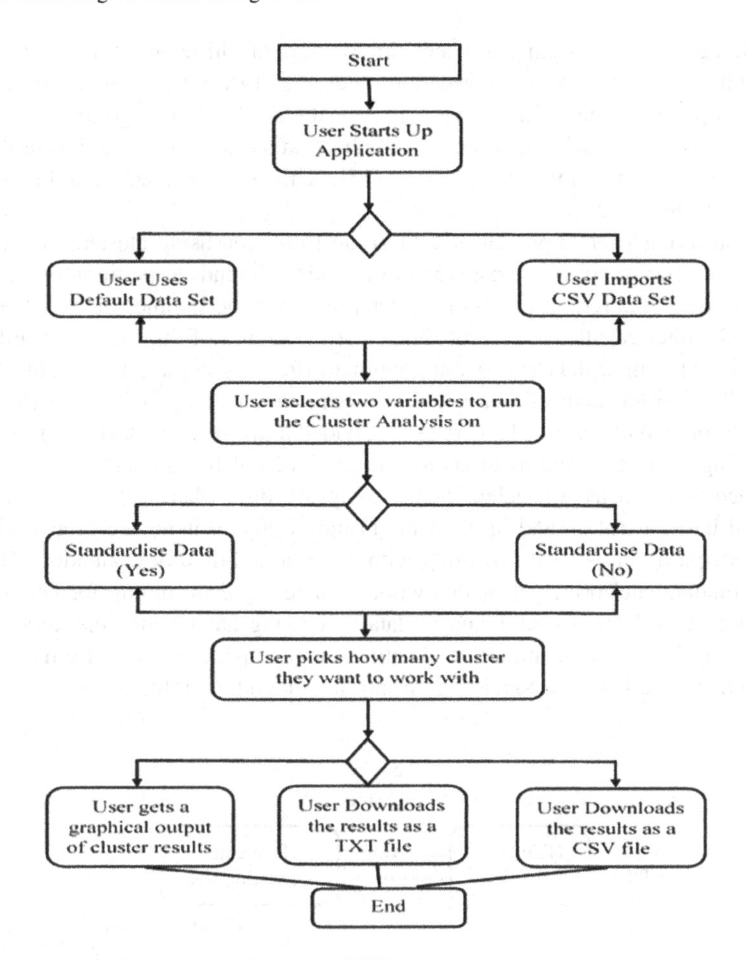

Fig. 4 Steps in M-Clustering evaluation in WEKA

improvisation required for handling Big Data analysis in an effective way [12] when one can build clustered information of nodes.

This is achieved by introducing a new technique called parallel clustering where the system identifies grouping of data collection and the groups are converted into multiple parallel processing steps where cluster nodes are created on their own.

This parallel clustering helps in achieving effective clustering process in Data mining in CRM. Clustered evaluation is a highly used technique in mining solution, in which many of existing clustering algorithms are in-efficient in handling extremely higher volume of arbitrarily shaped distribution data and datasets with multi-dimension relation.

At the same time, statistics-based cluster evaluation methods always treats with high cost for computational evaluation [13] in cluster evaluation which prevents clustering algorithms to be an efficient solution usable practice.

Although, many visualization techniques used in cluster evaluation are used predominantly as means for information rendering [14], rather than for inspecting how data pattern varies with the variations of the fields of the algorithms.

To add to this, the impreciseness of visualization has limitation that it is usable in contrasting grouping information only [15]. This is explained in a high-level architecture in Fig. 5.

M-Clustering is an approach of data group formation using clustered nodes, as the data in every group share patterns of same kind. It tends to partition by itself in the data space in a region group or clusters, in which the samples in the tables are allocated, either calculative or probability-wise. The aim of this step is identifying every sets of similar dataset, in some optimal fashion as compared in Table 1.

Parallel Cluster analysis is a solution model which helps to discover the sub-structure of a dataset in which data is divided into several clusters. The term "clustering" is to describe methods for grouping of unlabeled data.

Hence this is a useful technique for analyzing data where resembling set are grouped into a partition and in the same fashion different items are grouped in its own parts/partition when it is dealing with the cost and time of evaluation. This is also a fundamental principal of this work in handling data mining for better performance in evaluating results and in data processing having multiple iteration of evaluation. This is the solution of evaluating a unique pattern of data from different perspectives and helps in concluding it into a value added information.

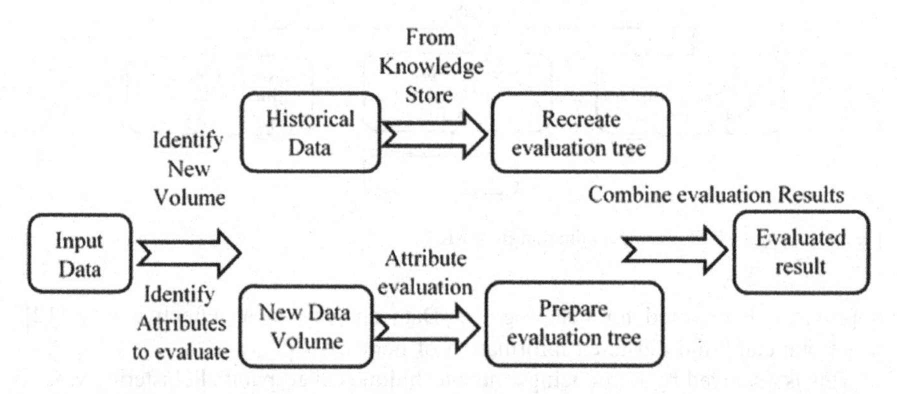

Fig. 5 High-level architecture of parallel clustering system

Table 1 Functional comparison between traditional and customized clustering algorithm

Functional aspect	K-means cluster/spectral cluster	Customized M-Cluster
Clustering method	Distance based clustering [16]	Tree based clustering (graph)
Data volume handling	Any size of data can be handled [17]	Any size of data can be handled
Data attribute volume	Lower number of attributes [18]	large number of attributes
Cluster depth	User specified [19]	Auto calculated
Hierarchical	Depth based [20]	Depth and node based
Cluster assignment	Probabilistic [20]	Ordered node based

6 Conclusion

Though Data mining in CRM can be addressed with proven solutions using existing algorithm like K-Means, this work primary focusses on Data mining technique in CRM where there are frequent volume of data change and re-evaluation is done with every dataset addition.

The primary goal of this work is to define a customized clustering approach for Data mining which can be efficiently used for field like CRM where frequent dataset change occurs and re-evaluation is executed. This helps in higher Return of Investment (RoI) as the processing efficiency increases during each cycle of re-evaluation of mining. The M-Clustering algorithm involves partitioning a given dataset into some groups of data whose members are similar in some way.

The usability of this cluster analysis can be applied in different CRM evaluation for Lead generation, Market analysis and advertisement improvement. Since the data volume is higher in such area, using a historical algorithm like K-Mean clustering algorithm would suffer by the problem of time complexity due to volume and frequent change in data collected from additional source. This has been overcome by M-Clustering and the experimental evaluation proves it to be efficient in terms of time (processing and execution time during higher data growth) and costs (lower costs in terms of less resource usage in execution without using complex system).

The experimental evaluation measures processing efficiency in terms of processing time when data volume increases in both dimensions (number of fields to process and number of records to process). The processing rate comparison is shown in Table 2.

Table 2 Experimental results

Evaluation data		WEKA algorithm evaluation		
No. of attributes × No of records	Total data size (Columns × Rows)	Attribute classification and Evaluation (in milliseconds)		
		Simple K-means (No clustering)	Hierarchical clustering	M-Cluster algorithm
16 Columns × 690 Rows	11,040	25	268	16
217 Columns × 510 Rows	110,670	469	531	453
27 Columns × 4627 Rows	124,929	739	904	516
72 Columns × 10,108 Rows	727,776	953	890	878
61 Columns × 39,644 Rows	2,418,284	16,968	18,903	16,281

M-Clustering is a technique that can be applied in different mining solution. There are various techniques similar to this available but when it is compared with cost and time efficiency, as shown in above table, this algorithm evaluates in better productivity and optimal in evaluation. This work can be further extended to multi-dimensional parallel clustering for big data based data mining solutions.

Example sources for dataset for above evaluation:

http://repository.seasr.org/Datasets/UCI/arff/
https://dreamtolearn.com/doc/2HDNJH3XJU6CVGKZ7SDM4MCSW
http://kdd.ics.uci.edu/databases/20newsgroups/20newsgroups.html
http://kdd.ics.uci.edu/summary.data.application.html

References

1. Suryanarayana, S. V., Venkateswara Rao, G., & Veereswara Swamy, G. (2016). Role of the scaling factor in spectral clustering algorithm. In *International Conference on Inventive Computation Technologies (ICICT)*, Vol. 3 (pp. 1–6). New York: IEEE.
2. Munir, M., et al. (2016). Comparative study of clustering techniques in data mining. *International Journal of Computer Science and Information Security, 14*(11), 332.
3. Yang, J., & Leskovec, J. (2015). Defining and evaluating network communities based on ground-truth. *Knowledge and Information Systems, 42*(1), 181–213.
4. Yoo, I., Alafaireet, P., Marinov, M., Pena-Hernandez, K., Gopidi, R., Chang, J. F., & et al. (2012). Data mining in healthcare and biomedicine: A survey of the literature. *Journal of Medical Systems, 36*(4), 2431–2448.
5. Zelnik-Manor, L., & Perona, P. (2005). Self-tuning spectral clustering. *Advances in Neural Information Processing Systems*, 1601–1608.
6. Zhao, Q., Shi, Y., Liu, Q., & Fränti, P. (2015). A grid-growing clustering algorithm for geo-spatial data. *Pattern Recognition Letters, 53*, 77–84.
7. Zimek, A., & Vreeken, J. (2015). The blind men and the elephant: On meeting the problem of multiple truths in data from clustering and pattern mining perspectives. *Machine Learning, 98*(1-2), 121–155.
8. Tanuja, V., & Govindarajulu, P. (2017). Application of trajectory data clustering in CRM: A case study. *International Journal of Computer Science and Network Security (IJCSNS), 17*(1), 137.
9. Soni, R. (2013). Visualization of behavioral model using WEKA. *International Journal of Science, Engineering and Computer Technology, 3*(3), 90.
10. Saranya, N., Rajesh Kanna, A., & Arunesh, P. K. (2017) Towards mobile users satisfaction in telecommunication networks using data mining classifiers. ISBN: 978-93-86171-18-4.
11. Induja, S., & Eswaramurthy, V. P. (2016). Review on CRM through data mining techniques and its tools. *IJMCS, 4*(6), 95–98.
12. Miyan, M. (2017). Applications of data mining in banking sector. *International Journal of Advanced Research in Computer Science, 8*(1), 108–114.
13. Chaudhari, B., & Parikh, M. (2012). A comparative study of clustering algorithms using weka tools. *International Journal of Application or Innovation in Engineering & Management (IJAIEM), 1*(2), 154–158.
14. Prabha, K., & Rajeswari, K. (2014). A hybrid clustering algorithm for data mining. In *International Conference on Information and Image Processing (ICIIP-2014)* (pp. 237–238). ISBN 978-93-83459-16-2.

15. Niu, S., Wang, D., Feng, S., & Yu, G. (2009). An improved spectral clustering algorithm for community discovery. In *HIS'09. Ninth International Conference on Hybrid Intelligent Systems, 2009*. Vol. 3 (pp. 262–267). New York: IEEE.
16. Chezhian, V., Thanappan, S. U., & Ragavan, S. M. (2011). Hierarchical sequence clustering algorithm for data mining. *Proceedings of the World Congress on Engineering, 3,* 223–234.
17. Marinova–Boncheva, V. (2008). Using the agglomerative method of hierarchical clustering as a data mining tool in capital market. *International Journal "Information Theories & Applications, 15*, 382–386.
18. Niu, S., Wang, D., Feng, S., & Yu, G. (2009). An improved spectral clustering algorithm for community discovery. In *Ninth International Conference on Hybrid Intelligent Systems, 2009. HIS'09*. Vol. 3 (pp. 262–267). New York: IEEE.
19. Chen, H.-L., Chen, M.-S., & Lin, S.-C. (2009). Catching the trend: A framework for clustering concept-drifting categorical data. *IEEE Transactions on Knowledge and Data Engineering, 21*(5), 652–665.
20. Qian, X., & Wang, X. (2009). A new study of DSS based on neural network and data mining. In *International Conference on E-Business and Information System Security, 2009. EBISS'09* (pp. 1–4). New York: IEEE.

Digital Data Preservation—A Viable Solution

Krishnan Arunkumar and Alagarsamy Devendran

Abstract Almost all the artifacts we have in digital only format is susceptible to loss because of the media deterioration, where they are stored. Even if we argue that migrating digital data on newer media at regular intervals will solve this issue. We have an even more important issue of digital data being inaccessible or not readable. This happens if software interpreting the data becomes obsolete. In such case the data is lost, as a bit stream is meaningless unless we can interpret them. Digital data preservation is a long and still open research area. There are various solutions proposed and implemented till date. All the solutions can be broadly classified into two categories (migration and emulation), based on the strategy to ensure longevity. We studied various approaches and strategies done till date for digital data preservation and propose a new framework—a combination of migration and emulation for digital preservation with fewer dependencies on future technology.

Keywords Digital data longevity · Digital data preservation · Data longevity

1 Introduction

In an interview with BBC at February 13, 2015—father of the Internet Vint Cerf, caused a stir when he expressed his concern that today's digital content might be lost forever. He also mentioned that if technology continues to outpace preservation tactics, future citizens could be locked out of accessing today's digital content enter a "digital dark age".

K. Arunkumar
ppltech, Chennai, Tamil Nadu, India
e-mail: emailarunkumar@gmail.com

A. Devendran (✉)
Dr. M.G.R. Educational and Research Institute, Chennai, Tamil Nadu, India
e-mail: devendran.alagarsamy@gmail.com

V. E. Balas et al. (eds.), *Data Management, Analytics and Innovation*,
Advances in Intelligent Systems and Computing 808,
https://doi.org/10.1007/978-981-13-1402-5_10

What is "digital dark age"?
It is the situation, where we have lot of digital data (documents and multimedia) but not able to read or interpret them. This could happen, if the format used to represent the digital data becomes obsolete.

2 Challenges in Digital Data Preservation

2.1 Core Problem

In digital preservation paradigm one of the most difficult challenges is to maintain the interpretability of data across changing technologies. A sequence of bits is meaningless if it cannot be decoded and transformed into some meaningful representation.

Example Documents stored as Word Star, a 1978 format, are as good as unreadable today.

2.2 Modern World Problem

Digital data exists as static content (like books, images, video, music) or dynamic content. In today's world most of the content in internet is rendered dynamically.

How do we preserver this type of content? Can we free these contents and preserver? How to exactly reproduce the dynamic content?

Dynamic content generation involves reproduction of the entire set of software executions. Preserving the entire set of "software execution" is a complex problem of aligning many moving parts over time. Over time lot of things are changing, hardware, OS, libraries, configurations, preferences, location and many.

2.3 Secured Documents Problem

Consider the case of encrypted documents and keys distribution for accessing them. Let us consider the case of storing encrypted documents in a distributed environment of many servers for long term. There is lot of research work happening in this area [1].

To achieve secrecy and longevity, we can have secret stored in multiple servers and have a constraint that a minimum number of shares among them must be required to reconstruct the secret. From a long-term archival point of view, any retrieval of such secured document needs some piece of software along with the data for construction of secret key from k or more pieces. Hence, software for key generation also needs to be preserved along with the platform on which it is developed.

2.4 Standards for Digital Data Preservation

Earlier research work on digital preservation was carried out by the Commission on Preservation and Access and the Research Libraries Group (RLG) in 1994. This task force summarized the basic requirements of preservation and issued a report (D. Waters and J. Garrett 1996). Between 1996 and 2002, many libraries, archives, and data-centers started to create digital repositories for their digitized information. The Consultative Committee for Space Data System (CCSDS) proposed an infrastructure model the Open Archive Information System (OAIS). The term OAIS also refers to the ISO OAIS Reference Model (ISO 14721:2003). This reference model is defined by recommendation CCSDS 650.0-B-1 of the CCSDS to standardize digital preservation practice and provide a set of recommendations for preservation program implementation.

3 Related Works

There has been significant amount of research work happened over the past 2 decades to address this problem. All of the long-term access strategies proposed can be broadly categorized under Migration and Emulation strategies.

3.1 Emulation

Here the system changes the environment, so that the original environment is created. Key thing is that the original digital item for preservation is intact and never modified. It is rendered by the application in the emulated environment.

Emulation has steadily improved over the last decade. At National Library of the Netherlands has used a UVC [2] (Universal Virtual Computer) system to make images available, and among other things via the EU project Planets developed the Dioscuri emulation system that can emulate x86 machine architecture. The project Preserving Virtual Worlds, which focuses on the preservation of games and interactive fiction, and the EU project KEEP (Keeping Emulation Environments Portable), also work with emulation. The status of this research is that there are promising prototypes, but products that can function in a preservation system are still far away.

Pros There are some types of records, for example interactive records, whose functionality cannot be preserved by the migration strategy. For such cases emulation is the only possible solution.

Cons Costs of such system are really high. Usually the following tasks are performed for data preservation:

- store bit streams
- preserve digital medium
- refresh/copy data to new media
- preserve integrity data during copy step
- preserve original application program, which was used to access data

Along with these steps for a proper working of emulation, we need to design and run emulator programs to mimic the behavior of old hardware platforms, OS and supporting libraries to run preserved application.

3.2 Systems Using Emulation Strategy

Dioscuri [3]—Dioscuri is x86 hardware emulator (written in java). It is OAIS complaint. It emulates various components of the hardware architecture as individual emulators and interconnecting them in order to create a full emulation process.

OLIVE [4] (Open Library of Images for Virtualized Execution)—Olive is a collaborative project for long-term preservation of software, games, and other executable content.

How they do it? Entire virtual machine along with the "executable content" is preserved along with metadata about the software and games preserved. VMs are stored with lot of metadata, so that users have lot of options to search and run. Virtual machine is then streamed using Internet Suspend/Resume technology. Users download the respective VM and run the preserved software and games from their local machine.

3.3 Limitations of Emulation Based Systems

Scalability
The ideas work well for artifacts from the pre-Web era. Considering the facts of huge number of web sites, the proposals of dedicated resources for the above methods need to be reconsidered.

Bit Preservation
Technical and Economic challenge of keeping the bits safe at huge scale cannot be simply ignored.

Intellectual Property
Digital copyright law on the data being preserved need poses lot of constraints to the emulation based solution.

Security Threats
Following are the major known threats on the virtualization technology.

- Hyper-jacking
- VM Escape
- VM Hopping

Technology Dependency
There are still a lot of dependency on the current technologies especially existence of JVM (Java Virtual Machine) in future.

3.4 Migration

Here the system changes the actual digital object, so that it adapt to the new environment to be migrated. This is akin to "recompilation". An earlier model could be portability through source code (for, e.g., a C program that is retargeted (re-compiled) from MIPS to x86-64); if source is not available but a binary is, then we may need to use the emulation method unless reverse assembly is feasible. However, the environment is typically complex and only part of this environment may be retargeted (firmware is almost always binary and proprietary).

This is an efficient approach as there is no interpretive overhead. But there is a startup cost and may be the "full system" needs to be converted. This latter aspect may be problematic if some migrated part is never used again.

Conversion to Standard Formats Another popular strategy is to migrate digital objects from many different formats to standard formats. For example, "Text documents" in several commonly available word processing formats to standards like SGML (ISO 8879). Images to standard file format and standard compression algorithms (e.g., JFIF/JPEG).

You can observe technical people used to migration of data from one data format to another format (especially documents). In the last decade this has become a primary interest to many long-term archival systems because of the less cost intensive task involved in this. There are various factors for this: Today fewer formats, each formats are used relatively longer and they are relatively better documented.

Definitely "Data" itself cannot be used. It requires a system that can interpret the data format in order to reproduce the data content. The more complex the data content and data format become, the greater the dependence on systems to interpret the data format. The migration strategy is thus by no means perfect, but currently in technical and financial terms the most simple and secure preservation strategy. Hence, used in practice by most preservation institutions.

Costs of Migration Apart from the regular bit stream protection, we have additional cost for building a monitor system which will

- Identify obsolete formats and of which the content must be migrated to other formats.
- Actual migration of data from one format to another.

3.5 Systems Using Migration Strategy

LOCKSS (Lots of Copies Keep Stuff Safe) [5] LOCKSS Program, from Stanford University Libraries and network, provides open-source digital preservation tools to preserve and authoritative access to digital content. It is OAIS complaint. It was started in 1999 @ Stanford University Libraries. Currently, there are 10,000+ e-journal titles from 520 publishers use LOCKSS. Libraries (United States and United Kingdom) that participate in the Global LOCKSS Network pay a small annual fee to join and get service.

Internet Archive Currently there are 361 billion web pages; 1,406,181 movies; 121,293 music concerts; 1,737,905 recordings; 5,279,432 texts are available.

Portico Porticos mainly involved in E-Journal Preservation Service. Currently there are 27,727,531 items available for preservation.

3.6 Limitations of Migration Based Systems

None of the above experiments really speaks about long-term preservation and also they heavily depend on current technologies. We are not very sure what future technologies will be? How, what and where information will be stored or processed? The argument within the community is like with continuous effort on migrating data at regular intervals will suffice.

Apart from the amount of effort in terms of finance and time it needs few more pressing disadvantages of migration strategy are

- Possible loss of information
- It requires manual/semi-manual checking of results
- Results are unpredictable
- It must be repeated every few years
- Error propagation

3.7 Insurable Storage Services [6]

This works deals with cost aspect for digital preservation and proposes an economically sustainable solution. The proposed solution creates a market place bringing two parties, people storing data for preservation and technology partners.

4 Our Contribution

We build a more concrete solution by a combination of migration, emulation, and encapsulation for digital preservation with less dependency. We looked at digital data being preserved and in retrieving the meaningful information from them in various scenarios and come up with a new concept of "Eternal Binary". A detailed discussion is given in the following section.

4.1 Digital Preservation Framework

Core Idea Packing a basic machine emulator (written in some abstract language), a simple Operating System to support the execution of Application along with the data to support the rendering of the same in future. Our claim is that in future technology, it will be easy to create an interpreter for this basic machine emulator packed along with data. With that interpreter we can emulate the basic machine, OS, Application and hence render data stored for longevity.

Eternal Binary With our system in place, every reader application (nothing but a native binary file specific to architecture) can be stripped and converted as "Eternal Binary" and from that moment all the data that are viewable via that reader can be viewed in future technology with the "Eternal Binary" being created through our system. This "Eternal Binary" application along with the "basic machine emulator", "simple OS" can render the data.
There are two ways to ensure the "Eternality" of the Application:

1. Create a distributed self-healing system to maintain all the "Eternal Binaries". So that any reference in future for viewing related data can be taken from it.
2. Always pack the "Reader Application" (Eternal Binary version), "basic machine emulator" and "simple OS" along with data. As of now, "Reader application developers" are responsible for creating the stripped version of the Application binary with given specification of basic machine and simple OS details.

In our experiments we created two such Applications, for reading and displaying ASCII encoded English txt file and "TSCII" encoded Tamil txt file. Later, we need an automated solution for converting any binary for a given architecture to an "Eternal Binary" file.

Even though our solution falls under "Emulation" category, our approach of stripping the minimal OS and basic machine emulation just to support the running of a single application only makes it unique and has the following key advantages:

Simplicity Proposed solution needs an interpreter, which can be easily implemented in future technology (compared to the complexity of hypervisors in virtualization technology.)

Security Proposed solution does not have the entire OS virtualized and hence exposing the security flaws of guest OS is avoided. Also, the code is open source which can be inspected for security loopholes.

Beyond 0s and 1s Proposed solution is not limited to a computers based on binary system (0s and 1s), even if the future technology has some other format of representing data. Creating an interpreter will be relatively easy with specification given.

5 Proposed Solution

5.1 Basic Machine Emulator—Von Neumann Architecture (MIPS R2000 ISA)

Almost all the computers we have today are based on the basic design "Von Neumann Architecture". As per "Von Neumann Architecture", all computers have

- Four main sub-systems: Memory, ALU, Control Unit, I/O
- Program resides in memory
- Program instructions are executed sequentially

Our emulator is built on Von Neumann Architecture. A detailed design/description for the same is out of scope of this document. Here, Execution of instruction happens in instruction cycle. This is the basic operation cycle of a computer.

5.2 Encapsulation/Packing

The digital data to be preserved (For our experiment we used data in "txt" files, both English language [ASCII] and Tamil language [TSCII]), basic machine emulator, simple operating system, application binary (simple "C" language program for reading and printing data stored in "txt" file). We believe carrying metadata about the requirement of interpreter and how to get display should be packed like a "readme.txt". This will help technologist of future to decode with their setup. This particular metadata is always kept updated (in latest standard format), may be every time a migration happens to avoid physical deterioration. A physical copy of the same is maintained with the digitally preserved artifact (Fig. 1).

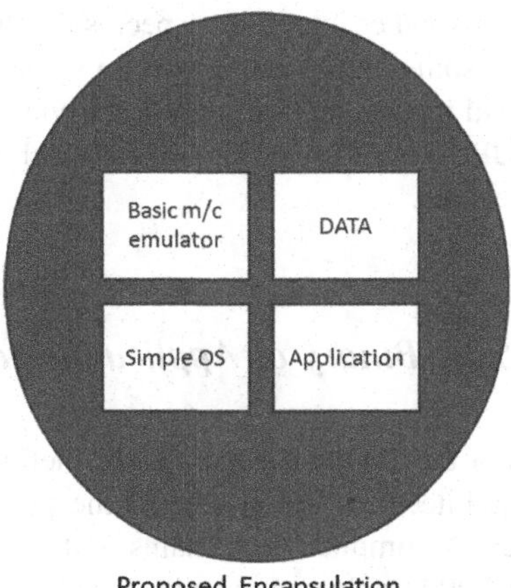

Fig. 1 Encapsulation and architectural stack

6 Experiments and Results

We created a POC (proof of concept) implemented in Java based on proposed framework for digital data stored in txt files. Hence, demonstrated the proposed framework will solve issues in digital data longevity. Following sections explains in details about the components of POC used in our experiments.

6.1 Interpreter (in Future Technology)

In our experiment we use applet-viewer, which can run in any standard browser. It will download the Basic machine emulator, OS, Application, data and render accordingly.

6.2 Base Machine Emulation

In our experiment we choose the following components:

Processor MIPS R2000 ISA implementation [7]

Co-Processor CP0, CP1 Interrupt/Exception handling and floating point operations

RAM Array of bytes (128 MB)

System Clock @ every nano-second (1 GHz Clock)

Once the base emulator is ready we hardcoded simpler programs like add, gcd for correctness of basic flow of instructions.

Another important aspect is the display module. Display is not a simple mapping of some region in memory to some patterns. It is not that simple. Basic VGA card and its programming are not trivial. We have used an open-source implementation (JPC Java X86 PC Emulator, which uses code from Bochs emulator) for our display system.

6.3 Binary of Application (as MIPS R2000 Machine Code)

For the test we needed a basic "hello world" program compiled for the MIPS R2000 architecture. We also need the program to be statically linked. We tried various cross compiler tool chains and got the basic program cross compiled for MIPS R3000 architecture with Linux. We ran the binary with QEMU-MIPS [8] (MIPS + Linux setup) and got required "Hello World" output.

6.4 Cross Compiler Tool Chains

With cross compiler—we can create executable code for any platform. For example: we can create executables for Linux/ARM on Linux/x86 based architecture.

For our experiments, we compiled for Linux/MIPS R3000 using "gcc binutils" tool chain. Details of how cross compiler works, the process to build a tool chain for a given target architecture and compile/build using the tool chain are out of scope of this document.

6.5 QEMU [8]

QEMU ("Quick EMUlator")—is an open-source processor emulator. It emulates CPU through dynamic binary translation and provides a set of device models, enabling it to run a variety of operating systems.

We used QEMU-MIPS to test our programs compiled for MIPS architecture.

6.6 ELF Files and ELF Loader

All the binary files generated by gcc and its tool chains are in ELF format. To run them we need a module to parse, interpret and load them in appropriate areas in memory. For that we implemented an ELF Loader parsing and loading corresponding sections into main memory.

6.7 System Call

A user program makes a "system call" to get the operating system to perform a service for it, like reading from a file, start another process, etc. In general system calls provides the essential interface between a user process and the operating system.

In our experiment, we ended-up in implementing the SYSCALL handler for 350 + system calls with very few actual implementations as per the trace from QEMU-MIPS trace.

6.8 Experiment Setup—Summary

Base Machine Emulator

- 140 Instructions (MIPS R2000)
- 128 MB RAM
- System Clock
- Co-Processors 0,1 Exception Handling and Floating Point operations
- 32 Integer Registers, Memory Address Register (MAR), Memory Data Register (MDR), Instruction Register (IR), Program counter (PC), HI, LO

Simple OS

- Interrupt handling and system calls implementation
- ELF Parser and Loader: To load and run single application
- File System (read from Actual data _le packed, write to Standard output device via VRAM)

Display

- VGA Card emulation
- Character mode initialization

Application

- C program with read, write to STD OUTPUT system call (statically linked and cross compiled for MIPS 3000 + Linux)

Data

- Simple txt file storing sentences in English encoded in ASCII format.

Package
All these information will be packed as a single JAR file and available for download/interpretation Interpreter. We depend on java's "applet-viewer" for downloading and interpreting the basic machine emulation, OS, Application along with data.

6.9 *Results and Analysis*

After the study of various contemporary approaches and strategies taken for digital preservation, we propose a new framework for digital preservation with very minimal expectation/assumption on future technology. We have come up with a proof of concept on how a data stored in our framework can be accessed in future after decades or even centuries. Of course, we assume the hardware availability and retrieval of bit streams is taken care and an applet-viewer like tool exists in future.

Digital data longevity is a deep recursive problem, which makes it difficult to come up with complete solution. We believe that packing entire computation system along with data and metadata for rendering is a reasonable solution. Our experiment on decoding the encapsulation with an applet-viewer, which is a part of every modern browser these days, proves our proposed framework works. Especially the second experiment we ran to retrieve non-Unicode TSCII encoded Tamil document. Retrieving such a document involves considerable time of either converting the document to Unicode document with some TSCII to Unicode conversion tool or by finding a font supporting TSCII encoding + installing it in the OS. If we cannot find them, then it is as good as lost data.

7 Conclusion

7.1 *What Needs to Be Done Further?*

Key differentiator of our proposal is combining Migration and Emulation strategy. We do a one-time migration to create an Eternal Binary out of application along with platform details. This step is to ensure the execution of application in future with very minimal dependency. That is, with the given data (0s 1s) stored privately/publicly and a reference to the Eternal binary stored in a common place. Retrieval of the data is guaranteed anytime anywhere in future.

After our experiments, we propose to build an open-source system like OLIVE to maintain all the Eternal binaries of possible applications generating digital data.

The system will have three main components

1. Interface for creating Eternal Binaries
2. Interface to link private/public data with available Eternal binaries
3. Interface to load and view data with Eternal binaries

7.2 *Motivation*

Scalability In the proposed solution, only the digital data needs to be preserved. This can be stored in a private or public cloud with different access privileges to ensure secured access over data. The application and platform to create or render the digital data are borrowed from the centralized repository, during the retrieval time. The complexity of the retrieval system is not a problem as we are going to maintain the eternal binary version of the same. With the implementation of bigger government projects like Aadhar, Digital India, e-Governance data preservation strategy ensuring security and scalability will be critical.

Intellectual Property As the data is separated and stored alone, every security feature available now can be applied.

Security Threats Proposed solution does not have the entire OS virtualized and hence exposing the security flaws of guest OS is avoided. Also, the code is open source which can be inspected for security loopholes.

References

1. Gupta, V. H., & Gopinath, K. (2006). An extended verifiable secret redistribution protocol for archival systems. In: *International Conference on Availability, Reliability and Security (ARES)*.
2. Van Diessen, R. J., & Lorie, R. A. (2005). IBM Research Report—UVC: A universal virtual computer for long-term preservation of digital information.
3. Rothenberg, J. (1999). Avoiding technological quicksand: Finding a viable technical foundation for digital preservation. A report to the Council on Library and Information Resources.
4. Gloriana, St. C., Linke, E., Satyanarayanan, M., & Bala, V. (2014). Collaborating with executable content across space and time. *ICST Transactions on Collaborative Computing*.
5. Dobson, C. (2003, February). From bright idea to beta test: The story of LOCKSS. Searcher 11.
6. Gopinath, K., & Simha, R. (2006). Insurable storage services: Creating a marketplace for long-term document archival. In *International Conference on Computational Science*, 3.
7. MIPS ISA. The MIPS R2000 Instruction Set manual. http://ti.ira.uka.de/TI-2/Mips/Befehlssatz.pdf.
8. Bellard, F. (2005). QEMU, a fast and portable dynamic translator. In *Proceedings of Usenix Annual Technical Conference*, Usenix Association.

Big Data Security Threats and Prevention Measures in Cloud and Hadoop

Manoranjan Behera and Akhtar Rasool

Abstract Big Data, collection of huge data sets is a widely used concept in present world. Although being stored and analyzed by Cloud services, it poses the greatest challenge of security threats, occurring in the exposure of enormous amount of data. This paper we are going to explain recent security risks, threats, and vulnerabilities with respect to Cloud Services, Big Data [with extra focus on EnCoRe system], Hadoop [with extra focus on HDFS] and throws light on issues dealing with big data analytics. It prominently makes use of recently developed approach called sticky policies and the existing security framework to improve security. This paper provides a literature review on security threats and privacy issues of big data, Hadoop concurrent processing. It also uses Verizon and Twilio which is familiar for its trustworthy implementation of the Hadoop using Amazon Simple Storage services (S3 services).

Keywords Big data · Cloud services · Hadoop · Security

1 Introduction

The stream computer science witnessed significant change in its domain due to widespread popularity of Web 2.0 in 2009. Later on, this became an inspiration for technology giants like (i) Microsoft and (ii) Google to offer browser based pursuit solutions. Amazon released {Elastic compute (EC2)/S3} in 2006 which was a landmark advancement in cloud computing but year 2009 marked the actual of cloud technology.

M. Behera (✉) · A. Rasool
Maulana Azad National Institute of Technology, Bhopal 462003
Madhya Pradesh, India
e-mail: nadantilttlia@gmail.com

A. Rasool
e-mail: akki262@gmail.com

V. E. Balas et al. (eds.), *Data Management, Analytics and Innovation*,
Advances in Intelligent Systems and Computing 808,
https://doi.org/10.1007/978-981-13-1402-5_11

The cost effectiveness, scalability of cloud based computing services has made its use across commercial, noncommercial and academic sector for offering services like storage, infrastructure, platform, and software, etc. With the widespread use of these services in past few years, their security is of prime importance now.

Big Data is considered by world over as the biggest innovation in computing [1], and now a familiar term to almost all due to recognition in sectors such as (i) health, (ii) administration, (iii) Learning, (iv) Transportation, (v) retail, (vi) manufacturing and (vii) personal location data.

Huge (big) data set is a word used for group of data sets of such complex size which makes it extremely difficult to be processed by application of commonly used data processing algorithms and platforms. Some examples of big data are (i) social networks, (ii) sensor networks, (iii) high throughput instrument, (iv) satellite and (v) streaming machine [2].

Hadoop released 1.0 in December 2011, equipped with security features which was an improvement from its previous versions. Thus emergence of Hadoop led to store, process and analyze big data as a cluster scale platform. This technology is being used by large corporations like (i) Adobe, (ii) Facebook and (iii) Spotify. International Data Corporation (IDC) estimates its growth as a significant software market with $130 billion share in 2016 [3].

2 Literature Review

In distributed programming, the security concern start working when enormous amount of private data stored in a database which is not encrypted format. When moving from similar type data to the different type data certain tools and technologies for huge data set is not often developed yet with more security [4]. It has been reported that big data advancement increases the threats to the existing security of the data. One issue with big data privacy is policy management and how to enforce it with this huge data and not affecting the performance [5]. Big Data security usually is to the use of the Big Data to implement solutions increasing security, reliability, and safety of a distributed system. Big Data privacy focuses on the protection [6]. There is no limit of ownership and right to some sensitive data that is still the risk of information leakage [7]. Hadoop security issue should not be given a priority facing in distributed and parallel processing of data. The diversity and the increased volume of data exchanged make the security problem [8].

3 Cloud Services

With development of cloud computing technology, concerns associated to data, security, and user privacy have intensified. We have perceived low user reliance in the different security initiatives undertaken by cloud service providers [9] as a

Table 1 Security threats with their relevance [10]

i.	Data breaches	91%
ii.	Data loss	91%
iii.	Insecure interfaces and APIs	90%
iv.	Malicious insiders	88%
v.	Service traffic hijacking	87%
vi.	Misuse of cloud services	84%
vii.	Shared technology vulnerabilities	82%
ix.	Denial of service	81%

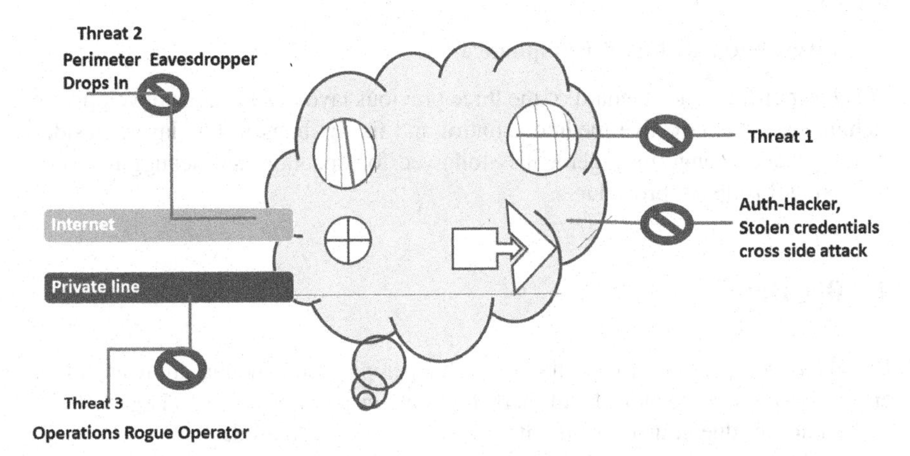

Fig. 1 Verizon cloud infrastructure [11]

reason for basic characteristics of shared resources of cloud, dearth of regulation over storage of data and openness by providers. Relevance of security threats in cloud services in present era as estimated by Cloud Security alliance (CSA) [10] is shown in Table 1. Figure 1 shows Verizon Cloud Infrastructure which uses a layered security model [11]. The four security layers are as follows:

A. **Base Security**

It refers to outer security which does not rely on technology. Verizon has started different initiatives like continuous video monitoring with 3 months support, biometric readable support system, and 24*7 on-site guard services. Algorithm such as "need to know" and "event by event" are used for access control. Along with it to strengthen this physical security, background check of employees and training on security updates are being emphasized.

B. **Logical Security**

This Security level is the second line of defence which is divided into different sub layers that work in an integrated manner to protect the whole infrastructure. They are

I. Compute sub-layer, II. Network sub-layer, III. Storage sub-layer, IV. Management sub-layer.

C. **Value-Added Security**

It contains some value-added security features comprising of Firewall and VPN potentialities, predesigned security solutions having uses of Big Data and its characteristics, which can detect security vulnerabilities and identify mitigation options.

D. **Governance, Risk and Compliance**

This layer ensures the standard of the three previous layers. Also agile development techniques and innovative methods control and fix the bugs in this layer. Besides this, high-end change management is followed for updation and debugging, verification and rollback procedures.

4 Big Data

Big data [1] is going to be the most happening phenomenon in near future encompassing both technical and marketing data inside it. It contains large volume of data and in due course of time its growth rate has increased.

4.1 Major Issues in Big Data

a. Management issue
b. Processing issues
c. Security issues
d. Storage issues

This paper mainly focuses on Security-Related-issues.

a. **Management Issue**

This issue pertains to gathering large volume of data from the organizations. This means the focus is on [12]

- High-data quality,
- data ownership,
- standardization,
- documentation and accessibility of data set.

b. **Storage Issues**

The quantity of data that has to be stored has always posed a problem. Take the case of most recent data explosion at social media. The biggest problem is there has been no new storage system. Current limits are 4 TB per disk. So 1 Exabyte would necessitate {Twenty-five thousand (25,000)} disks. Even if 1 Exabyte could be processed on a single computer system, it would not match the requisite number of disks [13].

c. **Processing Issue**

The analysis in big data operating involve higher unit sizes like Petabyte (10^15), Exabyte (10^18) or even in Zettabyte (10^18). For Example 1 K Petabyte involves total end-to-end processing time of 635 years. Hence the need of hour is requisite parallel processing and new analytics algorithm to provide desired result [11].

d. **Security Issue**

Both public and private database is equally vulnerable to data thefts due to lack of proper monitoring and tracking at the origin. Loopholes in security policy of both public and private organizations and lack of encryption at the time of storage of huge amount of data paves an easy path for the hacker to steal the required information. Apart from this, storage of data in presence of untrusted people and lack of development of proper tools and technologies with security certification has posed a great challenge with respect to the security issues [14].

There can be various type of data hacking. Like collection of data set along with copying it and storing it in common devices like USB drives {Pen drive} and hard disks. Also there can be hacking by sending threats like: {Denial of Service, Snoofing attack, Brute force attack}. The knowledge of key value pair of data makes the data more vulnerable. The solution lies at increasing the security tiers simultaneously with increase in data tiers and development of robust algorithm and efficient cryptographic framework techniques. Apart from this, landmark technologies in storage of data like Hadoop and NoSQL can solve many issues in big data. Hadoop is also in similar way vulnerable to security issues, as original Hadoop distribution is not designed keeping in view its use in enterprise data environment.

Solution of these issues need proper understanding of the problems and subsequent measures. They are

User Authentication and Access—Hadoop needs strict authentication of user and control of access rights. One possible solution for this problem is Apache Sentry.

Regulatory Requirements—Organizations with huge data system need extra regulatory requirement and system audits regarding generation and storage of records.

User Impersonation—Lack of strong protocol for user and service authentication in Hadoop increases the vulnerabilities in input of data. Kerberos and LDAP protocols are used against this weakness.

Protecting Data-At-Rest and Moving Data—Data encryption method is used for protecting Data at rest and Network encryption methods have been used for moving data. Although native Hadoop distribution system provides for the data encryption but later one is not included in this. So it requires the user to make that line of protection.

4.2 Hadoop Architecture

HDFS is nothing but Java portable file system. Whereas Hadoop cluster is in a group (Fig. 2). Commodity Hardware usage—{One Datanodes and more than one Namenode} provides redundancy in storage of vast pool of data with

(a) Low level of latency
(b) Operations like "Write Once" "Read Many Times" Getting done.

Remote Procedure calls (RPC)—As method for communication between the nodes.

Metadata like

- locations of blocks on data nodes,
- active data nodes,
- a bunch of other metadata,
- duplicates.

In addition to storage functions it also

- it decreases the dropping of data,

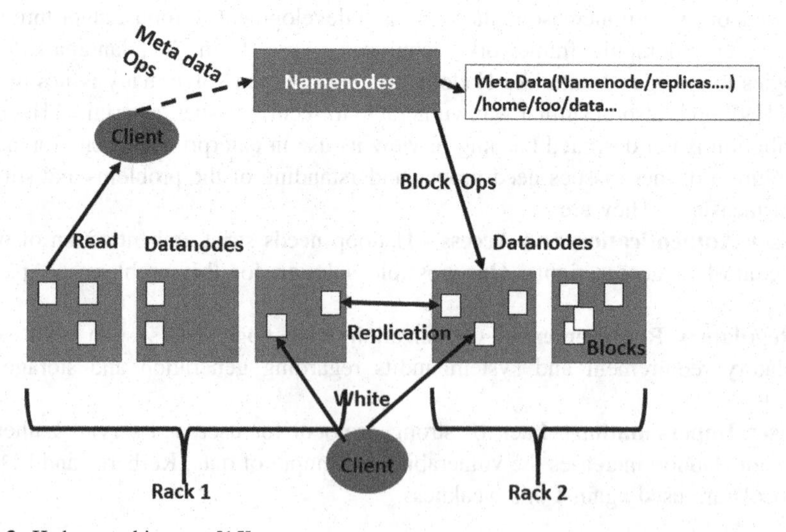

Fig. 2 Hadoop architecture [15]

- also decreases corruption of the file. In case of any lost or failed block, the namenode makes another carbon copy of the same block.

Any error or failure takes place in the node, periodic repairing can be done in case of any failure of the node.

HDFS permit more than one thousand nodes by a single Operator [15].

In case of replica of each block in various data nodes:

(a) Original data node as rack 1 (b) Replicated data node as rack 2.

4.3 HDFS Security Approaches

In Hadoop distributed file system {hdfs} security of data is ensured through below methods.

- Kerberos
- Bull Eye Algorithm Approach
- Namenode Approach

A. **Kerberos Mechanism**

Hadoop community has proposed and come to a conclusion that token based authentication is the appropriate approach to combine pluggable authentication providers, to improve upon single sign on for end user as desired and enforce centralized access control on the platform. This paper reflects on an innovative solution about implementation of token authentication based on the Kerberos pre-authentication framework. It proposes a pre-authentication mechanism for Kerberos that allows users to authenticate to Key Distribution Center (KDC) tossing a standard token [16].

The connection in HDFS is established by two methods.

- Remote Procedure Call—between client and Name node.
- Block Transfer—the link From Client to Data node.

Authentication of a RPC connection is done using Kerberos. Kerberos authenticated Connection is required for obtaining token by the user. We have TGT (Ticket Granting Ticket) or Service Ticketing used for authentication of name node in this mechanism. With each renewal of Kerberos,

- new TST
- ST

are being issued to all task. Finally, after getting request from any task, Kerberos Service Ticket using TGT is issued by—Key Distribution Center. The best part of Kerberos approach is that: ticket cannot be renewed even if the attacker steals (Fig. 3).

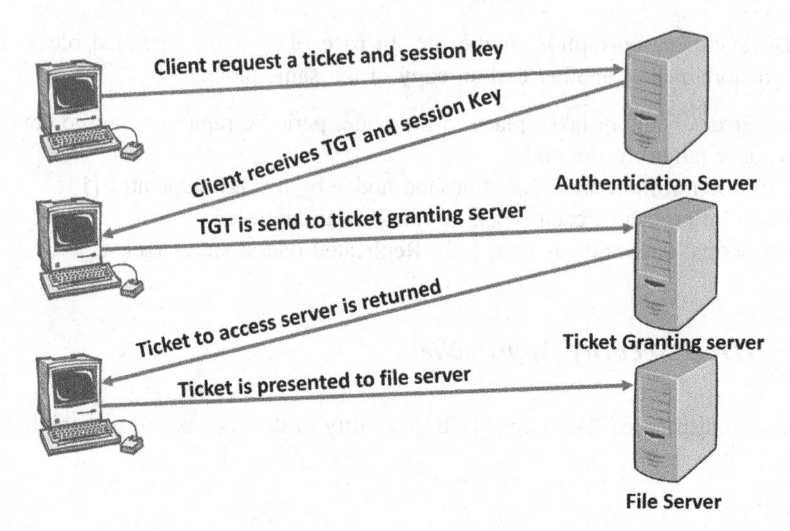

Fig. 3 Kerberos authentication [16]

B. **Bull Eye Algorithm Approach**

Hadoop is a technology to store sensitive data's like credit card number, passwords, bank details etc. So an algorithm, bull eye approach is used to enhance the security features in Hadoop. It analyzes all the information in 360° manner and data node of rack 1 is used for implementation [4].

Data's are stored in required blocks but this is solely allowed for that particular client. This algorithm forms a bridge between

(a) Real data node
(b) Replicated data node and also examine the relation between two racks.

But the authorized person permitted to

- The read or write about the data node.
- While implementation happens below the data node in both the racks, Data breaches and its complete checking is given immense importance Encrypted data—used for better protection in data node.
- This algorithm scans the data before getting allowed to enter into the blocks and also after entering both the rack 1, 2.
- Thus, this mostly concentrates only on the highly confidential data stored in the data nodes.

C. **Namenode Approach**

HDFS focuses on

- data streaming,
- storing large data sets,
- dealing with hardware failure.

But, it makes the number of system service and stored data equally not available. So, it is difficult for accessing the data in a perfectly safe way from this critical situation. Security in data availability is done. By using two Namenodes [4].

In HDFS,

- Namenode plays lead role
- Data Nodes assist namenodes running in the aforesaid cluster.

Hence, Name Node Security Enhance: [4]

(a) Name Node: An HDFS comprises of Single Namenode—for making namespace operations (like renaming, opening, closing of files and directories). A master server—for managing the file System namespaces and for regulating the access to files by clients.

(b) Data Node: The Data nodes functions with instruction from Name node and performs block creation, deletion, replication, and read write requests from file system clients.

4.4 Discussion

Hence we analyse that our work brings into picture different approaches for securing data in HDFS. In the first one, the basis is Kerberos in HDFS which is used to access a data block correctly and also by an authorized user. Whereas as in second one, the basis is Bull Eye Algorithm Approach which explains about security method from one node to another node and also scans nodes from all possible way to avoid attacks. In the third one, the basis is Namenode where security is obtained by replicating a name node to lower crashes for future references.

4.5 HADOOP Security Issue

When it comes to Hadoop security, we have seen many testing times, given its concurrent processing architecture. In the case of authentication vulnerabilities, this has led to launching central authentication via Kerberos systems. Now Hadoop framework despite forming an important role for traditional distributed systems also forms an integral part of a cloud environment. The basic construction of Hadoop makes a file to be split into blocks or chunks, further distributing over different nodes leading to its further processing at a time. But in the cloud technology background it becomes tough to exactly find out the specific node which holds that specific block of data. In the cloud it becomes hard to mark a specific node having a particular set of data.

This paper focuses on security architecture of Twilio. It is a more than one user communication media to ensure isolation of resources. There is still a possibility for physical theft of data, but is protected against using base security measures (Fig. 4).

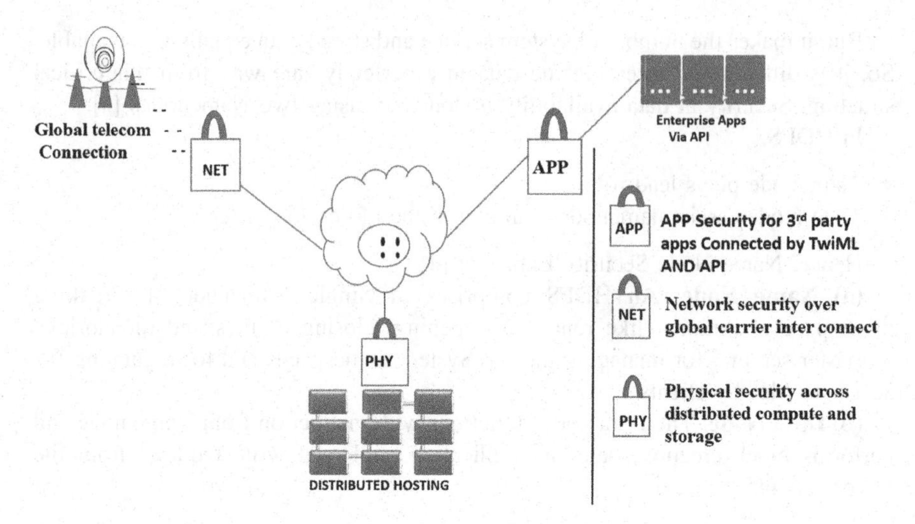

Fig. 4 Cloud security architecture named as Twilio [19]

5 Big Data Security: — {Risks, Vulnerabilities and Threats}

Although Big Data can have many advantages, it needs proper security measures. Storing and analyzing large volume of data extant a mammoth task for encryption. The cost incurred in encryption and decryption has impact on further uses of data. But an important point to note is that this encryption is require to neglect the mix of datasets to avoid risks and threats which has adds to possibility of further data loss. Analysis of the data should be done without decrypting it ensure security and privacy of the client (Fig. 5).

Different sets of data in a dataset are needed to be extracted and ply differently. Taking into consideration the high velocity of data, tracking access and data flow without fixed path of travel between two or more nodes can be a bigger issue. Ensuring control of access and data is necessary for protecting the data from the vulnerabilities of theft or loss. If corruption in one data set is allowed to go outside the proper bounds, it poses the same risk for other customer datasets or it may hamper the whole infrastructure of the cloud service provider. The Backup and Restore functions may negatively affect performance because of the presence of

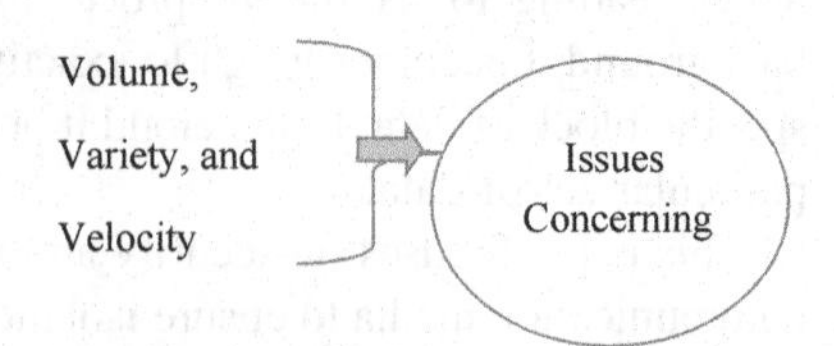

Fig. 5 Big data issues concerning

enormous amount and high velocity data. Hence if data loss is to be avoided this problem cannot be overlooked [17].

For the above-discussed purpose we will focus on sticky policy based framework in the subsequent paragraphs.

5.1 Sticky Policies

The security issues of big data have undergone transformation by design of Sticky Policy based framework. Lee and his co-authors have proposed framework with an innovative tool to help in securing big data application on the basis of data privacy management project called Encore. Encore implement sticky policy which apply constraints on data to define usage allowances. The architecture is given in Fig. 6.

There are two domains:

(a) Trusted authority. (b) Data center.

- The first one regulates identity, manage the sticky key and also manage policy engine; whereas
- The second one stores encrypted data.

In the first one

Identity and key management engine has the user information which comprises of

- Authentication and authorization information
- Privileges which it has for each piece of data, whereas;

Policy engine which comprises of various sub parts:

i. Policy Controller: ii. Policy Portal: iii. Policy negotiation: iv. Policy Update: v. Enforcement: vi. Policy store.

This Engine (policy) functions as the core of the engine which has further two functions:

(a) Controlling the data accessed
(b) Keeping track of parties privileges.

In the Policy Engine,

(a) Policy Portal acts as entry path to the engine

- receives data access requests from sender.
- gives final responses back to user.

(b) Policy Controller regulates the data for

- making a decision about the rejection of the entreaty or
- forwarding it to the responding part.

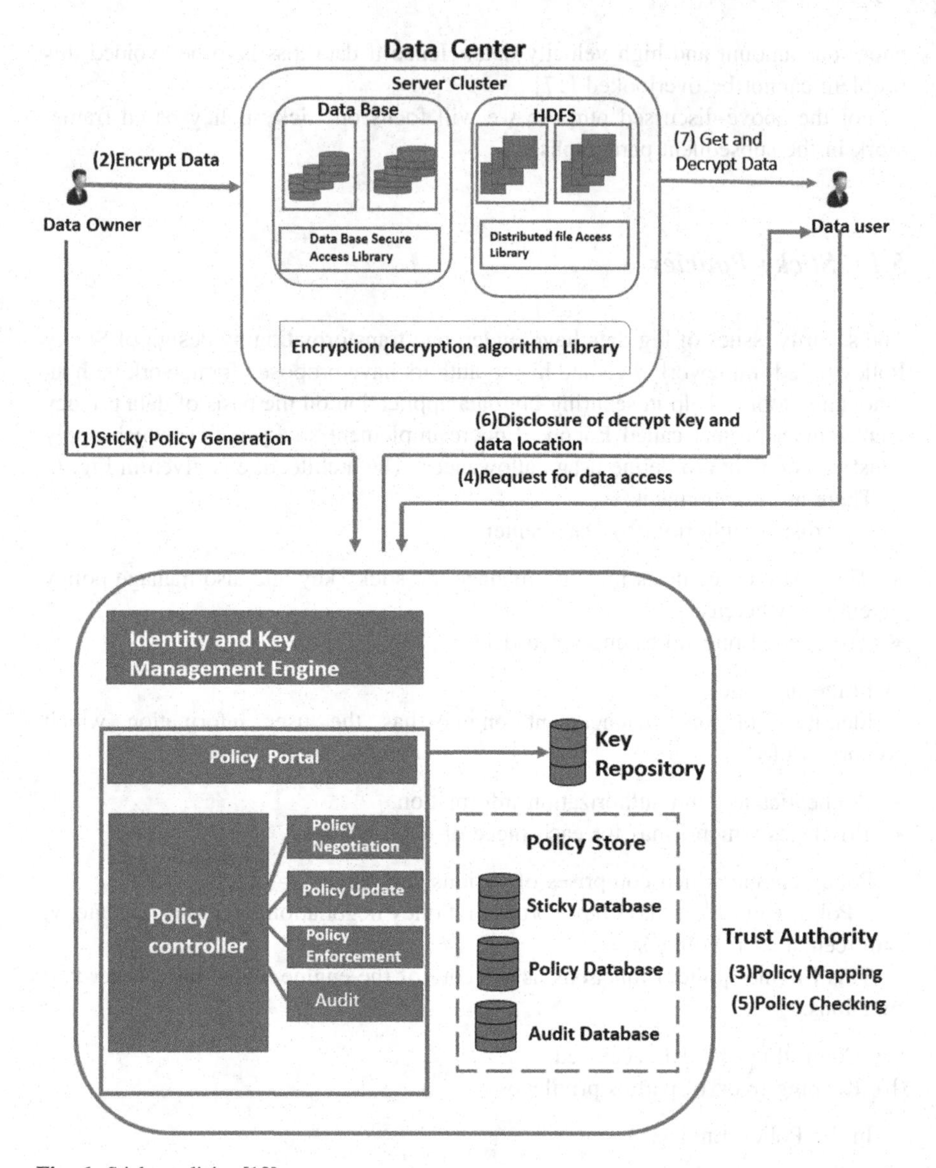

Fig. 6 Sticky policies [18]

(c) Policy negotiation asks

- Scheme,
- match policies to the database of users.

(d) Policy update—updates the security policies.

(e) Enforcement component

- confirms the fulfillment of required sticky policies by user
- keeps audit logs of all the activated policies and accessed data.

(f) Policy store keeps

- Data mapping
- And access it.

Then the stored data is encrypted which can in future be made accessible after policies have been given approval [18].

Sticky Policies gives an ideal framework to get more guarded transactions. This is used in private classified data like: (i) ATM Pin (ii) medical records (iii) law enforcement record (iv) highly confidential data {defence secret data, government organizations, investigation reports etc}.

We find that it is quite obvious for security be made a priority in the initial phase of design and implementation. But still there are chances that new vulnerabilities are sure to be identified. Further, a system to completely eliminate cyber criminals is yet to be designed.

On a concluding note, we can say that, In future Big Data security field is certainly going to play a more important role in the technical era as Cloud Services are increasing in popularity and are used in every sphere of life like right from personal data backup to large scale storage and analysis of big data in corporations.

6 Conclusion

In light of growing significance of big data collection and analysis in present day world, there has been a great leap in the volume of data produced. Thus we see many technologies getting developed to collect, analyses, and store this data but this comes with a price as they have vulnerabilities, risks and security threats associated with them. What is seen as breach in data across various platforms today is justification that security cannot be just skipped in developing the Big Data technology. Hence a lot of research is going on security issues in initial design and implementation phase and taking steps to restrain them.

Exactly, this paper tries to deduce current security threats in cloud services, big data and Hadoop which can be done by systemically bringing into effect any one of the approaches or by effective combination of the three in HDFS. Furthermore, a holistic system to fully stop cyber criminals is yet to be developed. Thus on a concluding note we can say that the Big Data security field is certainly going to play a major role in times to come.

References

1. Lv, Z., Song, H., Basanta-Val, P., Steed, A., & Jo, M. (2017, August). Next-generation big data analytics: State of the art, challenges, and future research. *IEEE Transactions on Industrial Informatics, 13*(4).
2. Guo, K., Tang, Y., & Zhang, P. (2017, January 29). Crowdsourcing semantic fusion for heterogeneous media big data in the internet of things. *Information Fusion, 37*, 77–85. Elsevier [Online]. Available: www.sciencedirect.com.
3. Goepfert, J., & Vesset, D. (2016, October). Worldwide semiannual big data and analytics spending guide. IDC [Online]. Available: http://www.idc.com/getdoc.jsp?containerId=prUS41826116.
4. Saraladevi, B., Pazhaniraja, N., Victer Paul, P., Saleem Basha, M. S., & Dhavachelvan, P. (2015). Big data and Hadoop—A study in security perspective. In *2nd International Symposium on Big Data and Cloud Computing (ISBCC'2015)*, Elsevier.
5. Al-Shomrani, A., Fathy, F., & Jambi, K. (2017). Policy enforcement for big data security. IEEE.
6. Matturdi, B., Xianwei, Z., Shuai, L., & Fuhong, L. (2014). Big data security and privacy. IEEE.
7. Mengke, Y., Xiaoguang, Z., & Jianqiu, Z. (2016). Challenge and solutions of information security issues in the age of big data. In *IEEE Transaction*.
8. Abouelmehdi, K., Beni-Hssane, A., Saadi, M., & Khaloufi, H. (2016). Big data emerging issues: Hadoop security and privacy. IEEE.
9. Singh, A., & Chatterjee, K. (2017, February 1). Cloud security issues and challenges. *Journal of Network and Computer Applications, 79*, 88–115. Elsevier [Online]. Available: www.sciencedirect.com.
10. Alliance, C. (2013). *The notorious nine cloud computing top threats*. Cloud Security Alliance, Tech. Rep.
11. Nejad, E. S., Majma, M. R., Izadpanahi, B., Natanzi, S. B. H., & Navaei, H. R. (2015). Infrastructure of data centers for transferring big data traffic: A survey research. In *International Conference on Technology, Communication and Knowledge (ICTCK)*, November 11–12, 2015, IEEE.
12. Paryasto, M., Alamsyah, A., Rahardjo, B., & Kuspriyanto. (2014). Big-data security management issues. In *IEEE, International Conference on Information and Communication Technology*.
13. Sookhak, M., Gani, A., Khan, M. K., & Buyya, R. (2017, February 20). Big data storage in cloud computing. *Information Sciences, 380*, 101–116. Elsevier.
14. Abouelmehdi, K., Beni-Hssane, A., Khaloufi, H., & Saadi, M. (2016). Big data emerging issues: Hadoop security and privacy. In *IEEE 5th International Conference on Multimedia Computing and Systems (ICMCS)*.
15. Jie, C., Dongjie, C., & Bangming, H. (2014). Research on big data information retrieval based on Hadoop architecture. In *IEEE Workshop on Electronics, Computer and Applications*.
16. Valliyappan, V., & Singh, P. (2016). Protecting the apache Hadoop clusters with Hadoop authentication process using Kerberos. In *Proceedings of 3rd International Conference on Advanced Computing, Networking and Informatics*, pp. 151–161. Springer.
17. Buyya, R., Calheiros, R. N., & Dastjerdi, A. V. (2016, June). Security and privacy in big data. Elsevier [Online]. Available: www.sciencedirect.com.
18. Li, S., Zhang, T., Gao, J., & Park, Y. (2015). A sticky policy framework for big data security. In *IEEE First International Conference on Big Data Computing Service and Applications*.
19. Sharif, A., Cooney, S., Gong, S., & Vitek, D. (2015). Current security threats and prevention measures relating to cloud services, Hadoop concurrent processing, and bigdata. In *IEEE International Conference on Big Data*.

Optimized Capacity Scheduler for MapReduce Applications in Cloud Environments

Adepu Sree Lakshmi, N. Subhash Chandra and M. BalRaju

Abstract Most of the current-day applications are data centric and involves lot of data processing. Technologies like hadoop enable data processing with automatic parallelism. Current-day applications which are more data intensive and compute intensive can take advantage of this automatic parallelism and the methodology of moving computation to data. In addition to it the Cloud computing technology enables users to establish the required clusters with required number of nodes instantly. Cloud computing has made easy for the users to execute large data applications without any requirement to establish/maintain the infrastructure. As cloud gives readily installed infrastructures, using hadoop on cloud has become common. The existing schedulers are very effective in static cluster environments but lack performance in virtual environments. The purpose of this work is to design an effective capacity scheduler for MapReduce applications for virtualized environments like public clouds by making scheduling decisions more intelligent using the characteristics of job and virtual machines.

Keywords Big data · Cloud computing · CloudSim · Hadoop MapReduce · Virtual machine

A. Sree Lakshmi (✉)
Geethanjali College of Engineering and Technology, Hyderabad, Telangana, India
e-mail: adepu.sreelakshmi@gmail.com

N. Subhash Chandra
CSE Department, CVR College of Engineering, Hyderabad, Telangana, India
e-mail: subhashchandra.n.cse@gmail.com

M. BalRaju
CSE Department, Swami Vivekanandha Institute of Techology, Hyderabad, Telangana, India
e-mail: drraju.jb@gmail.com

V. E. Balas et al. (eds.), *Data Management, Analytics and Innovation*, Advances in Intelligent Systems and Computing 808,
https://doi.org/10.1007/978-981-13-1402-5_12

1 Introduction

Hadoop [1] uses MapReduce framework where the given application executes in parallel on multiple nodes. The data of application is divided into multiple chunks of equal size and distributed to multiple nodes for processing. Map task perform the map function of the application on a split of data and generates the key value pairs which are further hashed and shuffled to an appropriate reduce task. Reduce task performs reduce logic on the output (Key/Value pairs) received from different mappers.

There are different schedulers provided in hadoop package which are pluggable. The schedulers provided in hadoop package are FIFO, Capacity scheduler, Fair scheduler. The default scheduler used in hadoop 2.0 is capacity scheduler. Our papers [2] can be referred for description of schedulers provided in hadoop package. All the schedulers provided in hadoop are optimal in fixed cluster environment. When executed in cloud environments it does not give optimized performance as the cloud is completely virtualized and each physical machine is shared by multiple virtual machines. Our paper [3] gives the study of performance of virtual machines for MapReduce applications in public cloud environment Amazon EMR. The execution time for hadoop MapReduce jobs can be further optimized by making scheduling decision more intelligent. If scheduling of MapReduce tasks considers the characteristics of virtual machine along with the job characteristics, we can make optimal scheduling.

Capacity Scheduler [4] in hadoop works in the form of hierarchy of queues and jobs are always placed in leaf queues. The available resources can be shared by multiple queues according to the configuration given in capapcityscheduler.xml file. In this paper we have improved the capacity scheduler by embedding the knowledge of virtual machine state and the resource requirements of the job. The proposed work is done in 3 modules: 1. Job resource requirements classification 2. Virtual machine state classification 3. Scheduling based on the above two classifications. Our experiments show that there is an improvement of 18–25% in the execution time of the hadoop jobs on virtualized environments. Less execution time is a major factor in cloud environments as users pay as use for the infrastructures. Reducing the execution time when executed on public clouds would reduce the costs incurred for cloud resources usage.

2 Background and Related Work

Capacity Scheduler is the default scheduler in Hadoop 2.0 [5]. One of the biggest advantages of Hadoop is multi-tenancy feature which enables multiple tenants to share the cluster in such a way that resources are allocated to their applications in a timely manner with respect to capacities allocated. Capacity scheduler enables sharing of large cluster among multiple users giving minimum capacity guarantee. It ensures that a single job/user/queue does not monopoly the resources available in the cluster.

Capacity scheduler mainly uses the concept of job queues and ensures capacity guarantees for queues. It uses hierarchy of queues where jobs are submitted to leaf queue. Certain capacity of resources is allocated to queues which can be used by the jobs submitted in the child leaf queues under it. The capacity guaranteed for the queues is configurable. Free resources can be allocated beyond the capacity of the queue also if not utilized by the other queues in the cluster. When an underutilized queue needs resources, the resources released by the over utilized queue are allocated to them. Hence elasticity is provided so that underutilized capacity of a queue can be provided to queues that need more resources beyond its allocated capacity.

Once the first job is submitted to the queue all the resources are given to it. When the job of other queue is submitted then further released resources are allocated to both the queues in accordance to the capacities guaranteed to the queues. Capacity scheduler does a static way to scheduling based on the resource requirements of the job and resources available in the cluster. Though it gives best run times in cluster environments. It needs some improvements when executed in virtualized environments.

Other works done in the direction of schedulers have done optimizations in different directions to decrease the execution time of the MapReduce applications. The authors of [6] designed a context aware scheduler for Hadoop1.x which improves the overall throughput of the system by leveraging the cluster heterogeneity. It schedules the tasks by context, i.e., job characteristics and resource characteristics of the nodes in the cluster. The authors neglected the IO bound shuffle phase where map outputs are moved to the reducer. Their work mainly classifies the node statically to determine whether a node is fast or slow. SAMR [7] self adaptive MapReduce scheduling algorithm in heterogeneous environment adapts to the continuously varying environment automatically by calculating the progress of tasks dynamically. It splits the given job into many fine grained tasks and while executing these fine-grained maps and reduce tasks stores the historical information on every node. Based on the historical information SAMR will adjust the time weight of each stage of the map and reduce tasks.

The authors of [8] designed a fine-grained and dynamic MapReduce task scheduling scheme for the heterogeneous cloud environment. Their work is also based on collecting historical and real time online information from each node in the cluster and selects the appropriate parameters in order to identify slow running tasks. Nodes are statically classified as high performance and low performance irrespective of the load on peer virtual machines running on the same physical node.

Research Work in [9] has a scheduling framework that takes into account the actual resource requirements of the job. Their scheduler classifies tasks into schedulable and non-schedulable classes based on whether a job will overload a node. They used only CPU utilization but other job features need to be considered relatively. Scheduling related works by different researchers [10–19] in the direction of dynamic resource allocation try to optimize the execution times in different directions like cost, IO time, Network IO, reduce operation time, etc.

3 Proposed Work

In virtual environments each physical machine has multiple virtual machines hosted on it and shares the resources in the physical machine and accesses them through hypervisor. When a Hadoop cluster is created in a virtualized environment the performance is not similar to that of dedicated cluster environment. An optimized scheduler is required for virtualized environments to schedule the Hadoop MapReduce applications. A better performance can be obtained by including the information related to job characteristics and VM characteristics in scheduling decisions. When a job is scheduled to a VM its performance is very much affected by the other virtual machines running on the same physical host. This is more effective in the case of disk performance as every read/write operation done by a VM goes through hypervisor and the disk I/O is shared by multiple VMs. This has more effect in Hadoop jobs as the map/reduce application input/output are done to disk. The amount of CPU and IO required by every job has different performance in different VMs depending on the current CPU and IO usage of the VM and the other VMs existing in the physical machine.

If job characteristics is understood to be whether it is CPU-intensive or IO-intensive then a proper VM can be chosen to give better performance of the MapReduce application execution. Scheduling can be made dynamic which includes the features of job, current VM characteristics. Our proposed work is to design a dynamic capacity scheduler which schedules MapReduce applications considering the job and VM characteristics. As the amount of data processed except the last map task is same and the map function applied is same, understanding the job characteristics is easy after execution of few map tasks, which can be used in scheduling the future map tasks of that job. Similarly if the current state/characteristics of every virtual machine is regularly monitored and a tag is set regularly based on the amount of CPU utilization and IO Utilization of each VM. If a VM is tagged to be using more IO then that VM is more appropriate for a map task which is less IO-intensive.

To simulate our work, we used CloudSimEx [20] which is a MapReduce simulator and an extension of CloudSim simulator [21]. CloudSimEx has simulation for both cloud environment and MapReduce execution. But as our scheduler is an improvement over current capacity scheduler, we built the existing scheduling model of capacity scheduler on CloudSimEx. To model capacity scheduler the following changes are made to the CloudSimEx tool.

1. Limited number of virtual machines, the number of virtual machines required by user is taken as parameter through cloud.yaml file as the cloud is usually rented by specifying the number and type of machines required.
2. In cloudSimEx a scheduling plan is built for all cloudlets considering the virtual machines available in the data centres and all cloudlets are submitted at a time. Cloudlets are mapped to VMs initially itself. As number of VMs is restricted based on user request, initial binding of cloudlets VM's is removed. It is being modelled in an incremental fashion. Cloudlets are submitted initially to VMs based on the capacity guaranteed (**Algorithm 1**).
3. Job submissions are made through queues where each queue is associated with capacity guarantee (queue_capacity). If queue_capacity is [60, 40] then 60% of resources are assigned to queue1 and 40% of resources to queue2. If nVMs = 5 then 3 VMs are allocated to queue1 and 2 VMs for queue2. Cloudlets are modelled to assign to nVMs as per the user provided queue_capacity.

4. Submit cloudlet method is overridden to submit cloudlet to a specified VM so that once the task assigned to slave node running on that VM is completed it can assign a new task to it within the constraints of capacity Scheduler (**Algorithm 2**).
5. Cloudlet return is being modified in order to invoke submitCloudlet(VM vm) to submit a cloudlet to the VM according to queue capacity allotment.
6. Job_submitted_count[] array is used to maintain the number of cloudlets submitted for each queue to enable the capacity allocation constraints. No of resources allocated to job in queue 0 is indicated in job_submitted_count[0].
7. job_index[] array to hold the index of each job submitted to cloudletList. This is being used to enable to move to next job in the queue easily instead of comparing with every task in the cloudletList. As every job is divided into tasks executed in cloudlets, when a task tag does not match with VM tag, the remaining tasks of the same job need not be compared with the VM tag, and we can directly compare with the cloudlet of next job. job_index[0] = 0 which is the index of cloudlet of the first job submitted and job_index [1] is assigned with the index of the first cloudlet of the second job submitted.

submitCloudlets() is invoked only once during the initial scheduling to VMs. First cloudlet is submitted and job_submitted_count is updated to reflect the current allocation done to the queue to which the job request belongs. If that request allocated resource (VM) count is less than the guaranteed queue capacity the next index also belongs to the same job request otherwise the index is made to point to next job request to fulfil its guaranteed queue capacity.

Algorithm1 submitCloudlets()

```
1  q_c = Queue_Capacity;
2  count, index, q= 0;  lastindex = job_index[q] – 1;
3  for each vm in VmsCreatedList
        3.1 cloudlet = get cloudlet at index
        3.2 if (cloudlet instanceof ReduceTask &&
            !isAllMapTaskFinished(cloudlet.getCloudletId())) then continue;
        3.3 r = request to which cloudlet belongs
        3.4 set vmid of cloudlet to vm
        3.5 put(cloudletid,vmid) into scheduling plan
        3.6 send CLOUDLETSUBMIT event;
        3.7 cloudletsSubmitted++;  count++;
        3.8 job_submitted_count[q] = count;
        3.9 for  i = q to job_index.length
                  3.9.1  job_index[i] = job_index[i] – 1;
        3.10      lastindex = job_index[q];
        3.11      if (!((count < q_c[q]) & (index < lastindex))) then
                   3.11.1 index = lastindex
        3.12      count = 0;  q++;
        3.13      lastindex = job_index[q];
        3.14      endif
        3.15      add cloudlet to CloudletSubmittedList
        3.16      remove cloudlet from CloudletList
4  end for
```

SubmitCloudlets(int Vmid) is invoked for submission of cloudlets as each VM is ready to accept a new task on finishing the previously allocated task. Index is made to point to the queue which has less allocated resources than the guaranteed capacity. If the task is a reduce task and its corresponding map tasks are not completed then cloudlet from next queue is considered. If the queue is last queue tried and no more cloudlets next to it in the cloudletlist we simply return as in step 4.4.2, otherwise point to the next queue. But if the selected task is a map task and within the capacity guaranteed for the corresponding queue then that task is submitted to the VM.

Algorithm2 submitCloudlets(int vmid)

```
1  if (getCloudletList().size() > 0) then
2  done = false;
3  index = 0;
4  for each vm in VmsCreatedList
   4.1 if (vm.getId() == vmid) then
       4.1.1 q = 0;
       4.1.2 if ((job_submitted_count[q] < q_c[q]) & (job_index[q] > 0)) then break;
       4.1.3 else q++;
       4.1.4 break;
       4.1.5 endif
   4.2 if (q != 0) then vindex = job_index[q -1];
   4.3 cloudlet = get cloudlet at index
   4.4 if (cloudlet instanceof ReduceTask &&
           !isAllMapTaskFinished(cloudlet.getCloudletId())) then
       4.4.1 if (q == q_c.length - 1) index = job_index[q - 1];
       4.4.2 if (index = = 0) return;
       4.4.3 endif
       4.4.4 else
             4.4.4.1     index = job_index[q]; q = q + 1;
       4.4.5 endif
       4.4.6 cloudlet = get cloudlet at index
       4.4.7 endif
       4.4.8 r = request to which cloudlet belongs
       4.4.9 set vmid of cloudlet to vm
       4.4.10put(cloudletid,vmid) into scheduling plan
       4.4.11send CLOUDLETSUBMIT event
       4.4.12cloudletsSubmitted++;
       4.4.13job_submitted_count[q]++;
       4.4.14if (q == job_submitted_count.length - 1) then job_index[q] - = 1;
       4.4.15if (q > 0)
       4.4.16if (job_index[q] == job_index[q -1])
             4.4.16.1   t = job_index[q];
             4.4.16.2   job_index[q] = 0; job_index[q - 1] = 0;
             4.4.16.3   for (int i = q + 1; i < job_index.length; i++)
              4.4.16.3.1        job_index[i]= job_index[i]–t;
```

```
        4.4.17endif
        4.4.18endif
        4.4.19else
        4.4.20for (int i = q; i < job_index.length; i++)
                4.4.20.1   job_index[i]= job_index[i]– 1;
        4.4.21endif
        4.4.22getCloudletSubmittedList().add(cloudlet);
        4.4.23getCloudletList().remove(cloudlet);
    4.5        endif
    4.6        endfor
5   endif
```

Our work is a contribution to CloudSimEx with the scheduling techniques of Hadoop. We implemented the scheduling algorithms FIFO, Capacity Scheduler of Hadoop MapReduce applications in CloudSimEx. This contribution enables people working in the area of scheduling, load balancing of Hadoop jobs for virtualized environments to simulate their work. Our proposed work of designing a dynamic capacity scheduler is simulated in CloudSimEx by including the job characteristics and VM characteristics in scheduling decisions. When a cloudlet is returned, submit cloudlet decides about which job is to be given to the VM based on the job tag and VM tag within the constraints of queue capacity.

Jobs are categorized to be CPU-intensive, IO-intensive and every job is associated with a 2-digit tag which indicates the intensiveness of the job. MSB indicates the CPU-intensiveness and LSB denotes IO-intensiveness. Tag with value 01 indicates a job which is IO-intensive and value 10 indicates a job which is CPU-intensive. Similarly every VM is associated with 2-bit tag to indicate the current load of CPU and IO of the host on which the virtual machine is deployed. Cloudlets can be submitted with reference to the tag of jobs and virtual machines. Jobs can be assigned a tag by user if the intensiveness of the job is known or can be calculated after few map tasks get executed. When a virtual machine tag indicates as IO heavy then a task which is more CPU-intensive and less IO-intensive would be appropriate choice.

Algorithm submitCloudlets()
// CapacitySchedulerLoadAware

```
1. q_c = QueueCapacity;
2. index,q= 0; lastindex = job_index[q] – 1;
3. for each vm in VmsCreatedList
4. cloudlet = first cloudlet in cloudletList;
      5.1 done = false;
      5.2 for idx=0 to job_index.length
      5.3 q = idx;
      5.4 if (idx == 0) then index = idx
      5.5 else index = job_index[idx – 1]
      5.6 if (job_submitted_count[q] < q_c[q]) then
         5.6.1 cloudlet = Cloudlet at index
      5.7 if (cloudlet instanceof ReduceTask &&
          !isAllMapTaskFinished(cloudlet.getCloudletId())) then continue;
      5.8 r= request to which cloudlet belongs
      5.9 if (((int) r.job.getTag() & (int) vm.getTag()) == 0) then
            5.9.1 set vmid of cloudlet to vm
            5.9.2 put(cloudletid,vmid) into scheduling plan
            5.9.3 send CLOUDLETSUBMIT event
            5.9.4 cloudletsSubmitted++;
            5.9.5 done=true;
            5.9.6 break;
      5.10      end if
      5.11      end if
6 end for
7 if(!done)
      7.1 cloudlet= first cloudlet in cloudletList;
      7.2 r = request to which cloudlet belongs
      7.3 q = 0;
      7.4 set vmid of cloudlet to vm
      7.5 put(cloudletid,vmid) into scheduling plan
      7.6 send CLOUDLETSUBMIT event,
      7.7 cloudletsSubmitted++;
8 end if
9 job_submitted_count[q] = job_submitted_count[q]+1;
10 for (int i = q; i < job_index.length; i++)
      10.1      job_index[i] – = 1;
11 lastindex = job_index[q];
12 if (!((job_submitted_count[q] < q_c[q]) & (index < lastindex)) ) then
      12.1      index = lastindex; q++; lastindex = job_index[q];
13 end if
14 add cloudlet to CloudletSubmittedList
15 remove cloudlet from CloudletList
16 end for
```

4 Evaluation and Results

Simulation.properties:
cloud.file = Cloud.yaml
experiment.files = test3.yaml
machines = ≪4,5,6,7,8≫
mtype = large-aws-us-east-1

MapReduce jobs used: MapReduce_9_2.yaml, MapReduce_15_2.yaml, MapReduce_30_2.yaml, MapReduce_50_1.yaml, MapReduce_100_3.yaml

Experiment file is specified as yaml file in which jobs are considered as part of queues in the order of submission. Different combinations of MapReduce applications are being submitted and the execution times of CapacityScheduler and CapacitySchedulerLoadAware by varying number of virtual machines. One of the sample of experiment file is as below (Fig. 1).

```
Expertiment:test3.yaml
!!org.cloudbus.cloudsim.ex.mapreduce.Experiment
workloads:
 - !!org.cloudbus.cloudsim.ex.mapreduce.Workload
   [
   CapacitySchedulerLoadAware, Public,
     {
      GOLD: 100.0,
      SILVER: 60.0,
      BRONZE: 0.0
     },
   [
   #[Submission Time, Deadline, Budget, Job, user class]
   !!org.cloudbus.cloudsim.ex.mapreduce.models.request.Request
     [200000, 120, 2.5, MapReduce_50_1.yaml, GOLD],
 !!org.cloudbus.cloudsim.ex.mapreduce.models.request.Request
     [200000, 120, 2.5, MapReduce_100_3.yaml, GOLD],
 !!org.cloudbus.cloudsim.ex.mapreduce.models.request.Reques
     [200000, 120, 2.5, MapReduce_15_2.yaml, GOLD],
 ]]
```

Test case	nVMs ($N = 4$)		nVMs ($N = 5$)		nVMs ($N = 6$)		nVMs ($N = 7$)	
	CS	CSLA	CS	CSLA	CS	CSLA	CS	CSLA
MR-30-2, MR-9-2	101.88	87.05	81.78	68.47	68.52	58.272	59.094	57.67
MR-50-1, MR-30-2	313.57	240.57	255.82	187.52	224.89	157.9	201.369	137.49
MR-50-1, MR-100-3	504.17	444.42	398.1	353.18	316.19	192.6	212.959	198
MR-100-3, MR-50-1	530.75	488.97	439.21	347.13	371.79	319.91	325.07	305.14

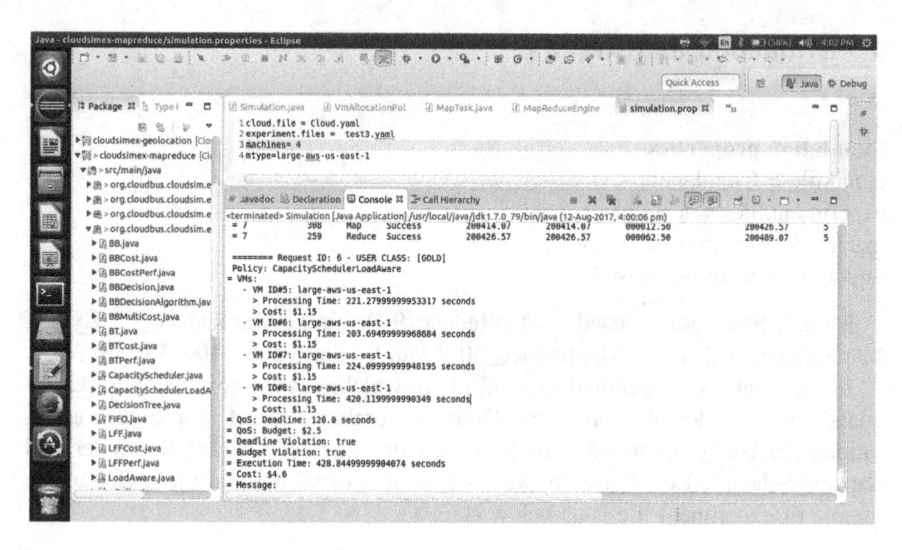

Fig. 1 CapacitySchedulerLoadAware execution

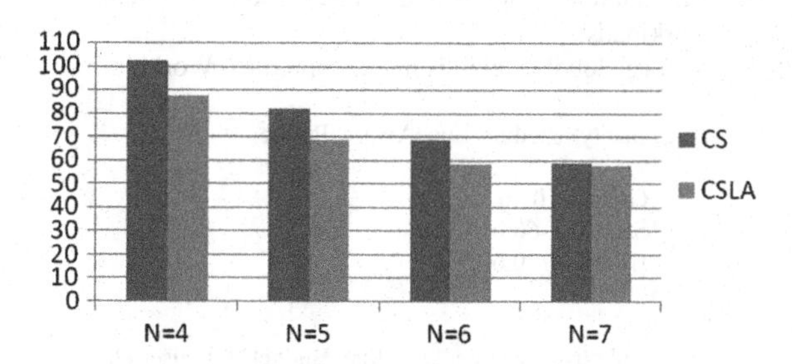

Fig. 2 MapReduce 30-2, MapReduce 9-2

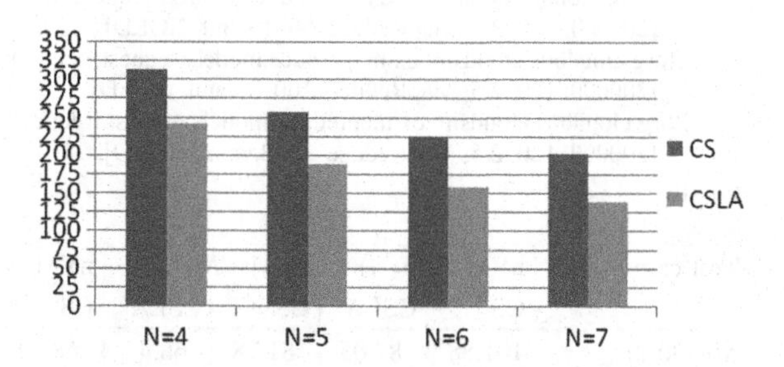

Fig. 3 MapReduce 50-1, MapReduce 30-2

The above results Figs. 2, 3, 4 and 5 indicate an improvement in execution time of the MapReduce jobs on an average of 20%. In few cases there is no major difference in the execution times as the jobs submitted are appropriate to the VM characteristics. There is an improvement brought in the job exection time where MapReduce applications are scheduled in appropriate to VM characteristics (Fig. 6).

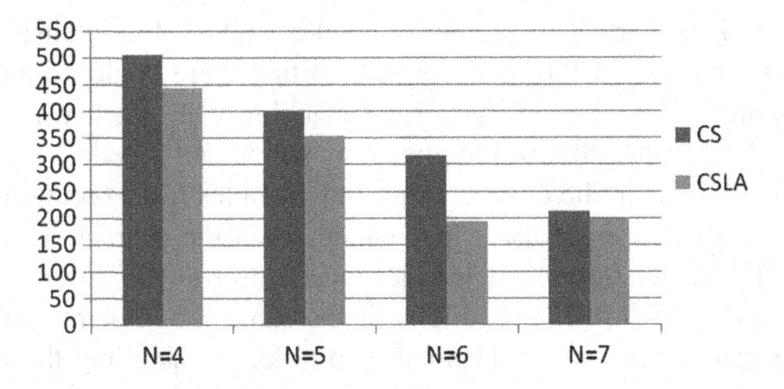

Fig. 4 MapReduce 50-1, MapReduce 100-3

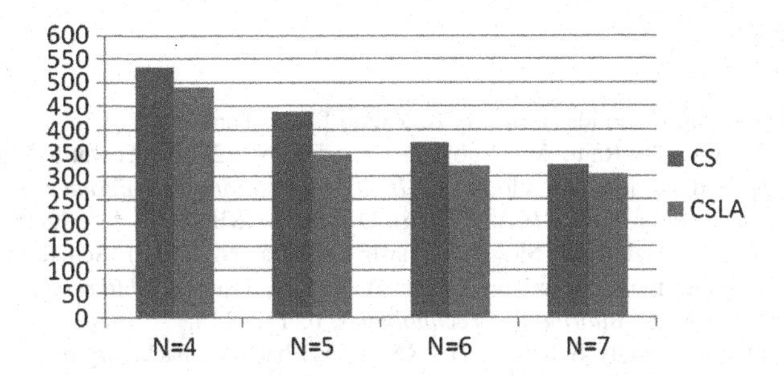

Fig. 5 MapReduce 100-3, MapReduce 50-1

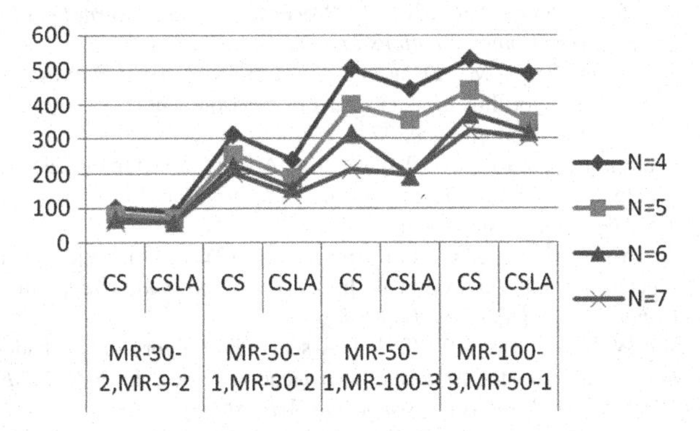

Fig. 6 CapacityScheduler versus CapacitySchedulerLoadAware

5 Conclusion

The MapReduce applications which automatically use the advantage of availability of multiple nodes by executing map and reduce in parallel can further be optimized if the scheduler logic includes a concept of learning of jobs and machine characteristics. If a virtual machine is loaded with CPU-intensive task then scheduler can schedule an IO-intensive task so that there could be an appropriate usage of resources available on the virtual machine. Our work was to design an optimal scheduler for MapReduce applications which can schedule balanced set of workloads to all nodes in the cluster. It not only enables fast execution of the tasks but can also be used to make users understand about the characteristics of the job submitted to the cluster. Job characteristics enable user to use proper configuration of Hadoop cluster for further executions of the same job. Our primary goal is to reduce the execution time of the Hadoop map tasks by including the behaviour of map task and virtual machine in scheduling decisions.

References

1. Hadoop The definitive guide, O'Reilly & Yahoo Press, Tom White.
2. Sree Lakshmi, A., BalRaju, M., Subhash Chandra, N. (2016). Towards optimization of hadoop map reduce jobs on cloud. In *IEEE International Conference on Computing, Analytics and Security Trends (CAST 2016)*, Dec 2016. ISBN: 978-1-5090-1338-8.
3. Sree Lakshmi, A., Bal Raju, M., & Subhash Chandra, N. (2015). Scheduling of parallel applications using map reduce on cloud: A literature survey. *International Journal of Computer Science and Information Technologies, 6,* 112–115.
4. Apache Hadoop Capacity Scheduler YARN: https://hadoop.apache.org/docs/r2.4.1/hadoop-yarn/hadoop-yarn-site/CapacityScheduler.html.
5. https://hortonworks.com/blog/understanding-apache-hadoops-capacity-scheduler/.
6. Kumar, A. K., Krishna, V., Voruganti, K., & Prabhakara Rao, G. V. (2012). CASH: Context aware scheduler for Hadoop. In *ICACCI '12 Proceedings of the International Conference on Advances in Computing, Communications and Informatics.*
7. Chen, Q., Zhang, D., Guo, M., Deng, Q., & Guo, S. (2010). SAMR: A self-adaptive mapreduce scheduling algorithm in heterogeneous environment. *Computer and Information Technology, International Conference*, 2736–2743.
8. Mao, Y., Qi, H., Ping, P., & Li, X. (2016). FiGMR: A fine grained mapreduce scheduler in the heterogeneous cloud. In *Proceedings of the IEEE International Conference of Information and Automation*, Ningbo, China, August 2016.
9. Deshmukh, S., Aghav, J. V., & Chakravarthy, R. (2013). Job classification for mapreduce scheduler in heterogeneous environment. In *2013 International Conference on Cloud & Ubiquitous Computing & Emerging Technologies.*
10. Wylie, A., Shi, W., Corriveau, J. P. (2016). A scheduling algorithm for hadoop mapreduce workflows with budget constraints in the heterogeneous cloud. In *2016 IEEE International Parallel and Distributed Processing Symposium Workshops.*
11. Kang, H., Chen, Y., Wong, J. L., Sion, R., & Wu, J. (2011). Enhancement of Xen's scheduler for MapReduce workloads. In *Proceedings of the 20th international symposium on High performance distributed computing*, New York, NY, USA, pp. 251–262.
12. Yazdanov, L., Gorbunov, M., & Fetzer, C. (2015). EHadoop: Network I/O aware scheduler for elastic MapReduce cluster. In *2015 IEEE 8th International Conference on Cloud Computing.* https://doi.org/10.1109/cloud.2015.113.
13. Ehsan, M., Chandrasekaran, K., Chen, Y., & Sion, R. (2016). Cost-efficient tasks and data co-scheduling with affordhadoop. *IEEE transactions on cloud computing.* https://doi.org/10.1109/tcc.2017.2702661.

14. Das, R., Singh, R. P., Patgiri, R. (2016). Mapreduce scheduler: A 360-degree view. *International Journal of Current Engineering and Scientific Research (IJCESR), 3*(11), ISSN (print): 2393–8374, (online): 2394–0697.
15. Kim, S., Kang, D., & Choi, J. (2015). I/O characteristics and implications of big data processing on virtualized environments. *Applied Mathematics & Information Sciences An International Journal, 9*(2L), 591–598.
16. Kim, S., Kang, D., Choi, J., & Kim, J. (2014). Burstiness-aware I/O scheduler for MapReduce framework on virtualized environments. In *2014 International Conference on Big Data and Smart Computing (BIGCOMP)* (pp. 305–308). https://doi.org/10.1109/bigcomp.2014.
17. Tian, W., Li, G., Yang, W., & Buyya, R. (2016). HScheduler: An optimal approach to minimize the makespan of multiple MapReduce jobs. *The Journal of Supercomputing, 72*(6), 2376–2393. https://doi.org/10.1007/s11227-016-1737-4.
18. Wang, X., Shen, D., Yu, G., Nie, T., & Kou, Y. (2013). A throughput driven task scheduler for improving mapreduce performance in job-intensive environments. In *2013 IEEE International Congress on Big Data* (pp. 211–218).
19. Yao, Y., Wang, J., Sheng, B., Lin, J., Mi, N. (2014). HaSTE: Hadoop YARN scheduling based on task-dependency and resource-demand. In *2014 IEEE International Conference on Cloud Computing*. 978-1-4799-5063-8/14.
20. http://nikgrozev.com/2014/06/08/cloudsim-and-cloudsimex-part-1/.
21. http://www.cloudbus.org/cloudsim/.

14. [illegible] (2016). MapReduce scheduling: [illegible] *International [illegible] Journal of Computing and Security [illegible]* [illegible] ISSN [illegible] 8571 (Online) 2394-0697.
15. [illegible] S., Kang, H., & Chen, [illegible] (2015). [illegible] characteristics and applications of big data processing on virtualized [illegible] computers. *Applied Mathematics & Information Sciences An International Journal* [illegible]
16. Kim, [illegible] Lee, [illegible], Choi, [illegible] & Kim, J. (2013). [illegible] aware I/O scheduler for MapReduce framework on virtualized environments. In *2013 [illegible] International Conference on Big Data and Smart Computing (BIGCOMP)* (pp. [illegible]). [illegible] 10.1109/bigcomp.2013. [illegible]
17. Tian, [illegible], & Buyya, R. (2016). [illegible] approach to minimize [illegible] the MapReduce jobs. *the Journal of Supercomputing* [illegible]
18. Wang, [illegible], Shen, [illegible] (2013). [illegible] for minimizing [illegible]. In *[illegible] International Conference on [illegible]* (pp. [illegible]).
19. [illegible], Wang, J., Cheng, [illegible] (2013). [illegible] scheduling [illegible] computing [illegible] [illegible] 2014.
20. [illegible] scheduler [illegible]

Big Data Analytics Provides Actionable Insights into Ship Emissions

Frank Cremer and Muktha Muralee

Abstract Atmospheric emissions such as NO_x from ship engines have a drastic impact on the environment. Controlling them is crucial for maintaining a sustainable growth for any logistics company. The Port of Rotterdam (The Netherlands) is using big data analytics to gain actionable insights into these emissions. Our case study deals with the implementation of the emission calculations and reporting implemented in Hadoop. In the analytical setup we introduce the method for estimating emissions based on recorded ship position data and information about its engines. We present a flexible approach that stores intermediate results allowing different levels of aggregation. These results are visualized in a Geographical Information System (GIS). We present some selected results followed by conclusions and recommendations.

Keywords Atmospheric emissions · Emission model · GIS · Hadoop Mapreduce

1 Introduction

Big data solutions like implementation based on frameworks such as Hadoop provides insights that were previously not possible. This has been demonstrated in many web-based applications such as recommendation lists for online shops and suggested friends for Facebook and LinkedIn. However, in this case study for the Port of Rotterdam we introduce a novel application of big data technology: to calculate aerial emission from ship traffic in such a way as to provide actionable insights.

F. Cremer (✉)
Geomatik, The Hague, The Netherlands
e-mail: frank@geomatik.nl

M. Muralee
MFasize, The Hague, The Netherlands
e-mail: m.muralee@mfasize.com

V. E. Balas et al. (eds.), *Data Management, Analytics and Innovation*,
Advances in Intelligent Systems and Computing 808,
https://doi.org/10.1007/978-981-13-1402-5_13

In this paper, we first explain why this study is relevant to the Port of Rotterdam. Next, we introduce the analytical setup used to model the emission per ship, based on each individual ship characteristics, its actual route. We then describe how it is aggregated into actionable results and visualized. Then, we show the implementation in software and discuss some selected results. At the end of the paper we present the conclusions and the future scope of this work.

2 Impact of Ship Emissions

The Port of Rotterdam is located near the metropolitan city of Rotterdam and is close to protected and sensitive nature areas. Since it is one of the ten largest ports in the world, it has a very intensive ship traffic. This ship traffic is the source of large amounts of atmospheric emissions of NO_x, CO_2, SO_x and particulate matter (PM) [1], even though less than other sources (such as road traffic and agriculture). This work is comparable to other studies [2, 3] in terms of emission calculations, but presumably one of the first to use big data technology to calculate emissions exhaustively at high temporal resolution.

For the Port of Rotterdam authority, it is a necessity to manage the ship emissions in order to be able to grow while reducing the footprint on the environment. The emission per terminal is restricted by its environmental license. Hence, it is imperative that the Port Authority have information about the emission attributed to each terminal. Only then operational decisions can be made to reduce the emission and/or incentives be given to less polluting ships.

Up until now the emission was calculated on a large grid, by RIVM, a government body. The results were not at a fine-grained level to be useful for determining emissions per terminal. Hence, the Port decided to implement their own model based on their dataset of ship positions, which was stored in Hadoop. The following sections of this paper show the implementation of the emission model in Hadoop and its results.

3 Analytical SETUP

In this section, we describe the analytical model setup to determine the ship emissions and to provide aggregated results that can be visualized and used as input for other models, for example, the deposition model. The first part deals with the emission model, the second part describes the stored ship position data. How data is aggregated is described in the third part and the visualization is described in the last part.

An overview of the processing from ship position data to storing intermediate results for further aggregation and visualization is shown in Fig. 1. Examples of the input and output data are shown for illustration. There are three steps involved: (1) creating tracks from individual position updates, (2) matching these tracks with

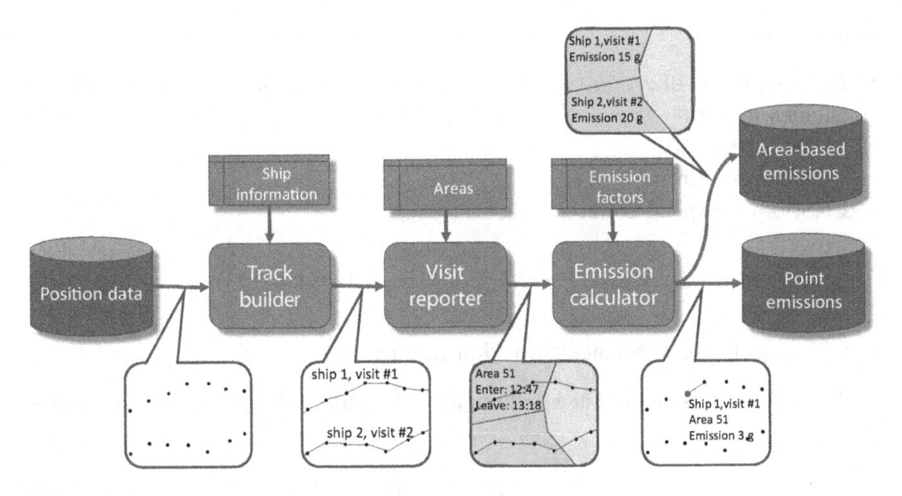

Fig. 1 Overview of the processing steps

predefined areas and (3) calculating the emission and storing these for these areas and separately for segments (two consecutive points).

3.1 *Emission Model*

The emission model is a model that estimates the various emissions into the air from sea ships. It is modeled, as opposed to measured, since as of date ships do not have sensors for measuring it in real time.[1]

The model uses as input the speed, the duration and the navigational status of the ship. The navigational status of a ship can be three states:

- Cruising at sea
- Maneuvering in the port
- Anchored or at a berth

For each state, different configurations of the main engine and auxiliary engines are used. The speed of the ship determines how many engines are used (if there is more than one) and to what percentage each engine is used.

The emission from the main engine during cruising at sea and maneuvering is given by the following formula:

$$E_{\mathrm{ME,subst}} = \mathrm{EF}_{\mathrm{subst}}\left(n_{\mathrm{eng}}\right) * t * \mathrm{CRS}(v) * \mathrm{CEF}_{\mathrm{subst}}(v), \tag{1}$$

[1]This model is common practice for aerial emission estimation and in use by government bodies such as RIVM.

with

- $E_{ME,subst}(n_{eng})$—the emission (in g), which depends on n_{eng}, the number of engines in use (-)
- t—the time (in s)
- v—the speed (in m/s)
- CRS(v)—the speed correction factor (-)

$$\mathrm{CRS}(v) = \left((v/v_{cruise})^3 + 0.2\right)/(1 + 0.2) \quad (2)$$

- $\mathrm{CEF}_{subst}(v)$—the substance correction factor (-)

When the ship is moored at a berth or at anchor, the main engine is considered to be off, so

$$E_{ME,subst} = 0 \quad (3)$$

For the auxiliary engine, the formulas are more straightforward

$$E_{AE,subst} = \mathrm{EF}_{AE,subst,<state>} * t, \quad (4)$$

with

- $E_{AE,subst}$—the emission for the auxiliary engine (in g)
- $\mathrm{EF}_{AE,subst,<state>}$—the emission factor for the auxiliary engine, given for the states "anchor", "berth", "sea" and "maneuver" (in g/s).

The emission factors for each individual ship have been provided by TNO (National research institute). TNO has determined these factors based on the ship data (such as engine, cruise speed, gross tonnage) obtained from the Lloyds register of ships.

3.2 *Ship Position Data*

The Port maintains a continuously updated dataset of all the ship traffic in its area. This data set is used operationally for traffic management and stored in Hadoop for traffic analyses. The data set consists of the ship position along with another 60 parameters (such as ship name, identifier, length, speed, course, etc.) and is stored every 10 s. Given an average number of ships in the range of a thousand ships, this gives a total of about 10 million records per day. It is obvious that processing these many records (especially for a whole year) is way beyond the reach of traditional solutions.

Based on the emission model (as described in part A) and the ship positions, the emission for the ship between any subsequent 10 s is calculated. The distances and speeds are calculated based on the coordinates of the subsequent time intervals.

These emission results are stored back into Hadoop. With a 10 s interval for measurements, we have created a very detailed emission model. However, this model is too fine-grained to provide any insights, hence we need to aggregate the data as described in the next paragraph.

3.3 Aggregating Results (Port/Area/Berth Visits)

There are four different selections within the emission data set that are of importance to the Port

- Total emission from activities within the Port.
- The emission attributed to the individual terminals, e.g., container terminals, refineries, etc.
- The emission per administrative area, e.g., a port basin.
- The emission per visit per ship. A visit is defined as a ship entering the port's operational area at 60 km from the entrance of the port, visiting one or more terminals and then leaving outside of the 60 km from the entrance.

An overview of the administrative areas is shown in Fig. 2. These areas range from 60 km out in sea (with designated anchor areas) to the port area itself (at a length of more than 40 km). In Fig. 3 a selected area of the port is shown. In this figure the port basins are visible (side arms of the river) as well the berths at the terminals. For illustration, a track of ship—actually its individual positions—is shown. This ship enters the port from the sea, sails up to the river and into the port basin called Maasvlakte to moor at the berth of terminal. After unloading and loading (not visible here) the ship leaves again the port to set sail for (presumably) the next port.

For each of these selections, the emissions can be aggregated in different ways:

- A uniform grid.
- The grid used by RIVM for yearly country-wide emission reporting.
- A grid which is finer (i.e., has smaller cells) closer to sensitive areas.
- Emissions per administrative area.
- Emissions for complete port visits attributed to specific berths belonging to the terminals.

Due to these many combinations of selections and aggregations, a flexible computational storage structure is necessary. The following intermediate data is stored:

- Visit information, containing the following:

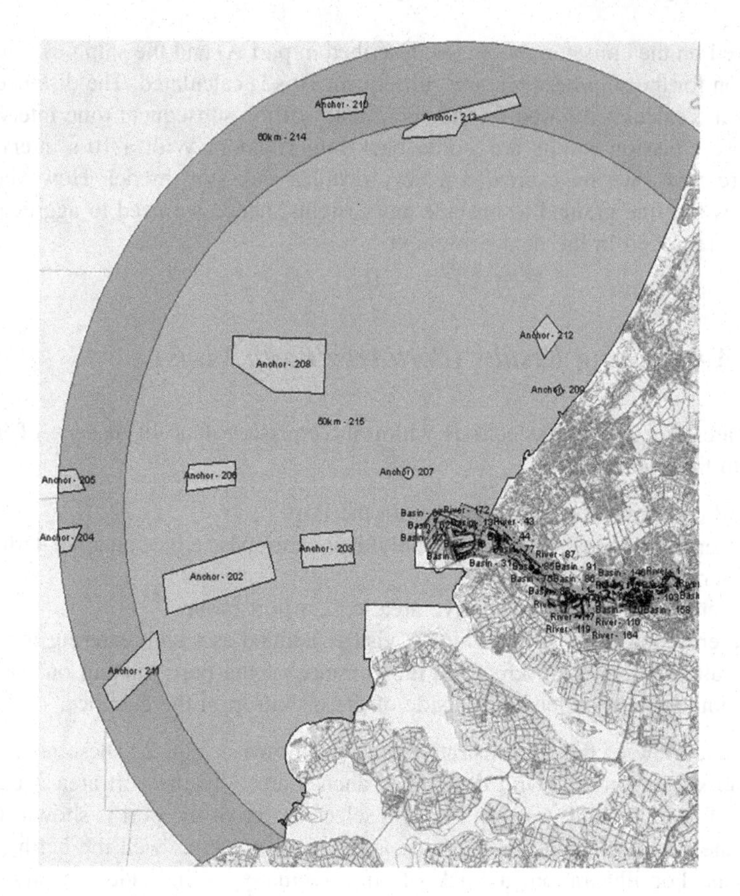

Fig. 2 Overview of the administrative areas

- Unique identifier for the ship.
- Unique visit identifier.
- Per area:
  - Area type (e.g., port area, anchor area, port basin and berth).
  - Area identifier.
  - Timestamp of entry.
  - Timestamp of exit.

- Segment emission data (10 s interval), containing the following:
  - Timestamp.
  - Unique ship identifier.
  - Visit identifier.
  - Navigational status (anchor/berth, cruising at sea or maneuvering).
  - Emission for the main and auxiliary engines for the eight different emission substances.

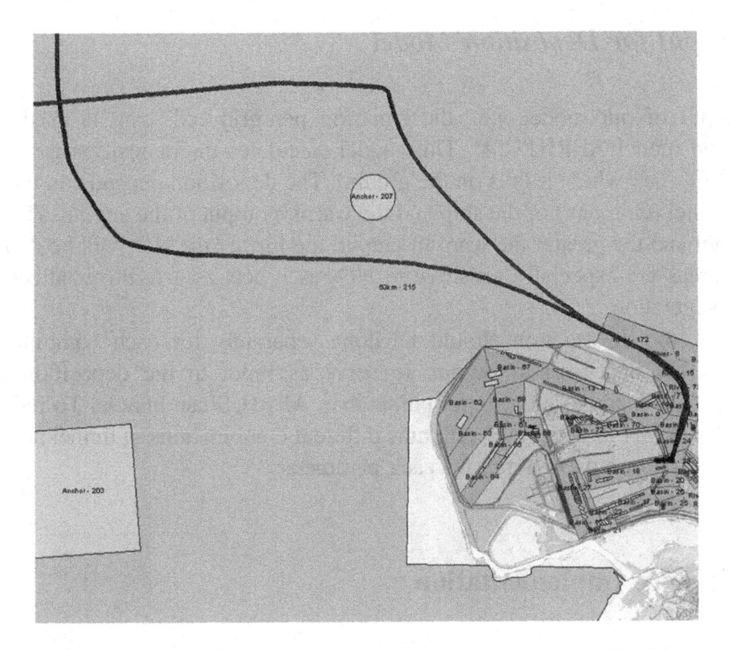

Fig. 3 Selection of the river, port basin and terminal area, with the track of a ship

- Funnel height.
- Warmth output.

Using these two data sets, it is possible to query the data to obtain the required combinations of selections and aggregations. For instance, the emissions of all the ships visiting a specific terminal is calculated in two steps. First, all the visit identifiers are determined by selecting the visits that have as type "berth" and as area identifier the required berth identifier. Second, a selection is made of all segment emissions which is aggregated onto the required grid.

3.4 Visualization

Insights into emissions is not only obtained by the numerical values of the emission estimates, but also through visualization. Obviously, any statistics can be put into graphs, e.g., comparing terminals or the same terminal over several years. However, in this paper, we limit the results to how we visualize the data set geographically.

Although the individual segment emission can be easily displayed geographically, there are too many points to provide any insights. Therefore, the visualization uses the emission per grid cell. The output of the aggregation is a set of 8 different emissions per grid cell. These grid cells are converted into a geographical database and visualized into a geographical information system (GIS).

3.5 *Input for Deposition Model*

The output of our model, i.e., the emission per grid cell, can be used for the deposition model AERIUS [4]. This model calculates the deposition for the substance NO_x, i.e., where it falls on the ground. The deposition depends on the height of the funnel (chimney) of the ship and the warmth output of the engine. The higher the funnel and the greater the warmth output, the further the NO_x will be deposited. Nature areas are especially sensitive to NO_x as it acts as a fertilizer affecting the natural vegetation.

Ideally, the aggregation should be done separately for each combination of funnel height and warmth output, to serve as input to the deposition model. However, this leads to more input points than AERIUS can handle. To reduce the number of points for each grid cell, nine different combinations of funnel height and warmth output are used (three for each parameter).

4 Software Implementation

The analytical setup as described in the previous section has been implemented using various software components. In this section, we describe how it is implemented.

The software implementation is divided into three parts: (1) the analytics for modeling the emissions and determining ship visits, (2) the selection and aggregation of these results and (3) the visualization of the results. For our work, we have used the following components:

- Hadoop filesystem for storage of (1) and (2)
- Custom developed MapReduce [5] Java code for (1) and in part for (2)
- Hive [6] (the Hadoop SQL engine) queries for (2)
- FME for converting the results from (2) in a geographical format
- ArcGIS for geographical visualization of the results (3).

The custom developed Java code has been developed using common IT standards. Especially for the emissions, model test cases have been created and the results have been validated externally by TNO.

5 Selected Results

In this section, some selected results are shown to demonstrate the insights into the emission calculations. Figure 4 gives an overview of the total NO_x emissions per grid cell for all ships visiting the port (as the actual numbers are not relevant for this

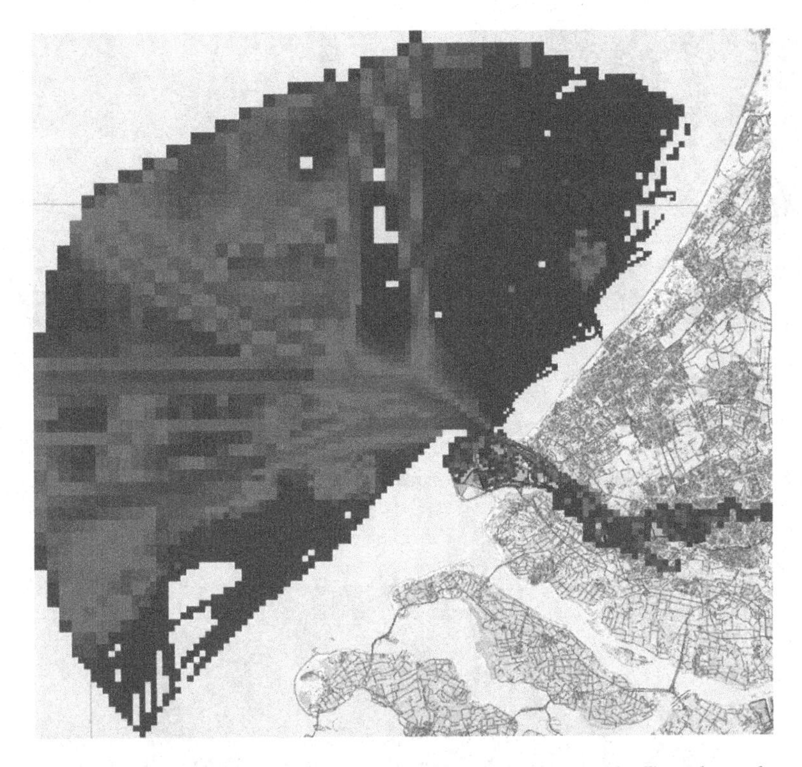

Fig. 4 Overview of the NO_x emission from ship traffic that visits the port. The color scale ranges from green at low emission, via yellow to red at high emissions

paper, the scale has been omitted). Although the emission is not normalized by the grid size, travel patterns can clearly be identified.

Figure 6 gives similar results for ships that do not visit the port from seaside (there is some residual traffic between the port and the neighboring ports in Dordrecht and Moerdijk). In this graph, there is clearly ship traffic passing along Rotterdam at seaside (please note that the scales are not the same as in Fig. 4).

Figure 5 shows the emission per administrative area. The emission is normalized by its area to provide emissions in kg per m^2. Obviously, these normalized emissions appear to be high for berths at terminals as ship need to spend some time at the same location for loading and unloading. The emission model allows to estimate the savings in emissions when onshore power supply is used, so that ships do not need to run their auxiliary engines while being moored at the berth (Fig. 6).

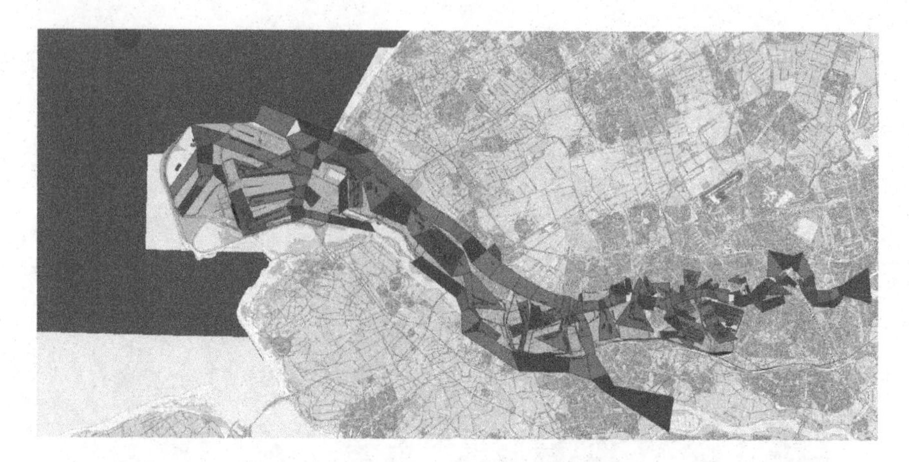

Fig. 5 The emission per area normalized to area size

Fig. 6 Overview of the NO_x emission from ship traffic that does not visit the port. The color scale ranges from green at low emission, via yellow to red at high emissions

6 Conclusions

With the approach followed in this paper, we have demonstrated that actionable insights can be gained by deploying big data methodologies for emissions calculations for the Port of Rotterdam. Specifically, the Port now has insights in the emissions per terminal, the emission close to sensitive areas, and the emission per ship per visit to the operational area of the port. A fine-grained emission model has been deemed essential for providing these results.

Based on these results actions can be defined to reduce the ship emissions while increasing the throughput of cargo, such as providing electricity for moored ships and/or providing incentives (e.g., by reducing the mooring fees) for cleaner ships and/or enforcing change or routes to diminish the impact on sensitive areas.

7 Future Scope

The results obtained using this methodology are more accurate than has previously accomplished by less fine-grained models, due to (1) better position estimates (2) shorter time intervals and (3) using emission factors for individual ships. Therefore, these results could be used for the Dutch National report on emissions. Providing these inputs for Rotterdam area should give a better estimate country wide. Moreover this methodology could be used for other Dutch areas as well, e.g., the Port of Amsterdam and the North Sea even further improving the estimates. Obviously, the same approach can be used for other countries.

A recommendation for improvement is for AERIUS to be implemented to run on Hadoop, thereby lifting the restriction for the number of input points. As the deposition model is additive, i.e. the total deposition is the sum of the deposition of each point emission, it should be able to parallelize the computations.

Acknowledgements We would like to acknowledge the Port of Rotterdam for the work carried out in this project. We would also like to state on record our appreciation to TNO for providing the emission model and the ship-specific emission factors.

References

1. Denier van der Gon, H., & Hulskotte, J. (2012). Methodologies for estimating shipping emissions in The Netherlands. The Netherlands Research Program on Particulate Matter, Report 500099012, ISSN: 1875-2322 (print), ISSN: 1875-2314 (on line).
2. Chen, D., Wang, X., Li, X., Lang, J., Zhou, Y., Guo, X., & Zhao, Y. (2017). High-spatiotemporal-resolution ship emission inventory of China based on AIS data in 2014. *Science of the Total Environment*, December 2017.
3. Cullinane, K., Tseng, P.-H., & Wilmsmeier, G. (2016). Estimation of container ship emissions at berth in Taiwan. *International Journal of Sustainable Transportation, 10.*

4. Duyzer, J., Zandveld, P., & Lohman, W. (2015). Doelmatigheidsonderzoek AERIUS Calculator (betaversie 8) en Monitor (versie 2014). TNO Rapport 2015 R10211.
5. Dean, J., & Ghemawat, S. (2004). MapReduce: Simplified data processing on large clusters. In *OSDI'04: Sixth Symposium on Operating System Design and Implementation*, San Francisco, CA, December, 2004.
6. Shaw, S., Vermeulen, A. F., Gupta, A., & Kjerrumgaard, D. (2016). *Practical Hive—A guide to Hadoop's data warehouse system*. Berkeley: Apress.

Part III
Artificial Intelligence and Data Analysis

Optimizing Deep Convolutional Neural Network for Facial Expression Recognitions

Umesh Chavan and Dinesh Kulkarni

Abstract Facial expression recognition (FER) systems have attracted much research interest in the area of Machine Learning. We designed a large, deep convolutional neural network to classify 40,000 images in the dataset into one of seven categories (disgust, fear, happy, angry, sad, neutral, surprise). In this project, we have designed deep learning Convolution Neural Network (CNN) for facial expression recognition and developed model in Theano and Caffe for training process. The proposed architecture achieves 61% accuracy. This work presents results of accelerated implementation of the CNN with graphic processing units (GPUs). Optimizing Deep CNN is to reduce training time for system.

Keywords Convolutional neural network · Deep learning · Graphical processing unit (GPU)

1 Introduction

1.1 Background

Facial expression recognition have found applications in technical fields such as Human–computer Interaction (HCI) which detect people's emotions using their facial expressions and security monitoring [1]. Use of Machine learning is powerful approach to detect and classify images. To improve their performance, it is necessary to collect larger datasets, as well as need to build powerful models. The weakest point of machine learning is that it cannot do feature engineering. The limitations of machine learning in many cases learned model does not generalize well. An algorithm can only work well on data with assumption of the training data. The biggest drawback is it is time consuming for learning with large datasets with powerful model. Deep Learning (DL) is a new advancement in area of machine

U. Chavan (✉) · D. Kulkarni
Walchand College of Engineering, Sangli, Maharashtra, India
e-mail: umesh.chavan@walchandsangli.ac.in

V. E. Balas et al. (eds.), *Data Management, Analytics and Innovation*,
Advances in Intelligent Systems and Computing 808,
https://doi.org/10.1007/978-981-13-1402-5_14

learning research whose motivation is moving closer to the objective of Artificial Intelligence (AI). Convolutional neural networks (CNNs) are a special kind of DL method. CNNs are useful in the area of computer vision. Facial expression recognition (FER) techniques detect people's emotions using their facial expressions. We build a model for FER using deep CNN. CNNs have much fewer parameters as compared to neural networks. So they are easier to train. More training time in CNN with large dataset is major bottleneck. We designed model in Theano framework [2] and exploited Graphics Processing unit (GPU) computation. The result shows that it is achieving 2–5 times speedup with training on GPU. Also shows that it achieves 61% accuracy. Our intention is to exhibit the performance and scalability improvement for FER using deep CNN.

2 Convolution Neural Network (CNN)

A CNN [3] is an advance in neural network evolution. It consists of sequences of one or more convolutional layers (CLs). CL's are mostly with pooling layers (PLs). PLs are succeeding by one or more fully connected layers (FCs), FCs are just as standard neural network. Accurate and correct feature extraction is necessary. This is the base for CNN. Input to a neural network is fed from output of feature extractor. It is challenging work to select a "suitable" feature extractor. It cannot adapt to network configuration. It is not part of learning procedure. These layers arranged in feed-forward structure as shown in Fig. 1. In a CNN only a small region/subset of input layer connects neurons in hidden layer. These regions are referred as Local receptive fields. The local receptive field is translated across an image to create a feature map. It can use convolution to implement this operation efficiently. So it's called as a CNN. The typical neural network has parameters—weights and biases [4]. The model learns these values during the training process and it's continuously updates with each new training examples. However, in the case of CNN, weights and bias values are the same for all hidden neurons in a given layer. This means that all hidden neurons detect the same feature such as an edge or

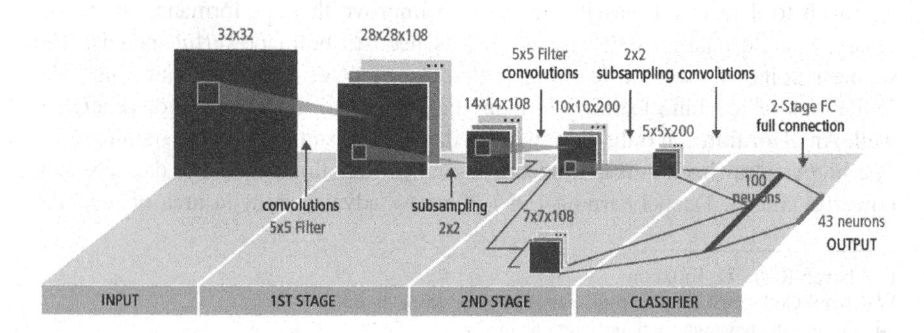

Fig. 1 A typical CNN model

blobs in different regions of the input image. This makes network tolerant to translation of object in an image. Activation step applies the transformation to the output of each neuron by using activation functions Rectified Linear Unit commonly known as a ReLUs [5]. Most DL network use ReLU for hidden layers. The power of ReLU is it trains much faster is more expressive than other alternatives—logistic function. Also ReLU prevents the gradient vanishing problem. It takes output of neurons and maps it the highest positive value or if the output is negative the function maps it to zero. We can further transform the activation step by applying a pooling step. Pooling reduced the dimensionality of the feature map by condensing the output of small regions of neurons into a single output. The CNN with its layer(s) is briefly explored in [6].

3 GPU (Graphics Processor Units) Programming

GPUs are massively parallel numeric processors. It is programmed in C with extensions for GPU programmers. It has application programming interfaces for programmers. It takes advantages of heterogeneous computing systems that contain both CPUs and massively parallel GPU's. For a GPU developer, the computing environment consists of a host that is traditional CPU and one or more devices that are processors with massive number of arithmetic units. GPU is a typically known as device in CUDA. Use of GPUs together with a CPU is GPU-accelerated computing. It accelerates deep learning applications. This work presents results of accelerated implementation of the deep CNN in graphic processing units (GPUs) for FER.

3.1 CUDA Programming Architecture

CUDA architecture allows the programmer to write one code that will run on both CPU and GPU. In CUDA GPU is referred as a device and CPU is referred as Host. CUDA assumes that the device and the host have their own separate memories where they store data. The function that executes on GPU is known as a Kernel. The kernel is invoked and executed by 100 s or even 1000 s of threads at a time. The CPU launches the kernels with a specific syntax to let GPU know how many threads should be used. The kernel function reserves memory in the device on board global memory. It takes one of the device's pointers as the first argument and the second argument is how many bytes to reserve. This function copies data from the host and device memories. CUDA provides a scalable approach to express parallelism. The CUDA is suitable for large amount of data and having lots of computations like image convolution. It achieves high throughput. CUDA uses thousands of threads executing in parallel, all these threads are executing the same function which is known as a kernel. Programmer can organize threads into blocks.

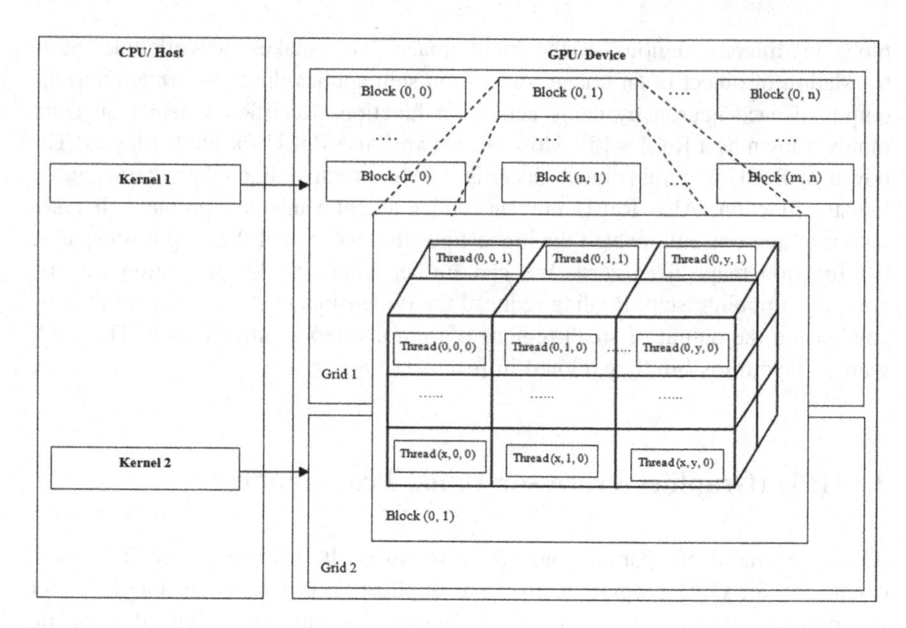

Fig. 2 CUDA programming model

These thread blocks are again arranged into grids of multiple thread blocks. All thread blocks at the same time work together through shared memory. A thread block has its own ID for its grid. Streaming microprocessors (SMs) are the part of GPU that actually runs this kernel. SMs are the heart of the architecture. They perform computations which have their own control units, execution pipelines, caches and registers. A typical CUDA architecture is shown in Fig. 2.

3.2 System Design

We built a CNN that has two convolution layers and two fully connected (FC) layers. In the first convolution layer, have 20, 5 × 5 filters with pooling. In the second convolution layer, we had 20, 5 × 5 filters and also pooling. In all convolution layers ReLU activation function is used. In the first FC layer it have 500 neurons and in second FC layer have 300 neurons. In both FC layers same as in the convolution layer we used ReLU activation function. Also we used softmax as loss function. We trained the network for varying number of epochs on each run (for 2, 10, 30, 50, 70 epochs) and with batch size of 30 samples. We cross-validated the hyper-parameters.

3.3 CNN Model

For an image with a size $M \times N$ of and kernels with a size of $m \times n$, the convolution is represented as

$$z_{ij}^{(k)} = \sum_{s=0}^{m-1}\sum_{t=0}^{n-1} w_{st}^{(k)} x(i+s)(j+t) \tag{1}$$

Here w is the weight of the kernel, which is the model parameter. Above Eq. (1) is for one kernel. For multi-convolution layers, the equation is

$$z_{ij}^{(k)} = \sum_{c}\sum_{s=0}^{m-1}\sum_{t=0}^{n-1} w_{st}^{(k,c)} x_{(i+s)(j+t)}^{(c)}(i+s)(j+t) \tag{2}$$

Here, c denotes the channel of the image. If the number of kernels is k and the number of channels is c, we have $w \in R^{k \times c \times m \times n}$. Then we see from the Eq. (2) that the size of convolved image is $(M - m + 1) \times (N - n + 1)$. After the convolution, all the convolved values will be activated by the activation function. We will implement CNN with the ReLU (rectified Linear Unit) function. With the activation we have

$$a_{ij}^{(k)} = h(z_{ij}^{(k)} + b^{(k)} = \max(o, z_{ij}^{(k)} + b^{(k)})) \tag{3}$$

where $b \in R^{k}$, a one-dimensional array. Next is the max-pooling layer. The propagation can simply be expressed as

$$y_{ij}^{(k)} = \max(z_{(l_1 i+s)(l_2 j+t)}^{(k)}) \tag{4}$$

Here l_1 and l_2 are the size of pooling layers and $s \in [0, l_1], t \in [0, l_2]$.

Usually l_1 and l_2 are set to same sizes (2 or 4). The simple Multilayer Perceptron Network (MLP) follows after sequences of convolutional layers and pooling layers to classify data. MLP can accept one-dimensional data. Output of CLs and PLs are two-dimensional, we need to flatten the down sampled/pooled data as preprocessing to adapt it to input to input layer of MLP. The error from input layer of MLP is back-propagated to the max-pooling layer, and this time it is un-flattened to two dimensions to be adapted properly to the model. Max-pooling layer simply back-propagates error to its previous layer as max-pooling layer does not have parameters. The equation can be expressed as

$$\frac{\partial E}{\partial_{(l_1 i+s)(l_2 j+t)}^{(k)}} = \begin{cases} \frac{\partial E}{\partial y_{ij}^{(k)}} & \text{if}, y_{ij}^{(k)} = a_{(l_1 i+s)(l_2 j+t)}^{(k)} \\ 0 & \end{cases} \tag{5}$$

Here E denotes the evaluation function, the error is then back-propagated to the CL, and with it can calculate the gradient of the weight and bias. Gradient of bias is represented as

$$\frac{\partial E}{\partial b^{(k)}} = \sum_{i=0}^{M-m} \sum_{j=0}^{N-m} \frac{\partial E}{\partial a_{ij}^{(k)}} \frac{\partial a_{ij}^{(k)}}{\partial b^{(k)}} \tag{6}$$

$$\partial_{ij}^{(k)} = \frac{\partial E}{\partial a_{ij}^{(k)}} \tag{7}$$

$$c_{ij}^{(k)} = z_{ij}^{(k)} + b^{(k)} \tag{8}$$

Then we get,

$$\frac{\partial E}{\partial b^{(k)}} = \sum_{i=0}^{M-m} \sum_{j=0}^{N-m} d_{ij}^{(k)} \partial_{ij}^{(k)} \frac{\partial a_{ij}^{(k)}}{\partial c_{ij}^{(k)}} \frac{\partial c_{ij}^{(k)}}{\partial b_{ij}^{(k)}} = \sum_{i=0}^{M-m} \sum_{j=0}^{N-m} d_{ij}^{(k)} h'(c_{y}^{(k)}) \tag{9}$$

In the same way the gradient for weight (kernel) is

$$\frac{\partial E}{\partial w_{st}^{(k,c)}} = \sum_{i=0}^{M-m} \sum_{j=0}^{N-m} d_{ij}^{(k)} h'(c_{y}^{(k)}) x_{(i+s)(j+t)}^{(c)} \tag{10}$$

When we think for multi-convolutional layers it is necessary to calculate the error of convolutional layers.

$$\frac{\partial E}{\partial w_{st}^{(c)}} = \sum_{k} \sum_{s=0} \sum_{t=0} \frac{\partial E}{\partial z_{i(i-s)(j-t)}^{(k)}} \frac{\partial z_{i(i-s)(j-t)}^{(k)}}{\partial x_{ij}^{(c)}} = \frac{\partial E}{\partial z_{i(i-s)(j-t)}^{(k)}} w_{st}^{(k,c)} \tag{11}$$

so, the error can be expressed as

$$\frac{\partial E}{\partial x_{ijst}^{(c)}} = \sum_{k} \sum_{s=0}^{M-m} \sum_{t=0}^{N-m} \partial_{(i-s)(j-t)}^{(k)} h'(c_{(i-s,j-t)}^{(k,c)}) \tag{12}$$

4 Experiments and Results

All benchmark in this paper were performed in machine having computation platform with (1) CPU: AMD Phenom II X4 B97—processor; (2) GPU is GeForce GTX520, compute capability 2.1, 48 cores. The software platform is composed of: Ubuntu 14.04 Operating system, CUDA 7.5, Python with Theano. All training and testing are in single precisions.

4.1 Program Code

In the part of code snippet; the model is created in Python having two convolutional layers and one FC layer. The first CL has 32 filters of size 32 × 32. Second CL has 64 filters of 3 × 3 sizes. The object CNN is instantiated with parameters for CNN layers.

```
def main():
X, Y = getImageData()
model = CNN(convpool_layer_sizes=[(32, 3,3), (64, 3, 3),(96,3,3),
(128,3,3)], hidden_layer_sizes=[200],
)
model.fit(X, Y,epochs=3,batch_sz=30)
if __name__ == '__main__':
main()
```

4.2 Dataset

The experiment was conducted on the dataset provided by Kaggle [7] website for Facial Expression Recognition Challenge [8]. This dataset consists of 37,000—48 × 48 pixel gray-scale images of faces. Each image is labeled with one of seven expression categories: Fear, Happy, Sad, Angry, Disgusts, Surprise, and Neutral. We used a training set of 36,000 samples, a validation set of thousand examples.

The emotions are labeled in each image. Network is trained on the dataset, which comprises 48-by-48-pixel gray-scale images of human faces each labeled with one of seven expressions. Some samples images with labeled expression are shown in Fig. 3. There are variations in the dataset considerably in scale, rotation, and illumination.

4.3 Experiment

We built a CNN that had two convolution layers and two fully connected (FC) layers. In the first convolution layer, we had 20, 5 × 5 filters with pooling. In the second convolution layer, we had 20, 5 × 5 filters and also pooling. In all convolution layers ReLU activation function is used. In the first FC layer we had 500 neurons and in second FC layer we had 300 neurons. In both FC layers same as in the convolution layer we used ReLU activation function. Also we used softmax as our loss function. Figure 4 shows the architecture of this deep network. We trained the network for varying number of epochs on each run (for 2, 10, 30, 50, 70 epochs) and

Fig. 3 Example images from Kaggle dataset [7]

with batch size of 30 samples. We cross-validated the hyper-parameters (Learning rate, regularization, decay, epsilon, Batch size) as shown in Table 1. To make the model training faster, we exploited GPU-accelerated deep learning facilities on Theano [2] library in using Python.

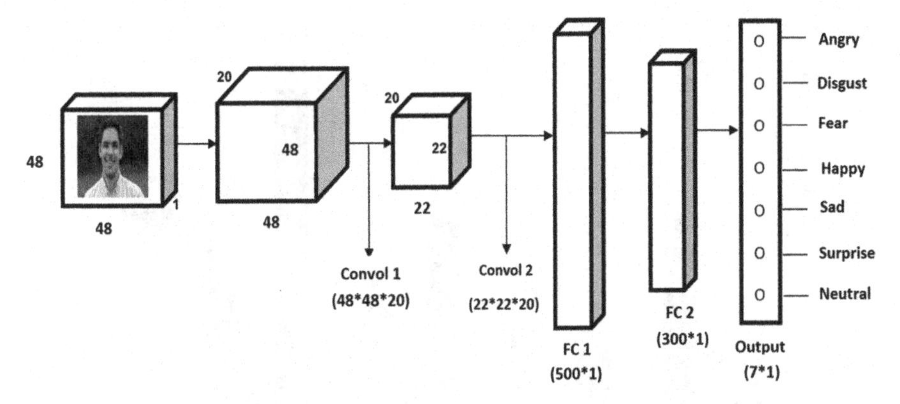

Fig. 4 FER CNN architecture

Table 1 The hyper-parameters for model

Parameters	Value
Learning rate	0.00001
Regularization	0.0000001
Decay	0.9999
Epsilon	0.001
Batch size	30

4.4 Results and Evaluation

The final validation accuracy we obtained is 61% for training with epochs =10. Training loss plot is shown in Fig. 5. The confusion matrix for classification is shown in Fig. 7. The performance in GPU speedup over Kaggle [7] FER 2013 dataset is shown in Table 2. We can see the performance improvement in speed of execution time of CPU and GPU training with different number of epochs which is shown in Table 2. The average speedup gain is approximately five times (Figs. 6 and 7).

5 Future Scope

The proposed work can be further extended by increasing the number of different expressions other than the six universal expressions (anger, fear, disgust, joy, surprise, sadness). The classification of other facial expressions may require the

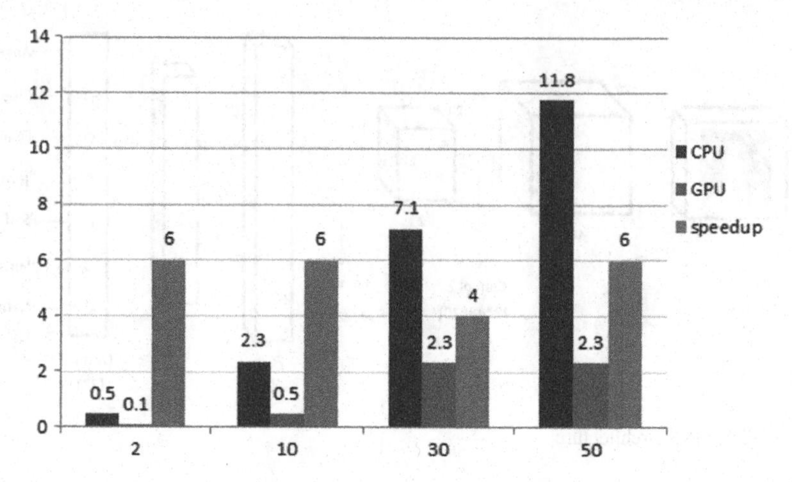

Fig. 5 Execution time (in h) and speedup with GPU

Table 2 Performance: execution time and accuracy

Epochs	Execution time (in minutes)		Speedup	Accuracy (%)
	CPU	GPU		
2	28	5	6	39
10	141	27	6	55
30	427	141	4	59
50	706	141	6	61

extraction and tracing of additional facial points and corresponding features. The system can be improved by using a wider training set so as to cover a wider range of poses and cases of low quality of images.

6 Conclusions

Although topology structure of convolution neural network is simple, it still needs a huge amount of work in calculation. NVIDIA GPU based on hardware architecture of stream processor has significant improvement in face expression recognition based on convolution neural network in support of programming model in CUDA. Compared with CPU, it has amazing advantages. Experiments show that stream

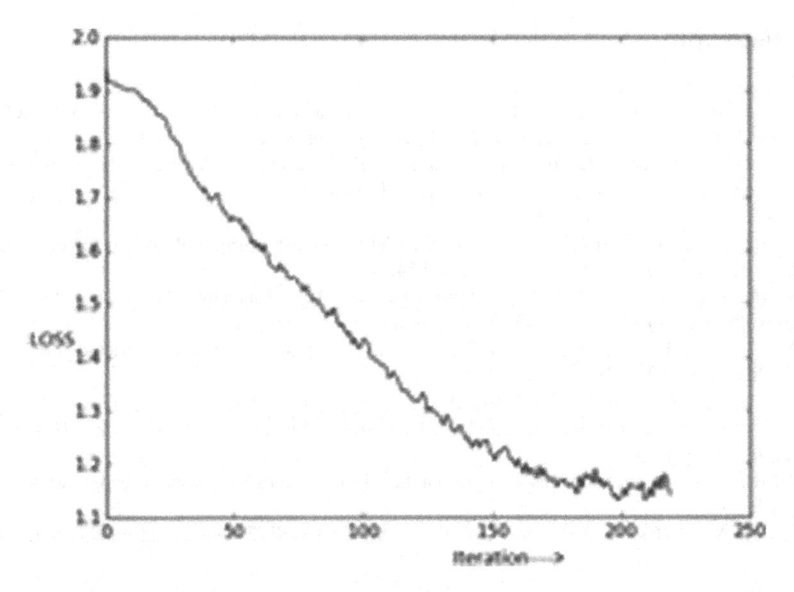

Fig. 6 Training loss

Fig. 7 Confusion matrix

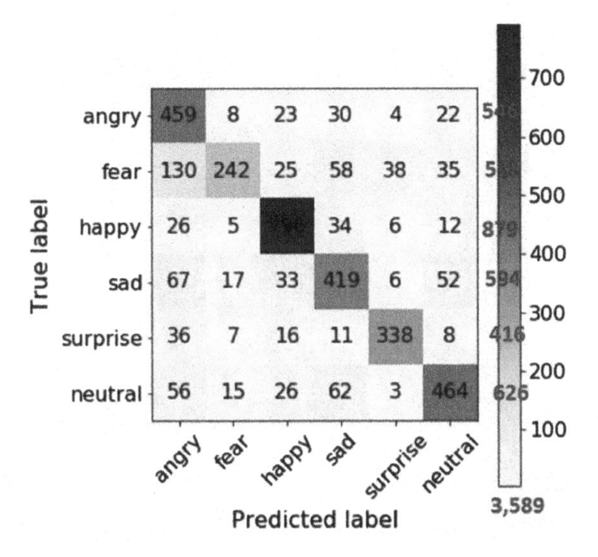

processor is suitable for convolution neural network. The results in this work show that GPUs are just as fast and efficient for deep learning. In this work, we evaluated their performance using different performance measurement and visualization techniques. Some of the difficulties with improving this is that images are very small and some cases it is difficult to distinguish which emotion is on each image.

Bibliography

1. Pantic, M., & Rothkrantz, L. J. M. (2004). Facial action recognition for facial expression analysis from static face images. *IEEE Transactions on Systems, 34*(3).
2. Bergstra, J., Breuleux, O., Bastien, F., Lamblin, P., Pascanu, R., Desjardins, G., et al. (2010). Theano: A CPU and GPU math compiler in Python. In *Proceedings 9th Python in Science Conference* (pp. 1–7).
3. Hinton, G. E., Osindero, S., & Teh, Y.-W. (2006). A fast learning algorithm for deep belief nets. *Neural Computation, 18*(7), 1527–1554.
4. Krizhevsky, A., Sutskever, I., & Hinton, G. E. (2012). Imagenet classification with deep convolutional. In *Advances in Neural Information Processing Systems.*
5. Nair, V., & Hinton, G. E. (2010). Rectified linear units improve restricted Boltzmann machines. In *ICML*, 2010.
6. Chavan, U., & Kulkarni, D. (2016). Accelerating learning performance of facial expression recognition using convolution neural network. *International Journal of Control Theory and Applications, 9*(43).
7. www.kaggle.com/c/challenges-in-representation-learning-facial-expression-recognition-challenge.
8. LeCun, Y., Cortes, C., & Burges, C. J. C. (1998). The MNIST database of handwritten digits.

Hybrid Kmeans with Improved Bagging for Semantic Analysis of Tweets on Social Causes

Mani Madhukar and Seema Verma

Abstract Analysis of public information from social media could yield fascinating outcomes and bits of knowledge into the universe of general conclusions about any item, administration, or identity. Social network data is one of the most effective and accurate indicators of public sentiment. Analysis of the mood of public on a particular social issue can be easily judged by several methods developed by the technicians. In this paper, analysis of the mood of society towards any particular news from the twitter post in form of tweets. The key objective behind this research is to increase the accuracy and effectiveness of the classification by the process of the NLP that is Natural Language Processing Techniques while focusing on semantics and World Sense Disambiguation. The process of classification includes the combination of the effect of various independent classifiers on one particular classification problem. The data that is available in the form of tweets on twitter can easily frame the insight of the public attitude towards the particular tweet. The proposed work is well planned to design as well as implement the best hybrid method that includes Hybrid Kmeans/Modified Kmeans (MKmeans) that involves clustering and Bagging for sentiment analysis. With this proposed idea one can easily understand the behavior of the public towards the post and further assist in the future policy making taking the results as the basis. At the end results are compared with the existing model with the motive of validating the findings.

Keywords Natural language processing (NLP) · Sentiment analysis
Social networking analysis · Social networking sites (SNS)

M. Madhukar (✉)
IBM India Pvt. Ltd., Bengaluru, Karnataka, India
e-mail: manimmadhukar@gmail.com

S. Verma
Banasthali Vidyapeeth, Banasthali, Rajasthan, India
e-mail: seemaverma3@yahoo.com

V. E. Balas et al. (eds.), *Data Management, Analytics and Innovation*,
Advances in Intelligent Systems and Computing 808,
https://doi.org/10.1007/978-981-13-1402-5_15

1 Introduction

Information era has developed over the years like no different technology. It has contributed in making unimaginable things conceivable. As this exploration proposition is being composed, something some place may get advanced, changing a few myths, breaking some taboos; Information innovation assumes its part some place in recording them. Sentiments or emotions as said have been the sole property of human heart and mind which has sounded good to us, people. Data Technology (IT) has taken it over from us and maybe has reshaped our feelings and conclusions as well. At this juncture of human history, innovation can help us in understanding ourselves. Sentiment analysis is one of the newly discovered measures that advertisers have found to quantify state of mind of a client via web-based networking media towards a brand or product or premium. The quantity of preferences and takes after on stages like Facebook and Twitter have made a considerable measure of footing by advertisers in coming to and mining these social stages. These suppositions which are found in input, discussions, or evaluates remarks are filling the Big Data universe with bunches of assorted data that is demonstrated gainful to different organizations for various causes [1].

Sentiment analysis is one of the viable methods for finding public sentiments. Different organizations regularly utilize on the web- or paper-based reviews to gather client remarks. Because of the development of informal communication destinations and applications, individuals tend to remark on their Facebook or tweet profile. Along these lines, the paper-based approach is not a proficient approach. Just a little client base can be come to and there is no certification that their answers in the review are straightforward or not. Here online networking becomes an integral factor. Facebook, Twitter and all other online networking locales are brimming with individuals' assessments about administrations, items, monetary, and political issues and so on they utilize, remarks about mainstream identities and considerably more. Henceforth mining tweets about different social issues from web-based social networking is a substantially more imaginative approach for estimation investigation. Sentiment Analysis helps us in understanding the ways the discussion has been hollowed, and what it has implied regardless of varieties in the translating parties. Each discussion passes on some type of estimation, which can extensively be named Positive, Negative, and Neutral. A great deal of work has been done on assumption examination, significantly in positive and negative marking [2].

As discussed, it is required to understand that sentiments are mere feelings, inclinations and not exact facts, however the feelings and inclinations direct us towards some real facts about the general mood of the sentiment conveyors, the masses, the people. An opinion or a view is generally classified as falling under one of the two opposing sentiment polarities; the opinions/sentiments are categorized as binary in nature—0/1 or true/false or can be understood as good/bad, like/dislike. This interpretation is generally referred to as Polarity or Semantic Orientation [2]. A considerable measure of research has been done on sentiment analysis from online networking, the vast majority of which concentrates on individuals' feeling towards different social issues. However, breaking down online networking information in this way gives a much-summed up thought. To make it more particular, feeling examination can be performed via web-based networking media information. Our approach is to discover the notions on different social issues.

There are several sentiments of the public towards the posts in SNS that includes: Positive and Negative along with the *n*-point scale that includes very good, good, satisfactory, bad and very bad [3]. Text mining is the famous way to analyze and understand the sentiment of people integrated with the content posted and these methods are: Machine Learning, Statistical/Quantitative Techniques or Natural Language Processing [4]. The sentiment analysis is of two kinds that is supervised or unsupervised. This paper includes one of the hybrid techniques that include Hybrid-Kmeans and improved bagging.

2 Related Work

In the research paper by Anguita et al. [5], a method to apply an approach of MLT, based on Maximal Discrepancy concept, to the problem of SVM model Selection has been described. The researcher has chosen Maximal Discrepancy (MD) method as it surpasses many resampling algorithms which are majorly used by practitioners and it is less susceptible to the availability of a good training set for the problem under investigation. The experimental result of this paper shows that when the number of samples is less than the dimensionality of the data in a small sample setting, use of machine-learning theory can outperform other concepts in selecting hyper parameters of a SVM.

In the research article by Joshi et al. [6], they have proposed a web-based system named as C-Feel-It which has been used for taping the emotions behind the posts from the micro blogging website twitter. Categorization of the post pertaining to a search string as positive, negative, or objective using input from four sentiments based repositories have undertaken and have given them an aggregate sentiment score that represented a sentiment snapshot for that search string. Architecture of the system consists of three parts. First, Tweet Fetcher which has been used for fetching tweets pertaining to the search string entered by the user. Second, Tweet Sentiment Predictor as the name suggested, analyzes sentiment of the one tweet at a time. Lastly, Tweet Sentiment Collaborator has given an overall prediction with respect to a keyword in the form of percentage of positive, negative, or objective content.

Mostafa et al. [7], in this paper we at first direct a factual examination on the contrasts between sentiment analysis of items and social issues. At that point, in view of our discoveries, we propose a way to deal with consider the part of verb as the most imperative term in communicating suppositions in regards to the social issues. Measurable and test comes about demonstrate that considering verbs is required and evident, as well as enhances the execution of assessment examination.

Wang et al. [8] conclusion grouping of tweets has been appeared to be a significant approach for tending to certifiable concerns, for example, forecast of race comes about. Tweet conclusion order execution might be contrarily influenced by components, for example, class irregularity, high dimensionality or boisterous preparing information. Information inspecting, gathering students and highlight determination are machine-learning procedures that can be utilized to address these issues

Revathy & Sathiyabhama [9] have built up a Semantic Sentiment Mining framework by joining both the govern-based approach and machine-learning calculation (Random Forest Model, SVM and Naive Bayesian) to group tweets. The half breed approach utilizes three diverse machine-learning models. The main model uses manage construct order situated in light of compositional semantic guidelines that recognizes articulation level extremity. The second one performs sense-construct characterization situated in light of WordNet faculties as highlights to Support Vector Machine classifier, while the third model performs element level examination in view of ideas acquired.

Saraswathi & Tamilarasi [10] have focused on movie reviews, investigating opinion classification of online movie reviews based on opinion/corpus words. They used SVM to classify as positive or negative feature sets from reviews extracted through the use of Inverse document frequency.

In the previous paper [11] the author has done analysis on social issues sentiment analysis that will automatically classify the issues into positive, negative, and neutral class label. Hybrid clustering with classification model framework is proposed and tweets on girl child issues are analyzed and the experimental results shown the proposed technique better than the existing framework.

3 Problem Formulation

The current scope of the proposed work is to identify and analyze Tweets from Twitter social platform to ascertain views of masses in India on social issues. Twitter is the biggest small-scale blogging administration with 200 million clients. Messages, called tweets, are restricted to 140 characters. This rouses unique methods for imparting, for example, shortening words, broad utilization of emojis, and the utilization of informal language expressions. The amount of public information present on twitter makes it a unique data source, with the challenge of overcoming the particular language used on it. During the Data Gathering, he Twitter messages to be characterized are recovered from its source. At that point, the Pre-handling stage happens; this stage is made by a set out of interior advances where the Twitter messages are disintegrated into information that is set up to be broke down by the classifier. Finally, the Classification stage, which is implemented through Hybrid Kmeans clustering and bagging, that takes some training data as input during its initialization, analyses the preprocessed twitter messages and returns the twitter messages arranged by assumption as either negative or positive.

The proposed framework has been thought to sort the messages into either positive or negative contingent upon sentiment they carry. Literature survey summarizes machine-learning approaches yearning accuracy of around 85 percent. The proposed system based on hybrid approach will surely improve the accuracy of the results.

4 Approach

Social Issues Sentiment Analysis automatically analyses social issues. It identifies the positive, negative, or neutral opinion. In this section, the proposed methodology clustering with Classification is explained in detail. The steps of the technique are as follows:

1. Data Collection
2. Data Preprocessing and Filtration
3. Clustering the data
4. Classification via Improved Bagged Learning
5. Building the model

The rest of the section explains these steps in detail

1. **Data Collection**

For the analysis of social issues tweets, data is collected The input data will be raw text from tweets on social causes in India, in particular on "JNU agitation", "Price rise". The motivation for the topics has been derived from the web-based interview of Twitter, India Director, Rishi Jaitly (2016) as the topics had the power to polarize the entire country and to shape opinion of common countrymen leading to sharp divide in Indian society. For creating the corpus of tweets, the tweets will be fetched from the Twitter database based on HashTags(#), using Twitter API for connecting and authenticating. The collected text is noisy and methods for cleaning and parsing of the data to form a corpus for further processing.

2. **Data Preprocessing and Filtration**

Preprocessing and feature extraction is a preliminary phase. Preprocessing includes three phases:

(a) Tokenization and parsing of words: In this phase, each tweet sentence splits into words of any natural processing language. For example for a sentence "I am a good girl", there are five words "I", "am", "a", "good", "girl" each word is represented as a token.
(b) Removal of stop words: Stop words are the words that contain little data so should have been evacuated. As by evacuating them, execution increments. Along these lines, at the season of pre-handling we have closed this stop word so every one of the words are expelled from our dataset.

(c) Stemming: It is characterized as a procedure to diminish the determined words to their unique word stem. For instance, “talked”, “talking”, “talks” as in view of the root word “talk”. We have utilized Snowball stemmer to diminish the determined word to their cause.

3. **Clustering the Data**

Applying the Hybrid Kmeans clustering algorithm on collected data. In the hybrid approach initially partitioning the dataset into sub tests and afterward applying the lessened emphasis way to deal with each subsample to discover the groups. This will spare part of time and enhance the execution for the extensive datasets.

Split the entire information into different subsamples.

At that point apply the calculation that will lessen the quantity of emphases.

Algorithm: Hybrid Kmeans Clustering

Input: Instances
Output: Clustered data

(1) Extract subsamples from the dataset
(2) For each extracted subsample
Randomly select k objects from dataset D as initial cluster centers.
Calculate the distance between each data object $di(1 \leq i \leq n)$ and all k cluster centers $cj(1 \leq j \leq k)$ as Euclidean distance $d(di, cj)$ and assign data object di to the nearest cluster.
(3) For each data object *di*, find the closest center *cj* and assign *di* to cluster center *j*;
Store the label of cluster center in which data object *di* is and the distance of data object *di* to the nearest cluster and store them in array Cluster[] and the Dist[] separately.
Set Cluster[*i*] = *j*, *j* is the label of nearest cluster.
Set Dist[*i*] = *d*(*di*, *cj*), *d*(*di*, *cj*) is the nearest Euclidean distance to the closest center.
(4) For each cluster $j(1 \leq j \leq k)$, recalculate the cluster center;
Repeat
(5) For each data object *di*
Compute its distance to the center of the present nearest cluster;
If this distance is less than or equal to Dist[*i*], the data object stays in the initial cluster;
Else
For every cluster center $cj(1 \leq j \leq k)$, compute the distance $d(di, cj)$ of each data object to all the center, assign the data object *di* to the nearest center *cj*.
Set Cluster[*i*] = *j*;
Set Dist[*i*] = *d*(*di*, *cj*);

(6) For each cluster center $j(1 \leq j \leq k)$, recalculate the centers;
Until the convergence criteria is met.
(7) Output the clustering results.

4. Classification via Improved Bagged Learning

The clustered information is classified by utilizing Enhanced Bagging approach which diminishes the variance of the forecasting by utilizing the data by using various combinations with repetitions to deliver multisets of size equals to the original data. The Bagging learning calculation is carried out to order the examples for each multi set and a model is generated. Also a vote is being identified for the generated model. The average of all the anticipated votes is thought to be the aftereffect of the classifier. This will order the information into three classes: positive, negative, and neutral.

In this algorithm dataset is sampled with replacement into ten datasets with same number of tuples using bagging. And then for every bootstrap sample Bagging classification is used as a base classifier and returns prediction results. At the end the final prediction is produced using average voting.

As compared to the single classifier, the improved classifier frequently has greater accuracy that derived from the original training data T. It will not be significantly more awful and is more powerful to the impact of noisy data. This combined model decreases the variance of the single classifier as the result performance and accuracy increases (Fig. 1).

Algorithm: Improved Bagging Procedure

Input:

D, a set of *d* training tuples;
k, the number of models in the ensemble;
a learning algorithm AdaBoost.

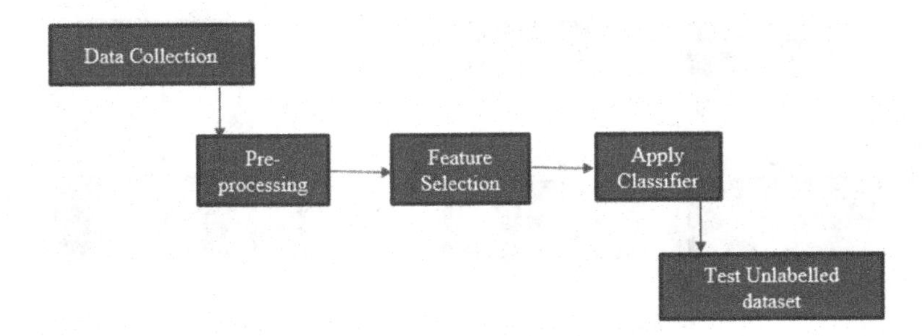

Fig. 1 Procedure of classification

Output: A classification model, M^*.

(1) for $i = 1$ to k do //create k models:
create bootstrap sample, *Di*, by sampling *D* with replacement;
use *Di* to derive a model, *Mi*;
end for

To use the composite model on a tuple, *X*:

(2) if classification then
let each of the *k* models classify *X* and return the majority vote;
(3) if prediction then
let each of the *k* models predict a value for *X* and return the average predicted value;

5. Building Model

The classifier model built by using proposed hybrid technique is evaluated and tested using *k*-fold Cross Validation approach (*k*-fold CV). In this, the training set is divided into *k* smaller sets. The procedure followed in *k*-fold CV approach is as follows:

A model is prepared utilizing $k - 1$ of the folds as preparing information; the subsequent model is approved on the rest of the piece of the information (i.e., it is utilized as a test set to figure an execution measure, for example, precision).

The execution measure assessed by *k*-crease cross-approval is then the normal of the qualities processed on repetitively (Fig. 2).

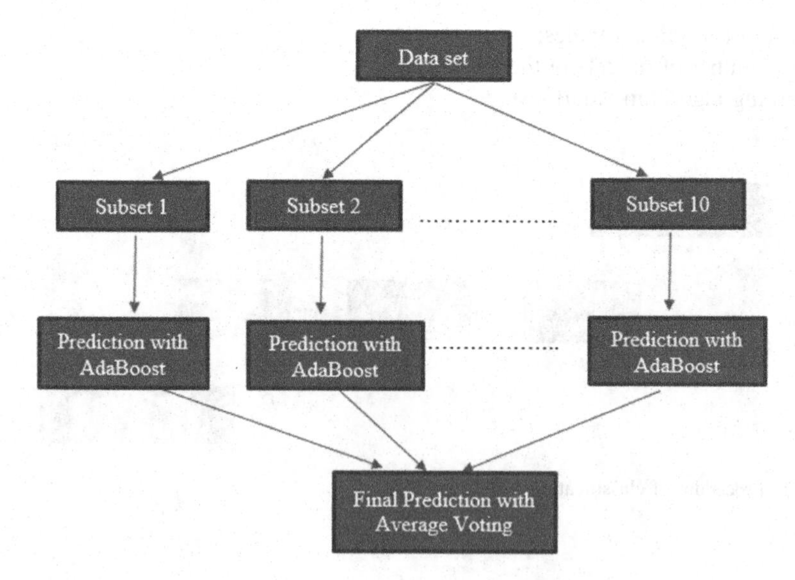

Fig. 2 Improved bagging

5 Results and Discussions

The Proposed Enhanced Bagging with Clustering algorithm is tested on JNU hash tags tweets for predicting the sentiments. The performance is analyzed based on below mentioned parameters.

Confusion Matrix Confusion matrix that is a matrix represented with two rows and two columns contains the value true positives, false positives, false negatives and true negatives.

Where $(n_{s \to s})$ is True Positive rate, $(n_{h \to s})$ is False Positive rate, $(n_{s \to h})$ is False Negative and $(n_{h \to h})$ is True Negative (Fig. 3; Tables 1, 2 and 3).

Figures 4 and 5 shown above gives the comparison of the classified instances correctly recognized and the class parameters by the proposed approach and the base approach. As the number of correctly classified instances in case of proposed approach is more, i.e., it has more correctly done the prediction of the tweets sentiments; concludes to perform better than the base Kmeans with SVM approach.

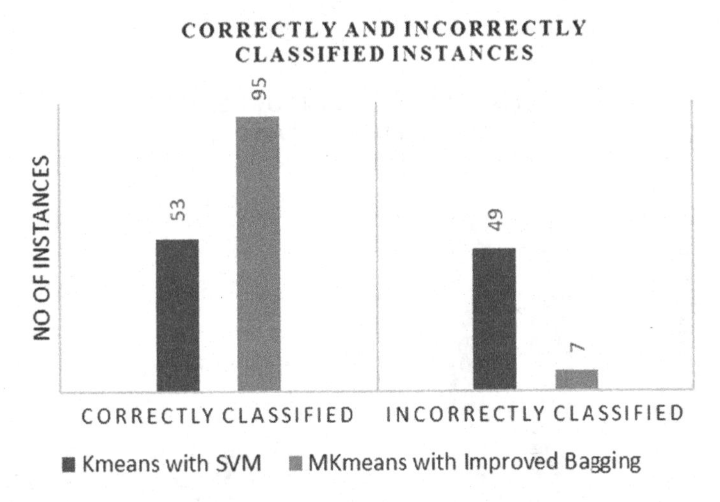

Fig. 3 Classified instances comparison of JNU tweets

Table 1 Table of confusion matrix for JNU tweets dataset

	Kmeans with SVM	(Hybrid-Kmeans/ MKmeans) Kmeans with improved bagging
Correctly classified $(n_{s \to s} + n_{h \to h})$	53	95
Incorrectly classified $(n_{h \to s} + n_{s \to h})$	49	7

Table 2 Table of confusion matrix for price rise tweets dataset

	Kmeans with SVM	(Hybrid-Kmeans/ MKmeans) Kmeans with improved bagging
Correctly classified ($n_{s \to s} + n_{h \to h}$)	241	269
Incorrectly classified ($n_{h \to s} + n_{s \to h}$)	33	5

Table 3 Representing the class parameters of the MKmeans with E-Bagging and Kmeans with SVM

Algorithms	Precision	Recall	*F* Measure
Kmeans with SVM	0.774	0.88	0.823
Hybrid-Kmeans/MKmeans with improved bagging	0.982	0.982	0.980

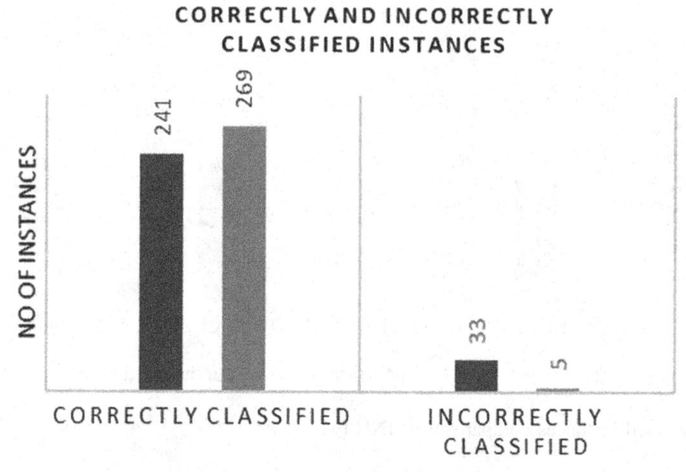

Fig. 4 Classified instances comparison of price rise tweets

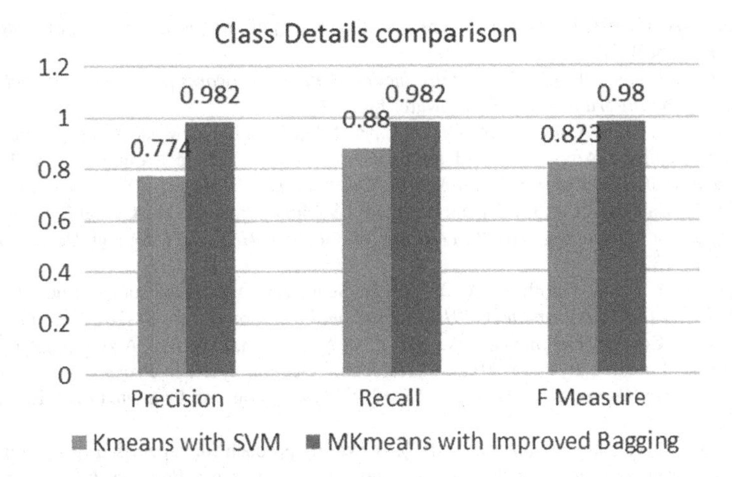

Fig. 5 Class details comparison of price rise tweets dataset

6 Conclusion

Sentiment analysis, as an interdisciplinary field that crosses natural language processing, artificial intelligence, and content mining, perceives suppositions of individuals in regards to an item, administration, protest, or social issues communicated in a given content. As of not long ago most if not all research in opinion investigation has been done on the items and administrations. Popular feelings with respect to social issues are so critical for government and by and large who include during the process of making decisions. This paper has concentrated on estimation examination of social issues including two datasets JNU tweets and Price Rise Tweets. We have proposed a strategy for slant examination of social issues. It removes the feelings from each sentence, builds correspondence conclusion structures, and afterward decides their introductions with respect to the social issue. The exploratory outcomes demonstrate that the proposed procedure performs superior than the base approach on the premise of different specified parameters. In future, the work can be stretched out by containing a bigger dataset and considering more occurrences which may coordinate in higher exact expectation investigation.

References

1. Pang, B., & Lee, L. (2004). A sentimental education: Sentiment analysis using subjectivity summarization based on minimum cuts. In *Proceedings of the 42nd ACL* (pp. 271–278).
2. Hatzivassiloglou, V., & McKeown, K. R. (1997). Predicting the semantic orientation of adjectives. In *Proceedings of 35th Annual Meeting of the Association for Computational Linguistics and Eighth Conference of the European Chapter of the Association for Computational Linguistics* (pp. 174–181).

3. Abbasi, A. (2010). Intelligent feature selection for opinion classification. *IEEE Intelligent Systems, 25*(4), 75–79.
4. Blais, A., & Mertz, D. (2001). *An introduction to neural networks pattern learning with back propagation algorithm*. Gnosis Software, Inc., July 2001.
5. Anguita, D., Ghio, A., Greco, N., Oneto, L., & Ridella, S. (2010). Model selection for support vector machines: Advantages and disadvantages of the machine learning theory. In *IEEE International Joint Conference on Neural Networks (IJCNN)* (pp. 1–8).
6. Joshi, A., Balamurali, A. R., Bhattacharyya, P., & Mohanty, R. (2011). C-Feel-It: A sentiment analyzer for micro-blogs. In *Proceedings of the ACL-HLT 2011 System Demonstrations* (pp. 127–132).
7. Karamibekr, M., & Ghorbani, A. A. (2012). Sentiment analysis of social issues. In *Social Informatics (SocialInformatics), 2012 International Conference* (pp. 14–16), December 2012.
8. Wang, H., Can, D., Kazemzadeh, A., Bar, F., & Narayanan, S. (2012). A system for real-time twitter sentiment analysis of 2012 us presidential election cycle. In *Proceedings of the ACL 2012 System Demonstrations* (pp. 115–120). Association for Computational Linguistics, 2012.
9. Revathy, K., & Sathiyabhama, B. (2013). A hybrid approach for supervised twitter sentiment classification. *International Journal of Computer Science and Business Informatics, 7*(1), 1–11.
10. Saraswathi, K., & Tamilarasi, A. (2014). Investigation of support vector machine classifier for opinion mining. *Journal of Theoretical and Applied Information Technology, 59*(2), 291–296.
11. Madhukar, M., & Verma, S. (2017). Hybrid semantic analysis of tweets on girl-child in India. *Engineering, Technology & Applied Science Research, 7*(5), 2014–2016.

Subspace Clustering—A Survey

Bhagyashri A. Kelkar and Sunil F. Rodd

Abstract High-dimensional data clustering is gaining attention in recent years due to its widespread applications in many domains like social networking, biology, etc. As a result of the advances in the data gathering and data storage technologies, many a times a single data object is often represented by many attributes. Although more data may provide new insights, it may also hinder the knowledge discovery process by cluttering the interesting relations with redundant information. The traditional definition of similarity becomes meaningless in high-dimensional data. Hence, clustering methods based on similarity between objects fail to cope with increased dimensionality of data. A dataset with large dimensionality can be better described in its subspaces than as a whole. Subspace clustering algorithms identify clusters existing in multiple, overlapping subspaces. Subspace clustering methods are further classified as top-down and bottom-up algorithms depending on strategy applied to identify subspaces. Initial clustering in case of top-down algorithms is based on full set of dimensions and it then iterates to identify subset of dimensions which can better represent the subspaces by removing irrelevant dimensions. Bottom-up algorithms start with low dimensional space and merge dense regions by using Apriori-based hierarchical clustering methods. It has been observed that, the performance and quality of results of a subspace clustering algorithm is highly dependent on the parameter values input to the algorithm. This paper gives an overview of work done in the field of subspace clustering.

Keywords Clustering · Subspace clustering · High-dimensional data

B. A. Kelkar (✉)
KLS Gogte Institute of Technology, Belgaum, Karnataka, India
e-mail: kelkar.ba@sginstitute.in

B. A. Kelkar
Sanjay Ghodawat University, Atigre, Kolhapur, Maharashtra, India

S. F. Rodd
Department of Computer Science and Engineering, Gogte Institute of Technology, Belagavi, Karnataka, India
e-mail: sfroddgit@git.edu

V. E. Balas et al. (eds.), *Data Management, Analytics and Innovation*,
Advances in Intelligent Systems and Computing 808,
https://doi.org/10.1007/978-981-13-1402-5_16

1 Introduction

Clustering is an essential data mining task for summarization, learning, and segmentation of data. It has been applied for target marketing, machine learning, pattern recognition, and statistics. Clustering is an exploratory data analysis task and aims to discover groups of similar objects called as clusters from input data set. The objects belonging to the same cluster must be highly similar whereas objects from different clusters must be highly dissimilar. Desired properties of the clustering algorithm are completeness, stability, homogeneous and significant results and efficiency.

1.1 The Curse of Dimensionality

Data analytics and machine learning is an evolving area. The grand challenge in this research lies in dealing with ever-increasing amounts of high-dimensional data gathered from multiple sources and different modalities. Bellman [1] refers to the combinatorial explosion that is observed in a data mining task implied due to processing of large number of dimensions as curse of dimensionality. This is due to the fact that, high dimensionality increases the computational complexity and memory requirements. It can adversely degrade underlying algorithm's performance. Clustering is usually done based on distance notations like the Euclidean distance and due to increased dimensionality distance between data points become meaningless. Additional dimensions spread the data points further apart as shown in Fig. 1a. With one dimension, half of the points were in a unit bin. If second dimension is added, data gets stretched as shown in Fig. 1b and the points get spread out further, pulling them apart, resulting in only about a one fourth of the points into a unit bin. Further addition of a third dimension again spreads the data and a unit bin holds only a few points as shown in Fig. 1c. When the dimensionality of the data becomes too large, the points then are all almost equidistant [2, 3] and distance between the points tend to zero as shown in Fig. 2. Hence large amount of

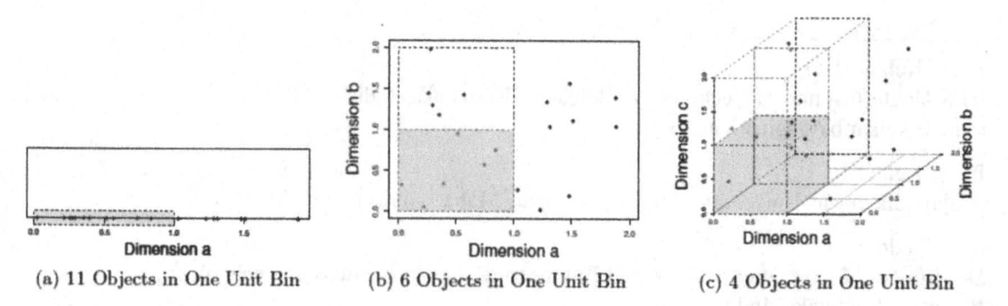

(a) 11 Objects in One Unit Bin (b) 6 Objects in One Unit Bin (c) 4 Objects in One Unit Bin

Fig. 1 The curse of dimensionality [1]—data becomes extremely sparse with increasing dimensions

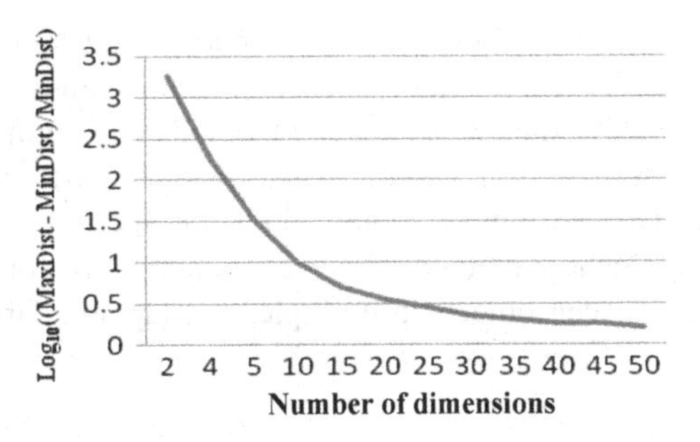

Fig. 2 Distance between data points is no longer meaningful with increased dimensions

data objects are required to satisfy a given density threshold. This fact badly affects performance of clustering algorithms as cluster membership is mainly determined based on distance between and density of data points.

The curse of dimensionality has many aspects. First, in a dataset, all attributes may not contribute to define a certain cluster. Rather the clusters may be present in subspaces. Second, a different subset of attributes may be involved in defining different subspace clusters. Hence a global feature selection procedure may not be applicable to identify attributes contributing to subspace clusters. Third, two subspace clusters might be overlapping, i.e., data point belonging to one subspace cluster $C1$ can be member of another subspace cluster $C2$. Hence subspace clustering requires appropriate feature selection methods which are different from the methods for traditional clustering based on density or partitioning of data.

In high-dimensional data, not all of the attributes are important for good clusters. Some of the attributes may be simply "noise". The problem is further worsened by the fact that, objects may be related in different subsets of dimensions in different ways and also due to fact that, some of the attributes might also be correlated. Keeping these facts in view, approaches like feature transformation and feature selection have been suggested. Feature transformation methods uncover latent structure in datasets to create combinations of the original attributes and summarize given dataset in fewer dimensions. When the number of irrelevant attributes is large, these methods are rendered irrelevant as they preserve the distance between the objects. As the new features are combination of original features, it is difficult to interpret them. Feature selection is one of the dimensionality reduction techniques and is often applied as a preprocessing step to remove noisy features. It identifies most relevant attributes for the data mining task at hand. This is achieved by evaluating various feature subsets using some criterion. These methods are further classified as: (i) global versus local where global methods find features from complete dataset whereas local methods find features relevant for each individual cluster. (ii) wrapper (with feedback) versus filter (blind-without feedback) where the filter approach selects features based on criteria such as pair wise constraints, mutual information, Laplacian score, chi-square test, etc. then evaluate the attributes, rank them before applying selection criteria. Wrapper methods formulate the problem as a search problem. Different combinations of the features are prepared.

These combinations are evaluated and a comparison with other combinations is done. Combinations of features are scored based on model accuracy using a predictive model. Embedded methods like regularization methods for feature selection are based on learning which features best contribute to the accuracy of the model. The learning is done while the model is being created. However the feature selection methods have a critical limitation that, they cannot uncover relations between objects in multiple, overlapping sub-dimensional spaces.

2 Subspace Clustering

In subspace clustering, clusters are identified in subset of attributes. Subspace clustering algorithms can be considered as an extension to feature selection methods which identify most relevant attributes by evaluating various feature subsets using some criterion. The clustering process first identifies the projections in which clusters may reside and then applies a clustering algorithm in identified subspace. A search method is required to identify subsets of attributes and then they are evaluated based on certain criteria. In subspace clustering object similarity is measured based on the selected attribute subset. For given a database DB with a set Dim of dimensions, clustering result can be denoted as a set $C = \{(C_1, A_1), \ldots, (C_k, A_k)\}$ where $C_i \subseteq$ DB and $A_i \subseteq$ Dim. Figure 3 illustrates an example of subspace clustering.

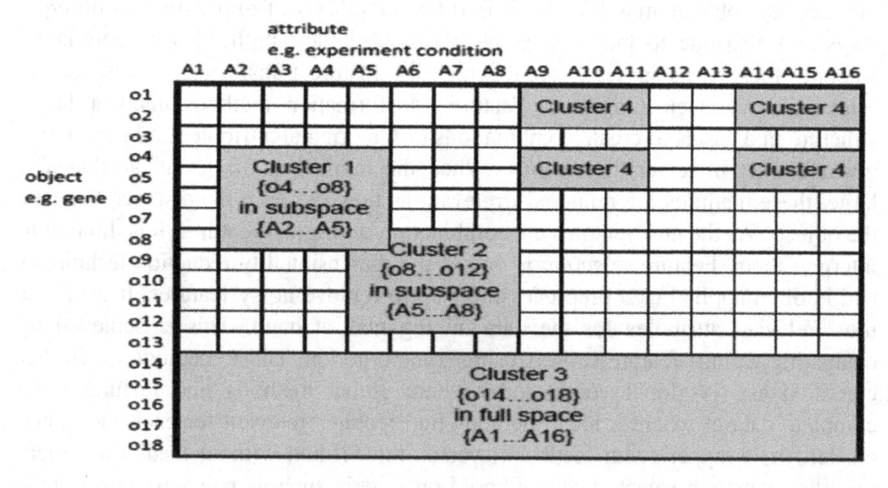

Fig. 3 Example for subspace clustering

3 Classification of Basic Subspace Clustering Approaches

The desirable property of any subspace clustering algorithm is that, it should identify all possible sets of subspace clusters. Also it must be ensured that, the outcome of clustering process must produce the same set of clusters during every run. Efficiency is another aspect of subspace clustering algorithms. The algorithms can be made to handle large data by applying proper heuristics to prune non-significant results. The results of subspace clustering should be easily interpretable. There are three major variants of subspace clustering, viz. the grid-based, window-based, and density-based. The homogeneity of attributes can be identified based on similarity between objects, density of objects, etc., depending on the criteria applied by clustering method.

3.1 Grid-Based Subspace Clustering

In this approach, data space is divided into axis-parallel cells [4]. Then the cells containing objects above a predefined threshold value given as a parameter are merged to form subspace clusters. Number of intervals is another input parameter which defines range of values in each grid. Apriori property is used to prune non-promising cells and to improve efficiency. If a unit is found to be dense in $k-1$ dimension, then it is considered for finding dense unit in k dimensions. If grid boundaries are strictly followed to separate objects, accuracy of clustering result is hampered as it may miss neighboring objects which get separated by string grid boundary. Clustering quality is highly dependent on input parameters.

3.2 Window-Based Subspace Clustering

Window-based subspace clustering [5] overcomes drawbacks of cell-based subspace clustering that it may omit significant results. Here a window slides across attribute values and obtains overlapping intervals to be used to form subspace clusters. The size of the sliding window is one of the parameters. These algorithms generate axis-parallel subspace clusters.

3.3 Density-Based Subspace Clustering

A density-based subspace clustering approach—SUBCLU is proposed by Kailing et al. (2004). It drops use of grids to overcome drawbacks of grid based subspace clustering algorithms. A cluster is defined as a collection of objects forming a chain which fall within a given distance and exceed predefined threshold of object count. Then adjacent dense regions are merged to form bigger clusters. As no grids are used, these algorithms can find arbitrarily shaped subspace clusters. Clusters are built by joining together the objects from adjacent dense regions. These approaches are prone to values of distance parameters. The effect curse of dimensionality is overcome in density-based algorithms by utilizing a density measure which is adaptive to subspace size.

3.4 Other Prominent Approaches

Overlapping cluster algorithms like CLIQUE [4], ENCLUS (ENtropy based subspace CLUStering) [6], MAFIA [7], SUBCLU [8], FIRES [9] try to enumerate all possible subspace clusters. When a data object belongs to many subspace clusters, the clustering is called overlapping. When each data object is member of a unique cluster or marked as outlier, the clustering is non-overlapping. Some of the non-overlapping approaches are PROCLUS [10], DOC [11], PreDeCon [12], etc. Lance et al. [13] classify subspace clustering as top-down- and bottom-up algorithms based on the strategy used to identify cluster subspaces. In bottom-up approach cluster discovery starts from individual attributes and then the subspaces grow to higher dimensional space. For pruning the search space APRIORI property of density is used. Candidate subspaces for the next higher level of dimension sets is formed only from the lower level dense regions. CLIQUE, OPTIGRID [14], DENCOS [5], MAFIA, SUBCLU, FIRES are some of the bottom-up approaches.

In top-down subspace clustering approach, all dimensions are initially part of a cluster and are assumed to equally contribute to clustering. In the subsequent iterations, importance of each dimension is recalculated and clusters are regenerated. This requires multiple iterations over full set of dimensions. The performance can be improved by making use of sampling technique. Due to top-down partitioning of the data, each data object can be member of a unique cluster. Some of the algorithms additionally identify outliers as a separate group. For meaningful results, parameter tuning is necessary. Further classification of this approach is per cluster weighting methods and per instance weighting methods. Few of the Top-down Algorithms are FIND-IT, ORCLUS [15], PROCLUS [10], COSA [16], δ-CLUSTERS [17]. Figure 4 presents a hierarchy of these two prominent classes of subspace clustering algorithms. Clustering oriented subspace clustering relies on predefined parameters such as the expected number of clusters, average dimensionality of clusters, etc. These algorithms try to optimize the solution and hence each data point is assigned to a cluster which results in assigning noise objects to some clusters.

In a dataset, when an object belongs to a cluster, the variance of the occurring values is less compared range of all other attributes. This geometrical intuition lead to identification of a cluster which contains data points which are densely clustered along relevant attributes. The resulting cluster is an axis-parallel subspace cluster.

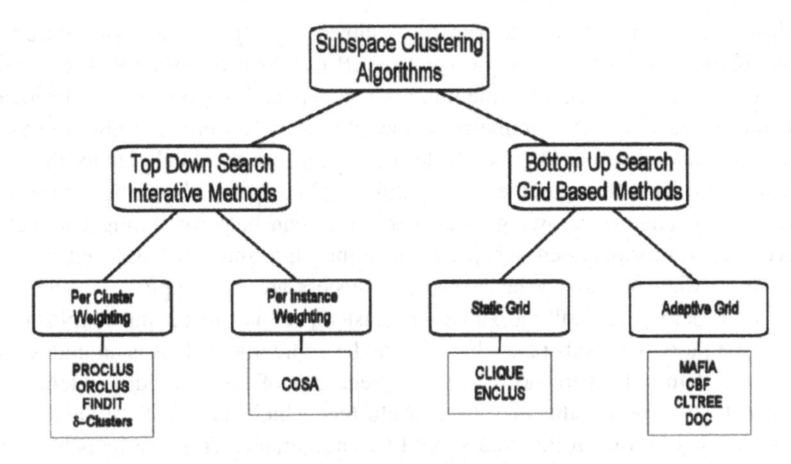

Fig. 4 Hierarchy of subspace clustering algorithms based on search strategy

Basic subspace clustering algorithms like CLIQUE, projected clustering algorithms fall under this category. The algorithms which result in arbitrary oriented subspaces, e.g., ORCLUS use the knowledge that, members of a subspace cluster are always close to the plane in which the subspace resides and use this information for cluster interpretation. Hard subspace clustering algorithms assume that all features have equal importance in forming a subspace whereas soft subspace clustering algorithms proceed by assigning a weight to each dimension based on its contribution to clustering.

4 Enhancements to Traditional Subspace Clustering

Although basic subspace clustering approaches look efficient in solving the clustering problem, they have certain major drawbacks. These algorithms can tackle quantitative 2-D data in format of object *X* attributes, but they are not customized to handle the data in 3-D format, i.e., having dimensions—objects, attributes and time stamp. Similarly, most of them cannot handle complex data such as categorical or streaming data. When the data is in 3-D format, it is rare that clusters can be found in every timestamp of the dataset when the data contains large number of timestamps and there is a need to develop efficient algorithms for mining 3D data. Distance measures applicable to numeric data cannot be applied directly to categorical data as they do not have natural order. Hence devising subspace clustering algorithms for categorical data is another challenge. Many real world data sets contain missing or erroneous values. Therefore, any subspace clustering algorithm working on real-world dataset must handle these datasets properly without affecting accuracy of the results.

Mostly all of the available subspace clustering algorithms work based on parameters values given by user at run time. It has been observed that, clustering output is very much sensitive to the input parameters and outcome varies drastically with minor changes in parameter values. Intuitively setting right values of parameters that will result in good clustering is very much difficult. Hence there is a need to overcome this parameter-sensitivity of subspace clustering algorithms. Domain knowledge or knowing data distribution can help for setting parameters. Sometimes, semi-supervised subspace clustering algorithms can be used to guide parameter setting process. When subspace clusters are overlapping, i.e., when an object may belong to multiple subspace clusters, it may result into explosion of clusters, i.e., too many subspace clusters may be enumerated. This is an undesirable solution as it may lead to too many interpretations of the same data. Hence it is desirable that, only significant subspace clusters which represent true and meaningful information of out the data should be enumerated. This can be achieved in two ways. First as a preprocessing step all significant subspaces can be mined, and then subspace clusters can be identified from these subspaces as in filter approach of feature selection [18]. In the second approach, what is significant in terms of subspace clusters is first defined and then the clustering algorithm mines these clusters directly.

5 Evaluation of Subspace Clustering

Evaluation of clustering output is a complex work. A clustering algorithm is evaluated in terms of execution time and quality of clustering results. Quality of clustering is defined in terms of compactness of a cluster and separation between different clusters and the same is true for subspace clusters. The motivation behind any clustering is to disclose the hidden information in the data as accurately as possible. Hence it is desirable to detect a minimum number of meaningful subspace clusters. There are various clustering quality indexes proposed in literature [19]. However there is lack of standardized guidelines for evaluation of clustering outcome. For a novice researcher, it is a dilemma which clustering quality index is to be used use for a particular dataset. Silhouette index, Simplified Silhouette index, Dunn index, Davies–Bouldin index, Isolation index, PBM index, Point-biserial index, RS index, Rand index are some of the indexes which can be used for the evaluation. In [2], the authors have analyzed some of the standard clustering quality measures and it reveals that, with increasing dimensionality different clustering quality indexes are affected in different ways and conclude that selecting a clustering quality index for high-dimensional data is nontrivial.

Liu et al. (2010) identify major criteria for evaluation of the clustering algorithms based on quality of the results produced namely, the results should be non-monotonous, the algorithm should be robust to noise, it should properly handle varying cluster density and skewed distributions of the data. Compactness of clusters and separation between the clusters are termed as internal clustering quality indexes. How well the data is partitioned is measured by external quality indexes. Müller et al. [20] present a common framework for evaluating major subspace clustering paradigms. Entropy, F1-measure and accuracy are some of the object based measures which mainly relate to (i) purity of clusters identified, (ii) an algorithm's power to discover hidden clusters and (iii) correctness of the algorithm in assigning objects to a cluster respectively. Relative non intersecting area (RNIA) is an object and subspace based measure to find the extent to which found sub-objects cover true sub-objects. Drawback of RNIA measure is that it cannot find if a true cluster is correctly covered by several found clusters or exactly one found cluster covers the true cluster. On the contrary, the clustering error (CE) is advancement over RNIA measure that maps each found cluster to at most one ground truth cluster and also each ground truth cluster to at most one found cluster. Intersection of sub-objects is determined for each such mapping of two clusters. After summing up the individual values give value I' which when substituted in place of I in the RNIA formula will give the CE-value. Thus CE-value penalizes the clustering results producing many smaller clusters. WEKA [21] is an open source framework containing various well known algorithms in clustering, classification, feature selection and association rule mining. It provides facility for visualization of the results. An open source framework OpenSubspace [22] can be used for evaluation of projected and subspace clustering algorithms in WEKA.

5.1 *Results Obtained from Earlier Work*

Müller et al. have systematically evaluated major paradigms of subspace clustering using OpenSubspace. The study highlights that, SUBCLU and CLIQUE have comparable F1 and Accuracy, but have to pay penalty in terms of RNIA and CE as they try to detect many clusters even more than the count of objects in the dataset as it tries to cover all of the data including noise. This also results into increased runtimes. SUBCLU does not even finish for the biggest real world data set, pendigits. The recent cell-based paradigms show best results with low runtimes. The distance-based approaches also face the problems of high runtimes. Clustering oriented approaches have easy parameterization as these settings decide on clustering output and they show reasonable runtimes. Cell-based approaches like CLIQUE and SUBCLU produce many more clusters in an attempt to achieve good results whereas clustering oriented approaches tend to produce comparatively few clusters.

Generally, high-quality results are paid with high runtime. But even in some algorithms meaningful results are not obtained within tolerable timeframe due to high runtimes even up to several days (for dimensionality >25). Hence practical application of such an algorithm with such high runtimes on high dimensionalities is infeasible. Hence a subspace clustering algorithm must have to find the trade-off between output quality and runtime. Also it is observed that cluster detection time increases with the number of objects. Several heuristics must be applied for having an efficient computation with acceptable accurate results. For neighborhood density computation, the density-based approaches have expensive database scans and hence they do not scale as dimensionality increases. DOC and MINCLUS are found to be good in handling noisy data. There are certain open issues in subspace clustering. Tuning parameter setting is a nontrivial task and usually guesswork is involved. Hence there is a need to have parameter-insensitive algorithms for subspace clustering. For time series data, there is need to identify proper search space pruning strategy. Appropriate post-processing methods for limiting output clusters, organizing the output clusters and formulating models to represent the output are necessary to uncover the information extracted from subspace clusters to useful knowledge.

6 Conclusion

High-dimensional data clustering is a challenging task which first requires formulating how a cluster needs to be represented. Many a times even though a dataset has lots of dimensions, only few of them are of importance for extracting knowledge and rest are noise. Subspace clustering algorithms solve this problem by finding clusters on subsets of attributes and objects. This has the advantage that, those patterns which may be missed by full dimensional clustering are also uncovered. A subspace clustering algorithm must ensure that the subspace projections must be dissimilar and at the same time must not be redundant. Performance of a subspace clustering algorithm is highly dependent on tuning parameters. It has been observed that, when dimensionality of the data increases, accuracy subspace clustering decreases with tremendous increase in runtime. Proper validation techniques must be applied to avoid spurious clusters. The quality evaluation of results obtained from subspace clustering algorithms is challenging as different subspace clustering approaches lead to different cluster characteristics and topologies. Fair and comparable evaluation based on objective evaluation measure of detected subspace clusters is of major importance. In synthetic datasets the best clustering is already known. But such a data might miss variations present in real-world data. Review of recent approaches for subspace clustering highlight that, cell-based approaches outperform in terms of efficiency and quality for low to medium dimensionality. It is also shown that instead of enumerating all subspace clusters which may contain many redundant clusters, outputting a few relevant clusters achieves best results. Further research direction in this field can be reducing database scans, automatic detection of clustering parameters based on data distribution, improving execution time and enhancements in existing algorithms to handle complex data.

References

1. Bellman, R. (1961). *Adaptive control processes*. Princeton: Princeton University Press.
2. Parsons, L., Haque, E., & Liu, H. (2004). Subspace clustering for high dimensional data: A review. *ACM SIGKDD Explorations, 6*(1), 90–105.
3. Francois, D., Wertz, V., & Verleysen, M. (2007). The concentration of fractional distances. *IEEE Transactions on Knowledge and Data Engineering, 19*(7), 873–886.
4. Agrawal, R., Gehrke, J., & Gunopulos, D. (1998). Automatic subspace clustering of high dimensional data for data mining applications. In *Proceedings of the ACM SIGMOD International Conference on Management of Data* (pp. 94–105).
5. Liu, G., Sim, K., Li, J., & Wong, L. (2009). Efficient mining of distance-based subspace clusters. *Statistical Analysis and Data Mining, 2*(5–6), 427–444.
6. Cheng, C.-H., Fu, A. W., & Zhang, Y. (1999). Entropy-based subspace clustering for mining numerical data. In *Proceedings of the Fifth ACM SIGKDD International Conference on Knowledge Discovery and Data Mining* (pp. 84–93).
7. Goil, S., Nagesh, H., & Choudhary, A. (1999). Mafia: Efficient and scalable subspace clustering for very large data sets. Technical Report CPDC-TR-9906-010, Northwestern University.
8. Kröger, P., Kriegel, H.-P., & Kailing, K. (2004). Density-connected subspace clustering for high-dimensional data. In *Proceedings of SIAM International Conference on Data Mining* (pp. 246–257).
9. Kriegel, H.-P. H., Kroger, P., Renz, M., & Wurst, S. (2005). A generic framework for efficient subspace clustering of high-dimensional data. In *IEEE International Conference on Data Mining* (pp. 250–257), Washington, DC, USA.
10. Aggarwal, C. C., Procopiuc, C. M., Wolf, J. L., et al. (1999). Fast algorithms for projected clustering. In *Proceedings of the ACM International Conference on Management of Data (SIGMOD)* (pp. 61–72), Philadelphia, PA.
11. Procopiuc, C. M., Jones, M., Agarwal, P. K., & Murali, T. M. (2002). *A Monte Carlo algorithm for fast projective clustering in SIGMOD* (pp. 418–427). USA.
12. Bohm, C., Railing, K., Kriegel, H.-P., & Kroger, P. (2004). Density connected clustering with local subspace preferences. In *Fourth IEEE International Conference on Data Mining, ICDM* (pp. 27–34).
13. Lance, P., Haque, E., & Liu, H. (2004). Subspace clustering for high dimensional data: A review. *ACM SIGKDD Explorations Newsletter, 6*(1), 90–105.
14. Hinneburg, A., & Keim, D. A. (1999). Optimal grid-clustering: Towards breaking the curse of dimensionality in high-dimensional clustering. In *VLDB* (pp. 506–517).
15. Aggarwal, C. C., & Yu, P. S. (2000). Finding generalized projected clusters in high dimensional spaces. In: *Proceedings of the ACM SIGMOD International Conference on Management of Data* (pp. 70–81).
16. Friedman, J. H., & Meulman, J. J. (2004). Clustering objects on subsets of attributes. *Journal of the Royal Statistical Society: Series B* (*Statistical Methodology*) (pp. 815–849).
17. Yang, J., Wang, W., Wang, H., & Yu, P. (2002). δ-Clusters: Capturing subspace correlation in a large data set. In *Proceedings of the 18th International Conference on Data Engineering* (pp. 517–528).
18. Dash, M., Choi, K., Scheuermann, P., & Liu, H. (2002). Feature selection for clustering – a filter solution. In *Proceedings of the IEEE International Conference on Data Mining* (ICDM02) (pp. 115–124).
19. Patrikainen, A., & Meila, M. (2006). Comparing subspace clusterings. *TKDE, 18*(7), 902–916.
20. Müller, E., Günnemann, S., Assent, I., & Seidl, T. (2009). Evaluating clustering in subspace projections of high dimensional data. *PVLDB, 2*(1), 1270–1281.
21. Weka 3: Data Mining Software in Java. (2014). Available: http://www.cs.waikato.ac.nz/ml/weka/.

22. OpenSubspace:Weka Subspace-Clustering Integration. (2014). Available: http://dme.rwth-aachen.de/OpenSubspace/.
23. Jaya Lakshmi, B., Shashi, M., & Madhuri, K. B. (2017). A rough set based subspace clustering technique for high dimensional data. *Journal of King Saud University-Computer and Information Sciences*.
24. Jaya Lakshmi, B., Madhuri, K. B., & Shashi, M. (2017). An efficient algorithm for density based subspace clustering with dynamic parameter setting. *International Journal of Information Technology and Computer Science, 9*(6), 27–33.
25. Tomašev, N., & Radovanović, M. (2016). Clustering evaluation in high-dimensional data. In *Unsupervised Learning Algorithms* (pp. 71–107). Berlin: Springer.
26. Zhu, B., Ordozgoiti, B., & Mozo, A. (2016). PSCEG: An unbiased parallel subspace clustering algorithm using exact grids. In *24th European Symposium on Artificial Neural Networks, Computational Intelligence and Machine Learning ESSAN16* (pp. 27–29), Bruges (Belgium).
27. Peignier, S., Rigotti, C., & Beslon, G. (2015). Subspace clustering using evolvable genome structure. In *Proceedings of the ACM Genetic and Evolutionary Computation Conference (GECCO 2015)* (pp. 1–8).
28. Kaur, A., & Datta, A. (2015). A novel algorithm for fast and scalable subspace clustering of high-dimensional data. *Journal of Big Data, 2*(1), 1–24.
29. Xu, D., & Tian, Y. (2015). A comprehensive survey of clustering algorithms. *Annals of Data Science, 2,* 165–193.
30. Sim, K., Gopalkrishnan, V., Zimek, A., & Cong, G. (2013). A survey on enhanced subspace clustering. *Data Mining and Knowledge Discovery, 26*(2), 332–397.
31. Liu, H. W., Sun, J., Liu, L., & Zhang, H. J. (2009). Feature selection with dynamic mutual information. *Pattern Recognition, 42*(7), 1330–1339.
32. Kriegel, H. P., Kröger, P., Zimek, A., & Oger, P. K. R. (2009). Clustering high-dimensional data: A survey on subspace clustering, pattern-based clustering, and correlation clustering. *ACM Transactions on Knowledge Discovery Data, 3*(1), 1–58.

Revisiting Software Reliability

Kavita Sahu and R. K. Srivastava

Abstract Reliability is an important issue for deciding the quality of the software. Reliability prediction is a statistical procedure that purpose to expect the future reliability values, based on known information during development processes. It is considered as a basic function of software development. A review-based research has been done in this work to evaluate the previously established methodologies for reliability prediction. In this paper, authors give a critical review related to successful research of reliability prediction. This paper also provides many challenges and keys of reliability estimation during software development process. Further, this paper gives a precarious discussion on previous work and identified factors which are important for reliability of software but still ignored. This work helps to developers for predicting the reliability of software with minimum risks.

Keywords Software reliability · Software development model · Reliability prediction · Soft computing techniques

1 Introduction

Software quality is the basic requirement to be fulfilled by a developers for user's satisfaction. This is because software quality plays a good role in developing high reliable software. The definition of reliability can be stated as the probability of software to perform under specified conditions without failing. ISO/IEC 25010:2011 product quality model defines reliability as the degree of performance for a system under specified conditions such as time and cost [1, 2]. The measurement of software reliability is degree of removal of errors. These errors or

K. Sahu (✉) · R. K. Srivastava
Department of Computer Science, Dr. Shakuntala Misra National Rehabilitation University, Lucknow, Uttar Pradesh, India
e-mail: kavi9839@gmail.com

R. K. Srivastava
e-mail: rks100664@gmail.com

V. E. Balas et al. (eds.), *Data Management, Analytics and Innovation*,
Advances in Intelligent Systems and Computing 808,
https://doi.org/10.1007/978-981-13-1402-5_17

faults are recognized during testing of software by software testers and thus removal of these is done by debugger and developers. This process of testing and debugging leads to more reliable software. Reliability modeling is the process of developing models to create error free and reliable software [3, 4]. Process of reliability modeling is called software reliability growth modeling (SRGM).

In past 30 years many SRGM's have been developed to ensure reliability and maintain quality in software [5]. But still we could not reach the level where a user can be satisfied with the reliability of software. In the proposed work we are hereby reviewing the already proposed reliability growth and prediction models and techniques used by these models. The area of reliability prediction covers techniques, metrics and models of how to develop a model which will be used for prediction and estimation of reliability [5–10]. This contains models for both the operational outline, to capture the projected procedure of the software, and models for the operational failure behavior. The latter category of models is then also used for prediction of the reliability in terms of failures. In this paper, a critical review is provided which is based on different soft computing techniques which are already been applied in the area of reliability prediction.

This paper has taken most challenging problems of reliability prediction in current years. The paper is mainly divided into four sections: second section emphasizes on software reliability. Third section describes the basics of reliability prediction. Fourth section describes reliability prediction related to soft computing techniques. At the end summary and discussion of the paper is given and then conclusion is described.

2 Software Reliability

With the ever-increasing role of computers in our daily lives, reliability of these machines has become an issue. Although software systems made in this era are designed to face failures in very extreme conditions. But still there is a missing link between the theory of reliability prediction and its application in real world. Testing for reliability of software is done during the development phase but that does not ensure the overall service life of software. Software, Once reliable, cannot be reliable over its whole life. Over the ages there have been multiple attempts made to predict the failure rate or the reliability of software, but none of them ensured to be applied to all kind of software. The facts related to these models can be uncovered during the review of the work. Hence an effort has been made to understand the pros and cons of the models given by different authors.

Reliability is defined as "The ability of software to function under specified circumstances for specific time duration". Software quality can be related directly with the software reliability, in both developers use mutual methods for its consideration and may involve input from each other [8, 11]. It pays attention on costs of failure which are caused by software downtime, repair tools and cost of assurance privileges. In software engineering, safety ordinarily highlights not cost, but stabilizing life and therefore deals only with specific risky software failure modes. The effect of reliability on different software characteristics are shown in Table 1.

Reliability is a significant factor of quality during software development process. Attacks and failures affect every characteristic of software which is clarified in Table 1.

3 Reliability Prediction

Reliability of software can be considered as a precise quantity [12]. It can either be one or zero means software will be either reliable or unreliable. Hence reliability prediction is an important activity to be considered during the early stages of software development. Reliability prediction is developing a methodology in early stages of development to predict the reliability in terms of its attributes such as defects, failures count, failure rate, etc. Prediction of reliability further helps in other activities of development such as evaluating the feasibility of planned requirements and delivers a stable origin for software design. Reliability of software may be enhanced by a focused defect removal, inspection, and test effort.

Figure 1 precisely describes the procedure to develop a new reliability prediction model. At first literature survey of available methods is done. Next is to evaluate the methods and compare them to traditional models. Design of experiments is done at level third. Testing of new model and regression analysis is done at last step in the procedure.

4 Reliability Prediction Related to Soft Computing

Software reliability growth and prediction modeling has been a thrust area of research in recent years. Software criticality and failure rates has been increased and thus giving the need to work more in the area of software reliability prediction and its modeling. Several statistical methods have been used in the modeling of reliability prediction but failed to give results which can be applied globally. Hence, soft computing methodologies are used for software reliability analysis, reliability optimization, etc. [13–18] because the results achieved from these methods are more precise and applicable to real world. This section focuses on description of application of soft computing technologies in software reliability prediction. Software reliability prediction focuses on developing and maintaining techniques by which software systems reliability can be quantitatively evaluated. In previous major models reliability of software was predicted through the data collected from past records such as failure counts on different time intervals.

Software reliability has remained a thrust area of research over the past 40 years, but still there are flaws in the modeling of software reliability. In the last two decades there has been lot of work done in the area of soft computing using the

Table 1 Effects of unreliable software on software characteristics

Unreliable software's proneness		Software characteristics							
		Quality	Efforts	Maintenance	Performance	Security	Cost	Usability	Efficiency
Attacks	Worms	✓	✓	✓	✓	✓	✓	✓	✓
	Virus	✓	✓	✓	✓	✓	✓	✓	✓
	Threats	✓	✓	✓	✓	✓	✓	✓	✓
Failures	Errors	✓	✓	✓	✓	✓	✓	✓	✓
	Faults	✓	✓	✓	✓	✓	✓	✓	✓
	Defects	✓	✓	✓	✓	✓	✓	✓	✓

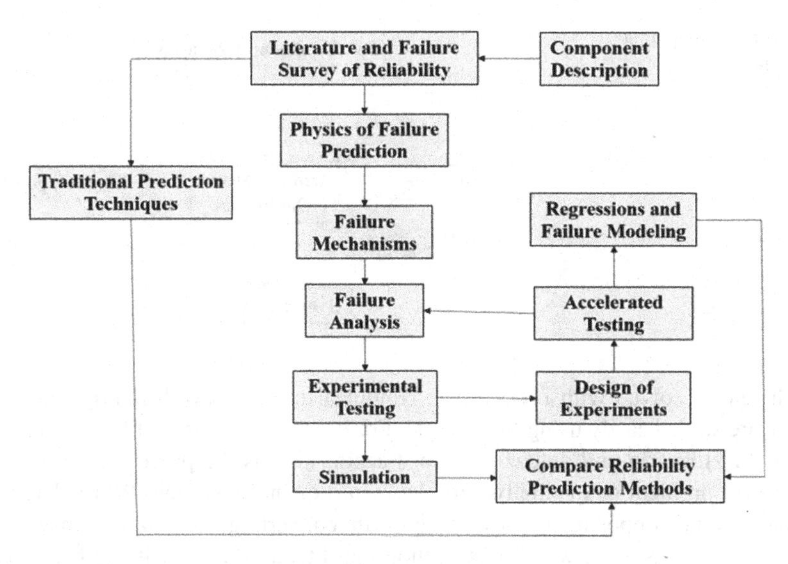

Fig. 1 Comparisons of reliability prediction methodology

fuzzy logic, neural network, genetic algorithms and their hybrid approaches [13, 19–21]. Both the statistical and soft computing techniques gives noteworthy results in predicting reliability but it varies as the type of data changes. There are various prediction techniques for software reliability, but before using these techniques, one must thoroughly go through different techniques according to their research question [9, 22–29]. Measurement of reliability in software is still in its initial stages. There is no good quantitative method for software reliability prediction which does not have any limitations. The study of different soft computing methodologies is as follows: fuzzy logic, probabilistic computing, neural networks, genetic programming and theory, evolutionary computing and probabilistic computing.

These soft computing techniques and its methodologies are applied to different areas such as computer science, robotics, engineering, and construction areas and many more [30, 31]. The predictive capability of soft computing methodologies helps in taking developing good models of these areas. Figure 2 presents soft computing techniques as fuzzy logic, neural networks, and genetic algorithm. The introduction of fuzzy logic, neural network, genetic algorithm and their hybrid techniques of reliability prediction with descriptions of relevant work related to each area is given below.

4.1 Fuzzy Logic for Reliability Prediction

Methodology of fuzzy logic is derived from fuzzy set theory which was first given by Zadeh in 1960. Fuzzy methodology defines linguistic values with reasoning which is in the form of numeric. Real-world problems that are full with complexity

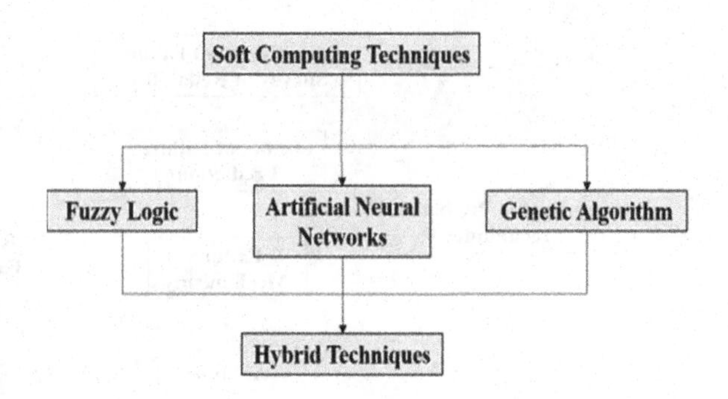

Fig. 2 Soft-computing techniques

can be easily solved with fuzzy logic. Nonlinear problems with multidimensional nature are solved easily using fuzzy logic. Fuzzy logic comprises of two broad steps that is fuzzification and defuzzification. Fuzzification is the process of converting real world problem or qualitative problems into quantitative one. While defuzzification is quite opposite to it, it deals with converting the fuzzify answers to real-world linguistic answers. After a thorough literature review it has been found that fuzzy logic is a well performing methodology for reliability prediction. Some of the pertinent works of fuzzy logic by previous researchers is given below:

Yuan [4] in his research predict the cumulative number of failures and further predict the reliability by using fuzzy subtractive clustering. The methodology of fuzzy clustering is combined with module order modeling. Both the methods work on different terms that is fuzzy subtractive clustering is used to predict number of faults while other is used to determine while those failures are fault vulnerable or not. In one of his work Khalaf Khatatneh [18] in 2009 proposed a model of predicting software reliability with the help of fuzzy logic. Proposed model was then applied on a custom set of data to certify that results were better than previous models. Xu [4] first proposed a model which uses the methodology of fuzzy nonlinear regression modeling technique for the prediction of the range in which a fault can be occurred in software. The proposed software was verified and validated using a real-time case study of full-scale industrial software and its modules. Sultan Alijahdali [8] researched on the usage of fuzzy logic in developing SRGM's for predicting reliability. The author here used Takagi-Sugeno technique for the training of proposed model. Authors represented the software failures using triangular fuzzy membership functions. He used the dataset of John Musa bell laboratories failures and shows his model efficiency.

Dilip Kumar Yadav et al. [23] presented a research paper on software defect prediction using fuzzy logic. To verify and validate their proposed model they used software metric for size estimation and three different metrics of requirement analysis using fuzzy logic. Predictive capability of presented the proposed approach was compared with existing models. Aleksandar Dimov and Sasikumar Punnekkat [21] presented their research that was based on fuzzy logic and prediction of reliability in component based software systems. They applied possibility theory for solving the uncertainty aspect which was based on fuzzy sets.

4.2 Artificial Neural Networks for Reliability Prediction

Today's world is more concerning on artificial intelligence, which is achieved by using the soft computing methods. Artificial neural network works on the methods of human brain neurons systems. Artificial neurons are trained and tested using different algorithms and methodologies in this method. A lot of work has been done in this field during last two decades. Reliability prediction and software reliability growth modeling is also done by various researchers. Some of these works are reviewed and detailed below:

Karunanithi et al. [10] was the first to use neural networks for the prediction of reliability by predicting its number of failures. The authors have taken execution time as the input of the neural network. Authors used different algorithms to train the networks that are Feed Forward neural networks and recurrent neural networks. Further in recurrent neural network Jordan neural network and Elman neural network were used for training and testing the network. Statistical methods were used to validate the results at the end.

Yu-Shen-Su et al. [6] proposed the software reliability model using neural networks. They also proposed a dynamic weighted combinational model by training it through back propagation algorithm. A Dataset of John Musa bell laboratory was used for implementation and RMSE values were calculated to show the results were better. Sultan Alijahdali and Khalid A. Buragga [7] presented a paper in which they used artificial neural networks for reliability prediction. The author used four algorithms to predict the number of failures that are a fuzzy inference system based on Takagi-Sugeno methodology, radial basis functions, Elman recurrent neural networks and multi-layer perceptron neural network.

Mohamed Benaddy and Mohamed Wakrim [22] presented a research on simulated annealing neural network based reliability modeling. In their work they proposed a failure count prediction model which is further based on the methodology of simulated annealing algorithm. Simulated annealing algorithm is used to train the neural network for predicting the increasing software failure. A dataset of John Musa bell laboratory was used for implementation and results. This paper also described the basic concepts of neural network and simulated annealing. Manjubala Bisi et al. [15] presented a paper based on feed forward architecture to predict software failures. The authors calculated the effects of encoding and different parameters. A comparison with existing statistical models was also done to improve efficiency. Gaurav Agrawal and Dr. V. K. Gupta [32] presented a paper on neural network based SRGM. They developed a SRGM which used error back propagation algorithm for learning. They also developed software and collect its failure datasets to predict reliability. Also NASA datasets were used that are JM1, PC1, KM1, KC1, and KC2.

4.3 Genetic Algorithm for Reliability Prediction

Genetic algorithms and genetic programming the other part of soft computing that was developed to decrease the local minima problem into the networks of neural methodology. Prof. John Holland and his students developed the solution to this by using genomes in their work. This methodology of optimization of problem using genomes was called genetic algorithm. This algorithm was developed at the University of Michigan during the 1960s and 1970s. The genetic algorithm works by creating a number of solutions which is called population in this algorithm. This population is then optimized by genetic operators and crossovers to get the best results out of it. Genetic algorithm works on the basic concept of neural network but reducing the problem of local minima which is generally occurred in neural networks.

Sultan Alijahdali and Mohammad E. El-Telbany [11] presented a research which was based on genetic algorithm. According to the authors Genetic Algorithm is proposed as a controlling soft computing algorithm and might overcome the uncertainties caused by other machine learning techniques. In this paper experiments were done to certify the theory by predicting the reliability of datasets and comparing it to other models result. In his other work Alijahdali et al. [12] presented the work of multi objective genetic algorithm for reliability prediction. The authors used genetic algorithms for predicting the software faults during the software testing process.

Oliveira [13] proposed the usage of genetic programming to predict software reliability. The author also used the genetic programming by using re-weighting in it. The algorithm of re-weighting uses recursion method which is to call itself many times with assigned weights to each example. Zainab Al-Rahamneh et al. [9] presented a research in which she conducted experiments to show that the proposed Genetic Programming model is way superior in comparison to other models like Schneidewind, Yamada S-Shaped, Generalized Poisson and NHPP reliability models.

4.4 Hybrid Techniques for Reliability Prediction

Soft computing methods are used for predicting reliability for many years of research. But there is always a lack of one single software reliability growth model that can predict reliability for all kinds of software. Researchers now are focusing on developing such systems by integrating different soft computing techniques to predict the reliability. As it is known that software changes its behavior when its code changes in testing phase so predicting reliability becomes more complex process. Here are some works that have been done recently in area of hybrid techniques for reliability prediction.

Chua et al. [24] presented a hybrid model for predicting reliability. They stated that back propagation neural network has some shortcomings like over fitting. Hence in their work they followed genetic algorithms search techniques with integration to Bayesian neural network methodology. Two examples were also presented to apply the model and showed better results than using only back propagation algorithm. Jin [33] presented a model which overcomes the limitations of using genetic algorithm. GA has problems of local minima and parameter convergence. Author integrated genetic algorithm and simulated annealing algorithm to overcome the limitation. He then applied it to support vector (SVM) for better results.

Mohanty et al. [26] presented an intelligent approach for the prediction of software reliability during the early stages of SDLC. They proposed a hybrid model which was a combination of genetic programming and group method of data handling. They also compared their model with different machine learning techniques like back propagation, counter back propagation, multiple linear regression, etc. Jung Hua Lo [34] used ARIMA and SVM in his hybrid model. This paper used ARIMA for linear model and SVM for nonlinear model of reliability prediction. Further he calculated Absolute error and Mean square error to prove that results were better. Jaydeep Pati and Shukla [27] presented a hybrid model in which they used ARIMA and ANN models to predict reliability. They used ARIMA for linear and ANN for nonlinear modeling. They used Levenberg–Marquardt algorithm with Bayesian Regularization to train neural networks [9]. The model was applied on three real time systems and it was found that results were better than using single methodology in prediction.

Bal et al. [35] presented a research paper on reliability prediction through two hybrid techniques that are radial basis function and feed forward neural network. The verification and validation of proposed models is done by using artificial neural networks and statistical methods. Wang and Zhang [36] presented a reliability model that was built by using a deep learning model based on the recurrent NN (RNN) encoder–decoder. This model was tested on 14 real-time datasets and the results were presented in the form of AE (Average Error) and AB (Average Bias). As per the literature reviewed by the author its model resulted well in respective to previous models.

5 Discussion and Findings

Software reliability is a key concern in this era, as reliable systems are more in demand due to the increasing number of faults per year. Reliability is also responsible for improving the quality of software. Reliability is composed of three attributes that are availability, fault tolerance, and maturity. Measurement and prediction of reliability helps in improving the service life of any software. Economic success of any organization depends on the reliability of software that it delivers. Hence software reliability growth and prediction models keep its importance in development organizations and software market.

Reliability prediction assists developer to maintain reliable software throughout its lifetime. Prediction in early stages of development provides better results than assessment of reliability at the end of development. Soft computing techniques thus emerge as a best solution to predict the number of faults in software and maintain the reliability of software. Some important Soft computing methods are Fuzzy Logic, Neural Network, Genetic Algorithms and Genetic Programming, etc., a brief summary of literature review preformed on prediction techniques of software reliability models has been summarized in this section as shown in Table 2.

This paper focuses on various software reliability growth and prediction models based on different soft computing techniques such as fuzzy logic, neural network and genetic algorithm proposed by different authors. By going through the literature we came to the various conclusions for improving the methodology of predicting software reliability. One of the inferences that can be drawn from the review of the above methodologies is that hybrid techniques of soft computing such as Fuzzy-neural, neural-genetic, Fuzzy-genetic, etc., give the best results when applied to the real-world problems.

6 Reliability Prediction Challenges

This paper gives some of the main objectives of future challenges that are as follows:

- Applicability of prediction models in real time scenarios
- Impacts of Environments Changes on reliability factors
- Integration of Past and Present Research on Reliability prediction models
- Check Development Tools Failure
- Ensuring Adequacy
- Re-evaluation of Role of Designers
- Understanding of the Market Issues
- Operational Practices
- Managing the Operational Impacts.

7 Conclusion

Reliability prediction is a mathematical process of estimating and predicting reliability in the early phases of software development. Measurement and prediction of reliability plays an important role in deciding the place of software industry in market. Choosing the right software reliability growth and prediction model for the development of the software is very much critical and the software developers

Table 2 Summary of SRGM and reliability prediction techniques

Sr. No.	Author(s)/ Reference(s)	Technology/Algorithm and year	Data of previous projects	Summarization
1.	Yu Shen Su, ChinYu Huang, Yi Shin Chen and Jing Xun Chen	Neural Network/2005	Real command and control project, John D Musa, Bell Lab	Used neural network to build a dynamic weighted combinational model. Calculated RA, AE and AB for showing the results
2.	Sultan Alijahdali et al.	Neural Network/2008	Dataset from DACS services at the Department of Defence	Calculated RMSE values from four different connectionist learning algorithms as MLP neural network, Radial basis function, Elman recurrent neural network, Takagi-Sugeno fuzzy inference system
3.	Mohamed Benaddy and Mohamed Wakrim	Neural Network/2012	Dataset from John Musa bell telephone laboratories	Simulated annealing is used for data optimization and neural network is trained. Also author calculated NRMSE values for showing the comparison between previous and proposed model
4.	Manjubala Bisi et al.	Neural Network/2012	18 different data sets (Military, Real time System, Real time Command ant Control, Online data Entry etc.)	A Feed Forward neural network with two encoding scheme has been proposed
5.	Gaurav Agrawal and Dr. V. K. Gupta	Neural Network/2015	NASA datasets JM1, PC1, KM1, KC1 and KC2	Author designed a SRGM that works in two phases. Error back propagation was used to train the neural network
6.	Khalaf Khatatneh and Thaer Mustafa	Fuzzy Logic/2009	Dataset from command and control applications, Musa, John D	Author developed model that can predict accurate results in the target database
7.	Aleksandar Dimov et al.	Fuzzy Logic/2010	Dataset from simple event based system	Based on fuzzy logic for reliability estimation of component based systems
8.	Sultan Alijahdali	Fuzzy Logic/2011	Real command and control project, Military and Operating system, John D Musa, Bell lab	Calculated Variance Accounted For (VAF) performance criterion. Based on Takagi-Sugeno technique

(continued)

Table 2 (continued)

Sr. No.	Author(s)/ Reference(s)	Technology/Algorithm and year	Data of previous projects	Summarization
9.	Yadav, Dilip Kumar et al.	Fuzzy Logic/2012	Twenty different datasets from real time systems	Author used software metric with fuzzy logic to predict reliability and compared lie result with existing models
10.	Sultan Alijahdali and Mohammed E. El-Telbany	Genetic Algorithm/2008	Data from three projects Military. Real System Control and Operating system	Measured the predictability of software reliability using ensemble of models winch performed better than the single model and also find that the weighted average combining method for ensemble has a better performance in a comparison with average method
11.	Sultan Alijahdali and Mohammed E. El-Telbany	Genetic Algorithm/2009	Data from three projects. They are Military, Real Time Control and Operating System	As far as the predictability of the single AR model and ensemble of AR models trained by GA algorithm over the trained and test data is concerned, the ensemble of models performed better than the single model. Also, found that the weighted average combining method for ensemble has a better performance in a comparison with average method
12.	Eduardo Oliveira Costa et al.	Genetic programming/ 2010	Dataset from John Musa bell telephone laboratories	Used genetic programming and boosting to improve the performance of GP. Less expensive techniques than GP and also there were similar results from both of them
13.	Zainab Al-Rahamneh et al.	Genetic programming/ 2011	CASRE tool is used for implementation	Calculated NRMSE values and used lilgp programming package for implementation in computer aided software estimation models (CASRE)
14.	C. G. Chua and A. T. C. Goh	Genetic algorithm + Bayesian neural network/2003	Data generated by two genera examples	Approach analyses nonlinear multi variant problems

(continued)

Table 2 (continued)

Sr. No.	Author(s)/ Reference(s)	Technology/Algorithm and year	Data of previous projects	Summarization
15.	C. Jin	Genetic algorithm + Stimulated annealing + Support vector machine (CSVM)/ 2011	Data of turbochargers and software failure is analyzed	A comparative study with different models is done with GA-SA-SVM model
16.	Jung Hua Lo	ARIMA + SVM/2011	General dataset used	Calculated MSE and MRE values
17.	Ramakanta Mohanty, V. Ravi, M. R. Patra	Genetic programming + Group method of data handling/ 2013	Data obtained from Musa and Iyer and Lee	Calculated NRMSE values which were better than other models
18.	Bonthu Kotaiah et al.	Neuro-fuzzy approach/ 2015	Java programming based evaluation	Evaluation criteria are RMSE values. Prediction using non parametric models in comparison to parametric model is better
19.	Jaydeep Pati and K. K. Shukla	ARIMA + ANN/2015	Real Time System Application (System 5), Military System Application (System 40), Word processing System Application (System SS3)	Calculated RMSE and MAE values. Good alternative solution to SRGM limitations
20.	Pravas Ranjan Bal, Nachiketa Jena and Durga Prasad Mohapatra	Feed forward neural network + Radial basis function/2017	Tested on three benchmark datasets	Software reliability Ensemble model based on two types Feed Forward Neural Networks and one Radial Basis Function Neural Network and Radial basis function Neural Network Ensembles (RNNE) model is developed and rested
21.	Jinyong Wang and Ce Zhang	Deep learning model based on RNN encoder–decoder/2017	DS1, DS2, DS3 and 11 others	Average Error (AE) and Average Bias (AB) is calculated

should choose a widely adaptable model for this purpose. In this paper we have investigated the performance of the previously developed software reliability growth and prediction models and their capabilities for predicting the software reliability. This paper presented the usages of the soft computing techniques for software reliability. An early phase software reliability prediction model is need of today's competitive world. This early phase reliability prediction model further helps developers in gaining the trust of user as well as in market of software.

References

1. Smidts, C., Stoddard, R. W., & Stutzke, M. (1998). Software reliability models: An approach to early reliability prediction. *IEEE Transactions on Reliability, 47*(3), 268–278.
2. Gokhale, S. S., & Trivedi, K. S. (1999). A time/structure based software reliability model. *Analysis of Software Engineering, 8,* 85–121.
3. Musa, J. D. (1999). *Software reliability engineering: More reliable software, faster development and testing*. McGraw-Hill.
4. Su, Y. S., Huang, C.-Y., Chen, Y. S., & Chen, J. X. (2005). An artificial neural-network-based approach to software reliability assessment. In *TENCON, IEEE Region 10* (pp. 1–6).
5. Hu, Q. P., Dai, Y. S., Xie, M., & Ng, S. H. (2006). Early software reliability prediction with extended ANN Model. In *Proceedings of the 30th Annual International Computer Software and Applications Conference* (pp. 234–239).
6. Su, Y.-S., & Huang, C.-Y. (2006). Neural-network-based approaches for software reliability estimation using dynamic weighted combinational models. *Journal of Systems and Software, 80*(4), 606–615.
7. Aljahdali, S. H., & Buragga, K. A. (2008). Employing four ANNs paradigms for software reliability prediction: An analytical study. *ICGST-AIML Journal, 8*(II). ISSN: 1687-4846.
8. Aljahdali, S. (2011). Development of software reliability growth models for industrial applications using fuzzy logic. *Journal of Computer Science, 7*(10), 1574–1580.
9. Al-Rahamneh, Z., Reyalat, M., Sheta, A. F., Bani-Ahmad, S., & Al-Oqeili, S. (2011). A new software reliability growth model: Genetic-programming-based approach. *Journal of Software Engineering and Applications, 4,* 476–481.
10. Karunanithi, N., Malaiya, Y. K., & Whitley, D. (1991). Prediction of software reliability using neural networks. In *Proceedings of the Second IEEE International Symposium on Software Reliability Engineering* (pp. 124–130), 1991.
11. Aljahdali, S. H., & El-Telbany, M. E. (2008). Genetic algorithms for optimizing ensemble of models in software reliability prediction. *ICGST-AIML Journal, 8*(I).
12. Aljahdali, S. H., & El-Telbany, M. E. (2009). Software reliability prediction using multi-objective genetic algorithm. 978-1-4244-3806-8/09/$25.00, IEEE, 2009.
13. Oliveira, E., Pozo, A., & Vergilio, S. (2006). Using boosting techniques to improve software reliability models based on genetic programming. In *ICTAI'06: Proceedings of the 18th IEEE International Conference on Tools with Artificial Intelligence*, Washington, USA, IEEE Computer Society, 2006.
14. Huang, C. Y., & Lyu, M. R. (2011). Estimation and analysis of some generalized multiple change-point software reliability models. *IEEE Transaction on Reliability, 60*(2), 498–514.
15. Bisi, M., & Goyal, N. K. (2012). Software reliability prediction using neural network with encoded input. *International Journal of Computer Applications (0975–8887), 47*(22).
16. Aljahdali, S., & Debnath, N. C. (2004). *Improved software reliability prediction through fuzzy logic modeling* (pp. 17–21). IASSE.
17. Cai, K. Y., Wen, C. Y., & Zhang, M. L. (1991). A critical review on software reliability modeling. *Reliability Engineering and System Safety, 32*(3), 357–371.

18. Khatatneh, K., & Mustafa, T. (2009). Software reliability modeling using soft computing technique. *European Journal of Scientific Research, 26*(1), 147–152. ISSN 1450-216X.
19. Zhang, Y., & Chen, H. (2006). Predicting for MTBF failure data series of software reliability by genetic programming algorithm. In *Proceedings of the Sixth International Conference on Intelligent Systems Design and Applications*, Washington, USA, IEEE Computer Society, 2006.
20. Costa, E. O., Pozo, A. T. R., & Vergilio, S. R. (2010). A genetic programming approach for software reliability modeling. *IEEE Transactions on Reliability, 59*(1).
21. Dimov, A. (2010). Fuzzy reliability model for component-based software systems. In *36th EUROMICRO Conference on Software Engineering and Advanced Applications* (pp. 39–46), IEEE.
22. Benaddy, M., & Wakrim, M. (2012). Simulated annealing neural network for software failure prediction. *International Journal of Software Engineering and Its Applications, 6*(4).
23. Yadav, D. K., Chaturvedi, S. K., & Misra, R. B. (2012). Early software defects prediction using fuzzy logic. *International Journal of Performability Engineering, 8*(4), 399–408.
24. Chua, C. G., & Goh, A. T. C. (2003). A hybrid bayesian back-propagation neural network approach to multivariate modeling. *International Journal for Numerical and Analytical Methods in Geomechanics, 27,* 651–667.
25. Kumar, R., Khan, S. A., & Khan, R. A. (2015). Durable security in software development: Needs and importance. *CSI Communications, 10*, 34–36.
26. Mohanty, R., Ravi, V., & Patra, M. R. (2013). Hybrid intelligent systems for predicting software reliability. *Applied Soft Computing, 13*(2013), 189–200.
27. Pati, J., & Shukla, K. K. (2015). A hybrid technique for software reliability prediction. In *ISEC'15*, February 18–20, 2015.
28. Sahu, K., Rajshree, Kumar R. (2014). Risk Management Perspective in SDLC. *International Journal of Advanced Research in Computer Science and Software Engineering, 4*(3), pp. 1247–1251, March, 2014.
29. Sahu, K., Rajshree. (2015). Stability: Abstract Roadmap of Software Security. *American In ternational Journal of Research in Science, Technology, Engineering & Mathematics, 2*(9), pp. 183–186.
30. Kumar, R., Khan, S. A., Alka & Khan, R. A. (2018). Measuring the Security Attributes through Fuzzy Analytic Hierarchy Process: Durability Perspective, ICIC Express Letters-An. *International Journal of Research and Surveys, 12*(6), June 2018.
31. Kumar, R., Khan, S. A., Alka & Khan, R. A. (2018), Security Assessment through Fuzzy Delphi Analytic Hierarchy Process, ICIC Express Letters-An *International Journal of Research and Surveys, 12*(10), October 2018.
32. Available Online at: https://www.iso.org/obp/ui/#iso:std:iso-iec:25010:ed-1:v1:en.
33. Jin, C. (2011). Software reliability prediction based on support vector regression using a hybrid genetic algorithm and simulated annealing algorithm. *The Institution of Engineering and Technology, 5*(4), 398–405.
34. Lo, J.-H. (2011). A study of applying ARIMA and SVM model to software reliability prediction. In *International Conference on Uncertainty Reasoning and Knowledge Engineering*, 2011, 978-1-4244-9983-0.
35. Bal, P. R., Jena, N., & Mohapatra, D. P. (2017). Software reliability prediction based on ensemble models. In *Proceeding of International Conference on Intelligent Communication, Control and Devices* (pp. 895–902). Singapore: Springer.
36. Wang, J., & Zhang, C. (2017). Software reliability prediction using a deep learning model based on the RNN encoder–decoder. *Reliability Engineering & System Safety*.
37. Kumar, R., Khan, S. A., & Khan, R. A. (2016). Durability Challenges in Software Engineering. *Crosstalk-The Journal of Defense Software Engineering*, 29–31.

Application of Classification Techniques for Prediction of Water Quality of 17 Selected Indian Rivers

Harlieen Bindra, Rachna Jain, Gurvinder Singh and Bindu Garg

Abstract Objective: In this study, prediction using classification techniques are used to predict the water quality of the 17 selected rivers in the year 2011 using their water quality in 2008 to interpret whether the water quality has improved or deteriorated. Methods/Analysis: For this prediction, we have used data mining classification techniques using Waikato Environment for Knowledge Analysis (WEKA) API to the dataset of selected 17 Indian rivers. The data used for prediction was created from ambient water quality of Aquatic Resources in India in 2008 and 2011. Data is obtained from data portal which was published under National Data Sharing and Accessibility Policy (NDSAP) and the contributor was Ministry of Environment and Forests Central Pollution Control Board (CPCB). Findings: Out of the four techniques used, prediction of classes, i.e. excellent, good, average and fair is best done by Naive Bayes followed by J48, SMO and REPTree technique.

Keywords Prediction using classification techniques · Weka · Data mining
Water quality · Indian rivers

1 Introduction

India is popularly referred to as the land of rivers since it has been blessed with several water bodies which not only enhance the beauty of the country but is also the source of livelihoods for a large number of people. They are the main sustainability source for people especially the farmers since the soil lands in proximity

H. Bindra (✉) · R. Jain · B. Garg
CSE Department, Bharati Vidyapeeth's College of Engineering, New Delhi, India
e-mail: harlieenbindra@gmail.com

R. Jain · G. Singh
CSE Department, Guru Tegh Bahadur Institute of Technology, New Delhi, India

V. E. Balas et al. (eds.), *Data Management, Analytics and Innovation*,
Advances in Intelligent Systems and Computing 808,
https://doi.org/10.1007/978-981-13-1402-5_18

to the rivers are nourished and fertile. For many holy reasons, these rivers are worshipped in India; specially "The Ganges" which is considered to be the holiest of all.

Indian Rivers not only nourish the flora and fauna but also attract tourist from all around the world and play an indispensable role in our economy. They are the witness of how the civilisation evolved but they are not only significant historically but also culturally and religiously. Even their inherent nature could not be altered by the dams. They still originate from the mountains and gush down the plains and valleys with the same force as several years ago. They nourish the plain with vitality and fertility.

But people day by day are forgetting the importance of rivers. The rivers now have fertilisers, pesticides and more different types of chemical products. A number of instances oil spills have disturbed the aquatic animals. The banks of the rivers are piled up with non-biodegradable wastes. But we need to understand that improving the unhygienic and dirty conditions of the rivers is not the sole responsibility of the government. We as the citizens of this nation should take special precautions and actions to improve the water quality of rivers. Even in western countries, the citizens themselves take measures to keep their rivers and river banks clean. We must strive to keep the best gift of nature clean and preserve its water quality.

For present study, 17 rivers were selected for prediction of water quality in 2011 using 2008 instances. The number of stations used to collect data for each river is mentioned in the parenthesis, which are

Beas (19), Satluj (20), Ganga (36), Yamuna (19), Brahmaputra (10), Dhansiri (7), Mahi (7), Narmada (6), Tapi (10), Mahanadi (14), Brahmani (11), Baitarni (5), Subarnarekha (6), Godavari (34), Krishna (22), Pennar (4), Cauvery (20). So, in total 250 instances were used for analysis.

In this paper, there are seven sections. Section 1 is the Introduction which is the current section. The Sect. 2 is Literature Review, the Sect. 3 is Materials and Methods, the Sect. 4 is Performance Comparison, the Sect. 5 is Result and Discussion, the Sect. 6 is Conclusion, seventh is Acknowledgement and the last section is References.

2 Literature Review

In paper [1], performance of CART, J48, REPTREE, Bayes Net and Naïve Bayes classification algorithms are compared by applying them to a dataset consisting of only 11 attributes, for predicting heart attacks. In the research work algorithms for prediction are applied using WEKA as it provides proficiency in analysing, discovering and predicting patterns. The results of the paper helped us in concluding that J48, CART and REPTREE shows the best results and there is not much difference in their performance factor. In paper [2], the author has compared the results of two decision trees ID3 and J48. The two techniques are applied to a dataset of students enrolling for MCA. The research work explains how tree based classification algorithms ID3 and J48 works and are used to analyse the data. From the results it can be concluded that ID3 decision tree algorithm shows an accuracy of 69.69% as compared to that of J48 which is 67.67%. In paper [3] the two data mining algorithms which are used for producing the classification model are Naive Bayesian Classifier algorithm and Decision Tree algorithms. These algorithms are applied on preprocessed student dataset. Decision Tree algorithms has an accuracy of 93.33% over 71.67% of Naive Bayesian Classifier algorithm. Hence decision tree algorithm proves to be better than naïve Bayesian classifier. In paper [4], the author has explained about the heart diseases and symptoms of heart attack. The paper has talked about various models that are developed using different data mining techniques. In paper [5] has bestowed satisfactory modifications for calculation of the water quality index (WQI). To calculate the general water quality index nine parameters are required but sometimes a few parameters are missing or unavailable, in that case the modified formula given in this paper helps user to calculate WQI.

National River Conservation Plan [6] was initiated with the launching of Ganga Action Plan (GAP) in 1985. In 1995 GAP was expanded to cover other rivers of the country. At a sanctioned cost of Rs. 5779.41 crore, NRCP, excluding the GAP-I, GAP-II and National Ganga River Basin Authority (NGRBA) programme presently covers polluted stretches of 40 rivers in 121 towns spread over 19 States running head, it will be shortened. Your suggestion as to how to shorten it would be most welcome.

3 Materials and Methods

For present study, the data set was created using the data that referred to the ambient water quality of Aquatic Resources in India in 2008 and 2011 [7]. This Dataset is released under "National Data Sharing and Accessibility Policy (NDSAP)" and the contributor is "Ministry of Environment and Forests and Central Pollution Control Board". The values of water quality parameters like Fecal Coliform, Temperature, Nitrate, Biochemical Oxygen Demand (B.O.D), pH, etc. were given in the data used. The data was published on the data portal on December 22, 2014 which was released under National Data Sharing and Accessibility Policy (NDSAP) [8] and the contributor was Ministry of Environment and Forests Central Pollution Control Board [9].

In classification [10] a set of objects is classified into a group so that objects in a group are more similar to each other.

3.1 Classification Techniques Used

3.1.1 Naïve Bayes [11]

Naïve Bayes Classifier is a part of probabilistic classifier based on application of Bayes' Theorem with strong independent presupposition/presumption between the features. This classifier algorithm presumes that in a given class attribute values are independent of other attributes values.

3.1.2 J48 [12]

J48 is the appendage od ID3. The features of J48 are decision tree pruning, keeping accounts for missing values, derivation of rules, etc. In WEKA, the implementation for JAVA open source algorithm C4.5 is done using J48. For tree pruning a number of options are provided by WEKA (data mining tool). Pruning could be employed as a mechanism for précising if there is case of over fitting. The aim of this algorithm is to progressively generalise the decision tree till accurate and flexible tree is obtained. Continuous and discrete attributes can be handled by this algorithm.

3.1.3 SMO (Sequential Minimal Optimization) [13]

SMO stands for Sequential Minimal Optimization, John Platt invented this algorithm in 1998 at Microsoft Research. This algorithm is mainly used for solving quadratic programming problem which emerges at time of programming support vector machine. Nominal attributes are transformed into binary ones on implementation of the model. Also, by default it normalises all attribute. The worst case running time for this is $O(n^3)$.

3.1.4 REPTree [14]

REP stands for Reduce Error Pruning. This algorithm is based minimising error surfacing form variance and calculating the information gain using entropy. REP Tree generates various trees in reordered iterations and uses regression tree logic. It splits the missing values into pieces of corresponding instances.

3.2 *Software Used*

Eclipse [15] open source IDE (Integrated Development Environment) was used to compile the code wrote using Waikato Environment for Knowledge Analysis (WEKA) [16] (developed by the University of Waikato, New Zealand) API. It is commonly used for data mining works, as it has a number of machine learning algorithms. It has tools for preprocessing, classification, visualisation, etc. Eclipse is an IDE mostly used for computer programming in Java language but it also supports many other programming languages.

3.2.1 General Algorithm

1. Training dataset is loaded.
2. The class index is set to the last attribute.
3. Number of classes is fetched.
4. Class values in the training dataset is printed.
5. Class string value using the class index is fetched.
6. Creating and building the classifier.
7. The test dataset is loaded.
8. The class index is set to the last attribute.
9. Looping through the new dataset to make predictions.
10. Fetching class value for current instance.
11. Fetching class string value using the class index Class's int value is used.
12. Instance object of current instance is fetched.
13. Calling classifyInstance, which returns a double value for the class.
14. Use the double value to get string value of the predicted class.

3.2.2 Code Used

The following is the code used for Naïve Bayes Classifier. The classifier can accordingly be changed as per the requirement but the rest of the code will remain the same.

```
import weka.classifiers.bayes.NaiveBayes;
import weka.core.Instance;
```

```
import weka.core.Instances;
import weka.core.converters.ConverterUtils.DataSource;
public class RiversClassification
{
public static void main(String args[]) throws Exception{
DataSource source = new DataSource("C:\\Users\\Bindra\\Desktop\\data set\\-claasification\\train2008.arff");
Instances trainDataset = source.getDataSet();
trainDataset.setClassIndex(trainDataset.numAttributes()-1);
int numClasses = trainDataset.numClasses();
for(int i = 0; i < numClasses; i++){
String classValue = trainDataset.classAttribute().value(i);
System.out.println("Class Value "+i+" is " + classValue);
}

//Classifier used is Naïve Bayes here
NaiveBayes nb = new NaiveBayes();
nb.buildClassifier(trainDataset);
DataSource source1 = new DataSource("C:\\Users\\Bindra\\Desktop\\data set\\-claasification\\test2011.arff");
Instances testDataset = source1.getDataSet();
testDataset.setClassIndex(testDataset.numAttributes()-1);
System.out.println("===================");
System.out.println("Actual,NB Predicted Class");
for (int i = 0; i < testDataset.numInstances(); i++) {
double actualClass = testDataset.instance(i).classValue();
String actual = testDataset.classAttribute().value((int)actualClass);
Instance newInst = testDataset.instance(i);
double predNB = nb.classifyInstance(newInst);
String predString = testDataset.classAttribute().value((int) predNB);
System.out.println(actual+", "+predString);
        }
  }
}
```

4 Performance Comparison

Below is the performance of various classification techniques on the used data set (Table 1).

Table 1 Percentage error in classification techniques applied for analysis of water quality of rivers

Classification technique	Incorrectly classified elements (%)
Naïve Bayes	46.667
J48	60
SMO	66.667
REPTree	73.333

Fig. 1 Output of the code for Naïve Bayes classifier

The table consists of two columns, techniques and error percentage. Naïve Bayes show the best result and maximum error is found in REPTree technique.

Below are the screenshots of the outputs when the code was run on Eclipse IDE using different classifiers (Figs. 1, 2, 3, and 4).

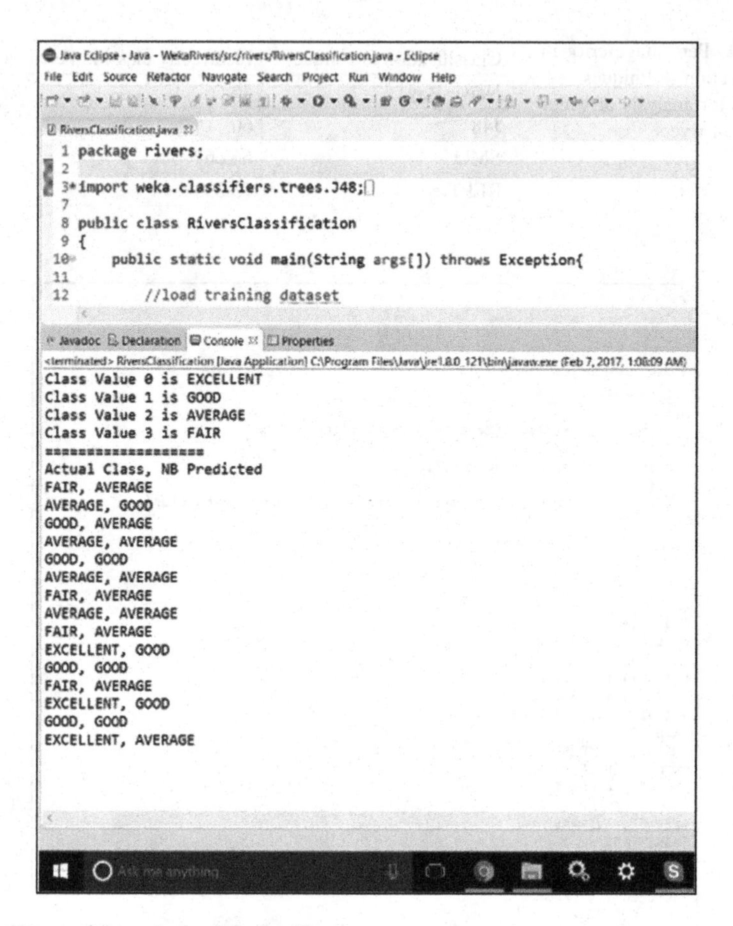

Fig. 2 Output of the code for J48 classifier

5 Result and Discussion

In this work, Naïve Bayes has proved to be the best technique with minimum error. The error percentage in the classification techniques we have applied in our analysis is high because the data set which is used as input is biased since in the dataset

Fig. 3 Output of the code for SMO classifier

number of instances in average and good water quality groups are greater than the number of instances in fair and excellent groups. Another reason for high error percentage is that the number of instances in excellent, good, fair and average water quality groups are not same.

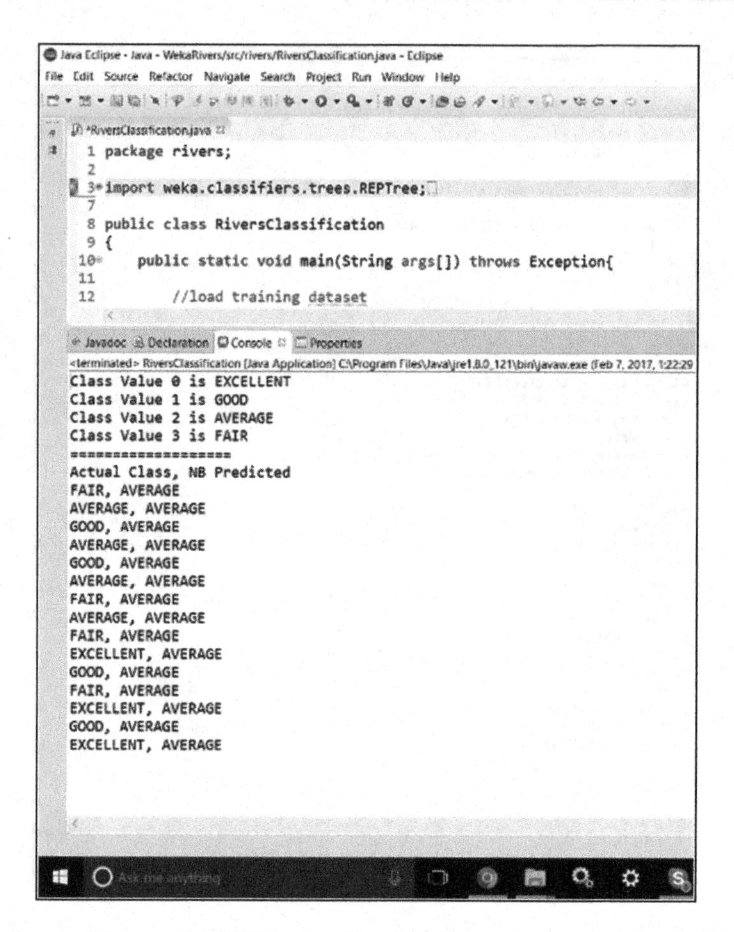

Fig. 4 Output of the code for REPTree classifier

6 Conclusion

Out of all the classification techniques, we have applied on our dataset, Naive Bayes has shown the results with least error.

In future, for effective and accurate analysis, some modifications need to be applied in these predefined classification techniques or new classification techniques need to be devised in order to form correct classes of such biased datasets.

Acknowledgements I profoundly thank Bharati Vidyapeeth's College of Engineering for constant support and encouragement.

References

1. Masethe, H. D., & Masethe, M. A. (2014). Prediction of heart disease using classification algorithms. In *Proceedings of the World Congress on Engineering and Computer Science 2014* (Vol. II), October 22–24, 2014, San Francisco, USA.
2. Saini, P., & Jain, A. K. (2013). Prediction using classification technique for the students' enrollment process in higher educational institutions. *International Journal of Computer Applications (0975–8887), 84*(14).
3. Padmapriya, A. Dr. (2012). Prediction of higher education admissibility using classification algorithms. *International Journal of Advanced Research in Computer Science and Software Engineering, 2*(11).
4. Sudhakar, K., & Manimekalai, M. Dr. (2014). Study of heart disease prediction using data mining. *International Journal of Advanced Research in Computer Science and Software Engineering, 4*(1).
5. Srivastava, G., & Kumar, P. (2013). Water quality index with missing parameters. *IJRET: International Journal of Research in Engineering and Technology, 02*(04), 609–614.
6. National River Conservation Directorate (NRCD) http://envfor.nic.in/division/national-river-conservation-directorate-nrcd. Date accessed on 30/9/2016.
7. Data Set https://data.gov.in/catalog/status-water-quality-india-2008-and-2011.
8. NDASP http://www.dst.gov.in/national-data-sharing-and-accessibility-policy-0.
9. Ministry of Environment and Forests https://data.gov.in/ministrydepartment/ministry-environment-and-forests.
10. Sujatha, M., Prabhakar, S., & Lavanya Devi, G. Dr. (2013). A survey of classification techniques in data mining. *International Journal of Innovations in Engineering and Technology (IJIET), 2*(4). ISSN 2319-1058.
11. Bhargavi, P., & Jyothi, S. Dr. (2009). Applying Naive Bayes data mining technique for classification of agricultural land soils. *IJCSNS International Journal of Computer Science and Network Security, 9*(8).
12. Patil, T. R., & Sherekar, S. S. Mrs. (2013). Performance analysis of Naive Bayes and J48 classification algorithm for data classification. *International Journal of Computer Science and Applications, 6*(2). ISSN 0974-1011.
13. Platt, J. C. (1998). *Sequential minimal optimization: A fast algorithm for training support vector machines* (Technical Report MSR-TR-98-14), April 21, 1998.
14. Kalmegh, S. (2015). Analysis of WEKA data mining algorithm REPTree, simple cart and RandomTree for classification of Indian News. *IJISET—International Journal of Innovative Science, Engineering & Technology, 2*(2). ISSN 2348–7968.
15. Eclipse IDE http://www.eclipse.org/users/. Date accessed on 11/2/2017.
16. Weka website (Latest version 3.6) http://www.cs.waikato.ac.nz/ml/weka/. Date accessed on 30/9/2016.

Trends in Document Analysis

Vaibhav Khatavkar and Parag Kulkarni

Abstract Document analysis is one of the emerging area of research in the field of Information Retrieval. Many attempts have been made for retrieving information from a document using various machine learning algorithms. A concept of context vector is frequently used in information retrieval from document/s. Context Vector is an vector, which is used for various feature selection from documents, automatic classification of text documents, Subject Verb Agreement, etc. This paper discusses, the attempts made in the field of Information Retrieval (IR) from document using context vector. It also discuss about pros and cons of each attempt. This paper propose a system which can give "context vector" of the document set using Latent Semantic Analysis which is the most trending method in document analysis. The system is tested on BBC news dataset and proves to be successful.

Keywords Information retrieval · Context vector · Document analysis
Machine learning · Subject verb agreement

1 Introduction

Due to digitization, the use of Internet has tremendously increased. Along with the usage of the Internet, many documents have been created. The structure of documents varies case to case. A document can be a web document, a text document, an image and so on. This survey is done with respect to text documents and web documents. Various machine learning algorithms have been applied depending on the type of document. Many attempts are made in document analysis; to be precise information retrieval (IR) of documents. Some of them use supervised, unsuper-

---

V. Khatavkar (✉) · P. Kulkarni
Department of Computer Engineering and Information Technology,
College of Engineering, Pune, Shivajinagar, Pune, Maharashtra, India
e-mail: vkk.comp@coep.ac.in

P. Kulkarni
e-mail: paragindia@gmail.com

V. E. Balas et al. (eds.), *Data Management, Analytics and Innovation*,
Advances in Intelligent Systems and Computing 808,
https://doi.org/10.1007/978-981-13-1402-5_19

vised and semi-supervised learning algorithms in their work. The attempt of using context vector is also done along with the above stated algorithms. For example, Harris et al. [1] analyse tweets by using twitter APIs and also carry out sentiment analysis of students using moodle. Haribhakta et al. [2] use a tool called "GATE" to extract necessary entities from a document for text categorization. Ye et al. [3] use multi-label classifier namely random k-label sets classifier to classify news articles in various categories. They also came across with terminology called "Reader Perspective" and "Writer Perspective".

The definition of context vector differs application to application. There are various methods to generate context vector. Some of them are term frequency (TF), Term Frequency and Inverse Document Frequency (TF-IDF), Similarity measures like cosine, euclidean distance. Many a times a semantic analysis is also considered while calculating Context Vectors. Khatavkar et al. in their work, [4], proposed "Context Vector Machine" which makes use of context vector for document classification. This paper states the use of context vector in IR of document. This paper is an attempt to find out context vector which is proposed in [4]. An attempt has been made by Khatavkar et al. in [5], to use context vectors in document identification by using Latent Semantic Analysis (LSA). This work is extension to their work. They shown the probability of finding the term in the dataset after the application of LSA algorithm. This paper takes a review of literature and an attempt is made to define context vectors for the given document set.

Section 3 gives usage of Context Vectors in document analysis. Section 4 discusses the proposed system followed by Experiments and results. Section 6 concludes the work.

2 Literature Review

When document analysis is considered the two basic concept from machine learning come into picture, which are "Document Clustering" and "Document Classification". In terms of machine learning when unsupervised approach is applied on the Documents then we call it Document Clustering while when supervised approach is applied on the Documents then it is called Document Classification. Researchers like Han et al. [6], Jain et al. [7], Steinbach et al. [8], Berkhin et al. [9] and Xu et al. [10] document clustering and clustering algorithms. On the other hand, Document Classification is nothing but to predict category of the document to a particular class [2, 11].

When we classify text using supervised approach, typically we use a large labelled dataset for training the system and then apply it on the unlabeled dataset. It is observed that use of the training dataset in this fashion is costly. Thus Nigam et al. [12] developed a model where a small labelled dataset and a large amount of unlabeled dataset are used for training. The model uses Expected Maximization (EM) algorithm along with naive Bayes classifier. It has been observed that the error rate in classification has been reduced. The recent approaches in text classification use two variants of Naive Bayes assumption. Mccallum et al. in [13] have discussed two variants of Naive Bayes viz. Multi-variate Bernoulli Model and Multinomial Model.

Applying these on five different datasets they concluded that the multinomial event model performs better than multi-variant Bernoulli. Nigam et al. in their paper [14] mentioned that they use Query-by-Committee (QBC) approach of active learning to address the same problem. They use EM and density-weighted pool-based sampling to reduce the need of labelled data. In a pattern recognition data clustering is the core method. Puzicha et al. [15] developed a system which has a systematic approach for clustering proximity which is also called similarity data. They have used deterministic annealing and mean-field approximation. In Information Retrieval (IR), the most frequently used models are Vector Space, Probabilistic Model, Inference Model, Term Frequency and Language Model. There are some limitations of using these models. Boughanem et al. [16] in their work proposed and experimented an IR model based on possibilistic logic. For finding similarity between documents, the most widely used algorithm uses Term Frequency (TF) and Inverse Document Frequency (IDF). However, there are some limitations of using TF-IDF. In order to overcome them Latent Semantic Analysis (LSA) and probabilistic version of Latent Semantic Analysis (pLSA) are used. These methods came up with an idea of representing probabilistic models in terms of graphical models. Salakhutdinov et al. [17] address the same using the fast inference model. They have used a class of two-layer undirected graphical models. Haribhakta et al. [18] have proposed unsupervised model to detect a topic of the document. They have used Term Frequency (TF) measure to find the keywords. Using standard dataset, they proved that their model is effective in terms of time required for the classification. Haribhakta et al. [2] had also used a GATE tool to extract necessary entities from a document for text categorization. They have used three measures for feature selection viz. Term Frequency (TF), Information Gain (IG) and Chi-square (X^2). They have applied it to standard datasets and shown the improvement in the accuracy of classification. Dou et al. [11] have worked on query facets which can be mined automatically. A query facet can be a word or a set of words which describe and summarise the query from different perspectives. They have developed a system called as QDMiner in which for a given query q, they retrieve top K results from the search engine and fetch all documents to form an input as a set of R. Then, the query facets are mined in following steps:

- List and Context Extraction,
- List Weighting,
- List Clustering,
- Facet and Item Ranking.

As a conclusion of their work, they conclude with aggregating frequent lists from free text, HTML tags, and repeat regions within top search. They also create two human annotated datasets and apply various metrics to them to evaluate the quality of query facet.

Even after applying clustering or classification on document, some times conceptual analysis is also done to find topic of a particular document. Huang et al. [19] worked on clustering of documents with active learning on Wikipedia. Conventionally, text documents are represented using Bag of Words (BOW) which lags because of not using semantic relationships. Therefore, they propose to represent documents using concepts rather than words. Meena et al. [20] reviews features for sentence scoring. Sentence Scoring is used in automatic text summarization. They have identified various sentence scoring features and used seven combinations of them. One them was found impressive. Chiang [21] proposes and experiments a fuzzy clustering algorithm which is used to discover the latent semantics in a text corpus from a fuzzy linguistic perspective. It was used to evaluate:

- Topic relevance in a document,
- Difference between other topics.

Similarly, thematic relationship among documents is also important since it represents theme of the sets of documents. Hatzivassiloglou et al. [22] focuses on finding common information between two small textual unit and extends it to find thematic group of the text across multiple documents. They use primitive and composite features to define similarity and then test with various algorithms in machine learning. Use of Latent Sematic Analysis (LSA) is done to map the documents and the terms in Latent Semantic Space. Hofmann in his paper [23] proposes a variant of LSA which uses probabilistic approach; it is called Probabilistic Latent Sematic Analysis (PLSA). He has fitted the corpus of text documents for training using Expectation Maximization (EM) algorithm. Hofmann, in his another paper [24], has described Cluster-Abstraction Model (CAM) which extracts hierarchical relations between groups documents and an abstractive organisation of keywords. He has derived annealed version of Expectation Maximization (EM) algorithm. Huang et al. [25] uses thematic representation of the Wikipedia documents. Wikipedia was used to create Bag of Concepts (BOC). It incorporates the semantic connections among concepts into a document similarity measure. Latent Sematic Indexing (LSI) and Independent Component Analysis (LCA) are used along with Term Frequency-Inverse Document Frequency (TF-IDF) and Cosine measure.

Some documents are web documents. In order to analyse them work related to ontological document analysis is done. Ontology plays an important role in web semantics. Lee et al. [26] have presented conceptual resonance and a parallel fuzzy mechanism for conceptual resonance computing. Typically for ontology construction, taxonomic aspect is the most frequently used in the knowledge acquisition process. Rushall et al. [27] have presented a system which learn from non-taxonomic relationship from the web documents. Dahab et al. [28] have developed a system called "TextOntoEx" which is a link between linguistic analysis and ontology engineering. The aim of this system is to find out potential interesting concepts and relationships between them automatically. Villaverde et al. [29] proposed a system which finds relationships using ontology learning based on non-taxonomy. They use Vector Space Model (VSM) for the discovery of non-taxonomic relationships, Stanford Part-of-Speech (POS) tagger and then association rule mining algorithm.

They then extract lexical items as connectors among related concepts. Many a times for finding out Word Sense Disambiguity (WSD) using unsupervised algorithm, pairwise similarity matching is been done. In [30], researchers, Abdalgar s et al. proposed an unsupervised algorithm which will compute semantic similarity between target word and context vector. The system was tested on three benchmark datasets namely: SemCor corpus, Senseval-2 dataset, Senseval-3 dataset. The system shows better results than pairwise similarity matching method.

In [31], researchers, Oh et al. worked on using supervised approach for Word Sense Disambiguation. The researchers represented each sense of a word as a vector in word space. Then during training found Contextual Words which were tagged and represented as Context Vectors.

Using local density of a word in the context sense vector are weighted. Now for the target word, its sense, static as well as dynamic, is represented and is considered as centroid of the context vectors in word space. The system was tested on English SENSEVAL dataset and produced relatively good results.

3 Use of Context Vector in Documents: Current Trend

When we have a look at literature review, we conclude that, Context Vectors in Document Classification, Weighting Schemes in Documents and Applications in various domains. Usage of Context Vectors is trending topic in document analysis.

3.1 Document Classification and Context Vectors

Jennifer Farkas in his work [32] makes an attempt of improving classification accuracy in automatic text processing. Central idea is to build a Artificial Neural Network (ANN) with weights and hierarchical relationships among terms. The ANN uses sigma function as "Context Vector" with back propagation algorithm. This ANN is called as NeuroFile. The process of creating context vector of a document, *v(D)*, is totally automated. The context vector of a document *D* is calculated using Eq. 1:

$$v(D) = \frac{w_i}{\|\sum w_i\|} \tag{1}$$

where $\sum w_i$ is sum of vectors of the form $w_i = \langle v_1, \ldots, v_n \rangle$, in which each coordinate v_i corresponds to a unique thesaurus term t_i. Here the researchers have used thesaurus. The thesaurus is based on the terms from the selected document space

Table 1 Condition of term and respective vector calculation

Condition of t_i in document D	V_i for document D
If thesaurus term t_i does not occur in D	$0 * \mathrm{IDF}_i$
If thesaurus term t_i is Leading Term in D	$1 * \mathrm{IDF}_i$
If thesaurus term t_i is Narrower Term in D	$0.5 * \mathrm{IDF}_i$
If thesaurus term t_i is Broader Term in D	$0.2 * \mathrm{IDF}_i$

and the hierarchical relationship between them. The hierarchical relationships for terms are built from the terms which are related with each other. Using thesaurus, the researchers found out Leading Term and related to it, Broader Terms, Narrower Terms, Related Terms. The formulation is given in Table 1.

After Calculating context vectors, the Inverse Document Frequency (IDF) of a term was calculated using Eq. 2:

$$\mathrm{IDF}(t) = \ln(n) - \ln(f_k) + 1 \tag{2}$$

where n is number of document set, and f_k is number of documents where term t occurs at least once. For more accuracy, stemming and stop word removal is performed before calculating the context vector.

The researchers used back propagation algorithm with three layers in feed forward neural network having optimal parameters. The evaluation function used is given in Eq. 3.

$$\mathrm{Eval}(g) = \frac{\max_i \text{ of correctly classified test documents}}{\text{Total number of test documents}} \tag{3}$$

with g ranging over 80 generations and i over 12 individuals for each generation.

The system was trained with 600 documents and tested by 400 documents. All 1000 documents were provided by the Transport data over five classes. The tests were performed on test documents with various cycles and iterations. The result for each class was calculated based on three important factors: recall, precision and correlation. The systems proved to be effective when compared with NeuroClass. The system is trained after 21 cycles with comparable global classification accuracy of 77.8%. Though systems shows good results, it do not address the problem with more number of documents like news datasets, social network datasets. The context of the document according to researcher are decided by the leading terms.

The researchers, David et al. in [27], developed a system called as Depict which enables user to visualise information using two technologies namely Context Vectors and Neural Networks.

In neural networks the researchers have used Kohonen Self Organising Maps (SOM) which is effectively used to represent information in suitable visualisation.

According to the researchers, the context vector is representation of terms, documents and queries with high dimensional vectors which consists or real value numbers or components. These vectors are unit vectors in high dimensional vector space. The closeness in the space will give closeness in the subject content. SOM, once trained with the context vectors can determine the context of the term, document and query during testing. It will give the direction vector of term, document or query during testing. Any point which lies on the this vector will be similar to the term, document or query which is used while testing.

Context Vector technology is highly effective since it can be used in text retrieval, text routing and image retrieval. When the researchers, tested the system on single processor the computation was highly intensive because of high dimension of the vectors given to SOM. In order to overcome this problem they have used SIMD numerical array processors (SNAP). After the usage of SNAP the computational intensity was less.

The overall system designed can be summarised in Algorithm 1. The advantage of this system is that it gives theme of a terms, words, document or any free text using context vectors. The system was tested over corpus of AP news wire data of 4 months span in 1990. If the data contains timestamp then the system can be used for temporal analysis. The temporal analysis is analysis of data over the period of given time frame.

The researchers, Zhang et al. in [33] combined Recurrent Neural Network Model (RNNM) with the Context Vector Feature. They train the system based on data of long span context vector. In order to improve the prediction, they have used Part-Of-Speech (POS) tagging and Topic Modelling. Skip-gram is used to convert words into vectors. This build a context vector in the system. The context vector is applied on RNNM and we get the results. The system is tested on standard dataset Penn Treebank. Significant improvements had been observed with the above system on Wall Street Journal speech recognition system.

Algorithm 1 *Overall System Design of SNAP*

Step 1: Take the train dataset as input,
Step 2: Perform preprocessing, i.e., Remove stop words and stem the words,
Step 3: Learn stem context vector with the preprocessed data in order to form database of stemmed context vectors,
Step 4: Take the test dataset as input,
Step 5: Perform preprocessing, i.e., Remove stop words and stem the words,
Step 6: Use of database of stemmed context vectors to learn document context vector from test dataset and find document context vector,
Step 7: Store the document context vector into a database called document context vector,
Step 8: Visualise document stem vector using SOM.

3.2 Weights and Context Vectors

Salehi et al. in [34] states the application of text window as context and term frequencies as weighing method. The context vector is a vector with the dimension of the vocabulary size. The vector will represent co-occurrence count of the word with the target word in vocabulary.

Sharma et al. in [35] uses Vector Space Model along with concept and context for IR. For calculating weights for context vector there are two ways: static and dynamic. Static method of calculating weights are typically term document matrix methods like Latent Semantic Indexing (LSI), co-occurrence, correlation, etc. While in dynamic method of calculating weights context and various data distribution techniques are used. The researchers in paper, focus on this point and had made an attempt to use dynamic weighting scheme. In the method proposed by researchers here, use google APIs in order to get web contents. After getting web contents, the concept and context vectors are build and updated. The weighting scheme proposed is given in Algorithm 2. The work proves to be effective since the model considers context vector dynamically.

Algorithm 2 *Dynamic Weighting Scheme for concept vector and context vector*

1. Data Filtration,
2. Target Concept vector and Target Context Vector,
3. Dynamic weight calculation,
4. Calculate relation between concept and context.

3.3 Applications in Which "Context Vector" Is Used

The researchers, Wan et al. in [36] used gloss vector, i.e., annotated vector to find the semantic relationship between the documents. They used high-dimensional vectors for organising concepts in the documents. The context vector used in their research used second-order co-occurrence vectors ANN used wordnet dictionary for finding semantic ontology of the document.

Cosine similarity was used to find out the similarity between the two objects/ term/documents. It is 0 for identical objects while 1 for totally non-identical objects. Other relatedness measures are compared and an experimental evaluation against several benchmark sets of human similarity ratings is presented. The Context Vector measure is shown to have one of the best performances.

In this paper, researchers focused on context vector measure, in which context consisted of the gloss synsets in XWN format and all related synsets in WordNet. The researchers define relatedness of synsets using direct semantic relation, but also consider all hypernyms as a part of context.

Having information about context is important so researchers used context vectors which reacted into closeness of relations between the words. The context vectors were build using second-order vectors in the concept the space. The context was measured in terms of the sense. The context vectors were compared with the concept vectors. The Context vectors proved to be more useful than the concept vectors for word sense in documents.

The context vector is also used in application like forecasting. Researchers, Rongali et al. in [37], state the use of same in the work. They developed an application which empowers the utility of company to allocate resources optimally. The company works in forecasting demands in district heating and cooling systems. They used data analytical approach in combination with the context vectors in order to predict the energy consumption in Northern Sweden. The accuracy of prediction was up to 87%.

In document processing, we often come across collation of documents or words. If collation of word similarity is considered then its hard task to measure and quantify the process of collation. In [38], researcher, Kaufman et al. developed a system called "Vectile System". In this system the words are represented by encoding the environment in they occur in texts. This technique is used in order to reduce the unnecessary linguistic analysis. This technique provides sufficient information for building context vectors based on Word Space. The system was tested on corpus of New York Times articles which used co-occurrence count and showed promising results in IR. The vector space model (VSM) is typically used for ranking the documents using vector space. In [39], researcher, Massimo, made an attempt to use context along with VSM in order to improve the accuracy of ranking. The context vectors can be any basic vector consisting of context, such as, time, space or word meaning. The ranking is done based on VSM, i.e., using inner product of two vectors. But additionally the projection and distance of vector is considered which is nothing but the context vector. Since the context will change the projection and its distance will change which changes ranking of the document.

In [40], the researchers, Erk et al. address the task of computing vector space for meaning of term frequency, which varies according to context. According to the researchers, the existing computational model are robust towards vector-based computation of considering the meaning of the sentence, is crucial. So they proposed an integrated model which will consider meaning of term/word as context vector and syntax of the document. The results showed that "the model outperforms the state of art for modelling the contextual adequacy of paraphrases".

The researchers, Thater et al. in [41], used first and second-order vectors to develop a model which is based on dependency trees. The dependency trees are obtained from parsing the English Giga-word corpus. The context is determined as combination of word vectors which are grammatically co-related. The systems performs two different tasks, namely, paraphrase ranking and word sense similarity.

Fig. 1 Proposed system

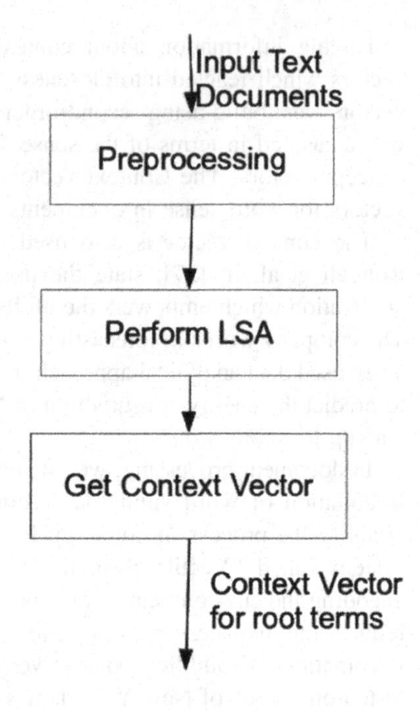

Maya Carrillo et al. in [42] represented context information for information retrieval using Bag of Concepts and Holographic Reduced Representation. They concluded with comparing results of their model with traditional vector space models. Their model gave 7% improvement in precision.

Milajevs et al. in [43] compared the quality of the distributional semantic NLP models against phrase-based semantic IR. They concluded with the observation that an IR model enriched with distributional linguistic information performs better in the long standing problem in IR of document retrieval where there is no direct symbolic relationship between query and document concepts.

4 Problem Statement and Proposed System

The problem statement can be defined as "to implement a system which can identify context vectors after LSA is applied on documents". It is depicted in Fig. 1.

The system will take document set as input and will preprocess the text data in document set. It will perform LSA on the preprocessed data. After application of LSA we will get the weight of each root term which can be stated as context vector.

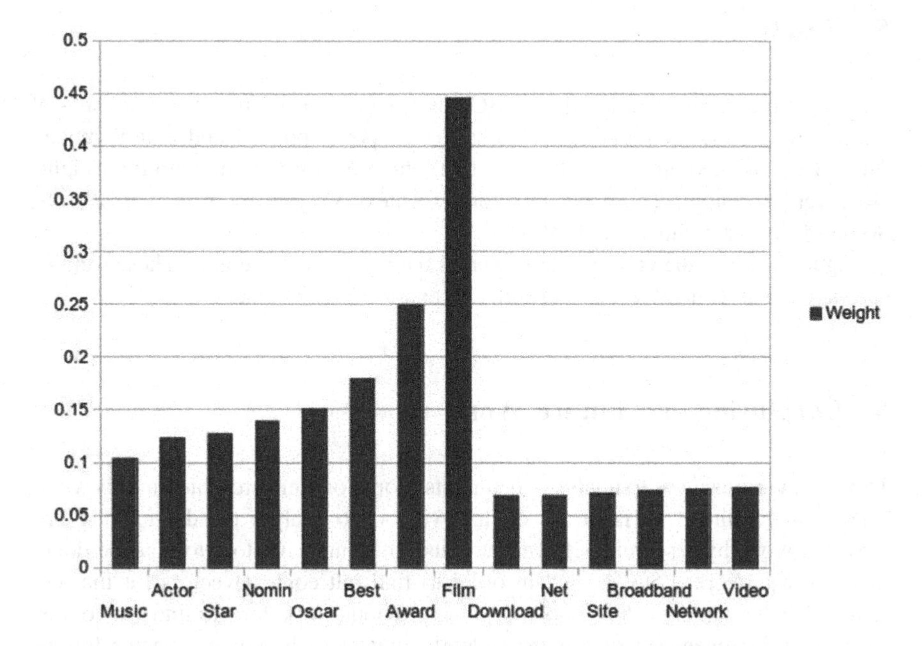

Fig. 2 Sample context vectors from BBC news dataset

Table 2 Context vectors

Term	Weight
Music	0.1036644819
Actor	0.1225405582
Star	0.1271796219
Nomin	0.1389425134
Oscar	0.150632962
Best	0.1792743336
Award	0.2492517543
Film	0.4451208228
Download	0.0682874384
Net	0.0683625163
Site	0.0703729296
Broadband	0.073135677
Network	0.0744756498
Video	0.07578611

5 Results

The experiments are performed on BBC news dataset [44]. The dataset consists of 2225 news article in domains: business, tech, sports, politics and entertainment. The LSA is applied on the whole dataset and the root words along with the weights are given as output. These weights form Context Vectors for root words/terms. Some of them are shown in Table 2.

Figure 2 shows the graph of terms verses their respective weights. These weights can be used as context vectors in order to analyse the document.

6 Conclusion and Future Work

There are various ways to analyse documents. Some of them are stated in this work. The most trending way is to use context vectors to analyse the documents. The previous work by researchers have made use of context vectors to analyse document. In this work, LSA is used in order to find out context vectors for the root terms in the documents. As compared to the previous work done to find out context vector for document set our method clearly mentions the context vector for the given set of documents. The context vectors can further be processed in order to find out the theme of documents correctly. The proposed system uses static dataset, future work can be to create context vectors dynamically.

References

1. Harris, S. C., Zheng, L., Kumar, V., & Kinshuk (2014). Multi-dimensional sentiment classification in online learning environment. In *IEEE Sixth International Conference on Technology for Education*.
2. Haribhakta, Y., Kalamkar, S., & Kulkarni, P. (2012). Feature annotation for text categorization. In *Cube 2012*.
3. Ye, L., Xu, R.-F., Xu, J. (2012). Emotion prediction of news articles from reader's perspective based on multi-level classification. In *Proceedings of the 2012 International Conference on Machine Learning and Cybernetics*.
4. Khatavkar, V., & Kulkarni, P. (2016). Context vector machine for information retrieval. *Atlantis Press, Advances in Intelligent Systems Research, 137,* 375–379.
5. Khatavkar, V., & Kulkarni, P. (2017). *Document context identification using latent semantic analysis*. Presented in 3rd International Conference on Computing, Communication, Control and Automation, August 17–18, 2017, Pune, MS, India. (To be published on IEEE).
6. Han, J., & Kimber, M. (2000). *Data mining: Concepts and techniques*. Morgan Kaufmann.
7. Jain, A. K, Murty, M. N., & Flynn, P. J. (1999). Data clustering: A review. *ACM Computing Surveys 31*(3), 264–323.
8. Steinbach, M., Karypis, G., & Kumar, V. (2000). *A comparison of document clustering techniques*. KDD Workshop on Text Mining (Vol. 400).

9. Berkhin, P. (2004). *Survey of clustering data mining techniques*. Available at http://www.accrue.com/products/rp_cluster_review.pdf.
10. Xu, R. (2005). Survey of clustering algorithms. *IEEE Transactions on Neural Networks, 16*(3), 634–678.
11. Dou, Z., Jiang, Z., Hu, S., Wen, J.-R., & Song, R. (2015). Automatically mining facets for queries from their search results. *IEEE Transactions on Knowledge and Data Engineering*.
12. Nigam, K., Mccallum, A., Thrun, S., & Mitchell, T. (1998). Learning to classify text from labeled and unlabeled documents. In *Proceedings of the Fifteenth National/Tenth Conference on Artificial Intelligence/Innovative Applications of Artificial Intelligence*.
13. Mccallum, A., & Nigam, K. (2003). A comparison of event models for naive bayes text classification. In *Proceedings of the Tenth Conference on European Chapter of the Association for Computational Linguistics* (Vol. 1).
14. Nigam, K., & Mccallum, A. (1998). Employing Em and pool-based active learning for text classification. In *Proceedings of the Fifteenth International Conference on Machine Learning*.
15. Puzicha, J., Hofmann, T., & Buhmann, J. M. (2000). A theory of proximity based clustering: structure detection by optimization. *Pattern Recognition 33*(4). 12.
16. Boughanem, M., Brini, A., & Dubois, D. (2009). Possibilistic networks for information retrieval. *International Journal of Approximate Reasoning (Special Section on Graphical Models and Information Retrieval) 50*(7). (Elsevier).
17. Salakhutdinov, R., Hinton, G. (2009). Semantic hashing. *International Journal Of Approximate Reasoning (Special Section On Graphical Models And Information Retrieval) 50*(7). (Elsevier).
18. Haribhakta, Y., Malgaonkar, A., & Kulkarni, P. Dr. (2012). Unsupervised topic detection model and its application in text categorization. In *Cube 2012*.
19. Huang, A., Milne, D., Frank, E., & Witten, I. H. (2008). Clustering document with active learning using wikipedia. In *Eighth IEEE International Conference on Data Mining*.
20. Meena, Y. K., & Gopalani, D. Dr. (2014). Analysis of sentence scoring methods for extractive automatic text summarization. In *International Conference on Information and Communication Technology for Competitive Strategies*.
21. Chiang, I.-J., Liu, C. C.-H., Tsai, Y.-H., & Kumar, A. (2015). Discovering latent semantics in web documents using fuzzy clustering. *IEEE Transactions on Fuzzy Systems*.
22. Hatzivassiloglou, V., Klavans, J. L., & Eskin, E. (1999). Detecting text similarity over short passages: Exploring linguistic feature combinations via machine learning. In *Proceedings of Joint SIGDAT Conference on Empirical Methods in Natural Language Processing and Very Large Corpora*.
23. Hofmann, T. (1999). Probabilistic latent semantic indexing. In *Proceedings of the 22nd Annual International ACM SIGIR Conference on Research and Development in Information Retrieval*.
24. Hofmann, T. (1999). The Cluster-Abstraction model: Unsupervised learning of topic hierarchies from text data. In *International Joint Conferences on Artificial Intelligence*.
25. Huang, A., Milne, D., Frank, E., Witten, I. H. (2009). Clustering documents using a wikipedia-based concept representation, advances in knowledge discovery and data mining. In *13th Pacific-Asia Conference, PAKDD 2009 Proceedings*. 11.
26. Lee, C.-S., Kao, Y.-F., Kuo, Y.-H., & Wang, M.-H. (2007). Automated ontology construction for unstructured text documents. *Data & Knowledge Engineering 60*(3). (Elsevier).
27. Rushall, D. A., & Ilgen, M. R. (1996). *DEPICT: Documents valuated as pictures visualizing information using context vectors and self organizing maps*. IEEE.
28. Dahab, M. Y., Hassan, H. A., & Rafea, A. (2008). TextOntoEx: Automatic ontology construction from natural English text. *Expert Systems with Applications 34*(2). (Elsevier).
29. Villaverde, J., Persson, A., Godoy, D., Amandi, A. (2009). Supporting the discovery and labeling of non-taxonomic relationships in ontology learning. *Expert Systems With Applications, 36*(7). (Elsevier); Montes-y-Gomez, M., & Villasenor-Pineda, L. (2009). *Representing context information for document retrieval*. LNAI. Springer, Berlin, Heidelberg.

30. Abdalgader, K., & Skabar, A. (2012). Unsupervised similarity-based word sense disambiguation using context vectors and sentential word importance. *ACM Transactions on Speech and Language Processing, 9*(1), Article 2.
31. Oh, J.-H., & Choi, K.-S. (2002). Word sense disambiguation using static and dynamic sense vectors. In *Proceedings of the 19th international conference on Computational linguistics* (Vol. 1).
32. Farkas, J. (1996). Improving the classification accuracy of automatic text processing systems using context vectors and back-propagation algorithms. In *CCECE'96*. IEEE.
33. Zhang, J., Qu, D., Li, Z. (2014). An improved recurrent neural network language model with context vector features. IEEE.
34. Salehi, M., Khadivi, S., & Riahi, N. (2014). Confidence estimation for machine translation using context vectors. In *2014 7th International Symposium on Telecommunications (IST'2014)*.
35. Sharma, D., & Jain, S. Dr. (2015). Context-based weighting for vector space model to evaluate the relation between concept and context in information storage and retrieval system. In *IEEE International Conference on Computer, Communication and Control (IC4-2015)*.
36. Wan, S., & Angryk, R. A. (2007). *Measuring semantic similarity using wordnet-based context vectors*. IEEE.
37. Rongali, S., Choudhury, A. R., Chandan, V., & Arya, V. (2015). A context vector regression based approach for demand forecasting in district heating networks. *Innovative Smart Grid Technologies—Asia (ISGT ASIA)*. IEEE.
38. Kaufmann, S. (2002). *Cohesion and collocation: Using context vectors in text segmentation*. CSLI, Stanford University, May 2002.
39. Melucci, M. (2006). Ranking in context using vector spaces. In *CIKM'06*. ACM.
40. Erk, K., & Padó, S. (2008, October). A structured vector space model for word meaning in context. In *Proceedings of the 2008 Conference on Empirical Methods in Natural Language Processing* (pp. 897–906).
41. Thater, S., & Furstenau, H., & Pinkal, M. (2010, July). Contextualizing semantic representations using syntactically enriched vector models. In *Proceedings of the 48th Annual Meeting of the Association for Computational Linguistics* (pp. 948–957).
42. Carrillo M., Villatoro-Tello E., López-López A., Eliasmith C., Montes-y-Gómez M., & Villaseñor-Pineda L. (2009). Representing context information for document retrieval. In: Andreasen, T., Yager, R. R., Bulskov, H., Christiansen, H., Larsen, H. L. (Eds.), *Flexible Query Answering Systems*. FQAS 2009. Lecture Notes in Computer Science, Vol 5822. Springer, Berlin, Heidelberg.
43. Milajevs, D., Sadrzadeh, M., & Roelleke, T. (2015). IR meets NLP: On the semantic similarity between subject-verb-object phrases. In *ICTIR, 15*. ACM 2015.
44. BBC news dataset available at: http://mlg.ucd.ie/datasets/bbc.html.

Comparison of Support Vector Machines With and Without Latent Semantic Analysis for Document Classification

Vaibhav Khatavkar and Parag Kulkarni

Abstract Document Classification is a key technique in Information Retrieval. Various techniques have been developed for document classification. Every technique aims for higher accuracy and greater speed. Its performance depends on various parameters like algorithms, size, and type of dataset used. Support Vector Machine (SVM) is a prominent technique used for classifying large datasets. This paper attempts to study the effect of Latent Semantic Analysis (LSA) on SVM. LSA is used for dimensionality reduction. The performance of SVM is studied on reduced dataset generated by LSA.

Keywords Document classification · Support vector machine · Latent semantic analysis · Dimensionality reduction · Singular value decomposition
Context vector

1 Introduction

Document classification sorts documents into classes on the basis of its contents [1]. By grouping documents, we are facilitating the job to manage large number of documents. In document analysis one interesting method is to extract context. Xu and Croft in [2] build a context. The context is build globally by using relation between the words in the document and locally using initial queries. The concluded with combining both of them which proves to be effective. With an increase in digital text, this field began spreading and gaining popularity. The number of document classification techniques began increasing gradually. These techniques are combined with different other processes or techniques to increase their efficiency.

V. Khatavkar (✉) · P. Kulkarni
Department of Computer Engineering and Information Technology,
College of Engineering, Pune, Shivajinagar, Pune, Maharashtra, India
e-mail: vkk.comp@coep.ac.in

P. Kulkarni
e-mail: paragindia@gmail.com

V. E. Balas et al. (eds.), *Data Management, Analytics and Innovation*,
Advances in Intelligent Systems and Computing 808,
https://doi.org/10.1007/978-981-13-1402-5_20

For example, a classification technique is merged with another classification technique or a classification technique is merged with a feature extraction technique and so on. Khatavkar and Kulkarni in [3] proposed a novel concept of "Context Vector Machine" which can be used for thematic analysis of documents. The system was proposed but not implemented. Khatavkar and Kulkarni in [4] implemented thematic analysis using LSA. In their work, they used "Context Vector" in order to get theme of a document. The clustering of themes of document set was missing. Sheikh et al. in [5] derived semantic context for Out of Vocabulary word. This motivates to develop a classification system after deriving the theme of document sets. Support Vector Machine is used for Classification and Latent Semantic Analysis for Theme Detection of document sets.

Section 2 gives details about the Datasets which are used in this work. Section 3 which explains how the data is preprocessed before processing for theme detection followed by Sect. 4 which give details about TF–IDF. Section 5 explains the algorithm used for theme detection. Section 6 deals with the explanation of SVM. Sections 7, 8, and 9 proposes system, the implementation of proposed system, its implementation and the results respectively. Section 10 concludes the work.

2 Datasets

Datasets play a very important role in machine learning. Winsser-Gross in [6] discussed about how dataset and speed of algorithm are dependent. The presence of high-quality training datasets has great effect on algorithms. This study explores BBC news dataset and 20 newsgroup dataset.

The **BBC news dataset** contains documents from BBC news website with five predetermined categories which are business, entertainment, tech, sports, politics. Each of these categories contains news articles related to the given category. Total there are 2225 documents present in the whole dataset [7].

The **20 Newsgroups dataset** consists of 19,999 news articles divided across 20 different categories/newsgroup which are alt.atheism, rec.autos, sci.space, comp.-graphics, rec.motorcycles, soc.religion.christian, comp.os.ms-windows.misc, rec.sport.baseball, talk.politics.guns, comp.sys.ibm.pc.hardware, rec.sport.hockey, talk.politics.mideast, comp.sys.mac.hardware, sci.crypt, talk.politics.misc, comp.windows.x, sci.electronics, talk.religion.misc, misc.forsale and sci.med. Some newsgroups are very closely related to each other, while others are highly unrelated [8].

3 Preprocessing

To represent a large set of document, raw text data is not useful. Preprocessing is applied on raw text data to convert it into numerical data and to get the most prominent features from the data. These features are typically words without special characters and numbers.

3.1 Tokenization

Document is a sequential collection of words. Words are separated by special characters viz., comma, blank space, etc. Each word is called a token. Tokenization is a process of finding the tokens (words) within a document and making its list. Further, working with tokens is easier and useful than working on raw data [9].

3.2 Removal of Stop Words

In information retrieval, the size of a dataset is usually very large. While taking all tokens in consideration, articles, prepositions, verbs will occur in large number but will be of no use in classification of document. Such features are called stop words. A database of such words is used to remove these trivial features.

3.3 Stemming

Stemming simplifies various words and their corresponding a noun, a verb form to its basic form. It reduces all related forms of a word to a common root form.

4 Term Frequency–Inverse Document Frequency Weighting

In document classification, a numerical measure is used to give importance to a feature (term/word). The measure represents an importance of feature to respective document. Term Frequency–Inverse Document Frequency (TF–IDF) is one of the most prominent techniques of weighting. TF–IDF takes into consideration that some words occur more in comparison with others but have less importance. For such features, the weight assigned is less. Higher TF–IDF weight indicates that the word is both important to the document and is uncommon. It interprets that the word is important to the document in classification. TF–IDF is given by the formula as stated in [10].

$$\mathrm{tf}-\mathrm{idf}(t,d,D)=\mathrm{tf}(t,d)\cdot\mathrm{idf}(t,D) \tag{1}$$

where

$$\mathrm{tf}(t,d) = 0.5 + 0.5 \cdot \frac{f_{t,d}}{\max\{f_{t',d} : t' \in d\}} \quad (2)$$

where

t term
d document
tf term frequency
t' any term
$f_{t,\ d}$ Number of occurrence of t in d and IDF is given as in [11]

$$\mathrm{idf}(t,D) = \log \frac{N}{|\{d \in D : t \in d\}|} \quad (3)$$

where

idf inverse document frequency
N Total number of documents
D Set of all documents

5 Latent Semantic Analysis

Latent Semantic Analysis (LSA) uses word occurrence to form a vector model. It uses the rule that if words occur multiple times close to each other, they must be semantically close and if they come close rarely, then they must be semantically distant. LSA uses Singular Value Decomposition to form Component verses Document Matrix. Onal Suzek in [12] proposed and experimented a system which used keyword extraction from the whole document corpora. The researcher compared tf–idf, LSA, and Metamap keyword extraction methods using the ROC curve in which tf–idf proves to be better. LSA can also be used for plagiarism detection with respect to context. Rajkumar and Karthik worked on the same in [13]. Marcolin and Luiz Becker in [14] explored LSA. The researchers worked on text mining from articles more that 42,079 published between the year 1979–2016.

They used LSA in R to investigate topics in the mined unstructured data. They concluded with the importance of application LSA in Big Data Hofmann in [15] explained probabilistic approach of LSA. It uses Bayesian Theory along with LSA. Zhu Zhiyuan in [16] explains the details of LSA. Dumais in [17] also gives overview of LSA.

The following subsection explain them in brief.

5.1 Singular Value Decomposition

Singular Value Decomposition (SVD) uses term by document matrix formed from training data. The values in the matrix are tf–idf weights. Now to reduce this sparse matrix to a compressed one, SVD is applied. SVD is given as:

$$M = U * S * V \tag{4}$$

The original matrix M is separated into three matrices—reduced rank term matrix U, singular value diagonal matrix S and document matrix V which is elaborated in [18, 19].

6 Support Vector Machine

Support Vector Machines (SVM) is one of the machine learning algorithms which uses supervised learning models for pattern recognition. SVM is often used for classification and regression analysis. In classification task, SVM leads to other methods because SVM provides a global solution for data classification. Researchers have used SVM in Text Categorization [20–22]. SVM gives a unique global hyperplane to separate data points for different classes. Abdiansah and Wardoyo in their work [23], discusses SVM and its implementation in LibSVM. They used two popular languages namely, C++ and Java, with three different datasets. They found that the complexity of SVM in LibSVM is O (n^3) and time complexity for C++ was faster than Java.

Assume a training set, (X_i, Y_i) for i = 0, 1, 2, …, n where X_i= (x_i, …, x_{id}) is a sample d-dimension and $y_i \in \{1, -1\}$ is a label given to a sample. The SVM's task is to find linear discriminant function $g(x) = w^T \cdot X + w_0$ so that, $w^T \cdot x_i + w_0 \geq +1$ for $i = 1, \ldots, n$ and $w^T \cdot x_i + w_0 \leq -1$ for i = 1, …, n. Solutions for these problems must satisfy the following equation:

$$y_i\left(w^T \cdot x_i + w_0\right) \geq +1 \tag{5}$$

Optimal linear function can be obtained by minimizing the following quadratic programming problems:

$$\min \frac{1}{2} \cdot w^T \cdot w - \sum_{i=1}^{n} \alpha\left(y_1\left(w^T \cdot x_i + w_0\right) - 1\right) \tag{6}$$

which will produce the following solutions:

$$w = \sum_{i=1}^{n} \alpha \cdot y_i \cdot x_i, \tag{7}$$

where $\{\alpha, i = 1, \ldots, n; \alpha \geq 0\}$ is Lagrange multipliers.

In order to make data linearly separate, feature space is mapped into high-dimensional space. The technique used to perform mapping function is called Kernel. Sometimes, SVM are also called Kernel Machines. The kernel function is $k = \chi \times \chi \to \mathbb{R}$ which takes two samples from input space and maps into a real number that indicates the level of similarity, $\forall x_i, x_j \in \chi$, then the kernel function must satisfy:

$$k(x_i, x_j) = \langle \phi(x_i), \phi(x_j) \rangle, \tag{8}$$

where ϕ is explicitly mapping from input space χ to the features of dot product of space H.

There are four basic types of kernels namely linear, polynomial, radial basis function and sigmoid. SVM can also be used for multiple class labeling. It is shown in Fig. 1. The example is about classifying samples into positives and negatives. A simple kernel function is used with single hyperplane. It classifies given samples into positives and negatives. The hyperplane selected is such that it classifies the samples correctly.

Another example of SVM is shown in Fig. 2. The example is about classifying samples into positives and negatives. A simple kernel function is used with single hyperplane. It classifies the given samples into positives and negatives. In this example, the hyperplane is such that it classifies some samples incorrectly.

The example of SVM, shown in Fig. 3, illustrates the use of multiple hyperplanes on same sample data.

Support vectors are critical elements of datasets because they are nearest points to the hyperplane and if these points are removed from the datasets, they would change the position of the dividing hyperplane. A hyperplane can be considered as

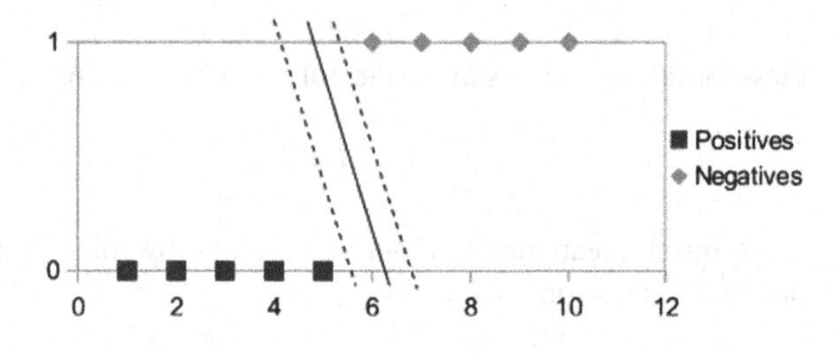

Fig. 1 Support vector machine: example 1

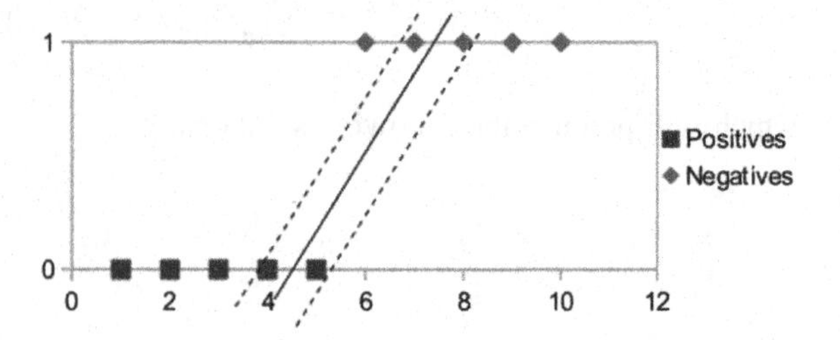

Fig. 2 Support vector machine: example 2

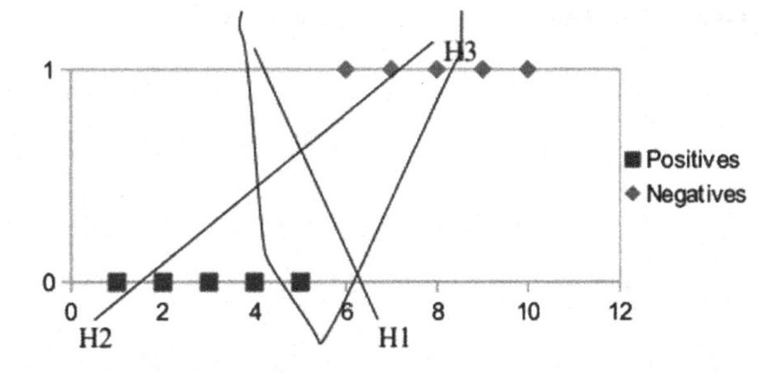

Fig. 3 Support vector machine: example 3

a line which straightly isolates and groups an arrangement of information. The more far a point lies from the hyperplane, it is more prominent that the points will be effectively arranged as per the rule. Thus, SVM picks such a hyperplane which maximizes the distance between the nearest data point on either side of the hyperplane or on the hyperplane itself. This distance is called a margin. In the light of the training information, a testing information point is arranged by allocating a side of the hyperplane. SVM is resistant to over fitting and has more probability of classifying the data properly as it has maximum margin.

7 Proposed System

This study proposes an algorithm given in Algorithm 1 to study the effects of dimensionality reduction on working of Support Vector Machine. The algorithm applies SVM classifier after application of LSA on the data. Before the LSA is applied the words/terms are stemmed. After stemming we get the root word of the stemmed word/term. All stop words are also removed from the data before stemming. The system is shown in Fig. 4.

Algorithm 1 *Extended LSA Algorithm*

1. The documents for training and testing in *.txt* format are categorized into their corresponding class folders.
2. The training dataset is loaded with predetermined categories.
3. The raw text dataset is then preprocessed as follows:
   (a) Tokenization is done which gives all the words as separate entities.
   (b) Stemming is done over the tokens.
   (c) Stop words are removed.
   (d) Occurrences are counted and frequency matrix is made.
4. Term Frequency and Inverse Document Frequency is calculated on the frequency matrix.
5. Further, latent semantic analysis is performed by calculating singular value decomposition (SVD) of the tf–idf matrix and thereby components are reduced.

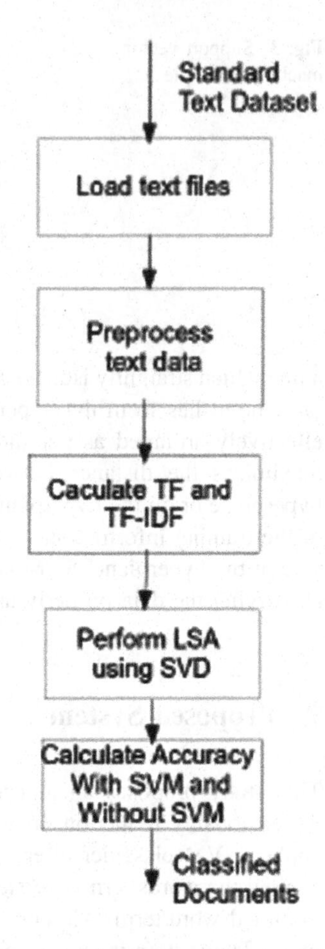

Fig. 4 Proposed system

The reduced components are variable based on the size of datasets. The size of components may vary from 50 to 100. LSA matrix is calculated which will have reduced components.

6. LSA matrix is then compared with available predetermined classes using SVM to find the resultant accuracy.

8 Implementation

For implementation, python language is used. And *sklearn* package is used for machine learning algorithms. The *sklearn* package is an open source Python library which provides implementation of many machine learning algorithms. Moreover, it has preprocessing algorithms, decompositions algorithms and many related techniques related to data mining and data analysis. Other important packages that are used are nltk for stemming and matplotlib and pyplot for plotting graphs. For converting raw text data into numerical data, *CountVectorizer* module from *sklearn* is used. It tokenizes the words, eliminates the stop words provided to it. For stemming, *SnowballStemmer* from *nltk* module is used [24–28].

9 Experimentation and Results

Confusion Matrix and accuracy are calculated for the training and test data. The results are as follows. The accuracy chart is given in Table 1 and Fig. 5.

Table 1 Results

Test data sets	Classification with LSA	Classification without LSA
BBC data (on train data)	98.47191	99.86516
BBC data (on test data)	84	88.0
20 newsgroup (on the train data)	94.08411	97.05455
20 newsgroup (on test data)	92.8	94.525

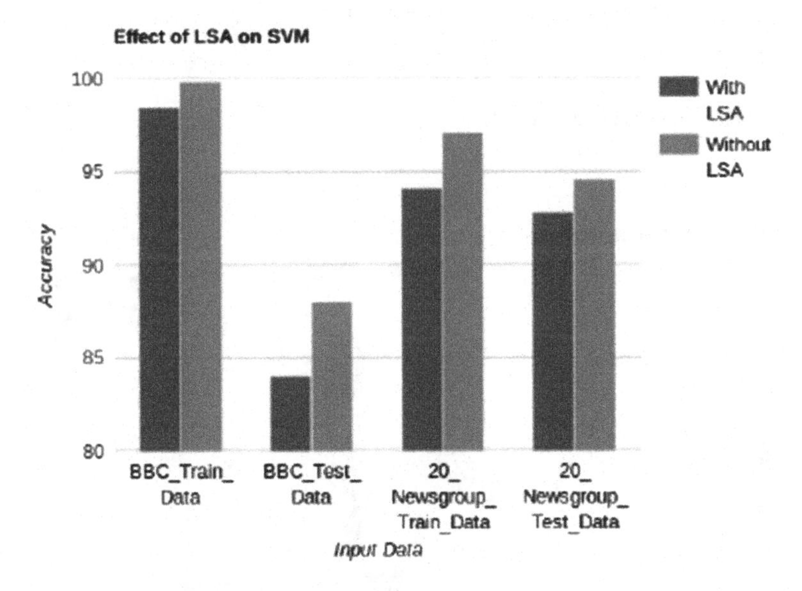

Fig. 5 Accuracy using different dataset

Table 1 shows that accuracy is decreased when LSA is used. The time difference in training the model in Fig. 6.

The time difference in predicting the test data from the model is shown in Fig. 7.

These graphs show that the time taken for training and testing are reduced drastically. It is observed from the graphs that SVM performs well without use of LSA than use of LSA.

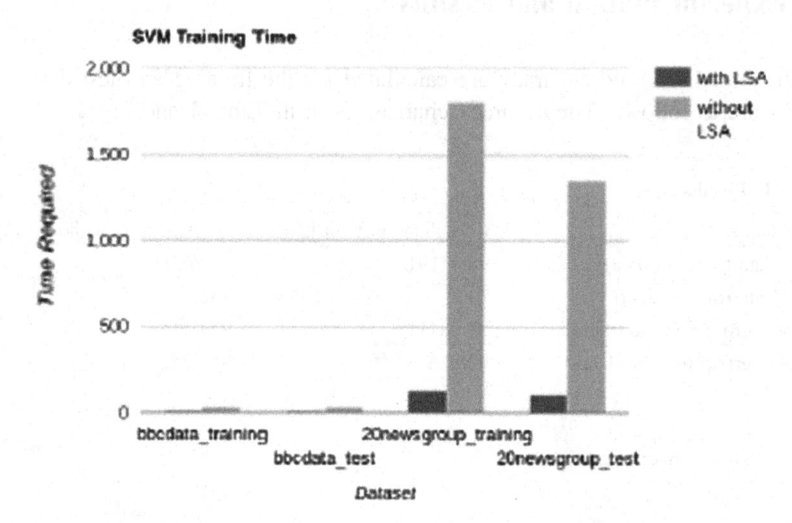

Fig. 6 Training time using different dataset

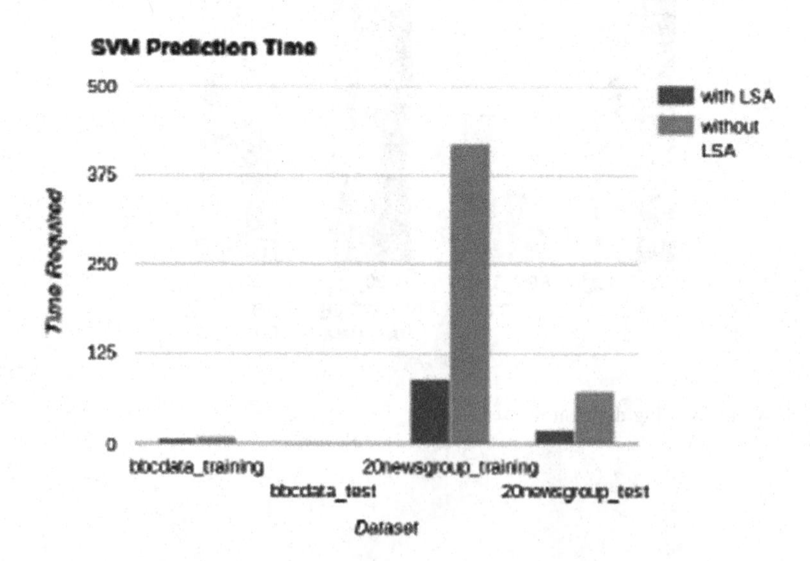

Fig. 7 Prediction time using different dataset

10 Conclusion

Accuracy graph, given in Fig. 5, shows that the accuracy decreases when SVM is combined with LSA. And when the dataset is more, the difference in accuracy is less. Further, time required for training and testing of model is reduced drastically when LSA is used.

SVM works well when the separation in the dimensions is high. When LSA is applied, the dimensions are cut to a very small number. Words which are close to each other are represented by a single component. This causes a loss of information. When these reduced components are fed to SVM for training, it gets very less number of features. Hence, the training time required is very less in comparison with the time required when LSA is not applied. Moreover, as the features are reduced for test data too, the time required for prediction is less too. This gives a very nice upgradation to SVM as SVM takes large time for training and hence is not preferred for very large dataset. Using LSA with SVM eliminates that shortcoming.

As the number of features are reduced, the dimensions get reduced too. SVM works well with higher dimensions and has methods to ignore outliers and handle high-dimensional input. Thus, when SVM is given low-dimensional input, it gives lesser accuracy.

It is therefore concluded that LSA should not be used with SVM when accuracy is the most important required feature. And in cases of large dataset, when a slight decrease in accuracy can be overlooked, LSA can be used to reduce the time.

Acknowledgements We acknowledge the help extended by Mr. Shubham Gatkal, Mr. Sandesh Gupta and Mr. Prathamesh Ingle for experimentation. We would also like to acknowledge the support and encouragements received from the authorities of College of Engineering, Pune. (COEP).

References

1. Camastra, F., Ciaramella, A., Placitelli, A., & Staiano, A. (2015). *Machine learning-based web documents categorization by semantic graphs*. ResearchGate Publisher. https://doi.org/10.1007/978-3-319-18164-6_8.
2. Xu, J., & Croft, W. B. (2017). Query expansion using local and global document analysis. In *ACM SIGIR Forum* (Vol. 51, No. 2), July 2017.
3. Khatavkar, V., & Kulkarni, P. (2016, December). Context vector machine for information retrieval. *International Conference on Communication and Signal Processing*.
4. Khatavkar, V., & Kulkarni, P. (2017). *Document context identification using latent semantic analysis*. Presented in 3rd International Conference on Computing, Communication, Control and Automation, August 17–18, 2017, Pune, MS, India. (To be published on IEEE).
5. Sheikh, I., Illina, I., Fohr, D., & Linar, G. (2016). Document level semantic context for retrieving OOV proper names. *ICASSP*.
6. Wissner-Gross, A. (2016). Datasets over algorithms, Edge.com. Retrieved January 8, 2016.
7. The BBC dataset available at: http://mlg.ucd.ie/datasets/bbc.html.
8. The 20 Newsgroup dataset available at http://qwone.com/~jason/20Newsgroups/.

9. The article about Tokenization available at https://nlp.stanford.edu/IR-book/html/htmledition/tokenization-1.html.
10. Manning, C. D., Raghavan, P., & Schutze, H. (2008). Scoring, term weighting, and the vector space model. In *Introduction to information retrieval (PDF)* (p. 100). https://doi.org/10.1017/cbo9780511809071.007. ISBN 978-0-511-80907-1.
11. Robertson, S. (2004). Understanding inverse document frequency: On theoretical arguments for IDF. *Journal of Documentation, 60*(5), 503–520. https://doi.org/10.1108/00220410410560582.
12. Onal Suzek, T. (2017). Using latent semantic analysis for automated keyword extraction from large document corpora. *Turkish Journal of Electrical Engineering & Computer Sciences.*
13. Rajkumar, K., & Karthik, K. (2017). Contextual plagiarism detection using latent semantic analysis. *International Research Journal of Advanced Engineering and Science, 2*(1), 214–217.
14. Marcolin, C. B., & Luiz Becker, J. (2016). Exploring latent semantic analysis in a big data (base). In *Twenty-second Americas Conference on Information Systems*, San Diego.
15. Hofmann, T. (2017). Probabilistic latent semantic indexing. In *ACM SIGIR Forum* (Vol. 51, No. 2), July 2017.
16. Zhu, Z. (2017). Latent semantic analysis, July 18, 2017. Available at http://www.professorbray.net/Teaching/89s-MOU/2017-summerTerm2/Papers/third\%20submissions/ZHUfiPaper2fiLSA.pdf10.
17. Dumais, S. T. (2005). Latent semantic analysis. *Annual Review of Information Science and Technology.*
18. DeAngelis, G. C., Ohzawa, I., & Freeman, R. D. (1995). Receptive-field dynamics in the central visual pathways. *Trends in Neuroscience, 18*(10), 451–458. https://doi.org/10.1016/0166-2236(95)94496-r. PMID 8545912.
19. Information about LSA and SVD is available at http://web.mit.edu/be.400/www/SVD/SingularfiValuefiDecomposition.htm.
20. Fatima, S., & Srinivasu, B. Dr. (2017, February). Text document categorization using support vector machine. *International Research Journal of Engineering and Technology (IRJET).*
21. Cortes, C., & Vapnik, V. (2003). Support-vector networks. *Machine Learning, 20*(3), 273–297.
22. Joachims, T. (1991). *Text categorization with support vector machines: Learning with many relevant features*. University at Dortmund Informatik LS8, Baroper Str. 30144221 Dortmund, Germany.
23. Abdiansah, A., & Wardoyo, R. (2015). Time complexity analysis of support vector machines (SVM) in LibSVM. *International Journal of Computer Applications.*
24. The python implementation of LSA was taken from http://mccormickml.com/2016/03/25/lsa-for-text-classification-tutorial/.
25. The python implementation for SVM was taken from https://pythonprogramming.net/linear-svc-example-scikit-learn-svm-python/.
26. The python implementation of LSA was taken from http://blog.josephwilk.net/projects/latent-semantic-analysis-in-python.html.
27. The python implementation of LSA was taken from https://gist.github.com/vgoklani/1267632.
28. The python implementation of LSA was taken from https://technowiki.wordpress.com/2011/08/27/latent-semantic-analysis-lsa-tutorial/.

Facial Recognition, Expression Recognition, and Gender Identification

Shraddha Mane and Gauri Shah

Abstract Face recognition has many important applications in areas such as public surveillance and security, identity verification in the digital world, and modeling techniques in multimedia data management. Facial expression recognition is also important for targeted marketing, medical analysis, and human–robot interaction. In this paper, we survey a few techniques for facial analysis. We compare the cloud platform AWS Rekognition, convolutional neural networks, transfer learning from pre-trained neural nets, and traditional feature extraction using facial landmarks for this analysis. Although not comprehensive, this survey covers a lot of ground in the state-of-the-art solutions for facial analysis. We show that to get high accuracy, good-quality data and processing power must be provided in large quantities. We present the results of our experiments which have been conducted over six different public as well as proprietary image data sets.

Keywords Machine learning · Deep learning · Face detection
Face recognition · Facial expression recognition · Gender identification
Hog descriptor · Logistic regression · Support vector machine · Random decision forest · Convolutional neural networks · Classification

1 Introduction

Over the last few decades, face recognition [1] and facial expression recognition have seen many commercial applications, and have become a popular area of research in the field of computer vision. Wide availability of low-cost desktop and embedded computing systems has created an enormous interest in automated processing of digital images and videos in applications such as biometric

S. Mane · G. Shah (✉)
Persistent Systems Ltd, Pune 411004, Maharashtra, India
e-mail: gauri_shah@persistent.co.in

S. Mane
e-mail: shraddha_mane@persistent.co.in

V. E. Balas et al. (eds.), *Data Management, Analytics and Innovation*,
Advances in Intelligent Systems and Computing 808,
https://doi.org/10.1007/978-981-13-1402-5_21

authentication, surveillance, human–computer interaction, robotics, and modeling techniques in multimedia management. Similarly, a facial expression recognition system has various applications such patient analysis in the medical field, human attentiveness in the field of advertising, human–robot interaction, etc.

Recent advances in automated face analysis, pattern recognition, neural networks, and other machine learning techniques have made it possible to develop automatic face recognition and facial expression recognition systems to address these requirements. A face recognition system automatically identifies the face present in an image or video. It can operate in one of the following two modes: face verification (or authentication), and face identification (or recognition). For verification, we have an input image and we need to check if this image matches one specific in a given database (one-to-one match). On the other hand, with identification we have to compare a given image with all the images in the database (one-to-many match). If we can restrict the set of images to compare, it is called a watch-list check (one-to-few match). A facial expression recognition system identifies the emotion of the user; in this paper, the emotion can be neutral or of six basic types—anger, disgust, fear, happiness, sadness, and surprise.

Although research in automatic face recognition has been conducted since the 1960s, this problem has been largely unsolved in its entirety mainly due to the lack of processing power. Recent years have seen significant progress in this area owing to advances in data collection, face modeling and analysis techniques as well as the availability of high-quality cheap hardware that can quickly run computationally intensive tasks. Systems have been developed for face detection and tracking, but reliable face recognition still offers interesting challenges in computer vision and pattern recognition research.

In this paper, we survey several different techniques for face recognition, facial expression recognition, and gender identification. Our goal is to compare these different approaches to compare the performance for different tasks. We evaluate (i) the public cloud-based platform—Amazon AWS Rekognition [2], (ii) other deep learning techniques such as Convolutional Neural Networks (CNNs) [3] and Transfer Learning [4] by using a pre-trained image model, and (iii) feature extraction [5] using facial landmarks [6].

2 Related Work

Although we have looked only at a small subset of techniques that can be used for facial analysis, there are many more solutions available to tackle these problems, some of which we summarize here. Face detection algorithms include those proposed by Rowley et al. [7], Viola and Jones [8], and Zhang and Zhang [9]. We have used the one proposed by Viola and Jones as it is widely used.

Facial features like skin color, gender, appearance, edges, facial marks, structure etc. can be extracted from facial images using different algorithms. Principal Component Analysis (PCA) [10] and Linear Discriminant Analysis (LDA) [11] capture the skin color, gender and general appearance information of a face. Gabor Wavelets [12], Local Binary Patterns (LBP) [13], Histogram of Oriented Gradient (HOG) [14] and Scale Invariant Feature Transform (SIFT) [15] are used to extract the facial structures that are relevant for recognition. These features are the most discriminative face features; to handle challenges like biologically identical twins, and faces across different age variations, we need to extract micro-level features on the face such as scars, moles, and facial marks for efficient face recognition. Adding depth information to color images also boosts the accuracy of facial analysis, however it adds to the computation, and requires special RGB-D sensors like Microsoft Kinect [16].

The reason behind the popularity of artificial neural nets is that the alternative process of feature extraction is time consuming, difficult, and requires domain experts. CNNs are stealing the show for the last few years. 2012 was the first year when neural networks grew to prominence for image processing as Alex Krizhevsky used them to win ImageNet competition that year. Standard CNNs like AlexNet [17], ZFNet [18], VGGNet [19], GoogleNet [20], and ResNet [21] are nothing but CNNs (with difference architectures) trained on ImageNet [22] data set. The inception module which uses multiple mini-models was introduced in GoogleNet. With transfer learning, these standard CNNs can be fine-tuned and used for feature extraction.

3 Data Sets

We used six different data sets to evaluate different techniques in facial analysis. In this section, we give the details of these data sets. Three of these data sets are public data sets available on the Internet, two are internal data sets obtained by us, and one is obtained from the Wikipedia and IMDb web sites.

3.1 Japanese Female Facial Expression (JAFFE)

The Japanese Female Facial Expression (JAFFE) [23] is a publicly available data set which consists of 213 grayscale images for seven facial expressions: happiness, sadness, surprise, anger, disgust, fear, and neutrality. The images are captured in a controlled environment with constant background and good illumination conditions. Expressions are posed by ten Japanese female models (Fig. 1).

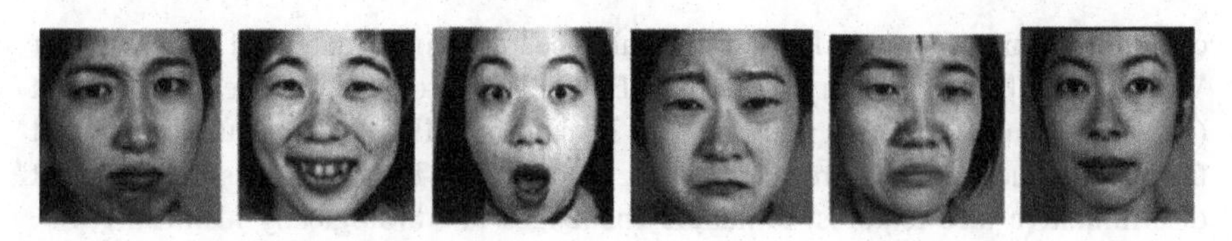

Fig. 1 Some sample images from the JAFFE data set [23]

3.2 Cohn-Kanade (CK)

The Cohn-Kanade (version 1) data set [24] is a publicly available expression data set containing a sequence of grayscale images taken in controlled environment. Each sequence begins with neutral expression and proceeds to a target expression. We manually separate all sequences into seven categories which are the same as the JAFFE data set (Fig. 2).

3.3 Persistent Systems Ltd. (PSL)

This data set was collated locally, and it consists of seven expressions posed by Persistent Systems employees. These colored images are captured using a webcam in a controlled environment. Due to privacy reasons, we omit giving samples from this data set but they are similar to the other samples given above.

3.4 FER2013

The FER2013 data set [25] was created as follows: 184 keywords related to human emotions (such as "enraged", "blissful", etc.) were used in the Google image search API (Fig. 3). The resulting images from this search were then cropped; if any images had incorrect labels, they were manually corrected. These images were then resized to 48×48 pixels and converted to grayscale. The distribution of the 35887 images in different classes is as follows:

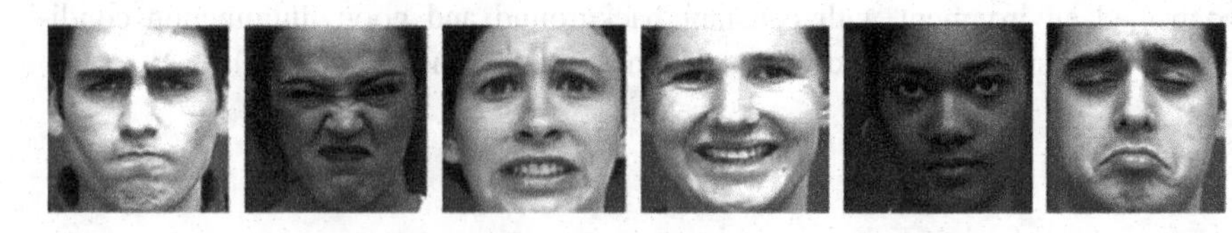

Fig. 2 Some sample images from the CK data set [24]

Fig. 3 Some sample images from FER2013 data set [25]

Fig. 4 Some sample images from the IMDb-Wiki data set [26]

Anger: 4953	Happiness: 6077	Surprise: 4002
Disgust: 5121	Sadness: 4002	Neutral: 6198
Fear: 8989		

3.5 *IMDb-Wiki*

The IMDb-Wiki data set [26] is the largest publicly available data set of face images with gender and age labels for training. The colored images are taken from the IMDb website as well as from Wikipedia, and related information such as age, gender, name, etc., is stored too (Fig. 4).

3.6 *Real-World Data Set*

This data set is captured by considering the real-world challenges like pose, illumination, expressions, background, etc. We obtained this data set as a part of a client project and the data has been suitably anonymized. There are two sets of training data and one set of test data. The first training data set (TS1) consists of nearly frontal color images, with only one image per user. The second training data set (TS2) contains one black-and-white, low-resolution photo ID for each of the participants. The test data set consists of multiple real world images per user, with different challenges as mentioned above (Fig. 5) (Table 1).

Fig. 5 Some sample images from the real-world data set

Table 1 A summary of all the data sets being used for our experiments

Data set	Description	Color	Resolution	#Images
JAFFE [23]	Japanese Female Facial Expression. Contains 7 standard expressions: happiness, sadness, surprise, anger, disgust, fear, neutrality. Controlled environment	Grayscale	100 × 100	213
CK [24]	Cohn-Kanade. Contains video sequences starting with a neutral expression and ending with a target expression. 7 standard expressions	Grayscale	100 × 100	460
PSL	Facial expression data with 7 expressions collected from Persistent Systems employees	Color	100 × 100	525
FER2013 [25]	Created using Google Search API. Cleaned up and labeled by humans. 7 expressions	Color	48 × 48	35 K
IMDb-WIKI [26]	Images takes from the IMDb and Wikipedia web sites. Stores related data such as name, age, and gender	Color	64 × 64	500 K+
Real-world data set	Captured using real-world conditions like pose and illumination. Two training data sets. TS1: one frontal colored image per user, TS2: one grayscale low-res photo ID per user	Color/ Grayscale	100 × 100 159 × 159 631 × 411	35

The resolution of the images from a few data sets has been changed; image resolution may differ in standard data sets

4 Methodologies

In this section, we give the details of the methodologies that were used in our experiments.

4.1 Convolutional Neural Networks

Convolutional Neural Networks (CNN) are a special type of Artificial Neural Network, specifically used for image processing. They are made up of different layers namely the convolutional layer, the pooling layer, and the fully connected layer. Typically, all the neurons in one layer perform one type of mathematical operation from which it gets its name (Fig. 6). We have used CNNs for facial analysis in the following experiments:

1. Facial Expression Recognition: We consider two data sets for facial expression (i) FER2013 data set, and (ii) all images combined from three data sets—JAFFE, CK, and PSL data sets. We train the network using 70% of the data, and use the remaining 30% of the data for testing.
2. Gender Identification: The IMDb data set has images labeled with gender. We use a similar 70–30 split for training and testing for gender identification with this data set.

4.2 Feature Engineering

When domain knowledge is required to extract features from data used by machine learning algorithms, that procedure is called feature engineering. Transforming input data into a reduced set of features is called *feature extraction* (Fig. 7). We use following methods for feature extraction:

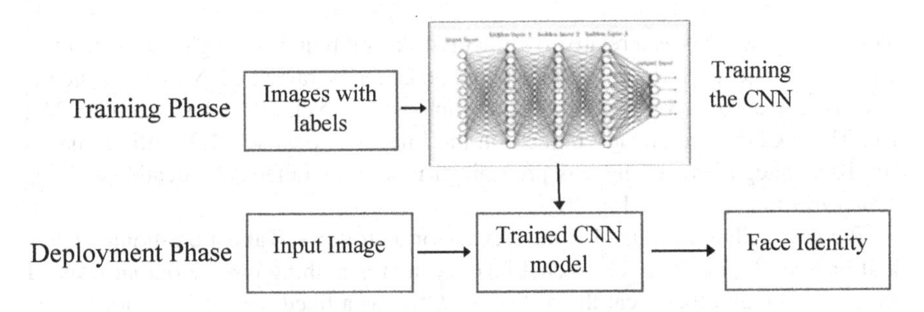

Fig. 6 Face recognition using convolutional neural networks

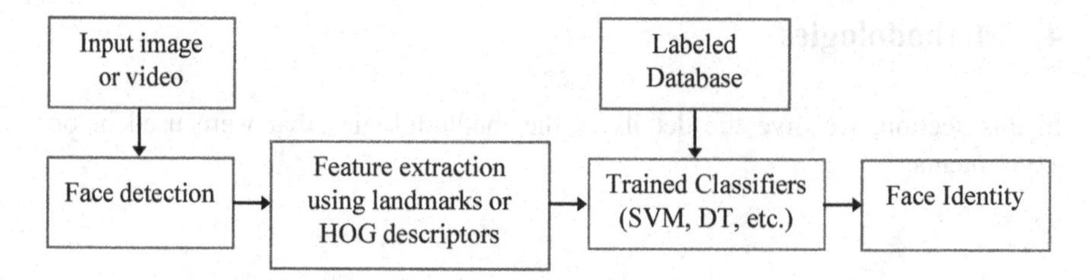

Fig. 7 Face recognition using feature engineering

- Facial Landmarks: The position of facial landmarks change as the expression on the face. We use 68 pre-defined landmarks for facial expression recognition [6]. Feature extraction is done using geometrical techniques on points on the face. For each such point, we get 4 features thus giving a feature vector of length $68 \times 4 = 272$.
- Histogram of Oriented Gradient (HOG): A HOG descriptor uses a single vector to describe the entire face instead of using multiple small feature vectors to describe smaller parts of the face. Typically, the face image is represented by feature vector of size 3780 by using the HOG descriptor.

Once we extract the features from an image, we can use any standard machine learning algorithm for pattern matching and classification. In particular, we tried the following combinations:

1. Facial expression recognition using facial landmarks and ML classification algorithms such as logistic regression and SVMs.
2. Facial expression recognition using HOG descriptors and classification using random decision forests.
3. Face recognition using facial landmarks and classification using logistic regression and SVMs.

4.3 Transfer Learning

Training a new CNN is a relatively difficult task as it requires a big data set and fast computing machines. So, in practice people take a pre-trained CNN model which is already trained on a very large data set such as AlexNet, VGGNet, Inception V3, etc. These CNNs are already trained on the ImageNet data set (1.2 million images for 1000 categories). Using this pre-trained model for other classification tasks is called *transfer learning* (Fig. 8).

We use a CNN as a fixed feature extractor as follows: Take a pre-trained CNN and remove the last fully connected layer (which is nothing but the output layer of the network), and then treat the rest of the CNN as a fixed feature extractor for the new data set. Once features are extracted using the pre-trained model, we can add a

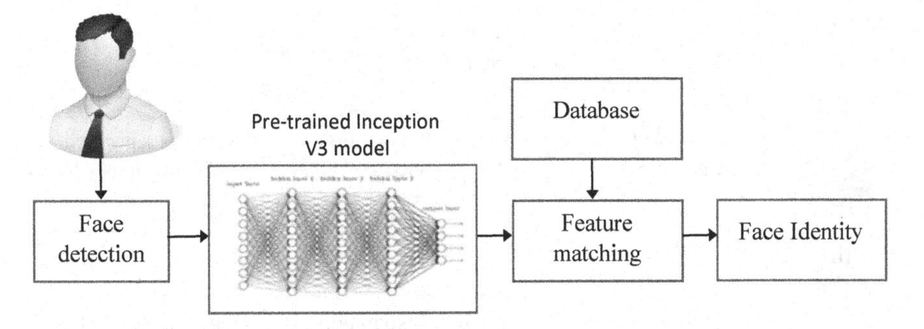

Fig. 8 Face recognition using transfer learning

classifier layer on top of it. Any linear classifier like a linear SVM or linear regression will serve our purpose. Additionally, we can fine-tune the CNN using our own data set.

For transfer learning, we do both facial expression recognition and face recognition using Inception V3 model and standard ML classifiers. In both these experiments, we get feature vectors of size 2048. For facial expression recognition, we use images from the JAFFE, CK, and PSL data sets. The real world data set is used for face recognition. These features are used to train different classifiers. We train six classifiers using these features and corresponding labels.

4.4 Amazon Rekognition

Amazon Rekognition is a tool for image analysis provided on the AWS cloud platform, which can be used for JPEG and PNG image formats. Amazon Simple Storage Service (S3) [27] which is a simple key-based object store, is used for effective data storage for Amazon Rekognition (Fig. 9). We have performed the following experiments using Amazon Rekognition API:

1. Facial Expression recognition, age prediction and gender identification using *DetectFaces*: The *DetectFaces* operation looks for key facial features such as eyes, nose, and mouth to detect faces in an input image.
2. Face Recognition using *IndexFaces* and *SearchFaces* operation: There are two stages here: first, we register the user in the system (registration stage), and then we match a test image with the data set images to predict the user (authentication stage).

We omit the details of these Amazon Rekognition APIs which can easily be found online [2].

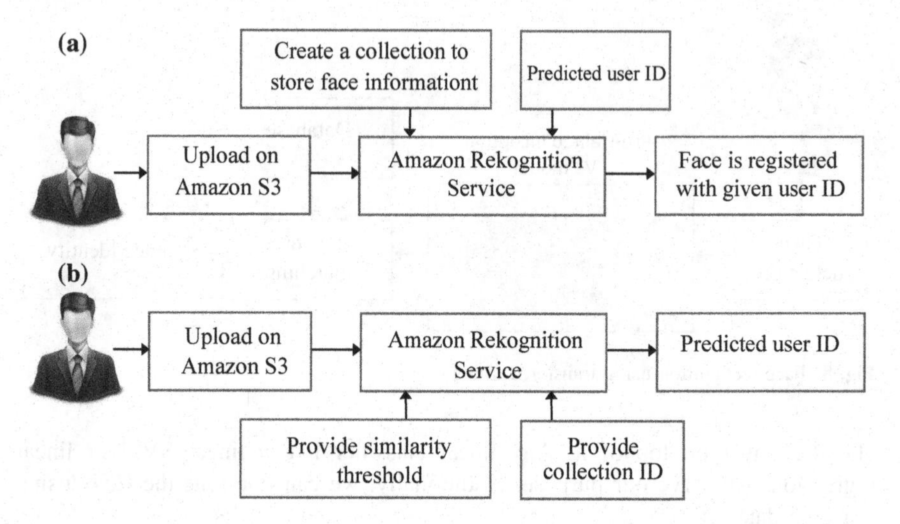

Fig. 9 **a** Face recognition using Amazon Rekognition: user registration. **b** Face recognition using Amazon Rekognition: user identification

5 Experiments and Results

We performed several experiments using the different data sets and methodologies explained above. We summarize our results in this section. We evaluated the following tasks:

1. Face Recognition using Amazon Rekognition, Transfer Learning, and Facial Landmarks.
2. Facial Expression Recognition using CNNs, Transfer Learning, Facial Landmarks, and HOG descriptors.
3. Gender Identification using CNNs.

5.1 Face Recognition

We evaluate face recognition using the real-world data set which has got two training data sets (TS1 and TS2), and one test data set. The test data set is the images of same users taken in the different poses and illuminations. In some cases, we do not use TS2 due to low resolution of the images.

5.1.1 Face Recognition with AWS Rekognition

In AWS Rekognition, the similarity threshold parameter (range is from 0 to 100) is the confidence with which the system should match a face; if the system matches a face with confidence less than the specified threshold, then it is not consider as a match. The following two tables show us the results for the two sets of training data. Note that we use a lower threshold with TS2 as the images have poor resolution. We see that with poor quality training images and a high threshold, the accuracy drops to as low as 69.23%. This low accuracy may not be acceptable in practice, for example when the face has to be identified for authentication and security purposes (Table 2).

5.1.2 Face Recognition with Transfer Learning

We use the pre-trained Inception V3 model for feature extraction from the real-world data set TS1, and then add a layer of classification for user identification. We tried different classifiers such as logistic regression, Naïve Bayes, etc. As the resolution of face images from TS2 was low, we did not use that data set to train our system. We get the highest accuracy using Naïve Bayes and SVMs; however overall the system does not perform very well which may be because the learnt model did not apply well to this data set, as there is only one image per user in the training data (Table 3).

Table 2 Face recognition results using AWS Rekognition with real-world data set

Training set	Similarity threshold	Accuracy (%)	Avg. time to upload image on S3 (s)	Avg. time to predict the user identity (s)
TS1	80	92.30	6.47	2.00
TS1	40	100.00	6.03	1.96
TS2	60	69.23	5.73	2.07
TS2	40	100.00	5.77	1.98

Table 3 Face recognition results using transfer learning with real-world data training set 1

	Classifier	Training time (s)	Testing time (s)	Accuracy (%)
1	Logistic regression	0.0229	0.0015	50.00
2	Naïve Bayes	0.0011	0.0026	57.69
3	Decision trees	0.0027	0.0001	23.07
4	Support vector machine	0.0007	0.0005	57.69
5	K nearest neighbor	0.0004	0.0007	15.38
6	Random decision forest	0.0269	0.0053	26.92

Table 4 Face recognition results using facial landmarks with real-world data training set 1

	Classifier	Training time (s)	Testing time (s)	Accuracy (%)
1	Logistic regression	0.0052	0.0006	7.69
2	Naïve Bayes	0.0013	0.0009	7.69
3	Decision trees	0.0010	0.0001	19.23
4	Support vector machine	0.0007	0.0002	3.84
5	K nearest neighbor	0.0005	0.0009	15.38
6	Random decision forest	0.0263	0.0052	19.23

5.1.3 Face Recognition with Facial Landmarks

Facial landmark detection method is used for feature extraction where we get a feature vector from 68 landmarks on the face. Using these features, several different classifiers are trained to identify the user. We use training data set TS1 but not TS2 as the resolution of the images is quite low. This method has very poor performance with the maximum accuracy around 19% (Table 4).

5.2 Facial Expression Recognition

5.2.1 Facial Expression Recognition with CNNs

CNNs are used to train the classifier for facial expression recognition. We use two data sets—FER2013 and JAFFE + CK + PSL—and train two different classifiers. Each data set contains the seven standard emotions. We split the data set into two parts for training (70%) and testing (30%) data. When we have a small number of images or lower resolution for training, we see that we do not get very high accuracy using CNNs as the training data is insufficient (Table 5).

5.2.2 Facial Expression Recognition with Transfer Learning

Once again for facial expression recognition, we use the pre-trained Inception V3 model for feature extraction, and different classifiers for user identification. We use a combination of three data sets—JAFFE, CK, and PSL. We restrict ourselves to using two or three emotions for classification to get better accuracy (Table 6).

Table 5 Facial expression recognition using CNNs with two different data sets

	Data set	Number of images	Accuracy (%)
1	FER2013	~35,000	66.00
2	JAFFE + CK + PSL	1195	54.00

Table 6 Facial expression recognition using Transfer Learning with two or three emotions

	Classifier	Accuracy of 2-class classifier (happy, neutral) (%)	Accuracy of 3-class classifier (happy, neutral, surprise) (%)
1	Logistic regression	88.18	81.56
2	Naive Bayes	68.50	61.45
3	Decision trees	68.50	55.86
4	Support vector machine	79.52	58.65
5	K nearest neighbors	78.74	65.36
6	Random decision forest	74.01	63.12

5.2.3 Facial Expression Recognition with Facial Landmarks

Facial landmarks like nose, eyes, lips, eyebrows, and jawline play an important role in facial expression recognition. We got 68 landmarks for a face which are then used to extract features (please see the details in Sect. 4). Different classifiers are applied on these features, and the results are compared. The data set is same as we used above with transfer learning (Table 7).

5.2.4 Facial Expression Recognition with HOG Descriptors and Random Decision Forests

In this experiment, HOG descriptors are used for feature extraction. Each subject from the JAFFE data set has three or four images per expression. Two images of each expression per subject are used to train the Random Decision Forest, and the remaining images are used to test the performance of the system. This system provides 97.5% accuracy on the JAFFE data set.

Table 7 Facial expression recognition using facial landmarks with 2 and 3 emotions

	Classifier	Accuracy of 2 class classifier (happy, neutral) (%)	Accuracy of 3 class classifier (happy, neutral, surprise) (%)
1	Logistic regression	96.85	92.09
2	Naive Bayes	85.82	74.01
3	Decision trees	85.3	79.09
4	Support vector machine	65.35	32.76
5	K nearest neighbors	70.86	55.36
6	Random decision forest	92.91	84.74

5.2.5 Facial Expression Recognition with Different Number of Emotions

In order to evaluate the effect of more training data on accuracy, we chose to focus on a fewer classes by choosing those emotions which have a larger number of samples. With just two classes and the largest number of training samples, we get the highest accuracy of 96.85% (across four different methods of classification) which is significantly higher than the highest accuracy of 66.01% when we include all the emotions.

5.3 Gender Identification

The IMDb data set has around 500 K images with gender labels. As thousands of images are available, CNNs (deep learning) are used to predict the gender. We divided the IMDb data into two parts—training and testing data sets using a 70–30 split. The CNN was trained on 70% of data, and the accuracy was measured on the remaining 30% of data. With this setup, we get an accuracy of 96% for gender identification (Table 8).

Table 8 Summary of all the experiments conducted with highest possible accuracy

Experiment	Method	Data set	Accuracy (%)	Num. emotions
Face recognition	AWS Rekognition	Real world TS1 TS2	100.0	–
Face recognition	Transfer learning	Real world TS1	57.69	–
Face recognition	Facial landmarks	Real world TS1	19.00	–
Emotion recognition	CNNs	JAFFE + CK + PSL	54.00	7
Emotion recognition	Transfer learning	JAFFE + CK + PSL	88.18	2
Emotion recognition	Facial landmarks	JAFFE + CK + PSL	96.85	2
Emotion recognition	HOG descriptors	JAFFE	97.50	2
Gender identification	CNNs	IMDb	96.00	–

6 Conclusions

We have evaluated several techniques in facial analysis primarily focusing on face recognition and emotion detection. In summary, Amazon Rekognition has the best performance for both face recognition and facial expression recognition. It can handle challenges like pose, illumination, expression, age, accessories, etc. According to security requirements, we can change the threshold in user identification system—the higher the confidence with which a face must be identified, the higher the threshold should be.

Convolutional Neural Networks also perform well when the data set is large enough for adequate training. For example, in case of gender identification, where we have $\sim$500 K images, the accuracy is 96%; however, for facial expression recognition where we have only 1200 images, the accuracy is significantly lower at 66%. In the case of facial expression recognition, we see that both CNNs and Facial Landmarks perform well when we use fewer emotions as we get more training data per emotion, and the accuracy of the system improves significantly.

Feature extraction using facial landmarks performs better than CNNs as well as transfer learning for facial expression recognition.

Acknowledgements We would like to thank our colleagues—Dr. Siddhartha Chatterjee and Aashis Tiwari—at Persistent Systems for valuable discussions and feedback during this work.

References

1. Facial recognition system. (2017, October). https://en.wikipedia.org/wiki/Facial_recognition_system.
2. Amazon Rekognition. (2017). https://aws.amazon.com/rekognition/.
3. Karpathy, A. Convolutional neural networks (CNNs/ConvNets). http://cs231n.stanford.edu/.
4. Karpathy, A. Transfer learning. http://cs231n.github.io/transfer-learning/.
5. Feature engineering. (2017, September). https://en.wikipedia.org/wiki/Feature_engineering.
6. Kazemi, V., & Sullivan, J. (2014). One millisecond face alignment with an ensemble of regression trees. In *The IEEE Conference on Computer Vision and Pattern Recognition (CVPR)*, June 2014, pp. 1867–1874.
7. Rowley, H. A., Baluja, S., & Kanade, T. (1998, January). Neural network-based face detection. *IEEE Transactions on Pattern Analysis and Machine Intelligence, 20*(1), 23–38.
8. Viola, P., & Jones, M. (2001). Rapid object detection using a boosted cascade of simple features. In *Proceedings of the 2001 IEEE Computer Society Conference on Computer Vision and Pattern Recognition. CVPR 2001*. Kauai, HI, USA: IEEE.
9. Zhang, C., & Zhang, Z. (2010, June). *A survey of recent advances in face detection* (Technical Report). Redmond, WA 98052: Microsoft Research.
10. Principal component analysis. (2017, October). https://en.wikipedia.org/wiki/Principal_component_analysis.
11. Linear discriminant analysis. (2017, September). https://en.wikipedia.org/wiki/Linear_discriminant_analysis.
12. Gabor wavelet. (2015, October). https://en.wikipedia.org/wiki/Gabor_wavelet.
13. Local binary patterns. (2017, September). https://en.wikipedia.org/wiki/Local_binary_patterns.

14. Dalal, N., & Triggs, B. (2005). Histograms of oriented gradients for human detection. In *IEEE Computer Society Conference on Computer Vision and Pattern Recognition*, July 2005. San Diego, CA, USA: IEEE.
15. Scale-invariant feature transform. (2017, September). https://en.wikipedia.org/wiki/Scaleinvariant_feature_transform.
16. Microsoft Developer Guide. Microsoft Kinect. (2017). https://developer.microsoft.com/en-us/windows/kinect.
17. Krizhevsky, A., Sutskever, I., & Hinton, G. E. (2012). ImageNet classification with deep convolutional neural networks. In *NIPS'12 Proceedings of the 25th International Conference on Neural Information Processing Systems—Volume 1*, Lake Tahoe, Nevada, December 03–06, 2012, pp. 1097–1105.
18. Zeiler, M. D. & Fergus, R. (2013). Visualizing and understanding convolutional networks. In *CoRR*.
19. Simonyan, K., & Zisserman, A. (2014). Very deep convolutional networks for large-scale image recognition. In *CoRR*.
20. Szegedy, C., Liu, W., Jia, Y., Sermanet, P., Reed, S., Anguelov, D., et al. (2015). Going deeper with convolutions. In *Computer Vision and Pattern Recognition (CVPR)*. IEEE.
21. He, K., Zhang, X., Ren, S., & Sun, J. (2016). Deep residual learning for image recognition. In *IEEE Conference on Computer Vision and Pattern Recognition (CVPR)*, June 2016. Las Vegas, NV, USA: IEEE.
22. Stanford Vision Lab, Stanford University, Princeton University. (2016). ImageNet.
23. Lyons, M. J., Akamatsu, S., Kamachi, M., & Gyoba, J. (1998). Coding facial expressions with Gabor Wavelets. In *Proceedings, Third IEEE International Conference on Automatic Face and Gesture Recognition*, April 14–16, 1998, pp. 200–205. Nara Japan: IEEE Computer Society.
24. Kanade, T., Cohn, J. F., & Tian, Y. (2000). Comprehensive database for facial expression analysis. In *Fourth IEEE International Conference on Automatic Face and Gesture Recognition*.
25. Challenges in Representation Learning: Facial Expression Recognition Challenge. (2013). https://www.kaggle.com/c/challenges-in-representation-learning-facial-expression-recognition-challenge/data.
26. Rothe, R., Timofte, R., & Van Gool, L. (n.d.). IMDB-WIKI—500 k + face images with age and gender labels.
27. Amazon Web Services, Inc. Amazon S3. (2017). https://aws.amazon.com/s3/.

Survey on Hybrid Data Mining Algorithms for Intrusion Detection System

Harshal N. Datir and Pradip M. Jawandhiya

Abstract Security is one of the most major concern issue arises in computer and internet technology. To conquer this problem, Intrusion Detection System (IDS) is the challenging solution in network systems. Such system is used to detect the known or unknown attacks made by intruders. Data mining methodologies like, clustering, classification, etc., plays a very important role in design and development of such IDS. They makes such system more effective and efficient. This paper describes some recent hybrid data mining based approaches used in development of IDS. We also describe the hybrid classification approaches used in IDS. Such Hybrid classifiers are any mixture of basic classifiers such as, SVM, Bayesian classifier, Neural network classifier, etc.

Keywords Intrusion detection · Data mining · Machine learning
Hybrid classification · Clustering · Intruders · Computer and network systems

1 Introduction

From last few years, computer network security become challenging and point of interest for many researchers. The various industrial organizations realize that, information and network security is most important part for protecting their information. Intrusion is an security attack, which is caused by either any successful or unsuccessful attempt for breaking the integrity, confidentiality or availability of information or its resource. All kind of informatics industries have faced such kind of information attacks or intrusions. To such kind of problems there is a solution called

H. N. Datir (✉)
Department of Information Technology, Sipna College of Engg & Tech,
Amravati, Maharashtra, India
e-mail: harshaldatir809@gmail.com

P. M. Jawandhiya
Department Computer Sci & Engg, PLIT, Buldhana, Maharashtra, India
e-mail: pmjawandhiya@rediffmail.com

V. E. Balas et al. (eds.), *Data Management, Analytics and Innovation*,
Advances in Intelligent Systems and Computing 808,
https://doi.org/10.1007/978-981-13-1402-5_22

as Intrusion Detection Systems (IDS). Intrusion Detection System is a key technique in information security area. It plays a very important role for detection of different kind of intrusions or attacks and maintains the security of networking systems.

It observes and analyze all kind of events occurs or arise in computer network systems for solving security problems. IDS provides following three important functions for security of information such as:

- Monitor Function
- Detect Function
- Respond Function.

IDS maintains the management of security systems with the help of non-expert staff with user-friendly interface [1].

Following are some services provided by IDS:

- Observe and analyze the activities of computer and network system
- Auditing of configurations and vulnerabilities of systems
- Data file integrity evaluation
- Estimations of malicious and non-malicious activities.

1.1 IDS Classification

The basic challenge in the field of network and system infrastructure is authorization of user identity and data access policy without violating privileges. Intruders are either insider threats or outsider threats, dangerous to the systems. IDS systems classify and deals with the malicious use of resources of network and computer systems. It ensures the security of network with identification of its violation of policies. The main aim of such system is to protect the system from malicious activities automatically.

Detection based IDS are discussed here:

- **Anomaly Detection**:

Anomaly IDS is used to detect anomalies if any deviation occur in the normal system. It is very helpful for fraud detection, network based intrusion detection and unusual event detection. All these are very hard to find. It is also called as behavior based detection as it is related to variations in user behavior.

Advantage: Able to detect novel or unknown attacks based on audit data.

Disadvantage: Malicious activity that falls within normal usage patterns is not detected.

- **Misuse Detection**:

Misuse IDS is used to detect the behavior with traffic analysis along with analysis and comparison of several rules. It can be used on the basis of matching signature pattern approach. It is also known as signature based detection system; alarms are generated if signature is wrong or modified.

Advantage: Able to generate accurate results along with minimum false alarms.

Disadvantage: Only detect known attacks.

1.2 IDS with Classification

Data Mining is the process of removing hidden procedure, previously unknown and valuable knowledge from big amount databases. Data mining focusing on various issues such as feasibility, efficacy, scalability and efficiency. Thus, data mining helps to detect patterns in the set of data and also use patterns to detect future intrusions in same data.

Data mining is one of the important techniques of classification. IDS with classification can classify all incoming network traffic in common or malicious class. Generally classification is used for anomaly detection. The fundamental steps are

1. It accepts collection of item as input.
2. Maps the items into predefined groups or classes define by specific qualities.
3. After mapping, it outputs a classifier that can correctly predict the class to which a fresh item belongs.

2 Literature Survey

A hybrid intrusion detection model is proposed in [2]. This model is a combination of base feature selection classifier. Each classifier uses partial original feature space and data mining classifier. This system basically combines feature selection approach to improve the detection rate and data mining technique for minimum false alarms. It is used for both anomaly and misuse detection. This hybrid model performs better with low FPR on normal computer utility and High DR with malicious activities. This hybrid system is better than system with individual feature selection classifier. It is effective for accurate intrusion detection (Fig. 1).

Three classifier named as: ANN-based, Naïve-Bayes, and DT classifier are merge in [3] for intrusion detection system. This system is named as hybrid-multistage IDSs. The combination algorithms had been implemented empirically to form four hybrid classifiers: MLP-based, SVM-based, Naïve-Bayes-based and J48-DT-based classifier. The multistage-MLP and hybrid-multistage-J48 achieves 100% recognition rate for normal and abnormal types of attacks. The cost of processing time and hardware complexities are minimum. Following attacks are detected with this proposed system:

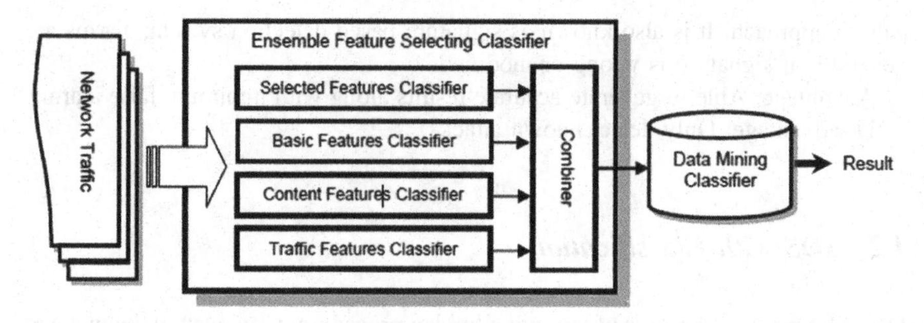

Fig. 1 Hybrid model for intrusion detection [2]

1. Denial-of-service (DoS)
2. Probing (PRB)
3. Remote-to-Local (R2L) access
4. User-to-Root (U2R) access.

A multiple-level hybrid classification model is proposed in [4]. This model combines decision trees classifier and Bayesian clustering. System is tested on KDD Cup 1999 Data and compare with existing approaches like MADAM ID and multiple level tree classifier. This system effectively and efficiently detects intrusions with very low false negative rate of 3.37%. False alarm rate is also very low. System combines concept of both supervised and semi supervised learning approach such as classification and clustering respectively (Fig. 2).

Another multiple-level hybrid classification model is proposed in [5] which combine decision tree classifier with enhanced heuristic clustering technique. It has low false negative rate up to 2.7% and low false alarm rate up to 9.1%. Enhanced fast heuristic clustering (EFHCAM) method used in this system performs effectively and effectively to detect both known and unknown intrusions (Fig. 3).

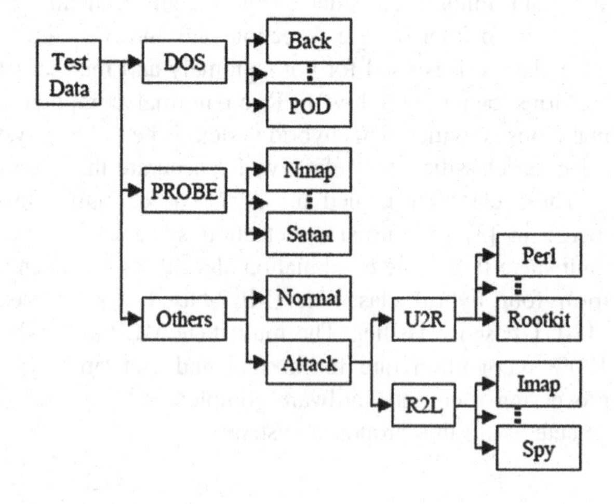

Fig. 2 Multiple level hybrid classifier [4]

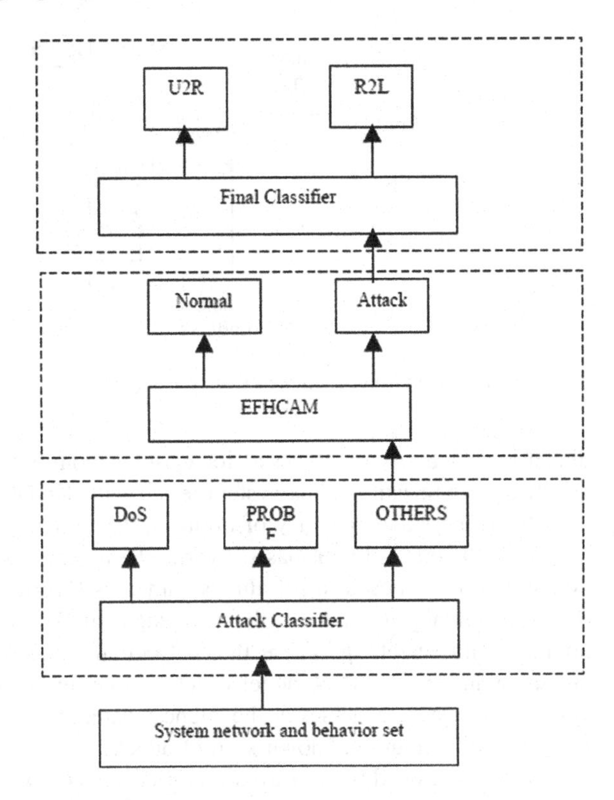

Fig. 3 Multi-level hybrid classifier with enhanced C4.5 for IDS [5]

Authors in [6] given a technique for intrusion detection which is based on classification attack. The present IDS makes use of all the present features, i.e., 41 features in network. Authors made use of AGAAR which reduces the redundant and irrelevant features. For classification of attacks in NSL-KDD a novel GPLS model is utilized. For tanning of the classifier they used full features of 20%-NSL-KDD. Classifiers are trained with 19.51% of the features of dataset. By minimizing datasets dimensionality the accuracy of the classifier is increased.

A new hybrid intrusion detection method is developed by the authors in paper [7]. This method consolidates the advantages present in anomaly as well as misuse detection. Authors have used K-Means clustering algorithm with the hybrid classifier which minimizes the false alarm rate present in anomaly detection also made use of naïve Bayes classifier merged with k-Nearest Neighbor for the propose of detection of intrusions. The real life data set has a very small variation in normal and anomalous data due to which algorithms used for calcification can misclassifies them which can cause the misclassification of few records. In developed technique by making use of some sort of fuzzy-based algorithms authors are able to overcome this issue.

Neural network is applied to the field of intrusion detection in [8]. It solves the problem of low rate, poor performance, etc. This paper develops a hybrid classifier, which is the combination of KPCA, RBFNN and PSO. KPCA is used to reduce the

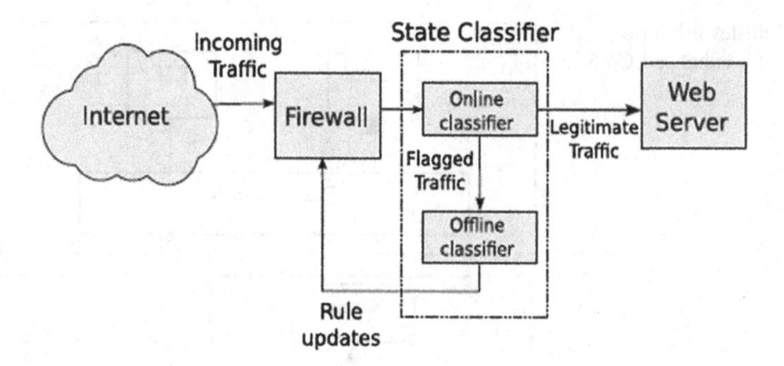

Fig. 4 Flow of hybrid intrusion detection system [11]

dimensions, RBF is purely used for classification and PSO is used for EBFNN parameter optimization. This system tested with KDDCUP99 data set and shows the effectiveness and accuracy of intrusion detections.

Another neural network based hybrid classification system for intrusion detection system is proposed in [9], this is named as Hybrid Neural Network Classifier (HNAA). Initially it combines the advantages of GA and LM algorithm fir global optimal point searching. These three algorithms effectively adjust the input and output parameters of ANN model which will be further used in intrusion detection system. This system achieves the highest detection rate for intrusion detections including known and unknown kind of attacks.

Decision trees and Naive Bayes classifiers are combine used in [10] for intrusion detection in computer or network systems. This system reduce the false negative rate with accurately identifying all kind of intrusions. This system can be extended for attack classification in future.

A hybrid approach for adaptive network intrusion detection is proposed in [11]. HMM is used for intrusion detection and experiment is performed on DARPA data set. System incorporated HMM model along with NB model into a hybrid model for intrusion detection (Fig. 4).

3 Research Directions

Challenges arises in the process of intrusion detection systems are listed here:

1. Reduce the number of false positives which increase efficiency of IDS
2. IDS should work with maintaining precision and recall very high
3. IDS should have low false positive rate and low false negative rate
4. Time required for huge amount of data training should be minimum
5. Improve classification accuracy in IDS
6. Use of multiple classifier to improve the effectiveness and efficiency of IDS

7. Implement such IDS along with heterogeneous operating systems and configurations is very challenging task in real-time environment
8. Study of standard datasets for evaluation of real-time IDS
9. Use of various feature reduction techniques instead of feature selection in the process of data reduction, which will lead to decrease the computational complexity
10. Implement IDS which can perform misuse detection and anomaly detection combine.

4 Comparative Analysis

This survey concludes that Good quality of IDS having combinations of multiple classifiers can detect both known and unknown types of intrusions. As we compare PROBE, U2R and R2L attack, performance of new hybrid intrusion detection method which is combination of K-Means + KNN + Naïve Bayes developed by the authors in paper [7] is better than other proposed hybrid classifier models. Whereas accuracy in detection rate of DOS attack is improved in model developed [8] which is combination of Kernel Principal Component Analysis (KPCA) + Particle Swarm Optimization (PSO) + Radial Basis Function Neural Network (RBF).

5 Conclusion

The Network Intrusion Detection System is a latest technique related to network security. Many advanced technologies related to intrusion detection systems proposed till date, are described in this paper. From this survey we conclude that, data mining techniques plays very important role in the successful execution if intrusion detection systems. Many classifiers are used in literature to implement the IDS, such as Bayesian classifier, neural networks, Support vector machine, Decision trees and K-Means, etc., are used. In this survey we studied various models that have used the concept of hybrid classification systems for intrusion detection. Various models are analyzed on the basis of false positive rate, false negative rate, false alarm, accuracy of detected intrusions.. The IDS should detect the latest types of attacks to prevent the attack at an early stage. Also system can generate immediate alert for admin or user related to occurrence of intrusions.

References

1. Wagh, S. K., Pachghare, V. K., & Kolhe, S. R. (2013). Survey on intrusion detection system using machine learning techniques. *International Journal of Computer Applications, 78*(16).
2. Chou, T.-S., & Chou,T.-N. (2009). Hybrid classifier systems for intrusion detection. In *CNSR'09 Seventh Annual Communication Networks and Services Research Conference*. IEEE.
3. Arumugam, M., et al. (2010). Implementation of two class classifiers for hybrid intrusion detection. In *2010 International Conference on Communication and Computational Intelligence (INCOCCI)*. IEEE.
4. Xiang, C., & Lim, S. M. (2005). Design of multiple-level hybrid classifier for intrusion detection system. In *2005 IEEE Workshop on Machine Learning for Signal Processing*. IEEE.
5. Rajeswari, L. P., & Kannan, A. (2008). An intrusion detection system based on multiple level hybrid classifier using enhanced c4. 5. In *International Conference on Signal Processing, Communications and Networking, ICSCN'08*. IEEE.
6. Hedar, A.-R., et al. (2015). Hybrid evolutionary algorithms for data classification in intrusion detection systems. In *16th IEEE/ACIS International Conference on Software Engineering, Artificial Intelligence, Networking and Parallel/Distributed Computing (SNPD)*. IEEE.
7. Om, H., & Kundu, A. (2012). A hybrid system for reducing the false alarm rate of anomaly intrusion detection system. In *1st International Conference on Recent Advances in Information Technology (RAIT)*. IEEE.
8. Xu, R., An, R., & Geng, X. F. (2011). Research intrusion detection based PSO-RBF classifier. In *2011 IEEE 2nd International Conference on Software Engineering and Service Science*. IEEE.
9. Xiangmei, L., & Qin, Z. (2011). The application of hybrid neural network algorithms in intrusion detection system. In *2011 International Conference on E-Business and E-Government (ICEE)*. IEEE.
10. Benferhat, S., & Tabia, K. (2005). On the combination of naive bayes and decision trees for intrusion detection. In *International Conference on Computational Intelligence for Modelling, Control and Automation and International Conference on Intelligent Agents, Web Technologies and Internet Commerce (CIMCA-IAWTIC'06)*, Vol. 1. IEEE.
11. Karthick, R. R., Hattiwale, V. P., & Ravindran, B. (2012). Adaptive network intrusion detection system using a hybrid approach. In *2012 Fourth International Conference on Communication Systems and Networks (COMSNETS 2012)*. IEEE.

An Efficient Recognition Method for Handwritten Arabic Numerals Using CNN with Data Augmentation and Dropout

Akm Ashiquzzaman, Abdul Kawsar Tushar, Ashiqur Rahman and Farzana Mohsin

Abstract Handwritten character recognition has been the center of research and a benchmark problem in the sector of pattern recognition and artificial intelligence, and it continues to be a challenging research topic. Due to its enormous application many works have been done in this field focusing on different languages. Arabic, being a diversified language has a huge scope of research with potential challenges. A convolutional neural network (CNN) model for recognizing handwritten numerals in Arabic language is proposed in this paper, where the dataset is subject to various augmentations in order to add robustness needed for deep learning approach. The proposed method is empowered by the presence of dropout regularization to do away with the problem of data overfitting. Moreover, suitable change is introduced in activation function to overcome the problem of vanishing gradient. With these modifications, the proposed system achieves an accuracy of 99.4% which performs better than every previous work on the dataset.

Keywords Data augmentation · Deep learning · Dropout · ELU
Neural network

A. Ashiquzzaman (✉) · A. K. Tushar
Computer Science and Engineering Department, University of Asia Pacific, Dhaka, Bangladesh
e-mail: zamanashiq3@gmail.com

A. K. Tushar
e-mail: tushar.kawsar@gmail.com

A. Rahman
Computer Science and Engineering Department, Bangladesh University of Engineering and Technology, Dhaka, Bangladesh
e-mail: ashiqbuet14@gmail.com

F. Mohsin
Department of Management Information Systems, University of Dhaka, Dhaka Bangladesh
e-mail: farzanamohsinmohona@gmail.com

V. E. Balas et al. (eds.), *Data Management, Analytics and Innovation*,
Advances in Intelligent Systems and Computing 808,
https://doi.org/10.1007/978-981-13-1402-5_23

1 Introduction

Automatic character recognition, also known as optical character recognition (OCR), has very high commercial and pedagogical importance and has been receiving a profound interest from researchers over the past few years. It is the key method in check reading, textbook digitalization, data collection from forms, and in other numerous kinds of automated data extraction. And this process is being applied to different languages all over the world.

Arabic is the fifth most spoken language in the world. And it is also one of the official languages of United Nation from 1973. Along with 290 million native speakers, it is spoken by 422 million speakers in 22 countries [1]. Arabic is the major source of vocabulary for languages Turkish, Uighur, Urdu, Kazakh, Kurdish, Uzbek, Kyrgyz, and Persian. And alphabets of many other languages like Persian, Syrian, Urdu, Turkish are quite similar to Arabic [2].

The past works of OCR of hand written alphabets and numerals were concentrated on Latin languages. Gradually other languages are also coming to the fore. But Arabic still has many unexplored fields of study. Maximum work done on this language was to detect printed characters [3]; however, detection of hand written alphabets is more challenging than printed characters. Some work has been done on Arabic handwritten digit recognition via patterns learned from other related languages [4]. The interesting point of note regarding Arabic is, though the words as well as characters are written from right to left, the numerals are written from left to right. Figure 1 depicts the ten Arabic numeral classes.

Das et al. [5] have devised a method to recognize handwritten Arabic numerals by multilayer perceptron (MLP) with a considerable accuracy. In [6] a convolutional neural network model is used with dropout based on the model [5] that achieves a better accuracy on the same dataset. In this paper, we propose a modification to the same method devised in [6]. We introduce data augmentation to make learning more robust against overfitting. Exponential linear unit (ELU) is introduced instead of Rectified Linear Unit (ReLU) to fix the vanishing gradient problem.

2 Background

The method devised in [5] to recognize Arabic numerals uses MLP as a classifier. It uses a set of 88 features; among them 72 are shadow feature and 16 are octant features. To train this MLP a huge data set of 3000 handwritten sample is used which is obtained from CMATERDB 3.3.1 [7]. Each of these pictures is scaled to the size of 32 $\times$ 32. For uniform training each pixel is rendered in gray scale.

The MLP used in [5] has a single hidden layer along with input and output layer. And this single hidden layer is enough to classify the given data set [8]. The setting of 3 layers for the MLP method is shown in Fig. 2. From Fig. 2 we can see that

Arabic Digit	English Digit	Image	Inverted Image
١	1		
٢	2		
٣	3	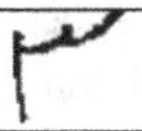	
٤	4		
٥	5		
٦	6		
٧	7	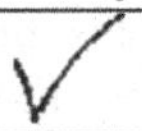	
٨	8		
٩	9		
٠	0		

Fig. 1 Handwritten Arabic digits and corresponding inverted images [6]

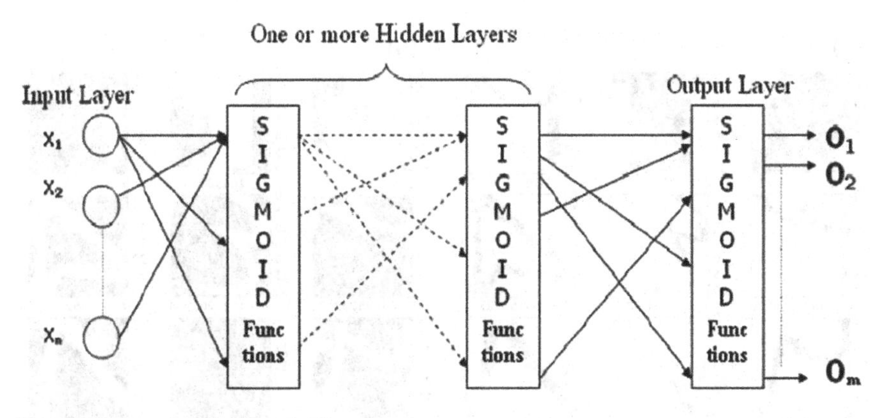

Fig. 2 MLP used by Das et al. [5]

sigmoid is used as nonlinearity after each layer. The width of the network is determined through trial and error to get a good bias and minimal variance. Supervised learning is undertaken and performed over 2000 samples. Simple back-propagation algorithm is used for training.

The result was evaluated by cross validation, and each time the numbers of neurons are varied to get a perfect bias-variance combination. By exchanging the number of neurons arbitrarily between training data set and test data set error can be brought very close to zero, but this will generate worse performance when tested on samples outside of sample dataset. This condition is known as "Overfitting" [9] and this is because the model gets too much acquainted to the training data set and cannot generalize over a broader spectrum as a result. The number of neurons in hidden layer is finally fixed to 54 that give an accuracy of 93.8% in recognizing Arabic numerals.

Das et al. split the total data set at a ratio 2:1 for training and testing. The images are normalized before feeding into the network. Before feeding data into MLP the images were transformed into a simple one dimensional vector.

The model described in [6] uses the same dataset as [5] but adopts the method of CNN for numeral recognition. The images are kept in the original form for CNN; however, colors of the images are inverted before feeding. Figure 3 shows the final condition of images before feeding in CNN model. Figure 4 describes the total model used by them. From Fig. 4 it is seen that, their model consists of two layers of convolution with kernels of size 5 × 5 and 3 × 3, respectively each followed by a 2 × 2 max-pooling layer. The fully connected network consists of a hidden layer with 128 neurons and one final layer with 10 neurons for each numeral class. They use ReLU as activation function and categorical cross entropy for error calculation. Softmax function is used to get the final result from output layer. Total dataset is split at ratio 2:1 for training and validation purpose. This model gives an accuracy of 97.4% over validation dataset.

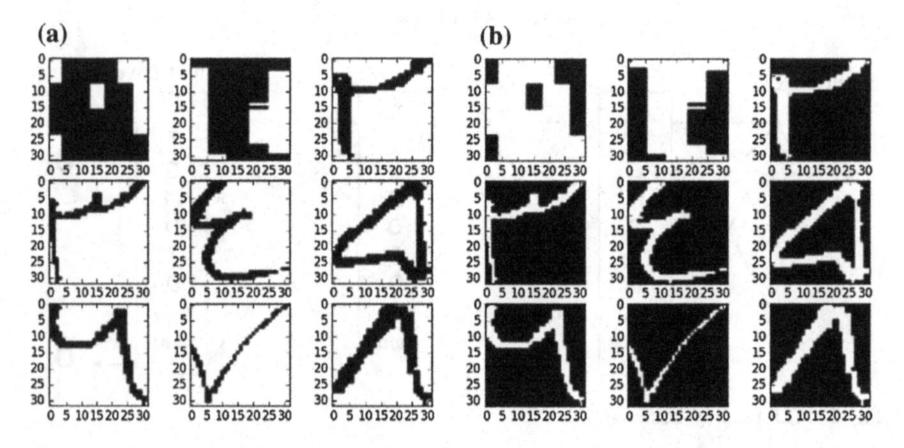

Fig. 3 **a** Original data. **b** Final data after processing

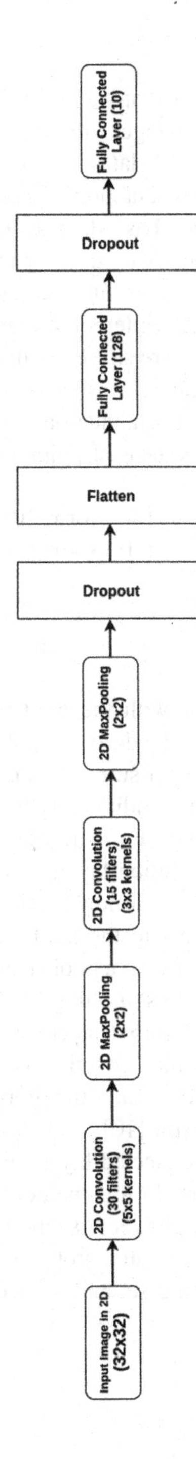

Fig. 4 CNN used in [6]

3 Proposed Method

In this section we discuss the changes that we bring to the method described in [6] to improve the accuracy. Two key changes are brought about—we introduce data augmentation to the CMATERDB 3.3.1 dataset, and change the activation function from ReLU to ELU. The pictorial presentation of our model is given in Fig. 5.

Data augmentation transforms our base data. In our case, it simply takes our dataset of images and transforms it by rotation, color variation or by adding noise. It makes our model more robust against over fitting and numerically increases the size of the dataset. We have apply ZCA-whitening as augmentation, images are also rotated in a range of 10°. The images are shifted both horizontally and vertically to an extent of 20% of the original dimensions randomly. The images are zoomed randomly up to 10%. During this augmentation process the points outside the boundaries are filled according to the nearest point.

$$\Big| f.'(x) = \begin{cases} 0 & \text{for}\, x<0 \\ 1 & \text{for}\, x\geq 0 \end{cases} \quad (1)$$

$$f'(\alpha, x) = \begin{cases} f(\alpha, x) + \alpha & \text{for}\, x<0 \\ 1 & \text{for}\, x\geq 0 \end{cases} \quad (2)$$

The ReLU is sometimes plagued with the gradient vanishing problem. From Eq. (1) it is seen that for region $x < 0$ the derivation of ReLU is 0, and due to this in this region the updating of weight vector stops. And this stops the learning process. The ELU function can prevent this condition as its derivative Eq. (2) does not become zero in any point on the curve. Furthermore, it assures a smooth learning.

In our model, we have four convolution layers, each with ELU as an activation function. The kernel size is determined through trial and error, and the window of 3×3 gives maximum accuracy. Next to the final layer of convolution, we have added a pooling layer doing max-pool with a pool size of 2×2. After max-pooling the convoluted images are flattened by squashing 2D convoluted data and fed into the fully connected layers. The final output layer contains 10 output nodes representing 10 classes of numerals. Softmax function is used to calculate final result from the output layer. This function calculates the probability of each class from the ELU value of each output layer neuron [10].

Both the convolution and fully connected layers have a dropout rate of 25% to prevent data over overfitting problem. In this method, in each epoch 25% neurons in each layers do not update their weight vectors, and the effect of these neurons are removed from network. If neurons are dropped, they are prevented from co-adapting too much with the training set and reduces overfitting.

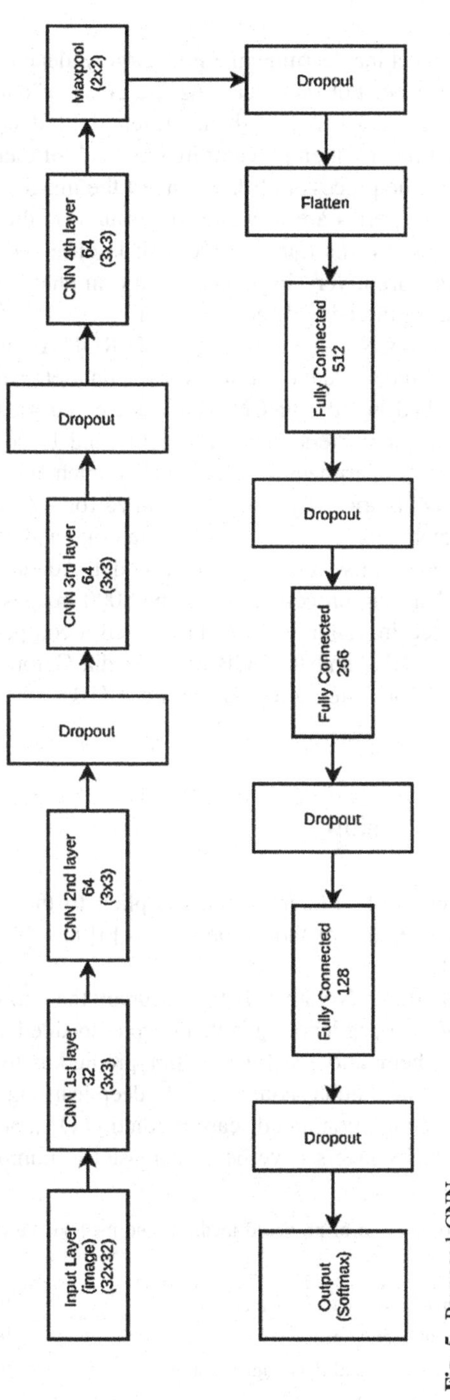

Fig. 5 Proposed CNN

4 Experiments

The success of any machine learning method largely depends on the size and correctness of dataset used. For deep learning the effect of data set is even more vital. The CMATERDB 3.3.1 Arabic handwritten digit dataset is used [7]. It contains 3000 separate handwritten numeral image. Each of them is 32 × 32 pixel RGB image. Similar to the process in [6], we invert the images before feeding into CNN. As a result the numerals are in white foreground on the backdrop of black background. For the past works done on OCR, it is observed through activation visualization that edges are a very important feature in character recognition, and black background makes the edge detection easier.

We train the proposed CNN model with CMATERDB Arabic handwritten digit dataset and test against the test data sample of same dataset. Similar to the process implementation described in [6], our CNN have increased 4; 977; 290 instead of total 75; 383 trainable parameters such as weights and biases. Model is implemented in Keras [11]. Experimental model was implemented in Python using Theano [12] and Keras libraries. The model is trained for 100 epoch with different kernel sizes. The batch size is 128 for both training and testing. Categorical crossentropy is used as the loss function in this model. Adadelta optimizer [13] is used to optimize the learning process. Among the 3000 images, 2500 are used for training and 500 is used in validation. We have used a computer with CPU Intel i5-6200U CPU @ 2.30 GHz and @ 4 GB ram. Nvidia Geforce GTX 625M dedicated graphics is used for faster computation, i.e. CUDA support for accelerated training is adopted.

5 Result and Discussion

Table 1 shows the result of different models applied to the same CMATERDB dataset. Method-1 denotes the method proposed in [5] and Method-2 denotes the method proposed in [6].

Our proposed method of CNN gives better accuracy than both the methods. The prominent attribute of the deep learning is that deep embedded layers recognize the features and cascade them into the final output prediction to cast classification cation. This is the attribute that has contributed to deep learning method being used in problems in image recognition and disease prediction [14], among others. In our proposed model, the CNN layers have been increased in number from the model

Table 1 Performance comparison of proposed methods and methods described in [6, 5]

Method Name	Accuracy
Method-1 (without dropout and data augmentation)	93.8
Method-2 (without data augmentation)	97.4
Proposed Method (with dropout and data augmentation)	99.4

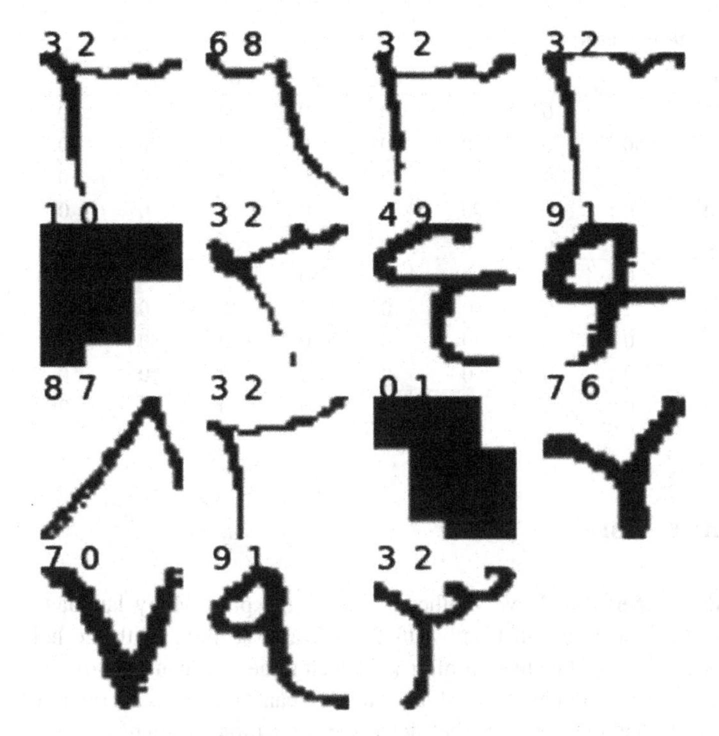

Fig. 6 Some misclassified images during training. Blue denotes actual class and red denotes predicted class

used by [6]. We have also added two other fully connected layers in later phase of the proposed model to enhance feature extraction and recognition. The result was significance improved from the previous methods because of the introduction of data augmentation during training. Data augmentation virtually makes the image transformed into new set of features for the neural network to detect and recognize. It also shifts and zooms the training images, making the image decentralized for the kernels to recognize them from various positions.

Figure 6 denotes some of the classified images during training. It demon-castrate some of classified images, which are actually impossible to recognize manually due to morphological decomposition of cure shape. Table 2 shows the class-wise classification of the test images, which denotes the final accuracy rate of 99.40%. The columns denote true classes and the rows denote predicted classes. The cell content denotes the count of predicted classes.

Table 2 Confusion matrix

	0	1	2	3	4	5	6	7	8	9
0	50	0	0	0	0	0	0	0	0	0
1	0	50	0	0	0	0	0	0	0	0
2	0	0	49	1	0	0	0	0	0	0
3	0	0	0	50	0	0	0	0	0	0
4	0	0	0	0	50	0	0	0	0	0
5	0	0	0	0	0	50	0	0	0	0
6	0	2	0	0	0	0	48	0	0	0
7	0	0	0	0	0	0	0	50	0	0
8	0	0	0	0	0	0	0	0	50	0
9	0	0	0	0	0	0	0	0	0	50

6 Future Work

Our model outperformed every other model in the past. Many languages do not have such high accuracy in Handwritten numeral detection. With the help of this model by applying Transfer learning [15] technique, we can improve the model strength of other languages. And this model can be taken as reference while working with Handwritten alphabet detection for Arabic language.

7 Conclusion

The OCR is a benchmark problem in pattern recognition and has wide commercial interest. The work is based on the dataset CMATERDB 3.3.1. Das et al. performed the first work done on this dataset while using MLP in recognition. This model gains an accuracy of 93.8%. The work in [6] is based on the work of [5] and uses CNN as its model. This model uses dropout regularization and ReLU as activation function, and for the final output softmax function is used in the output layer. This model achieves an accuracy of 97.3%. This paper proposes a modification to the model proposed in [6]. We add data augmentation to prevent overfitting, as well as change the activation function from ReLU to ELU to prevent vanishing gradient problem and make the learning more uniform. After adopting all these changes, the proposed model achieves an accuracy of 99.4% which is better than any of the previous works on this dataset.

References

1. World arabic language day, unesco. (2017). http://www.unesco.org/new/en/unesco/events/prizes-and-celebrations/celebrations/international-days/world-arabic-language-day/. Retrieved May 27, 2017.
2. Boucenna, A. (2006). Origin of the numerals. arXiv preprint math/0606699.
3. Amin, A. (1998). Off-line arabic character recognition: The state of the art. *Pattern recognition, 31*(5), 517–530.
4. Tushar, A. K., Ashiquzzaman, A., & Afrin, A., & Islam, M. (2017). A novel transfer learning approach upon hindi, arabic, and bangla numerals using convolutional neural networks. arXiv preprint arXiv:1707.08385.
5. Das, N., Mollah, A. F., Saha, S., & Haque, S. S. (2010). Handwritten arabic numeral recognition using a multi-layer perceptron. CoRR, abs/1003.1891. Retrieved October 30, 2017.
6. Ashiquzzaman, A., & Tushar, A. K. (2017). Handwritten arabic numeral recognition using deep learning neural networks. In *IEEE International Conference on Imaging, Vision & Pattern Recognition (icIVPR)*, pp. 1–4. IEEE.
7. Google code archieve—long-term storage for google code project hosting. https://code.google.com/archive/p/cmaterdb/downloads. Retrieved May 30, 2017.
8. Kubat, M. (1999). Neural networks: A comprehensive foundation by simon haykin, macmillan. *Knowledge Engineering Review, 13*(4), 409–412. ISBN 0-02-352781-7. Retrieved October 30, 2017.
9. Domingos, P. M. (2000). Bayesian averaging of classier and the overfitting problem. In *Proceedings of the Seventeenth International Conference on Machine Learning (ICML 2000),* Stanford University, Stanford, CA, USA, June 29–July 2, 2000, pp. 223–230. Retrieved October 30, 2017.
10. Hinton, G. E., & Salakhutdinov, R. R. (2009). Replicated softmax: An undirected topic model. In *Advances in Neural Information Processing Systems*, pp. 1607–1614.
11. Chollet, F. (2015). keras. https://github.com/fchollet/keras. Retrieved October 30, 2017.
12. Theano Development Team. (2016). Theano: A python framework for fast computation of mathematical expressions. arXiv e-prints, abs/1605.02688, May 2016. Retrieved October 30, 2017.
13. Zeiler, M. D. (2012). Adadelta: An adaptive learning rate method. arXiv preprint arXiv:1212.5701.
14. Ashiquzzaman, A., Tushar, A. K., Islam, M., & Kim, J.-M., et al. (2017). Reduction of overfitting in diabetes prediction using deep learning neural network. arXiv preprint arXiv:1707.08386.
15. Pan, S. J., & Yang, Q. (2010). A survey on transfer learning. *IEEE Transactions on knowledge and data engineering, 22*(10), 1345–1359.

Secured Human Authentication Using Finger-Vein Patterns

M. V. Madhusudhan, R. Basavaraju and Chetana Hegde

Abstract In any organization, providing a secured authentication system is a challenge. Here, we propose a secured authentication process using finger-vein patterns. Finger vein is a reliable biometric trait because of its distinctiveness and permanence properties. The proposed algorithm initially captures the finger-vein image and is preprocessed using Gaussian blur and morphological operations. Then features like number of corner points and the location of these corner points are extracted. The features fetched for an individual from database are compared against the extracted features. If the comparison satisfies predefined threshold value, then the authentication is successful. The simulation results of the proposed algorithm have produced the FAR as 2.78%, FRR as 0.09% and the overall performance as 99.96%.

Keywords Euclidean distance · Finger vein · Gaussian blur · Harris corner Patterns · Preprocessing

1 Introduction

Biometrics is a science of identifying a person using their physiological or behavioral characteristics [1]. If any behavioral or physiological characteristic of a person fulfils the requirements such as universality, permanence, collectability, and distinctiveness then it can be considered as a biometric trait [2]. Acceptability, circumvention and performance are the other concerns needs to be considered in a practical biometric system [3]. By taking all these requirements into consideration,

M. V. Madhusudhan
Sri Pillappa College of Engineering, Bengaluru, Karnataka, India

R. Basavaraju
Innominds Software Pvt. Ltd., Bengaluru, Karnataka, India

C. Hegde (✉)
RNS Institute of Technology, Bengaluru, Karnataka, India
e-mail: chetanahegde@ieee.org

V. E. Balas et al. (eds.), *Data Management, Analytics and Innovation*,
Advances in Intelligent Systems and Computing 808,
https://doi.org/10.1007/978-981-13-1402-5_24

biometric traits like hand veins [4], fingerprints [5], retinal patterns, handwritten signatures, ear patterns [6], electrocardiogram [7–9], Finger Knuckle Print [2, 10, 11], etc., are used extensively in areas where security is the major concern.

Finger-vein is an accepted biometric trait because it possesses the properties mentioned in [3]. Every individual including identical twins has a unique pattern of veins. As the individual grows, the vein size increases, but the number of veins and their position do not change from infancy. The vein structure is invisible to the naked eye as it is underneath the skin and hence one cannot spoof the system easily. As finger-vein images are captured without contact, it is more convenient and hygiene, which leads to high user acceptance.

In general, any finger-vein identification and/or authentication system involves four major steps viz. capture the image, image preprocessing, extracting the features and matching features for decision-making. The methods used extract the features from finger vein can be categorized into three categories viz. vein pattern-based methods, dimensionality reduction-based methods, and local binary-based methods. The vein pattern-based feature extraction methods have been explored in following ways: repeated line tracking [12], maximum curvature [13], Gabor Filter [14], mean curvature [15], region growth [16, 17], modified repeated line tracking [18], and Radon Transform [19]. In these methods, initially the vein patterns are segmented. Then the geometric shape or topological structure of vein pattern is used for feature matching. Usually in dimensionality reduction-based methods image will be transformed into low-dimensional space to classify. During transformation, they keep discriminating information and remove noises. Local binary-based methods are based on local area and the extracted features are in binary format. Some of the researchers have used near-infrared vein pattern [20], score fusion [21] etc. for identification purpose. A survey of various such techniques [22] has been done and compared. Identification based on finger-vein pattern is used even in multimodal biometric systems [23].

The content of this paper is presented in five sections. The related work is presented in Sect. 2. Section 3 contains architecture and model. Section 4 deals with implementation and performance analysis. Section 5 is about conclusions.

2 Related Work

This section briefly presents the work carried in the area of finger-vein biometrics.

The repeated line tracking, maximum curvature, region growth, and modified repeated line tracking—all these approaches use the cross-section of image to extract vein pattern. In repeated line tracking method [12] the line tracking starts from several positions to extract the finger-vein pattern from the image which are blurred. The major advantages of this method are ability to achieve better segmentation performance even for blurred image and it can be easily joined with any other hand-based biometrics. Low robustness and efficiency are its drawbacks. This method may degrade slightly during cold weather because of reduction in the clarity of finger veins.

The maximum curvature method [13], calculates the local maximum curvatures in cross-sectional profiles of vein to extract specific of finger-vein patterns. The vein image is viewed as a geometric shape in mean curvature method [15]. The group of pixels having negative mean curvature is seen as vein pattern. The ratio of such pixels is matched for authentication purpose. This method produced an equal error rate of 0.25%, which was significantly lower than the methods existing till then. In region growth method [16], it runs region growing operator on the different seeds to extract the vein pattern. Here it observes the symmetry and continuity of valleys in cross sectional profiles, which helps in extracting the finger-vein patterns by avoiding irregular shading and noise. The finger vein is segmented to obtain binary and skeleton image in [18]. The parameter is revised based on the width of vein computed using this skeleton image. These revised parameters are used to identify the locus space of vein. Then segmentation on this locus space of vein is extracted by using Otsu algorithm and hence proved that the said procedure performs better than traditional repeated line tracking algorithm.

3 Architecture and Modeling

The various steps involved in the authentication process are as shown in Fig. 1.

It is difficult for an organization to maintain a finger-vein image database for authentication of its employees. So, we suggest storing the features of finger-vein images instead of the images themselves. Every employee of the organization is assigned a unique ID. Various finger-vein features like coordinate positions of corner points of the vein image are extracted and are stored against the ID of every individual. This will be the initial setup of the database. During authentication, finger-vein image of a person is captured and the features are extracted from it. These features are compared with corresponding features stored in the database for a respective ID. The person can be authenticated, if comparison matches the threshold criteria, otherwise rejected.

3.1 Image Acquisition

To capture the finger-vein image, an infrared imaging device is used. The captured image is a gray scale image. Index finger images viz. I1 and I2 of two individuals are shown as sample images in Fig. 2.

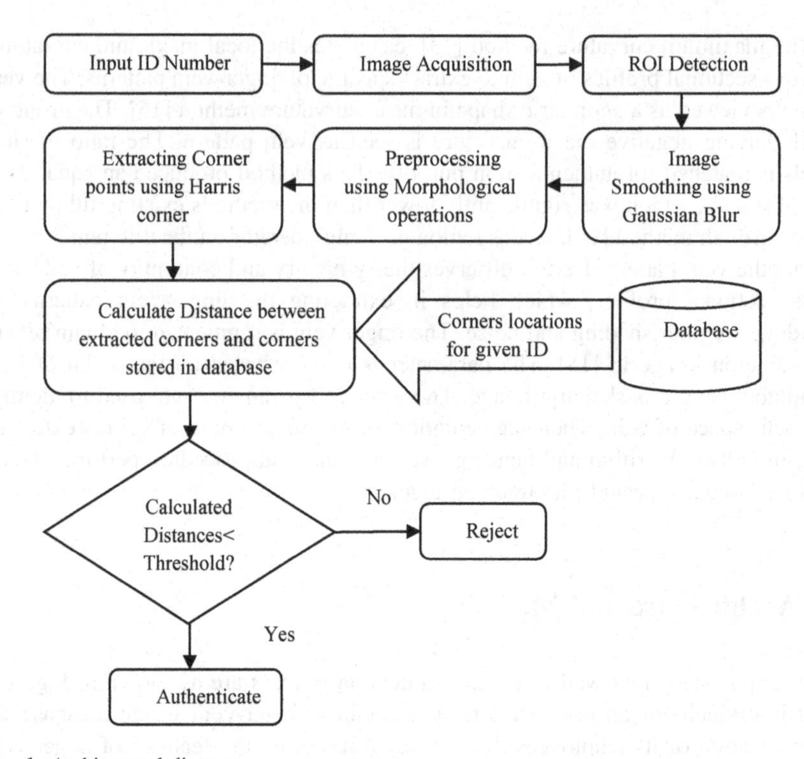

Fig. 1 Architectural diagram

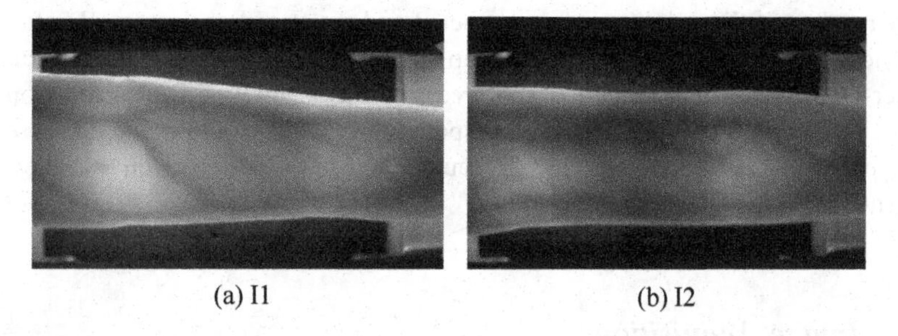

Fig. 2 Original images

3.2 ROI Detection and Smoothing

The region of interest (ROI) which contains the finger-vein pattern is extracted from captured image. Then image is resized to get a better clarity. The ROI is as shown in Fig. 3.

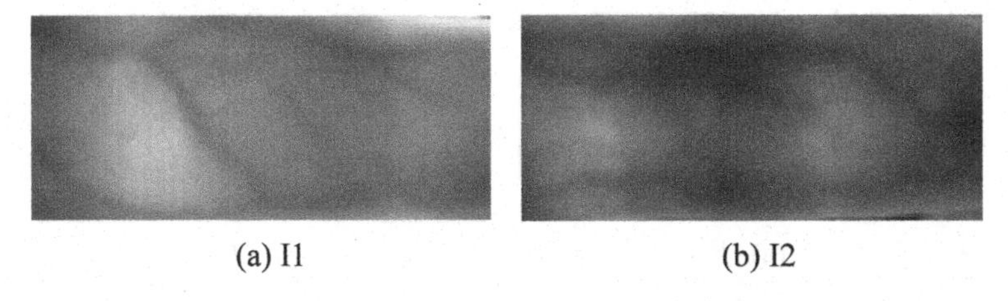

(a) I1 (b) I2

Fig. 3 ROI extracted from original images

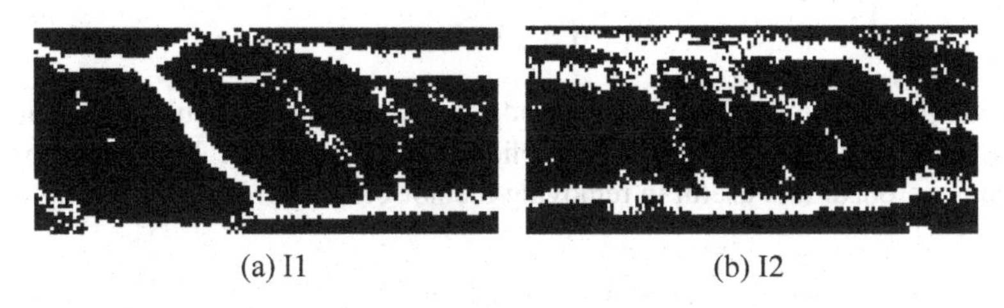

(a) I1 (b) I2

Fig. 4 Images after computing difference of gaussian

The resized image is smoothened by using Gaussian Blur, which uses below Gaussian function for calculating the transformation to apply to each pixel in the image. The Eq. 1 represents the two-dimensional Gaussian function.

$$G(x, y) = \frac{1}{2\pi\sigma^2} e^{-\frac{x^2+y^2}{2\sigma^2}} \tag{1}$$

Here, x represents the distance from the origin in horizontal axis, and y represents the distance from the origin in vertical axis, and σ represents the standard deviation of the Gaussian distribution. The Gaussian blur is applied on two different kernels of an image. There will be two resulting images, whose difference is computed. This is known as Difference of Gaussian (DoG). Figure 4 shows the computed DoG for the sample images I1 and I2.

3.3 Preprocessing

Preprocessing involves series of morphological operations. Any morphological operation requires two inputs viz. the input image and a structuring element deciding the nature of operation. Morphological opening is performed to remove possible noise in the image. Opening is performed by using Eq. 2.

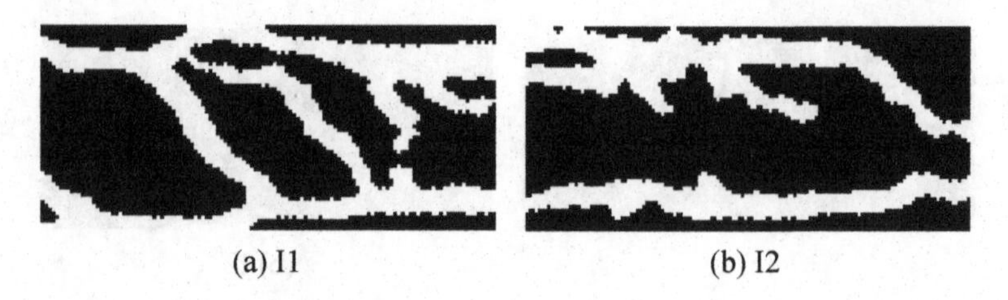

(a) I1 (b) I2

Fig. 5 Images after preprocessing and pattern extraction

$$A \circ B = (A \ominus B) \oplus B \tag{2}$$

Here, A represents an input image and B represents the structuring element.

When morphological erosion is applied on an image, the boundaries of the object are eroded. It is useful in removing the noise. Erosion is performed by using Eq. 3.

$$A \ominus B = \{z \in E | B_z \subseteq A\} \tag{3}$$

Dilation is used for joining the broken parts on an object in the image. Dilation is done using the equation

$$A \oplus B = \bigcup_{b \in B} A_b \tag{4}$$

Preprocessed images are the required patterns in the images. These are shown in Fig. 5.

3.4 Feature Extraction

The corners points on the hand-vein images are treated as features for authentication in this proposed technique. Chris Harris and Mike Stephens [24] finds the difference in intensity for a displacement of $(u,\ v)$ in all directions, which is expressed as below:

$$E(u,v) = \sum_{x,y} \underbrace{w(x,y)}_{\text{window function}} \underbrace{[I(x+u,y+v))}_{\text{shifted intensity}} - \underbrace{I(x,y)]^2}_{\text{intensity}} \tag{5}$$

The function $E\ (u,\ v)$ has to be maximized to detect the corners. Taylor's expansion is applied and derived on Eq. (5). The resultant equations are

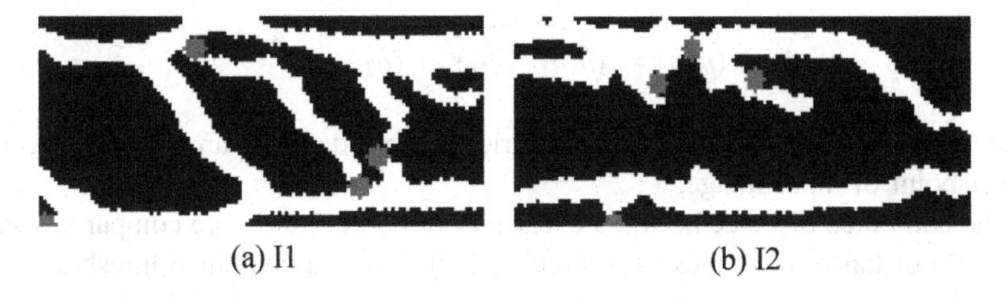

(a) I1 (b) I2

Fig. 6 Images with Harris corners

$$E(u,v) \approx [u \quad v]M\begin{bmatrix} u \\ v \end{bmatrix} \tag{6}$$

$$M = \sum_{x,y} w(x,y)\begin{bmatrix} I_x I_x & I_x I_y \\ I_x I_y & I_y I_Y \end{bmatrix} \tag{7}$$

Here, I_x and I_y are image derivatives in x and y directions respectively.
We can determine whether a window contain a corner or not, using equation

$$R = \det(M) - k(\text{trace}(M))^2 \tag{8}$$

- $\det(M) = \lambda_1 \lambda_2$
- $\text{trace}(M) = \lambda_1 + \lambda_2$
- λ_1 and λ_2 are the eigen values of M

The corners points are detected from preprocessed image by using Harris corner technique and it is as shown in Fig. 6.

3.5 Authentication

In this step, calculate the essential parameters based on which authentication is decided. The number of corners retrieved from the database for a given ID is compared with those of currently captured image. Then, the coordinate positions of these points are also compared using the formula for Euclidian distance as

$$d(p,q) = \sqrt{(q_1 - p_1)^2 + (q_2 - p_2)^2} \tag{9}$$

Here, $p = (p_1, p_2)$ is a corner point retrieved from database and $q = (q_1, q_2)$ is a corner point of input image.

The computed distance between every pair of corner points are compared with a threshold distance. If all these are found to be lesser than redefined threshold, then the authentication is successful, otherwise authentication failed. In the proposed algorithm the threshold value is taken as 35.

4 Implementation and Performance Analysis

The proposed algorithm given in Table 1 is implemented using OpenCV in Python 2.7.8 and is tested on Finger-vein images taken from Finger-vein database SDUMLA-FV built by Shandong University [25]. They used light transmission method to acquire the image. Database is constructed by taking six images of ring, index and middle fingers of both hands of 106 individuals. It contains 3816 images of 320 × 240 resolutions, presented in *.bmp* format.

Table 1 Algorithm for authentication

Inputs: I: Finger vein image N: ID number T: Threshold distance
Output: Authentication result
Procedure Authentication
Extract the ROI from I → ROI
Apply Gaussian blur on ROI with kernel k1 → I1
Apply Gaussian blur on ROI with kernel k2 → I2
Difference of Gaussian I3 = I2 − I1
Apply morphological operations on I3
Extract Harris corners C1={x_1, x_2, x_3, x_4} from I3
Fetch corners C2={y_1, y_2, y_3, y_4} for given N from database D = {} // Null set
For each x ∈ C1 and y ∈ C2
d_i = Euclidean distance (x_i, y_i) where i ∈ {1,2,3,4} D = D ∪ d_i
End
If every d_i < T where i ∈ {1,2,3,4} return True//Authenticated Else return False//Rejected End

Table 2 Confusion Matrix

		Actual	
		Accept	Reject
Tested	Accept	105	3
	Reject	1	11,127

For the experimental purpose, only the index finger images of 106 individuals are considered in this paper. The confusion matrix is generated based on simulation results of these 106 images, and is shown in Table 2.

The simulation results of proposed authentication algorithm produced 2.78% false accept ratio (FAR), 0.09% false reject ratio (FRR) and 99.96% overall system efficiency.

5 Conclusion

In this paper, we propose an efficient way of extracting corners as features from finger vein for authentication. In the proposed technique, we have extracted the corner points using Harris corner on preprocessed finger-vein image. The positions of corners are stored in database along with the unique ID assigned to every individual.

Authentication is decided by comparing distances calculated between each pair of fixed number of corners extracted for captured image and that in the database for given ID against the threshold value. If all distances are less than the threshold value, then authenticate the person, otherwise reject.

References

1. Lingyu, W., & Leedham, G. (2006). Near and far infrared imaging for vein pattern biometrics. In *IEEE International Conference on Video and Signal Based Surveillance (AVSS 06)* (p. 52).
2. Hegde, C., et al. (2011). FKP biometrics for human authentication using Gabor wavelets. In *Proceedings of International Conference IEEE—TENCON 2011* (pp. 1072–1076). Bali, Indonesia.
3. Jain, A. K., Ross, A., & Prabhakar, S. (2004). An introduction to biometric recognition. *IEEE Transactions on Circuits and System, 14*(1), 4–20.
4. Hegde, C., et al. (2009). Authentication of damaged hand-vein patterns by modularization. In *Proceedings of International conference IEEE TENCON-2009* (pp. 1–6), Singapore.
5. Liang, B., et. al. (2014). A novel fingerprint-based biometric encryption. In *Proceedings of International Conference on P2P, Parallel, Grid, Cloud and Internet Computing (3PGCIC).*
6. Yuan, L., Mu, Z., & Xu, Z. (2005, October). Using ear biometrics for personal recognition. In *International Workshop on Biometric Recognition Systems, IWBRS2005* (pp. 221–228), Beijing, China.

7. Hegde, C., et al. (2011). Heartbeat biometrics for human authentication. *Signal, Image and Video Processing Journal, Special Issue on Unconstrained Biometrics: Advances and Trends, 5*(3), 485–493. ISSN 1863-1703.
8. Singh, Y. N., & Gupta, P. (2009). Biometrics method for human identification using electrocardiogram. In M. Tistarelli, M. S. Nixon (Eds.), *ICB 2009* (LNCS, Vol. 5558, pp. 1270–1279). Springer, Heidelberg.
9. Simon, B. P., & Eswaran, C. (1997). An ECG classifier designed using modified decision based neural network. *Computers and Biomedical Research, 30,* 257–272.
10. Zhang, L., Zhang, L., & Zhang, D. (2009). Finger-knuckle-print: A new biometric identifier. In *Proceedings of IEEE International Conference on Image Processing*.
11. Hegde, C., et al. (2013). Authentication using Finger Knuckle Prints. *Signal, Image and Video Processing Journal, Special Issue on Image and Video Processing for Security, 7*(4), 633–645. ISSN 1863-1703.
12. Miura, N., & Nagasaka, A. (2004). Feature extraction of finger-vein pattern based on repeated line tracking and its application to personal identification. *Machine Vision and Applications, 15*(4), 194–203.
13. Miura, N., Nagasaka, A., & Miyatake, T. (2007). Extraction of finger-vein patterns using maximum curvature points in image profiles. *IEICE Transactions on Information and Systems, E90-D*(8), 1185–1194.
14. Kumar, A., & Zhou, Y. B. (2012). Human identification using finger images. *IEEE Transactions on Image Process, 21*(4), 2228–2244.
15. Song, W., Kim, T., Kim, H. C., Choi, J. H., Kong, H. J., & Lee, S. R. (2011). A finger-vein verification system using mean curvature. *Pattern Recognition Letter, 32*(11), 1541–1547.
16. Qin, H. F., Yu, C. B., & Qin, L. (2011). Region growth–based feature extraction method for finger-vein recognition. *Optical Engineering, 50*(5), 057208–057208.
17. Lee, E. C., & Park, K. R. (2009). Restoration method of skin scattering blurred vein image for finger vein recognition. *Electronics Letters, 45*(21), 1074–1076.
18. Liu, T., Xie, J. B., Yan, W., Li, P. Q., & Lu, H. Z. (2013). An algorithm for finger-vein segmentation based on modified repeated line tracking. *The Imaging Science Journal, 61*(6), 491–502.
19. Wu, J.-D., & Ye, S.-H. (2009). Driver identification using finger-vein patterns with Radon transform and neural network. *Expert Systems and Application, 36*(3), 5793–5799.
20. Kono, M., Ueki, H., & Umemura, S. (2002). Near-infrared finger vein patterns for personal identification. *Applied Optics, 41*(35), 7429–7436.
21. Nandakumar, K., Chen, Y., Dass, S. C., & Jain, A. K. (2008). Likelihood ratio based biometric score fusion. *IEEE Transactions on Pattern Analysis and Machine Intelligence, 30* (2), 342–347.
22. Mulyono, D., & Jinn, H. S. (2008). A study of finger vein biometric for personal identification. In *Proceedings of ISBAST* (pp. 1–8).
23. Sim, T., Zhang, S., Janakiraman, R., & Kumar, S. (2007). Continuous verification using multimodal biometrics. *IEEE Transaction on Pattern Analysis and Machine Intelligence, 29* (4), 687–700.
24. Harris, C., & Stephens, M. (1988). A combined corner and edge detector. In *Proceedings of the 4th Alvey Vision Conference* (pp. 147–151).
25. Yin, Y. L., Liu, L. L., & Sun, X. W. (2011). SDUMLA-HMT: A multimodal biometric database. In *The 6th Chinese Conference on Biometric Recognition, LNCS 7098* (pp. 260–268), Beijing, China.

Fuzzy-Based Machine Learning Algorithm for Intelligent Systems

K Pradheep Kumar

Abstract It has become essential to develop machine learning techniques due to the automation of various tasks. At present, several tasks need manual intervention for better reliability of the system. In this work, fuzzy-based approach has been proposed where systems are trained based on initial data sets. In several data sets, the data is either partially available or unavailable. When data sets need to be used on real time systems, the non-availability of data may lead to catastrophe. In this approach, a fuzzy-based rule set is formulated. The rule strength is used to determine the effectiveness. Rules with similar strengths are clustered together. The learning is carried out by determining a threshold for the formulated rule set. Based on the threshold computed, a modified rule set is formulated with rule strengths greater than the computed threshold. A semi-supervised learning approach that uses an activation function is employed. The fuzzy learning approach proposed in this work reduces the error by 20% compared to conventional approaches.

Keywords Threshold · Activation function · Learning rule matrix
Machine learning · Rule strength · Smart systems · Semi-supervised learning

1 Introduction

In today's world, there is a need to design gadgets and systems with automation. To enable automation of gadgets, it is essential that systems handle decisions independently based on the environmental conditions.

Systems should adapt to the external environment and handle tasks based on the underlying behaviour. Systems should also handle exceptional tasks. To fulfil the same, the system needs to be intelligent to take decisions in the absence of manual intervention.

K. Pradheep Kumar (✉)
BITS Pilani, Pilani, Rajasthan, India
e-mail: pradjourn@gmail.com

V. E. Balas et al. (eds.), *Data Management, Analytics and Innovation*,
Advances in Intelligent Systems and Computing 808,
https://doi.org/10.1007/978-981-13-1402-5_25

Machine learning is one way of implementing these features of Artificial Intelligence where systems are trained based on an initial data set.

The system forms the initial dataset by collecting the information from the external environment through sensors and actuators. System is trained to learn information and take decisions based on iteration of datasets. The time needed by the system to learn information and take decisions from the datasets should be finite and small so as to avoid failures that can lead to catastrophe.

In this work, a fuzzy rule matrix is constructed and the threshold of the entire rule set is also computed. The modified rule set is constructed by selecting rules whose rule strength is greater than the threshold. These are the rules used by the system to learn the data. Other rules are not considered.

The paper is organised as follows: Sect. 2 discusses the literature on machine learning algorithms. Section 3 explains the fuzzy model proposed, illustrates the same with an example and gives the algorithm. Section 4 explains the simulation environment and the results. Section 5 concludes the work and highlights the scope for the future work.

The block diagram of the entire procedure is shown in Fig. 1.

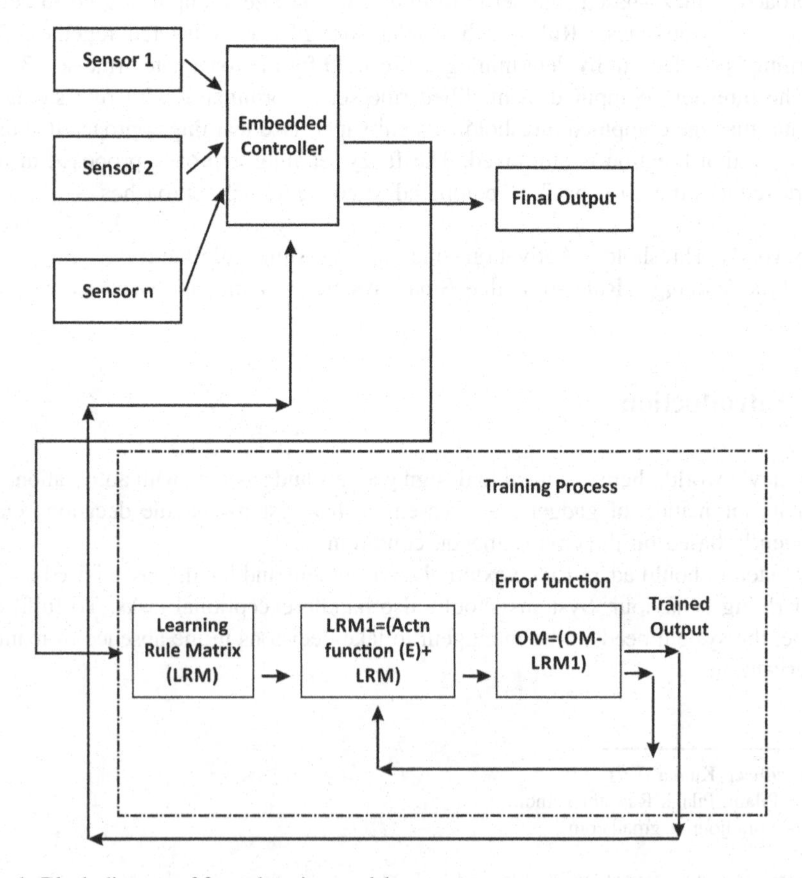

Fig. 1 Block diagram of fuzzy learning model

2 Literature Review

The most conventional machine learning strategy is the Naive Bayes classifier. Several literatures report the inaccuracies in Bayes' conditional independence. This is due to the suboptimal nature of the probability estimates as reported in [1–3]. Even when multidimensional tables are used to compute probabilities in real-time scenarios, many errors are observed as discussed in [2]. Support Vector method uses a hyperplane to classify the data. Based on the classification a number of support vectors are generated as explained in [4]. But when data sets are widely spread and have a large range of deviation, this method does not yield accurate results. The choice of the hyperplane is the key factor associated with the data classification as discussed in [5, 6]. Artificial neural networks give reliability but again this depends on the choice of the activation function used to train the data set as explained in [7, 8]. Linear and Logistic regression as explained in [9–11] uses a cost function and estimates the gradient. The reduction in the gradient is used to learn the datasets. The regression analysis depends on the effectiveness of the cost function.

Several deep learning neural network algorithms model a data set as a function using a mathematical model and train the same to interpolate or extrapolate the data value for a particular value. The Stochastic Gradient Descent technique as explained in [12] attempts to minimise the loss of the model by incorporating additional parameters and increments these proportional to the gradient estimated. But this makes the data noisy due and filters needs to be used to eliminate the noise associated. The method of steepest descent as discussed on [13] which when used on a very large data set requires an infinite number of iterations to get an optimum solution. It may so happen that an optimum solution may not exist also. Another approach which works in a different dimension is Reinforcement learning as explained in [14]. This method arrives at an optimal solution by trial and error in infinite time. But the time of training in an unsupervised scenario would be infinite and the optimal solution cannot be guaranteed. The drawbacks highlighted in the above techniques are overcome in the fuzzy-based analysis. In the fuzzy-based analysis a precise value for the rule strength are arrived and the optimal solution is arrived by training the same with an activation function in a finite period of time. It does not require any mathematical model and infinite number of trials to arrive at an optimal solution.

3 Proposed Work

In this work a fuzzy-based learning approach using Mamdani's engine as discussed by Venkat and Pradheep in [15] has been used. For each rule, a rule strength is computed. The rule strength is computed based on a set of sub-factors and their corresponding weights. For each linguistic variable used in the rule, the mid-value of the range is computed. The weighted average of each variable of the rule gives the rule strength. A typical rule is given as follows:

The Rule Strength is computed using the expression

$$\mathrm{RS} = \frac{\sum (\mathrm{Weight} * \mathrm{MidValue})}{\sum \mathrm{Weight}} \tag{1}$$

The rule strengths have been computed for all rules in the fuzzy rule set using the above expression. The threshold of the fuzzy rule set is computed using the following expression

$$T = \frac{\sum_{i=1}^{i=n} (\mathrm{Rulenumber} * \mathrm{RuleStrength})}{\sum_{i=1}^{n} \mathrm{Rulenumber}} \tag{2}$$

where n is the total number of rules.

The threshold computed is a measure of the upper bound till completion of the training of the system. The modified rule set is constructed by selecting rules, which have a rule strength greater than the computed threshold. After this a matrix is constructed in the form (RS − mT, RS, RS + mT), where $m = 1$ to n and n is the number of rules in the original rule set. The matrix is formulated for different values of m in such a manner that RS − T is positive. This criterion decides the number of columns of the matrix.

The matrix now is non-symmetrical as the number of rows would be greater than the number of columns. The matrix is split into a number of sub-symmetrical matrices by matching the number of rows with the columns.

The procedure has been illustrated with an example where RS − T is positive and RS − 2T, RS − 3T, etc. are negative. With only RS − T, RS and RS + T, the matrix has been formulated. Hence, the modified rule set should be split into 3 X 3 matrices based on the modified rule set. The actual output matrix is accepted as input. The error, which is the difference between the output and modified rule matrix, is computed. An activation function is used to iteratively train the modified rule set till the error is non-negative. The final value is the optimal rule set.

For computing the trained matrix for antivirus check, the different sub-factors and their corresponding weights are shown in Table 1.

The different Linguistic variables with their ranges and corresponding mid-values are indicated in Table 2.

A typical rule is shown below.

If (Phishing Check is Excellent) and (Spyware Check is Good) and (Malware check is Average) and (Trojan Check is Fair) and (Rootkit scan is Poor) Then (Antivirus software is Average).

The entire rule set comprises of 125 rules. Few sample rules from the entire rule set is shown below in Table 3.

Table 1 Weight for each sub-factor

S.No.	Sub-factor	Weight
1.	Phishing check	5
2.	Spyware check	4
3.	Malware check	3
4.	Trojan check	2
5.	Rootkit scan	1

Table 2 Linguistic variable with range and mid-value

S.No.	Linguistic variable	Range	Mid-value
1.	Excellent	0–19	9.5
2.	Good	20–39	29.5
3.	Average	40–59	49.5
4.	Fair	60–79	69.5
5.	Poor	80–100	90

The rule strength of the rule If (Phishing Check is Excellent) and (Spyware Check is Good) and (Malware check is Average) and (Trojan Check is Fair) and (Rootkit scan is Poor), then (Antivirus software is Average) is 36.2. The Rule strengths of the above rule base would be computed as indicated in Table 4.

Using the expression, the threshold has been computed as 62.02. Based on this threshold modified rule set is shown in Table 5.

Hence RS − T is only non-negative and the rest of RS − mT is negative. From the above discussion 14 (3 X 3 Symmetrical) sub-matrices have been formulated. The last row of the 14th matrix would be zeroes. The modified rule set is trained using the activation function $E = \left(\frac{1}{1-e^{-T}}\right)$ as shown in Table 6.

The modified rule matrix is LRM. The iteration matrix is LRM1. Given Output matrix is OM. The error is (OM − LRM1). The iteration is carried out till all elements of OM − LRM1 becomes negative.

The algorithm has been discussed below

- Form the fuzzy rule set
- Compute the rule strength and Threshold
- Form the Learning rule matrix (LRM) with selected rules (RS > T)
- Form modified rule matrix (RS − mT, RS, RS + mT)
- Accept the output matrix (OM)
- Start the iteration by computing Error E = (OM − LRM)
- Use activation function $E = \left(\frac{1}{1-e^{-T}}\right)$ to minimise the error.
- LRM1 = LRM + E
- If (LRM1 <= OM) then
    - Final output O1 = LRM1
      Else
    - L1: LRM1 = LRM + E

Table 3 Sample rule set

S.No.	Malware check	Spyware check	Rootkit check	Trojan check	Phishing check	Antivirus software
1.	Excellent	Excellent	Excellent	Excellent	Excellent	Excellent
2.	Good	Good	Good	Good	Excellent	Good
3.	Average	Average	Average	Average	Excellent	Average
4.	Fair	Fair	Fair	Fair	Excellent	Fair
5.	Poor	Poor	Poor	Poor	Excellent	Poor
6.	Excellent	Excellent	Excellent	Excellent	Good	Excellent
7.	Good	Good	Good	Good	Good	Good
8.	Average	Average	Average	Average	Good	Average
9.	Fair	Fair	Fair	Fair	Good	Fair
10.	Poor	Poor	Poor	Poor	Good	Poor
11.	Excellent	Excellent	Excellent	Excellent	Average	Excellent
12.	Good	Good	Good	Good	Average	Good
13.	Average	Average	Average	Average	Average	Average
14.	Fair	Fair	Fair	Fair	Average	Fair
15.	Poor	Poor	Poor	Poor	Average	Poor
16.	Excellent	Excellent	Excellent	Excellent	Fair	Excellent
17.	Good	Good	Good	Good	Fair	Good
18.	Average	Average	Average	Average	Fair	Average
19.	Fair	Fair	Fair	Fair	Fair	Fair
20.	Poor	Poor	Poor	Poor	Fair	Poor
21.	Excellent	Excellent	Excellent	Excellent	Poor	Excellent
22.	Good	Good	Good	Good	Poor	Good
23.	Average	Average	Average	Average	Poor	Average
24.	Fair	Fair	Fair	Fair	Poor	Fair
25.	Poor	Poor	Poor	Poor	Poor	Poor
26.	Excellent	Excellent	Excellent	Excellent	Excellent	Excellent
27.	Good	Good	Good	Excellent	Good	Good
28.	Average	Average	Average	Excellent	Average	Average
29.	Fair	Fair	Fair	Excellent	Fair	Fair
30.	Poor	Poor	Poor	Excellent	Poor	Poor
31.	Excellent	Excellent	Excellent	Good	Excellent	Excellent
32.	Good	Good	Good	Good	Good	Good
33.	Average	Average	Average	Good	Average	Average
34.	Fair	Fair	Fair	Good	Fair	Fair
35.	Poor	Poor	Poor	Good	Poor	Poor

Table 4 Rule strengths computed from rule set

Rule no.	Malware check	Spyware check	Rootkit check	Trojan check	Phishing check	Rule strength = ((5 * phishing check) + (4 * spyware check) + (3 * malware check) + (2 * trojan check) + (1 * rootkit check))/15	Threshold parameter Q = (Rule no. * rule strength)
1	9.5	9.5	9.5	9.5	9.5	9.50	9.5
2	9.5	9.5	49.5	9.5	9.5	12.17	24.33
3	9.5	9.5	9.5	9.5	9.5	9.50	28.50
4	29.5	9.5	9.5	9.5	9.5	13.50	54.00
5	9.5	9.5	9.5	9.5	9.5	9.50	47.50
6	9.5	29.5	9.5	9.5	9.5	14.83	89.00
7	9.5	9.5	9.5	9.5	9.5	9.50	66.50
8	9.5	9.5	9.5	9.5	9.5	9.5	76.00
9	9.5	9.5	9.5	29.5	9.5	12.17	109.50
10	49.5	9.5	9.5	9.5	9.5	17.50	175.00
11	9.5	9.5	69.5	9.5	9.5	13.50	148.50
12	9.5	9.5	9.5	49.5	9.5	14.83	178.00
13	9.5	9.5	29.5	9.5	9.5	10.83	140.83
14	9.5	9.5	9.5	69.5	9.5	17.50	245.00
15	9.5	29.5	29.5	29.5	29.5	25.50	382.50
16	29.5	29.5	29.5	29.5	29.5	29.50	472.00
17	9.5	9.5	90	9.5	9.5	14.87	252.73
18	29.5	29.5	69.5	29.5	29.5	32.17	579.00
19	9.5	9.5	9.5	9.5	29.5	16.17	307.17
20	9.5	69.5	9.5	9.5	9.5	25.50	510.00
21	29.5	49.5	29.5	29.5	29.5	34.83	731.50
22	29.5	29.5	29.5	29.5	49.5	36.17	795.67

(continued)

Table 4 (continued)

Rule no.	Malware check	Spyware check	Rootkit check	Trojan check	Phishing check	Rule strength = ((5 * phishing check) + (4 * spyware check) + (3 * malware check) + (2 * trojan check) + (1 * rootkit check))/15	Threshold parameter Q = (Rule no. * rule strength)
23	29.5	29.5	29.5	29.5	29.5	29.50	678.50
24	9.5	49.5	9.5	9.5	9.5	20.17	484.00
25	9.5	9.5	9.5	90	9.5	20.23	505.83
26	29.5	29.5	29.5	29.5	69.5	42.83	1113.67
27	29.5	29.5	29.5	49.5	29.5	32.17	868.50
28	69.5	9.5	9.5	9.5	9.5	21.50	602.00
29	49.5	29.5	49.5	49.5	49.5	44.17	1280.83
30	29.5	29.5	29.5	29.5	29.5	29.50	885.00
31	29.5	29.5	29.5	29.5	9.5	22.83	707.83
32	9.5	9.5	9.5	9.5	49.5	22.83	730.67
33	29.5	29.5	29.5	69.5	29.5	34.83	1149.50
34	49.5	49.5	9.5	49.5	49.5	46.83	1592.33
35	29.5	29.5	29.5	29.5	29.5	29.50	1032.50

Table 5 Modified rule set

Modified rule no.	Malware check	Spyware check	Rootkit check	Trojan check	Phishing check	Rule strength (RS) = ((5 * phishing check) + (4 * spyware check) + (3 * malware check) + (2 * trojan check) + (1 * rootkit check))/15	(RS − T)	(RS + T)
1	69.50	49.50	69.50	69.50	69.50	64.17	2.15	126.19
2	49.50	69.50	69.50	69.50	69.50	65.50	3.48	127.52
3	69.5	69.5	29.5	69.5	69.5	66.83	4.81	128.85
4	69.5	69.5	69.5	69.5	69.5	69.50	7.48	131.52
5	69.5	69.5	69.5	29.5	69.5	64.17	2.15	126.19
6	69.5	69.5	9.5	69.5	69.5	65.50	3.48	127.52
7	69.5	69.5	69.5	49.5	69.5	66.83	4.81	128.85
8	90	90	90	90	90	90.00	27.98	152.02
9	69.5	69.5	69.5	69.5	69.5	69.50	7.48	131.52
10	90	90	90	90	90	90.00	27.98	152.02
11	69.5	69.5	69.5	69.5	49.5	62.83	0.81	124.85
12	49.5	49.5	49.5	49.5	90	63.00	0.98	125.02
13	90	90	90	90	9.5	63.17	1.15	125.19
14	90	90	90	90	90	90.00	27.98	152.02
15	69.5	69.5	49.5	69.5	69.5	68.17	6.15	130.19
16	90	9.5	90	90	90	68.53	6.51	130.55
17	90	90	90	90	29.5	69.83	7.81	131.85
18	69.5	69.5	90	69.5	69.5	70.87	8.85	132.89
19	90	90	90	90	90	90.00	27.98	152.02
20	69.5	69.5	69.5	90	69.5	72.23	10.21	134.25
21	90	69.5	69.5	69.5	69.5	73.60	11.58	135.62
22	90	29.5	90	90	90	73.87	11.85	135.89
23	9.5	90	90	90	90	73.90	11.88	135.92

Table 6 Trained modified rule set

LRM				LRM1			OM			OM − LRM1		
Rule	RS − T	RS	RS + T	RS − T	RS	RS + T	RS − T	RS	RS + T	RS − T	RS	RS + T
1	2.15	64.17	126.19	62.66	124.68	186.70	5.67	72.34	132.65	−56.99	−52.34	−54.05
2	3.48	65.50	127.52	63.99	126.01	188.03	10.67	77.34	137.65	−53.32	−48.67	−50.38
3	4.81	66.83	128.85	65.32	127.34	189.36	15.67	82.34	142.65	−49.65	−45.00	−46.71
4	7.48	69.50	131.52	67.99	130.01	192.03	20.67	87.34	147.65	-47.32	-42.67	-44.38
5	2.15	64.17	126.19	62.66	124.68	186.70	25.67	92.34	152.65	−36.99	−32.34	−34.05
6	3.48	65.50	127.52	63.99	126.01	188.03	30.67	97.34	157.65	−33.32	−28.67	−30.38
7	4.81	66.83	128.85	65.32	127.34	189.36	35.67	102.34	162.65	−29.65	−25.00	−26.71
8	27.98	90.00	152.02	88.49	150.51	212.53	40.67	107.34	167.65	−47.82	−43.17	−44.88
9	7.48	69.50	131.52	67.99	130.01	192.03	45.67	112.34	172.65	−22.32	−17.67	−19.38
10	27.98	90.00	152.02	88.49	150.51	212.53	50.67	117.34	177.65	−37.82	−33.17	−34.88
11	0.81	62.83	124.85	61.32	123.34	185.36	55.67	122.34	182.65	−5.65	−1.00	−2.71
12	0.98	63.00	125.02	61.49	123.51	185.53	60.67	127.34	187.65	−0.82	3.83	2.12
13	1.15	63.17	125.19	61.66	123.68	185.70	65.67	132.34	192.65	4.01	8.66	6.95
14	27.98	90.00	152.02	88.49	150.51	212.53	70.67	137.34	197.65	−17.82	−13.17	−14.88
15	6.15	68.17	130.19	66.66	128.68	190.70	75.67	142.34	202.65	9.01	13.66	11.95
16	6.51	68.53	130.55	67.02	129.04	191.06	80.67	147.34	207.65	13.65	18.30	16.59
17	7.81	69.83	131.85	68.32	130.34	192.36	85.67	152.34	212.65	17.35	22.00	20.29
18	8.85	70.87	132.89	69.36	131.38	193.40	90.67	157.34	217.65	21.31	25.96	24.25
19	27.98	90.00	152.02	88.49	150.51	212.53	95.67	162.34	222.65	7.18	11.83	10.12
20	10.21	72.23	134.25	70.72	132.74	194.76	100.67	167.34	227.65	29.95	34.60	32.89
21	11.58	73.60	135.62	72.09	134.11	196.13	105.67	172.34	232.65	33.58	38.23	36.52
22	11.85	73.87	135.89	72.36	134.38	196.40	110.67	177.34	237.65	38.31	42.96	41.25
23	11.88	73.90	135.92	72.39	134.41	196.43	115.67	182.34	242.65	43.28	47.93	46.22

- If (LRM1 > OM)
    - O1 = LRM1
      Else
    - (LRM1 = LRM1 + E)
- Repeat iteration until LRM1 value is > OM.
- Goto L1.
- Compute RMSE, RRSE, RAE and MAE for final value of O1.
- End the procedure.

4 Simulation Results

The algorithm was simulated on Java platform with several IT-related issues like Antivirus, Data backup strategies, hardware maintenance. Simulations were carried out on fuzzy rule sets with about 500 rules.

- Root Mean Square Error (RMSE)
- Root Relative Square Error (RRSE)
- Relative Absolute Error (RAE)
- Mean Absolute Error (MAE).

4.1 Root Mean Square Error (RMSE)

It is the square root of the mean of the squared difference between the Output Matrix (OM) and Trained Learning Rule Matrix (LRM1). It is given by the equation

$$\mathrm{RMSE} = \sqrt{\frac{1}{n}\sum_{i=1}^{i=n} (\mathrm{OM}(i) - \mathrm{LRM1}(i))^2} \tag{3}$$

where n is the number of sample values of the data set under consideration.

4.2 Root Relative Square Error (RRSE)

It is defined as the normalised version of the total squared error to the total squared error of the prediction made.

$$\text{RRSE} = \sqrt{\frac{\sum_{i=1}^{i=n} (\text{LRM1}(i) - \text{OM}(i))^2}{\sum_{i=1}^{i=n} (\text{OM}(i) - A\sim)^2}} \tag{4}$$

where $A\sim$ is the average of the n sample values.

$$A\sim = \frac{1}{n}\sum_{i=1}^{i=n} \text{OM}(i) \tag{5}$$

4.3 *Relative Absolute Error (RAE)*

It is defined as the total absolute error normalised to the total absolute error of the prediction made.

$$RAE = \frac{\sum_{i=1}^{i=n} |(LRM1(i) - OM(i))|}{\sum_{i=1}^{i=n} |(OM(i) - A\sim)|} \tag{6}$$

where A $\sim$ is the average of the n sample values.

$$A\sim = \frac{1}{n}\sum_{i=1}^{i=n} OM(i) \tag{7}$$

4.4 *Mean Absolute Error (MAE)*

It is defined as the error difference for the n samples considered.

$$\text{MAE} = \frac{\sum_{i=1}^{i=n} (\text{OM}(i) - \text{LRM1}(i))}{n} \tag{8}$$

The results of the fuzzy-based approach proposed in this work have been compared with the four conventional machine learning algorithms listed below:

1. Naïve Bayes Classifier
2. Support Vector Method (SVM)

3. Regression Analysis
4. Artificial Neural Networks (ANN).

4.5 Comparison with Naïve Bayes Classifier

The results of the proposed fuzzy-based approach are compared with Naïve Bayes Classifier. The reduction in Error for the parameters RMSE, RRSE, RAE and MAE when compared with the Naives Classifier method are 21, 17, 18 and 27% respectively as indicated in Table 7.

4.6 Comparison with Support Vector Method (SVM)

The reduction in Error for the parameters RMSE, RRSE, RAE and MAE for the proposed work when compared with the Support Vector Method are 32, 10, 17 and 17% respectively as indicated in Table 8.

4.7 Comparison with Regression Analysis

The reduction in Error for the parameters RMSE, RRSE, RAE and MAE for the proposed fuzzy-based approach when compared with the Regression Analysis Method are 13, 4, 7 and 23% respectively as indicated in Table 9.

4.8 Comparison with Artificial Neural Networks (ANN)

The reduction in Error for the parameters RMSE, RRSE, RAE and MAE for the proposed fuzzy-based approach when compared with the Artificial Neural Networks method are 34, 21, 17 and 32% respectively as indicated in Table 10.

4.9 Summary

The reduction in error for the proposed work compared to conventional approaches is shown in Table 11.

The average reduction by using the fuzzy learning method for RMSE, RRSE, RAE and MAE are 25, 13, 15 and 25% respectively as indicated in Table 11.

Table 7 Comparison of parameters (Fuzzy learning model vs. Naïve classifier)

Rule	Naïve Bayes classifier				Fuzzy learning model				Reduction error (%)			
	RMSE	RRSE	RAE	MAE	RMSE	RRSE	RAE	MAE	RMSE	RRSE	RAE	MAE
1	0.41	0.56	0.62	0.28	0.24	0.41	0.41	0.19	41.93	26.79	33.87	32.14
2	0.47	0.6	0.65	0.35	0.32	0.42	0.45	0.24	31.91	30	30.77	31.43
3	0.51	0.64	0.68	0.41	0.37	0.47	0.51	0.29	27.45	26.56	25	29.27
4	0.54	0.68	0.74	0.45	0.41	0.54	0.55	0.34	24.07	20.59	25.68	24.44
5	0.58	0.71	0.75	0.49	0.47	0.59	0.59	0.39	18.97	16.9	21.33	20.41
6	0.62	0.74	0.78	0.57	0.51	0.62	0.64	0.42	17.74	16.22	17.95	26.32
7	0.66	0.77	0.81	0.62	0.55	0.65	0.69	0.46	16.67	15.58	14.81	25.81
8	0.71	0.81	0.84	0.65	0.59	0.71	0.73	0.49	16.9	12.35	13.1	24.62
9	0.75	0.84	0.87	0.74	0.63	0.75	0.77	0.53	16	10.71	11.49	28.38
10	0.79	0.87	0.91	0.78	0.67	0.77	0.81	0.57	15.19	11.49	10.99	26.92
11	0.84	0.91	0.94	0.81	0.71	0.81	0.84	0.61	15.48	10.99	10.64	24.69
12	0.87	0.94	0.96	0.88	0.75	0.84	0.88	0.64	13.79	10.64	8.33	27.27
13	0.91	0.97	0.98	0.95	0.81	0.89	0.91	0.71	10.99	8.25	7.14	25.26
Average	0.67	0.77	0.81	0.61	0.54	0.65	0.68	0.45	20.55	16.7	17.78	26.69

Table 8 Comparison of parameters (Fuzzy learning model vs. Support vector method)

Rule	Support vector method (SVM)				Fuzzy learning model				Reduction error (%)			
	RMSE	RRSE	RAE	MAE	RMSE	RRSE	RAE	MAE	RMSE	RRSE	RAE	MAE
1	0.57	0.52	0.61	0.32	0.24	0.41	0.41	0.19	57.89	21.15	32.79	40.63
2	0.61	0.55	0.64	0.36	0.32	0.42	0.45	0.24	47.54	23.64	29.69	33.33
3	0.65	0.59	0.67	0.39	0.37	0.47	0.51	0.29	43.08	20.34	23.88	25.64
4	0.68	0.62	0.71	0.41	0.41	0.54	0.55	0.34	39.71	12.9	22.54	17.07
5	0.72	0.65	0.74	0.43	0.47	0.59	0.59	0.39	34.72	9.23	20.27	9.3
6	0.74	0.68	0.77	0.46	0.51	0.62	0.64	0.42	31.08	8.82	16.88	8.7
7	0.78	0.71	0.81	0.51	0.55	0.65	0.69	0.46	29.49	8.45	14.81	9.8
8	0.82	0.74	0.84	0.54	0.59	0.71	0.73	0.49	28.05	4.05	13.1	9.26
9	0.85	0.76	0.87	0.57	0.63	0.75	0.77	0.53	25.88	1.32	11.49	7.02
10	0.88	0.79	0.91	0.61	0.67	0.77	0.81	0.57	23.86	2.53	10.99	6.56
11	0.91	0.84	0.93	0.71	0.71	0.81	0.84	0.61	21.98	3.57	9.68	14.08
12	0.94	0.89	0.94	0.81	0.75	0.84	0.88	0.64	20.21	5.62	6.38	20.99
13	0.97	0.93	0.97	0.91	0.81	0.89	0.91	0.71	16.49	4.3	6.19	21.98
Average	0.78	0.71	0.8	0.54	0.54	0.65	0.68	0.45	32.31	9.69	16.82	17.26

Table 9 Comparison of parameters (Fuzzy learning model vs. Regression analysis method)

Rule No.	Regression analysis (RA)				Fuzzy learning model				Reduction error (%)			
	RMSE	RRSE	RAE	MAE	RMSE	RRSE	RAE	MAE	RMSE	RRSE	RAE	MAE
1	0.39	0.42	0.51	0.31	0.24	0.41	0.41	0.19	38.46	2.38	19.61	38.71
2	0.42	0.45	0.54	0.35	0.32	0.42	0.45	0.24	23.81	6.67	16.67	31.43
3	0.45	0.49	0.59	0.39	0.37	0.47	0.51	0.29	17.78	4.08	13.56	25.64
4	0.48	0.54	0.62	0.44	0.41	0.54	0.55	0.34	14.58	0	11.29	22.73
5	0.51	0.61	0.65	0.49	0.47	0.59	0.59	0.39	7.84	3.28	9.23	20.41
6	0.55	0.65	0.69	0.55	0.51	0.62	0.64	0.42	7.27	4.62	7.25	23.64
7	0.58	0.69	0.72	0.57	0.55	0.65	0.69	0.46	5.17	5.8	4.17	19.3
8	0.62	0.73	0.75	0.61	0.59	0.71	0.73	0.49	4.84	2.74	2.67	19.67
9	0.65	0.76	0.78	0.65	0.63	0.75	0.77	0.53	3.08	1.32	1.28	18.46
10	0.7	0.81	0.81	0.71	0.67	0.77	0.81	0.57	4.29	4.94	0	19.72
11	0.79	0.84	0.85	0.76	0.71	0.81	0.84	0.61	10.13	3.57	1.18	19.74
12	0.89	0.88	0.91	0.83	0.75	0.84	0.88	0.64	15.73	4.55	3.3	22.89
13	0.94	0.92	0.94	0.91	0.81	0.89	0.91	0.71	13.83	3.26	3.19	21.98
Average	0.61	0.68	0.72	0.58	0.54	0.65	0.68	0.45	12.83	3.63	7.18	23.41

Table 10 Comparison of parameters (Fuzzy learning model vs. Artificial neural networks)

Rule No.	Artificial neural networks (ANN)				Fuzzy learning model				Reduction error (%)			
	RMSE	RRSE	RAE	MAE	RMSE	RRSE	RAE	MAE	RMSE	RRSE	RAE	MAE
1	0.57	0.59	0.61	0.32	0.24	0.41	0.41	0.19	57.89	30.51	32.79	40.63
2	0.61	0.63	0.64	0.43	0.32	0.42	0.45	0.24	47.54	33.33	29.69	44.19
3	0.65	0.67	0.67	0.47	0.37	0.47	0.51	0.29	43.08	29.85	23.88	38.3
4	0.69	0.71	0.71	0.51	0.41	0.54	0.55	0.34	40.58	23.94	22.54	33.33
5	0.73	0.75	0.74	0.54	0.47	0.59	0.59	0.39	35.62	21.33	20.27	27.78
6	0.78	0.81	0.77	0.57	0.51	0.62	0.64	0.42	34.62	23.46	16.88	26.32
7	0.81	0.84	0.81	0.64	0.55	0.65	0.69	0.46	32.1	22.62	14.81	28.13
8	0.84	0.87	0.84	0.69	0.59	0.71	0.73	0.49	29.76	18.39	13.1	28.99
9	0.87	0.89	0.87	0.77	0.63	0.75	0.77	0.53	27.59	15.73	11.49	31.17
10	0.91	0.91	0.89	0.84	0.67	0.77	0.81	0.57	26.37	15.38	8.99	32.14
11	0.94	0.94	0.91	0.87	0.71	0.81	0.84	0.61	24.47	13.83	7.69	29.89
12	0.96	0.97	0.94	0.91	0.75	0.84	0.88	0.64	21.88	13.4	6.38	29.67
13	0.99	0.99	0.97	0.94	0.81	0.89	0.91	0.71	18.18	10.1	6.19	24.47
Average	0.8	0.81	0.8	0.65	0.54	0.65	0.68	0.45	33.82	20.91	16.52	31.92

Table 11 Summarisation of results of comparison for error reduction

S.No.	Conventional approach	Reduction in error (%) for Fuzzy based approach			
		RMSE	RRSE	RAE	MAE
1	Fuzzy approach versus Naïve	21	17	18	27
2	Fuzzy approach versus SVM	32	10	17	17
3	Fuzzy approach versus Regression analysis	13	4	7	23
4	Fuzzy approach versus Artificial neural networks	34	21	17	32
Average		25	13	14.75	24.75

The average of these values would be 20%. Hence, the fuzzy learning method reduces error compared to conventional approaches by 20%.

5 Conclusion

A fuzzy-based learning approach has been proposed in this work. The fuzzy learning approach reduces the error by 20%. This approach computes the rule strength and chooses rules with rule strengths greater than the threshold limit for effective learning. The learning approach uses a predictive range for the linguistic variable, which is chosen before commencing the training process.

6 Future Work

The work can be extended by making this training process fully unsupervised by generating rule strengths using random function and then completing the entire process. The random unsupervised learning approach could later be implemented using Raspberry Pi as hardware for several applications like healthcare analytics, stock market, etc.

References

1. Taheri, S., Mammadov, M., & Bagirov, A. M. (2011). Improving Naïve Bayes classifier using conditional probabilities. In *9th Australian Data Mining Conference*, pp. 63–69.
2. Liu, H., Yin, X., & Han, J. (2005). An efficient multi-relational Naïve Bayesian classifier based on semantic relationship graph. In *MRDM*, pp. 39–49.

3. Ziddi, N. A., Cerquides, J., Carman, M. J., & Webb, G. I. (2013). Alleviating Naïve bayes attribute independence assumption by attribute weighting. *Journal of Machine Learning Research*, 1947–1988.
4. George, A., Poornachandran, P., & Kaimal, M. R. (2012). adsvm: Pre-processor plug-in using support vector machine algorithm for Snort. ACM 978-1-4503-1822-8/12/08, pp. 179–184.
5. Milchevski, A., Rozza, A., & Taskovski, D. (2015). Multimodal affective analysis combining regularised linear regression and boosted regression trees. ACM 978-1-4503-3743-4/15/10, pp. 33-39, AVEC'15.
6. Zikos, D., & Vandeliwala, I. (2015). Using binary logistic regression coefficients for the dynamic quantification of comorbidities. ACM 978-1-4503-3452-5/15/07 PETRA'15, pp. 12–14.
7. Plagge, M. (2013). Using artificial neural networks to predict first-year traditional students second year retention rates. ACM 978-1-4503-1901-0/13/14, ACMSE'13, pp. 13–18.
8. Trujillo, L., Martinez, Y., & Melin, P. (2011). How many neurons? A genetic programming answer. ACM 978-1-4503-0690-4-11-07/GEECO'11, pp. 175–176.
9. Gunasekara, M., Dharmaratne, A., & Sandaruwan, D. (2014). A feature-point based approach for pose variant face recognition. ACM 978-1-4503-2765-7/14/08, VINCI'14, pp. 242–243.
10. Kalles, D., Verykios, V. S., Feretzakis, G., & Papagelis, A. (2016). Data set operations to hide decision tree rules. ACM 978-1-4503-4304-6/16/08, PrAISe'16, pp. 12–20.
11. Ram, P., & Gray, A. G. (2011). Density estimation trees. ACM 978-1-4503-0813-7/11/08, pp. 627–635, KDD'11.
12. Senior, A., Heigold, G., & Ranzato, M. A. *An empirical study of learning rates in deep neural networks for speech recognition* (pp. 12–17). IEEE.
13. Wang, X. *Method of steepest descent and its applications* (pp. 1–3). IEEE.
14. Li, Y. *Deep reinforcement learning: An overview*. Thesis work: Arxiv, pp. 12–42.
15. Subramanian, D. V., Dr. & Kumar, K. P., Dr. (2016). Fuzzy based modelling for an effective it security policy management. In *SAI Computing Conference 2016*, 978-1-4673-8460-5/16, pp. 1–9. IEEE.

Exponential Spline Approximation for the Solution of Third-Order Boundary Value Problems

Anju Chaurasia, Prakash Chandra Srivastava and Yogesh Gupta

Abstract A general third-order boundary value problems (BVPs) are considered here, to find the approximate solution. An exponential amalgamation of cubic spline functions is used to form a novel numerical approach. Finite difference method supports the developed system to solve the problems slickly. Our method is convergent and second-order accurate. Numerical examples show that the method congregates with sufficient accuracy to the exact solutions.

Keywords Boundary value problems · Finite difference method
Splines · Third-order differential equation

1 Introduction

The solution of boundary value problems (BVPs) is of key importance for numerous scientific problems in many branches of science and engineering. Application of numerical methods to find the approximate solutions of these BVPs has been an active area of research in applied mathematics. A wide range of numerical approaches specifically tailored to approximate the solution of third-order BVPs is available today. An exploration of the literature on the solution of third-order BVPs using diverse numerical approaches can be comprehended as

---

A. Chaurasia (✉)
Birla Institute of Technology, Allahabad 211000, Uttar Pradesh, India
e-mail: anjuchaurasiya@rediffmail.com

P. C. Srivastava
Department of Mathematics, Birla Institute of Technology, Allahabad 211000, Uttar Pradesh, India
e-mail: prakash_bit123@rediffmail.com

Y. Gupta
Department of Mathematics, Jaypee Institute of Information Technology, Noida 201307, Uttar Pradesh, India
e-mail: yogesh.gupta@jiit.ac.in

V. E. Balas et al. (eds.), *Data Management, Analytics and Innovation*, Advances in Intelligent Systems and Computing 808,
https://doi.org/10.1007/978-981-13-1402-5_26

finite difference method [1–4], cubic spline method [5, 6] quartic spline method [7], quintic spline method [8], non-polynomial splines methods [9–13, 15], etc.

Authors in [1, 2, 4] considered the following system of third-order BVPs

$$u^{(3)}(x) = \begin{cases} f(x), & a \le x \le c, \\ p(x) + f(x)u(x) + r, & c \le x \le d, \\ f(x), & d \le x \le b, \end{cases} \tag{1.1}$$

with the boundary conditions (BCs)

$$u(a) = \alpha, \quad u'(a) = \beta_1, \quad u'(b) = \beta_2. \tag{1.2}$$

In these papers, authors tried to compute numerical solution of the above BVPs by means of finite difference method. Penalty functions as a supplemental tool supported the authenticity of the solution. Methods presented in these papers are found to be second-order convergent.

Authors in [5, 6] studied the same system of problems (1.1, 1.2) to find the approximate solution. They built up a technique based on uniform cubic spline functions of the form

$$T_n = \mathrm{Span}\{1, x, x^2, x^3\}$$

and established the consistency equations. Designated method had the convergence of order two with dominance over some other techniques such as finite difference, collocation, and splines.

The system of third-order BVPs was yet again deliberated by numerous authors with the treatment of non-polynomial spline functions to compute approximation to the solution. For solving the above proposed system (1.1, 1.2), authors in [9, 10, 14] looked for a non-polynomial spline method fabricated with quartic functions. In [9, 10], authors used the quartic spline function of the form

$$T_n = \mathrm{Span}\{1, x, x^2, \sin(kx), \cos(kx)\}.$$

Whereas in [14], authors dealt with a different form of quartic splines as

$$T_n = \mathrm{Span}\left\{1, x, x^2, \mathrm{e}^{(kx)}, \sin(kx)\right\}.$$

The discussed method was second-order convergent and facilitated the smooth approximation.

Authors [12], came up with their novel approach based on non-polynomial quintic spline functions to find the solution of above system (1.1, 1.2). They exercised the establishment of different order of methods for many third-order BVPs. Henceforth, numerical examples were tested by means of the proposed scheme along with improved end conditions. Rate of convergence are up to second and fourth.

Authors in [7], explicitly featured the development of a numerical technique in consequence of quartic spline functions to solve the special case of third-order BVPs given as

$$u'''(x) = \begin{cases} f, & 0 \leq x \leq \frac{1}{4} \text{ and } \frac{3}{4} \leq x \leq 1, \\ u + f - 1, & \frac{1}{4} \leq x \leq \frac{3}{4}, \end{cases} \tag{1.3}$$

with the BCs

$$u(0) = u'(0) = u'(1) = 0. \tag{1.4}$$

The above system of differential equations was solved by authors for $f = 0$ to test the pragmatism of the applied technique. Errors were computed for examples that evidence the better approximation than some existing methods in the nous of accuracy and computational work.

Authors in [8], proposed a particular form of third-order BVPs to lead the spline based numerical system

$$y'''(x) = f(x, y); \quad a \leq x \leq b, \tag{1.5}$$

subject to

$$y(a) = k_1, \quad y'(a) = k_2, \quad y(b) = k_3. \tag{1.6}$$

A structure was formed with the help of quintic spline functions to solve the third-order BVPs. This method provided accuracy up to fourth order.

Authors in [11, 15] numerically investigated the non-polynomial spline solutions of third-order BVPs of the form (1.5) with subsequent BCs

$$y(a) - A_1 = y'(a) - A_2 = y'(b) = A_3 = 0. \tag{1.7}$$

They employed with the non-polynomial splines based on quartic spline functions to yields the approximations. Order of the method in [11] was to be found two, where [15] assured the accuracy up to second and fourth order.

Once more, above specified BVP (1.5) along with BCs (1.7) was projected by authors in [13]. A framework was systematized with non-polynomial quintic spline functions to find the smooth approximations. Method was second and fourth order convergent while solving the third-order BVPs.

A general third-order BVPs of the form

$$Ly(x) = y'''(x) + a(x)y''(x) - b(x)y'(x) + c(x)y(x) = f(x), \tag{1.8}$$

with the boundary conditions

$$y(\delta_0) = \alpha_0, \quad y'(\delta_0) = \alpha_1, \quad y(\delta_1) = \alpha_2. \tag{1.9}$$

or

$$y(\delta_0) = \alpha_0, \quad y'(\delta_0) = \alpha_1, \quad y'(\delta_1) = \alpha_3.$$

were solved for linear and nonlinear cases by authors in [3]. They involved only four mesh points to customize the technique. They solved the examples for special cases of BVPs with different conditions. The method has the convergence up to order four.

Our paper focused on the solution of *general third-order boundary value problem* of the form

$$y^{(3)}(x) = f(x)y^{(2)}(x) + g(x)y^{(1)}(x) + h(x)y(x) + r(x), \quad -\infty \leq a \leq x \leq b \leq \infty, \tag{1.10}$$

with the boundary conditions:

$$y(a) = Z_1, \quad y'(a) = Z_2 \text{ and } y'(b) = Z_3, \tag{1.11}$$

where Z_i, i = 1, 2 and 3 are finite real constants. $f(x)$, $g(x)$, $h(x)$ and $r(x)$ are continuous functions on the interval $[a, b]$. To find the approximations to the solutions of above system (1.10, 1.11) through some different non-polynomial scheme, we used *exponential cubic spline functions* of the form

$$T_n = \text{Span}\left\{1, x, e^{(kx)}, e^{(-kx)}\right\},$$

where k is the free parameter. T_n reduces to cubic polynomial spline function in $[a, b]$, when $k \to 0$. The results obtained by our method will clearly indicate that this spline scheme produces more accurate numerical results than difference method in [16]. The paper at hand will cover the following aspects:

Section 2 introduces the development of our method. Section 3 describes the convergence study of the proposed scheme. In Sect. 4, we applied our scheme on two examples to assess the efficiency of the technique. Finally, paper is concluded in Sect. 5.

2 Mathematical Formulation

A grid of $N + 1$ equally spaced points x_i, is defined here which equally divides the interval $[a, b]$ into N parts. In each part, the mixed spline function $P_i(x)$ has the following form:

$$P_i(x) = a_i e^{k(x-x_i)} + b_i e^{-k(x-x_i)} + c_i(x - x_i) + d_i, \quad \text{for } i = 0, 1, 2\ldots, N. \tag{2.1}$$

where a_i, b_i, c_i and d_i are constants and k is free parameter which can be real or purely imaginary. Let $U(x)$ be the exact solution of above system of BVPs (1.10, 1.11) and S_i an approximation to $U_i = U(x_i)$ obtained by the segment $P_i(x)$ of the exponential spline function passing through the points (x_i, S_i) and (x_{i+1}, S_{i+1}).

Now, we assume

$$\begin{cases} P_i\left(x_{i+\frac{1}{2}}\right) = S_{i+\frac{1}{2}}, & P_i^{(1)}(x_i) = D_i, \\ P_i^{(2)}\left(x_{i+\frac{1}{2}}\right) = Q_{i+\frac{1}{2}}, & P_i^{(3)}\left(x_{i+\frac{1}{2}}\right) = T_{i+\frac{1}{2}} \end{cases} \tag{2.2}$$

to obtain the value of coefficients such as

$$\begin{cases} a_i = \frac{e^{-\theta}}{2k^2}\left[Q_{i+\frac{1}{2}} + \frac{1}{k}T_{i+\frac{1}{2}}\right], \\ b_i = \frac{e^{\theta}}{2k^2}\left[Q_{i+\frac{1}{2}} - \frac{1}{k}T_{i+\frac{1}{2}}\right], \\ c_i = D_i + \frac{\sinh(\theta)}{k}Q_{i+\frac{1}{2}} - \frac{\cosh(\theta)}{k^2}T_{i+\frac{1}{2}}, \\ d_i = S_{i+\frac{1}{2}} - \frac{h}{2}D_i - \left[\frac{1}{k^2} + \frac{h\sinh(\theta)}{2k}\right]Q_{i+\frac{1}{2}} + \frac{h\cosh(\theta)}{2k^2}T_{i+\frac{1}{2}}. \end{cases} \tag{2.3}$$

where $\theta = kh/2$, $i = 0, 1, 2, \ldots, N - 1$.

As cubic splines are continuous functions, it follows that

$$P_{i-1}^{(\mu)}(x_i) = P_i^{(\mu)}(x_i), \quad \mu = 0, 1, 2. \tag{2.4}$$

which gives the following consistency relations:

For $\mu = 0$,

$$\begin{aligned} D_i + D_{i-1} &= \frac{2\left(S_{i+\frac{1}{2}} - S_{i-\frac{1}{2}}\right)}{h} + \left(\frac{2(\cosh(\theta) - 1)}{k^2 h} - \frac{\sinh(\theta)}{k}\right)Q_{i+\frac{1}{2}} \\ &- \left(\frac{2(\cosh(\theta) - 1)}{k^2 h} + \frac{\sinh(\theta)}{k}\right)Q_{i-\frac{1}{2}} + \left(\frac{\cosh(\theta)}{k^2} - \frac{2\sinh(\theta)}{k^3 h}\right)\left(T_{i+\frac{1}{2}} + T_{i-\frac{1}{2}}\right). \end{aligned} \tag{2.5}$$

For $\mu = 1$,

$$D_i - D_{i-1} = \left(\frac{2\sinh(\theta)}{k}\right)Q_{i-\frac{1}{2}}. \tag{2.6}$$

and for $\mu = 2$,

$$Q_{i+\frac{1}{2}} - Q_{i-\frac{1}{2}} = \left(\frac{\tanh(\theta)}{k}\right)\left(T_{i+\frac{1}{2}} + T_{i-\frac{1}{2}}\right). \tag{2.7}$$

Adding Eqs. (2.5) and (2.6), we get

$$D_i = \frac{\left(S_{i+\frac{1}{2}} - S_{i-\frac{1}{2}}\right)}{h} + \left(\frac{1}{2k^2\cosh(\theta)} - \frac{\tanh(\theta)}{k^3 h}\right)\left(T_{i+\frac{1}{2}} + T_{i-\frac{1}{2}}\right). \tag{2.8}$$

From Eqs. (2.7) and (2.8), we get

$$D_{i+1} - 2D_i + D_{i-1} = \left(\frac{2\sinh(\theta)\tanh(\theta)}{k^2}\right)\left(T_{i+\frac{1}{2}} + T_{i-\frac{1}{2}}\right). \tag{2.9}$$

Substituting the value of D_i's from Eq. (2.8), we get the following relation

$$S_{i+\frac{1}{2}} - 3S_{i-\frac{1}{2}} + 3S_{i-\frac{3}{2}} - S_{i-\frac{5}{2}} = h^3\left[\alpha\left(T_{i+\frac{1}{2}} + T_{i-\frac{5}{2}}\right) + \beta\left(T_{i-\frac{1}{2}} + T_{i-\frac{3}{2}}\right)\right], \quad \text{for } i = 3, 4, \ldots, N-1. \tag{2.10}$$

where

$$\begin{cases} \alpha = \frac{\tanh(\theta)}{8\theta^3} - \frac{1}{8\theta^2\cosh(\theta)}, \\ \beta = \frac{\sinh(\theta)\tanh(\theta)}{2\theta^2} - \frac{\tanh(\theta)}{8\theta^3} + \frac{1}{8\theta^2\cosh(\theta)}, \\ \text{and } T_i = S^{(3)}(x_i). \end{cases}$$

For the ends of the range of integration, we can obtain three more equations. These three equations [17] are given by

$$\begin{cases} 8S_0 - 9S_{\frac{1}{2}} + S_{\frac{3}{2}} = -3hD_0 + \frac{3}{8}h^3\left[T_{\frac{1}{2}}\right], & \text{for } i = 1, \\ 2S_{\frac{1}{2}} - 3S_{\frac{3}{2}} + S_{\frac{5}{2}} = -hD_0 + \frac{1}{24}h^3\left[-T_0 + 12T_{\frac{1}{2}} + 12T_{\frac{3}{2}}\right], & \text{for } i = 2, \\ -S_{N-\frac{5}{2}} + 3S_{N-\frac{3}{2}} - 2S_{N-\frac{1}{2}} = -hD_N + \frac{1}{24}h^3\left[12T_{N-\frac{3}{2}} + 12T_{N-\frac{1}{2}} - T_N\right], & \text{for } i = N. \end{cases} \tag{2.11}$$

The solution of our system (1.10, 1.11) is based on the linear equations given by (2.10, 2.11). The local truncation errors t_i, $i = 1, 2 \ldots N$ associated with the above specified exponential cubic spline method, are given as follows:

$$t_i = \begin{cases} \frac{27}{1920} h^5 U_i^{(5)} + o(h^6); & \text{for } i = 1, \\ -\frac{1}{1920} h^5 U_i^{(5)} + o(h^6); & \text{for } i = 2, \\ (-2\alpha) h^5 U_i^{(5)} + o(h^6); & \text{for } 3 \le i \le N-1, \\ -\frac{1}{1920} h^5 U_i^{(5)} + o(h^6); & \text{for } i = N. \end{cases} \tag{2.12}$$

The third derivative specified in Eq. (1.10), is solved here by means of exponential non-polynomial spline method (2.10, 2.11) and approximate solutions are attained. Substituting $T_i = y^{(3)}(x)$ in Eq. (1.10), we get the following equation

$$T_i = f(x_i)y_i'' + g(x_i)y_i' + h(x_i)y_i + r(x_i).$$

Henceforward, finite difference method is used as a bridge to evaluate the first- and second-order derivatives by way of

$$y_i''(x_i) = \frac{y_{i+1} - 2y_i + y_{i-1}}{h^2} \text{ and } y_i'(x_i) = \frac{y_{i+1} - y_{i-1}}{2h}.$$

3 Convergence Analysis

The error equation of the methods (2.10, 2.11) can be written as:

$$AE = T, \tag{3.1}$$

where $E = (e_{i+1/2})$ is the error of discretization defined by $e_{i+1/2} = U_{i+1/2} - S_{i+1/2}$, $T = (t_i)$ and matrix A can be defined as

$$A = A_0 + hBF + h^2BG + h^3BH, \tag{3.2}$$

where, A_0 is nonsingular matrix, given by

$$A_0 = \begin{bmatrix} -9 & 1 & 0 & & & & \\ 2 & -3 & 1 & 0 & & & \\ -1 & 3 & -3 & 1 & & & \\ & -1 & 3 & -3 & 1 & & \\ & & \ddots & \ddots & \ddots & \ddots & \\ & & & \ddots & \ddots & \ddots & \\ & & & -1 & 3 & -3 & 1 \\ & & & & 1 & -3 & 2 \end{bmatrix}.$$

Matrices B, F, G and H are according [10]. From (3.1) and (3.2), we can have

$$E = A^{-1}T = \left\{A_0 + \left(hBF + h^2BG + h^3BH\right)\right\}^{-1}T. \tag{3.3}$$

Then,

$$\|E\| \leq \left\|\left(I + A_0^{-1}\left(hBF + h^2BG + h^3BH\right)\right\|^{-1}\left\|A_0^{-1}\right\|\|T\|.$$

Using

$$\left\|(I + A)^{-1}\right\| \leq (1 - \|A\|)^{-1};$$

we get

$$\|E\| \leq \frac{\left\|A_0^{-1}\right\|\|T\|}{1 - \left\|A_0^{-1}\right\|.\|(hBF + h^2BG + h^3BH)\|}. \tag{3.4}$$

Again, by using the fact as followed in [10],

$$\left\|A_0^{-1}\right\| \leq \frac{2}{81}(b-a)^3\left[1 + \frac{3h^2}{2(b-a)^2}\right]h^{-3}$$

and

$$\|BF\| = (48\alpha + 16\beta)\|F\|, \|BG\| = (12\alpha + 5\beta)\|G\| \text{ and } \|BH\| = 2(\alpha + \beta)\|H\|,$$

where,

$$\|F\| = \max_{a \leq x \leq b}|f(x_i)|, \|G\| = \max_{a \leq x \leq b}|g(x_i)| \text{ and } \|H\| = \max_{a \leq x \leq b}|h(x_i)|$$

and for λ_1, λ_2, $\lambda_3 < 1$ and $\lambda_1 + \lambda_2 + \lambda_3 = 1$;

$$\|F\| \le \frac{h^2 \lambda_1}{\omega(48\alpha + 16\beta)}, \|G\| \le \frac{h \lambda_2}{\omega(12\alpha + 5\beta)}, \|GH\| \le \frac{\lambda_3}{\omega}. \quad (3.5)$$

also,

$$\|T\| = (-2\alpha)h^5 M_5, \text{where } M_5 = \max_{a \le x \le b}\left|u^{(5)}(x)\right|.$$

So, by using Eq. (3.4), we obtain

$$\|E\| \le O(h^2). \quad (3.6)$$

4 Numerical Illustrations

To illustrate the use of exponential cubic spline, we consider two general third-order BVPs with a comparison over a difference method [16].

Problem 4.1

$$u^{(3)}(x) + xu^{(2)}(x) = -6x^2 + 3x - 6; \quad 0 < x < 1,$$
$$u(0) = 0, \quad u'(0) = 0 \text{ and } u'(1) = 0.$$

The theoretical solution for this problem is

$$u(x) = \frac{3}{2}x^2 - x^3.$$

The maximum absolute errors (MAE) in the solutions are tabulated in Table 1.

Problem 4.2

$$u^{(3)}(x) = 25\,u^{(1)}(x) - 1; \quad 0 < x < 1,$$
$$u(0) = \frac{r\left(-K + 2\tanh\frac{K}{2}\right)}{2K^3}, u'(0) = 0 \text{ and } u'(1) = 0.$$

Table 1 The maximum absolute errors in Problem 4.1

N	Our method	In [16]
128	3.7363×10^{-4}	3.0696×10^{-3}
256	9.3581×10^{-5}	6.1094×10^{-4}
512	2.3417×10^{-5}	1.4379×10^{-4}
1024	5.8568×10^{-6}	4.1723×10^{-5}
2048	1.4648×10^{-6}	1.6298×10^{-5}

Table 2 The maximum absolute errors in Problem 4.2

$r = 1$, $K = 5.0$.		
N	Our method	In [16]
64	3.7390×10^{-5}	3.6522×10^{-3}
128	1.1143×10^{-6}	4.3436×10^{-4}
256	4.5358×10^{-6}	5.1409×10^{-5}
512	3.5393×10^{-6}	9.9651×10^{-6}
1024	2.0973×10^{-6}	2.5145×10^{-6}

The theoretical solution for this problem is

$$u(x) = \frac{r\left(K(2x-1) - 2\sinh(Kx) + 2\cosh(Kx)\tanh\left(\frac{K}{2}\right)\right)}{2K^3}.$$

The MAE in the solutions are charted in Table 2.

5 Conclusion

There are a few articles that have addressed the approximate solutions of *general third-order boundary value problems* using spline functions. We presented a novel method based on exponential cubic splines to solve general linear third-order BVPs. Convergence analysis and solved examples show the feasibility of the method. Our method validates accuracy up to second order. Moreover, comparisons are made to assess the adeptness of the specified technique. The results obtained in tables clearly indicate that present method produces more accurate numerical results than difference method as developed in [16].

References

1. Said, E. A. (2001). Numerical solutions for system of third-order boundary value problems. *International Journal of Computer Mathematics, 78,* 111–121.
2. Noor, M. A., & Said, E. A. (2002). Finite-difference method for a system of third-order boundary-value problems. *Journal of Optimization Theory and Applications, 112,* 627–637.
3. Salama, A. A., & Mansour, A. A. (2005). Fourth-order finite-difference method for third-order boundary-value problems. *Numerical Heat Transfer, Part B: Fundamentals, An International Journal of Computation and Methodology, 47,* 383–401.
4. Noor, M. A., Said, E. A., & Noor, K. I. (2012). Finite difference method for solving a system of third-order boundary value problems. *Journal of Applied Mathematics, 2012,* 1–10.
5. Said, E. A., & Noor, M. A. (2003). Cubic splines method for a system of third-order boundary value problems. *Applied Mathematics and Computation, 142,* 195–204.
6. Said, E. A. (2008). A cubic spline method for solving a system of third order boundary value problems. *International Journal of Pure and Applied Mathematics, 43,* 677–686.

7. Noor, M. A., & Said, E. A. (2004). Quartic splines solutions of third-order obstacle problems. *Applied Mathematics and Computation, 153,* 307–316.
8. Khan, A., & Aziz, T. (2003). The numerical solution of third-order boundary-value problems using quintic splines. *Applied Mathematics and Computation, 137,* 253–260.
9. Islam, S. U., Khan, M. A., Tirmizi, I. A., & Twizell, E. H. (2005). Non-polynomial splines approach to the solution of a system of third-order boundary-value problems. *Applied Mathematics and Computation, 168,* 152–163.
10. Islam, S. U., Tirmizi, I. A., & Khan, M. A. (2007). Quartic non-polynomial spline approach to the solution of a system of third order boundary value problems. *Journal of Mathematical Analysis and Applications, 335,* 1095–1104.
11. Talaat, S., & Danaf, E. (2008). Quartic non-polynomial spline solutions for third order two-point boundary value problem. *International Journal of Mathematical, Computational, Physical, Electrical and Computer Engineering, 2,* 637–640.
12. Khan, A., & Sultana, T. (2012). Non-polynomial quintic spline solution for the system of third order boundary value problems. *Numerical Algorithms, 59,* 541–559.
13. Srivastava, P. K., & Kumar, M. (2012). Numerical algorithm based on quintic non-polynomial spline for solving third-order boundary value problems associated with draining and coating flow. *Chinese Annals of Mathematics, 33,* 831–840.
14. Li, L., & Gao, J. (2014). Non-polynomial spline method for a system of third order obstacle boundary value problems. *Journal of Mathematical Sciences, Advances and Applications, 30,* 49–69.
15. Salam, F. A. A. E., Sabbagh, A. A. E., & Zaki, Z. A. (2010). The numerical solution of linear third order boundary value problems using non-polynomial spline technique. *Journal of American Science, 6,* 303–309.
16. Pandey, P. K. (2017). A numerical method for the solution of general third order boundary value problem in ordinary differential equations. *Bulletin of the International Mathematical Virtual Institute, 7,* 129–138.
17. Said, E. A. (2004). Computational techniques for system of third order boundary value problems. *Journal of Nonlinear Functional Analysis and Applications, 9,* 305–314.

Customer Lifetime Value: An Ensemble Model Approach

Harminder Singh Channa

Abstract Customer lifetime value allows Banks and Financial Institutions to examine the worth of customers to the business, which provides important inputs to take informed marketing & retention decisions and better Customer Relationship Management. Traditional CLV approaches are primarily isolated at account level worthiness. Some Customer level CLV do take 360° view of the customer relationships but are more heuristic in nature or predicting the CLV based on historical CLV data using single model approach. In this paper, we have explored the existing solutions available to calculate the CLV and explained the rationale for not using with their respective limitations. The focus of the study was on retail banking sector, the proposal is to use whole gamut of existing marketing and risk predictive models for calculating the predicted CLV without taking the time value of money into consideration. It also discusses about the comparisons between the present and future CLV of the customer and how to check the overall health of the bank's business using calculated CLV.

Keywords CLTV · CLV · Customer lifetime value · Predicted customer value
Present customer value · CLTV limitations · Predictive model · Churn model
Survival model · Machine learning in banking · Customer potential
Time value · Present value · Customer relationship · Customer level CLTV

1 Introduction

CLV: Customer Lifetime Value in the most simplistic terms is the measure of the profitability of the customer during the lifetime. The limelight which CLV always receive is because all the banks are interested in focusing on the 'cream of the crop'

H. S. Channa (✉)
Analytics & Insights, Tata Consultancy Services, Bengaluru, Karnataka, India
e-mail: harminder.channa@tcs.com

V. E. Balas et al. (eds.), *Data Management, Analytics and Innovation*,
Advances in Intelligent Systems and Computing 808,
https://doi.org/10.1007/978-981-13-1402-5_27

customers to have sustaining mutually beneficial relationships with selected customers. Also, it exhaustively explains the relationship between the customer and the Bank.

This CLV soothsaying can vary from a subjective/heuristic approach to more sophisticated analytical techniques. There is no standard way of measuring CLV.

Bank and Data Sources

We were supporting an Indian bank[1] with strong regional presence and lot of data attributes were available to us to have a customized data lake. Besides model development for different bank departments, one of the prioritized goals was the calculate CLV for the customers.

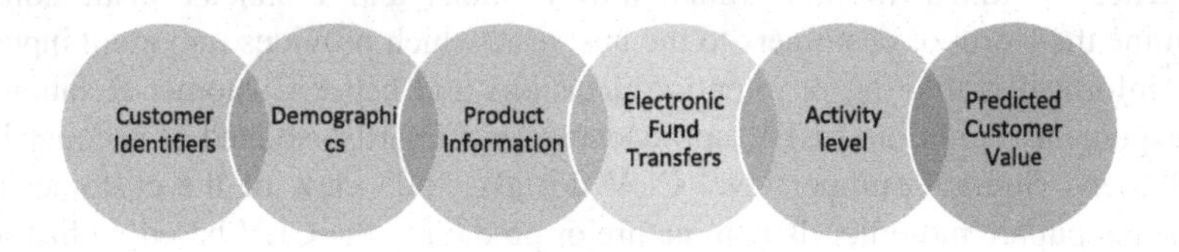

In order to have our own CLV measure for the retail banking exhaustive enough to encompass all relationships and potentials of a customer, we started adopting one of the available approaches but ended up devising a novel way to calculate CLV.

Various approaches are in place and are published, focusing on multiple aspects in the retail banking industry. We have explained some of those and then boiled down to the solution we have presented.

Three standard approaches/techniques were explored that are used in industry:

(1) Churn model based approach
(2) Survival analysis based approach
(3) Model for CLV.

2 Literature Review

2.1 *Approach 1: Churn Model Based Approach*

CLV is viewed as the present value of the future cash flows associated with a customer. It is basically the sum of the incomes obtained from a customer over the lifetime after deducting all the associated costs by taking the present value of the money.

The basic formula for calculating CLV for customer, i at time, t for a finite time horizon T is:

[1]Due to proprietary bank, we are keeping the bank anonymous.

$$\mathrm{CLV}_{i,t} = \sum_{t=0}^{T} \frac{\mathrm{profit}_{i.t}}{(1+d)^t} \text{ or } \mathrm{CLV}_i = \sum_{t=1}^{T} \frac{\mathrm{Revenue}_{i.t}}{(1+d)^t} - \sum_{i=1}^{T} \frac{\mathrm{Cost}_{i.t}}{(1+d)^t} \tag{1}$$

Now calculating CLV based on T that is equal to the term of term deposits or advances would be naïve. That is to assume all the customers on the banks books would complete the full terms. For Saving/Current deposits selecting T becomes little tricky. A rule of thumb is to calculate it for next 3–5 years or any other horizon, which business feels would not change substantially.

But this approach with some fixed horizon is little vulnerable to customer attrition or churn. So an improved approach is to incorporate the churn prediction.

Here is the mathematical model for CLV measuring of this research:

$$\mathrm{CLV} = \sum_{t=1}^{n} \frac{P_t(S_t X M_t)}{d^t} - \sum_{i=1}^{n} \frac{(P_t X D_t) + R_t)}{d^t} \tag{2}$$

where

- P_t The probability of continuous interaction of customer with the bank; P_t = 1 − C.R, and also C.R is churn probability.
- S_t The average amount of customer's accounts after subtracting by legal and liquidity saving rate; this amount of accounts inventory is the free deposits for retail banks.
- M_t The marginal profits for S_t.
- d_t Discount rate that is equal to: 1+ inflation rate.
- D_t This is the first group of costs that associated with the direct costs about the accounts.
- R_t This is the first group of costs that associated with the indirect costs. This group are including of costs such as: advertising and marketing costs, depreciation costs, administrative costs, other personnel costs, etc.
- N Number of periods.

2.2 Approach 2: Survival Analysis Based Approach

CLV based on survival function and time value of money.

Survival model is developed based on customer's past behavior and trends, to calculate the probability of a customer's survival for next "n" years. CLV is calculated based on historic and predictive Customer Lifetime Value for each customer.

$$\begin{aligned} \text{CLV} &= \text{CLV}_{\text{History}} + \text{CLV}_{\text{Future}} \\ \text{CLV}_{\text{Future}} &= \text{Survival}(t) * T * [\text{Monthly Potential}] \end{aligned} \tag{3}$$

In another way, CLV represents the net present value from profits, from a single customer. It can be represented as

$$\text{CLV}_{\text{customer\_i}} = \sum_{i=1}^{N} (\text{Revenue}_i \times \ \text{Survivalrate}_i - \text{Expenses}_i)/(1+r)^i \tag{4}$$

2.3 Approach 3: CLV Predictive Model

This approach is quite straightforward and is to develop a predictive model for Customer Lifetime value (CLV) using the most important variables from the banking industry.
This works like a standard model development approach with a dependent variable and lot of predictors. CLV is calculated for each customer in the historical data and used as a dependent variable. The dependent variable is the yearly profit obtained from the customer and is computed as taking the sum of the profits gained from each transaction, the assets and liabilities and the products/services used by the customers.

Various modeling techniques can be used to model CLV. This study for literature review did a comparison between least square estimation (LSE) and artificial neural network (ANN) in order to select the best performing forecasting tool to predict the potential CLV. Due to its higher performance; LSE based linear regression model is selected.

In addition to the common variables used in CLV prediction, monetary value and risk of certain bank services, as well as product/service ownership-related indicators, are also significant factors.

3 Proposed Work

As discussed above, we have first tried to adopt the approaches mentioned in the literature review but found that there was some room available to add value to the existing approaches. In this section, we first aimed to address the drawbacks of the approaches discussed through a simple but yet effective approach.

(1) Limitation with first two approaches:

Both the methods i.e. Churn & Survival based work well if CLV needs to be calculated at an account level. In order to take in every aspect of the customer relationship, the best we could do was to agglomerate individual CLVs at the customer level for different products. But it was missing out the entire customer potential. Customer value was also dependent on the future products/offers a customer can take up.

(2) Limitation with CLV predictive model:

Missing out on the customer potential and weighing so heavily on the historical predictors only, to predict the future CLV. It becomes a real challenge to squeeze in all the diverse customer information from all customer touch points/spheres into one model. This information ranging from demographics, experience, monetary and risk information to product/service ownership was so vast to be captured appropriately in one single model.

These limitations formed the basis of our research and drove us to work to encompass every aspect of the customer.

3.1 Predictive Models in Banks

From the literature review, one thing was certain that the predictive models, be it churn model or survival Cox proportional hazard model or Foreclosure model, do add a lot of value in CLV calculation/prediction.

Therefore, it became imperative for us to understand the current model repository a retail bank was equipped with and to assess whether we can leverage anything we have developed so far. The good part was that the retail banking industry has matured in the last few years and few banks managed to have evolved as a customer centric bank. This gave us a head start and following was how our banks model repository (Table 1) looked like for both Deposits & Advances[2] both marketing/risk models.

Table 1 Banks model repository

Marketing models	Risk models
Response & uplift model	Probability of Default & LGD Survival & Churn Model Payoff Model & Foreclosure

[2]Due to proprietary bank data, not all models were listed.

3.2 Usage of Predictive Models in Banks:

How banks use predictive models for marketing/Risk/Account management was an important area to look at.

We have observed, how much customer centric a bank can become, usage of models was mostly segregated or in isolation. What we mean here was various bank's departments mostly work in isolation.

- Marketing team's focus was always increasing the response rates and hence acquisitions. They have their own suite of models (like response, uplift, attribution) which they use without any knowledge what was happening to bank's NPAs
- Similarly Risk teams focus was to look into credit worthiness of the customers.

There was no problem working like this and that was a successful analytics model a bank can have but synergizing all departments would definitely yield better results.

Therefore, the available stock of models was good enough to have 360° view of the customers. Each model with individual probabilities predictions from all phases of customer lifecycle or customer potential and from all relationships are utilizing the entire data lake available.

3.3 Our Approach: Model Ensemble

Our approach was basically an ensemble approach where we were mainly "connecting the dots" to reveal a bigger picture where individual bank departments models were the dots.

Our definition of CLV is the measure of the profitability of the customer during the lifetime from all the relationships and future potentials of a customer without taking into account the time value of the money by only incorporating available predictive models

$$\mathrm{CLV}_{\mathrm{customer},i} = \mathrm{CLV}_{\mathrm{customer}_i,\mathrm{Potential}} + \mathrm{CLV}_{\mathrm{Fcustomer}_i,\mathrm{future}}$$

where

$\mathrm{CLV}_{\mathrm{customer}_i,\mathrm{Potential}}$ It was the customer potential. It was revealed by the marketing or acquisition models predicting the future products a customer can take up.

$\mathrm{CLV}_{\mathrm{Fcustomer}_i,\mathrm{future}}$ It is CLV of existing relationships: Using all the available risk and retention models as used in account level CLV calculations.

Our definition prefers to have a shorter horizon of the customer relationship from 6 to 12 months as business conditions would not change drastically in this period and because of this there was no need to incorporate the time value of money.

3.4 Demonstration

From the Deposit & Advances books, this was the snapshot of two customers with the following relationships with the bank.

Cust_num: Customer Number
Month Year: Month and year when snapshot was taken
Deposit Amt: represents the term deposit account amount
Saving Curr bal: represents the saving account current balance
HLoan Appr Amt: represents the Home Loan account approved amount
HLoan bal Amt: represents the Home Loan balance amount
Loan Approved Amt: represents Personal loan account approved amount
Loan Bal: represents the Personal Loan balance amount

STEP 1 CLV$_{,present}$ (CLV at present)

Cust_num*	Month Year	Age	Region	Deposit Amt	Saving Curr bal	HLoan Appr Amt	HLoan bal Amt	Loan Approved Amt	Loan Bal	Present Customer Value
ABC	Aug15	35	Region Central	Rs0	Rs6,500	Rs15,000	Rs12,000	Rs5,000	Rs1,000	Rs15,000
HSC	Aug15	65	Region East	Rs0	Rs2,500	Rs0	Rs0	Rs2,000	Rs1,500	Rs2,650
.										
.										

*Masking the bank account number.

Based on the present portfolio of customers (ABC), CLV$_{present}$ was calculated i.e.

$$\mathrm{CLV}_{\mathrm{ABC,present}} = (\text{Deposit Amt}) + (\text{Saving Curr bal}) + 70\% * (\text{HLoan bal Amt}) + 10\% * (\text{Loan bal})$$

$$\mathrm{CLV}_{\mathrm{ABC,present}} = \mathrm{Rs.}15{,}000$$

STEP 2 Adding predicted probabilities per customer

With lot of predicted models in place, we had made these predictions about the customer

Cust_num	Age	Region	Deposit Amt	Saving Curr bal	HLoan Appr Amt	HLoan bal Amt	Loan Approved Amt	Loan Bal	Present Customer Value	p1	p2	p3	p4	Dec 1	Dec 2	Dec 3	LGD
ABC	35	Region Central	Rs0	Rs6,500	Rs15,000	Rs12,000	Rs5,000	Rs1,000	Rs15,000	0.4	0.0055	0.06	0.15	3	10	2	0.15
HSC	65	Region East	Rs0	Rs2,500	Rs0	Rs0	Rs2,000	Rs1,500	Rs2,650	0.56	0.008	NA	0	2	9	NA	0.0001
.																	
.																	

*p*1 Represents the probability of existing customers buying a Term Deposit

*p*2 Represents the Churn probability of existing Saving customers within next 12 months

*p*3 Represents the payoff probability of existing Home Loan customers to foreclose

*p*4 Represents the predicted Loss Given Default of existing Personal Loans within next 6 months

Dec1 Represents the predicted decile of existing customers buying a Term Deposit

Dec2 Represents the predicted Churn decile of existing Saving customers within next 12 months

Dec3 Represents the predicted payoff decile of existing Home Loan customers to foreclose

LGD Represents the predicted Loss Given Default of existing Personal Loans within next 6 months.

STEP 3 Based on predicted probabilities/deciles per customer, predicted amounts are calculated, which finally converted into CLV

Cust_num	Age	Region	Present Customer Value	p1	p2	p3	p4	p5	Dec1	Dec2	Dec3	LGD	RsVal1	RsVal2	RsVal3	RsVal4	Predicted Customer Value
ABC	35	Region Central	Rs15,000	0.4	0.0055	0.06	0.15	0.002	3	10	2	0.15	Rs750	Rs6,500	(Rs8,400)	(Rs150)	(Rs1,300)
HSC	65	Region East	Rs2,650	0.56	0.008	NA	0	0.0005	2	9	NA	0.0001	Rs750	Rs2,500	Rs0	(Rs0)	Rs3,250

RsVal1 Represents the customer's potential to buy Term deposit. We have used average term deposit = Rs. 750

RsVal2 Represents the customer's potential loss as churning from Saving account, which was equivalent to the current saving balance. In the example, customer was not churning therefore, current balance i.e. Rs. 6,500

RsVal3 Represents the customer's potential loss on Home loan outstanding balance as prepaying the loan. Calculated as (−1)** × 70% of Rs. 12000 = −Rs. 8,400

RsVal4	Represents the customer's potential loss on Personal loan outstanding balance as customer will default. Calculated as (−1)** × LGD of Rs. 1,000 = − Rs. 150
Predicted customer value	It was the sum of the potential gain/loss per customer.

Calculated as $CLV_{ABC,predicted}$ = Rs. 750 + Rs. 6,500 + (−Rs. 8,400) + (−Rs. 150) = −Rs. 1,300.

**(−1) to subtract the loss

3.5 *Comparison Between Present and Future CLV*

Therefore, Customer who seemed very profitable as on Aug15 with $CLV_{ABC,present}$ = Rs. 15,000 has the predicted $CLV_{ABC,predicted}$ of – Rs. 1,300 as depicted in Fig. 1.

In order to pin point the concern areas, we calculated the percentage change in each of the customer products, which was represented in the Fig. 2.

Predicted CLV of the Bank: In the demonstration above, we had calculated the predicted CLV of one customer. When we replicated it across the Deposits and Advances book, we got astonishing results. The healthy and growing looking book based on the present value has shown inverted picture as shown in Fig. 3.

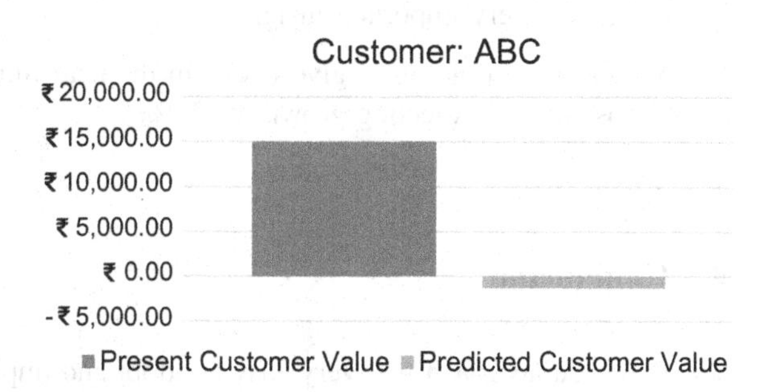

Fig. 1 Present versus predicted Customer value

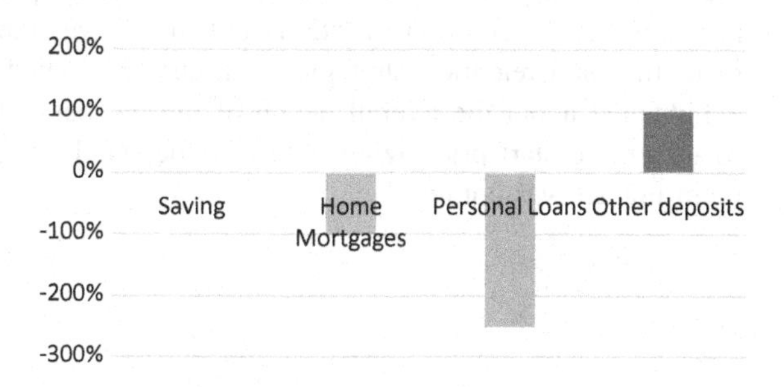

Fig. 2 Portfolio wise: percent change

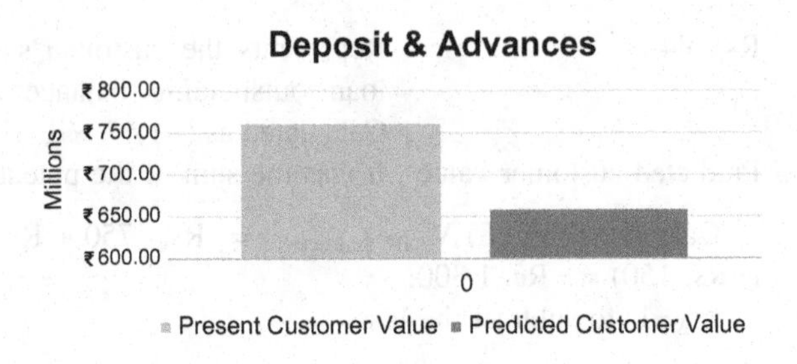

Fig. 3 Present versus predicted Customer value of deposits & advances

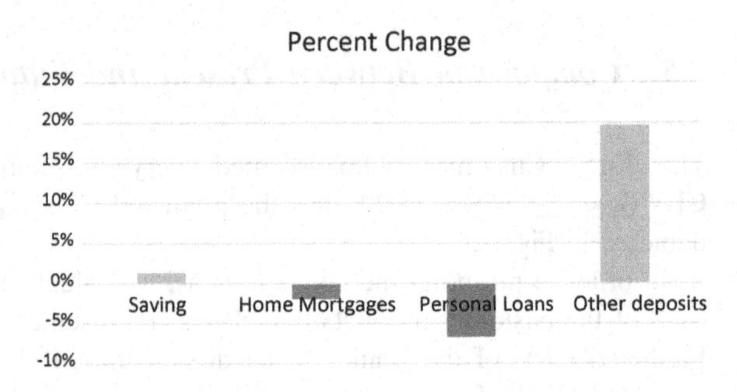

Fig. 4 Portfolio wise across bank: percent change

In order to pin point the concern areas, we calculated the percentage change in each of the customer products for the bank's Deposit and Advances, which was represented in the Fig. 4.

This gave a very important insights:

(1) Advances were going to give losses in the near future
(2) Deposits were expecting growth of 20%.

4 Conclusion

The proposed approach was very easy to adopt and implement and captures all the important parameters available in the retail banking space. It not only answered the problem CLV prediction questions but also gave important insights about how marketing and retention strategies at a customer level. The highlight was to get enlightened about the overall health of the business. It very easily explained the focus area to start prioritizing with to understand from which segment the maximum impact was coming.

References

1. Berger, P., & Nasr, N. (1998). Customer lifetime value: marketing models and applications. *Journal of Interactive Marketing, 12*(1), 49–61.
2. Kahreh, M. S., et al. (2014). Analyzing the applications of customer lifetime value (CLV) based on benefit segmentation for the banking sector. *Procedia—Social and Behavioral Sciences, 109,* 590–594.
3. Ramanathan, D. (2014–2015). How to become a CLV Aligned Organization: Fractal analytics.
4. Ekinci, Y., Uray, N., & Ülengin, F. (2014). Lifetime value in the bank industry: Sixth montreal industrial problem solving workshop August 17–21, 2015. *European Journal of Marketing; Bradford, 48*(3/4), 761–784.

Part IV
Advances in Network Technologies

A Cryptographic Algorithm Using Location-Based Service and Biometrics

Ridam Pal, Onam Bhartia and Mainak Sen

Abstract Modern advancement of different technologies led to the increasing computational power of each individual component of digital computers that threatens to crack many secure classical algorithms as they are based on mathematical assumption. Thus like authenticated users, hackers or intruders are also able to crack security system. So, researchers and scientists are moving to new directions as well as merging different types of algorithms. Different GPS-enabled devices like smartphone, PDA, etc. are easily accessible which also supports many different applications that extract patterns like iris, fingerprint, etc. Biometric features can be used along with location of intended receiver to develop a cryptographic algorithm. Different smartphone apps provide both locations and can extract the biometric features by which people can form new key. The focus of this paper is to examine that merging of two approaches is advantageous as it provides more security to data.

Keywords Location-based services · Biometric cryptosystem · Feature extraction · Image processing · Image segmentation

1 Introduction

Researchers are trying to improve the efficiency or hardness of cryptographic algorithm so that the attacker might face more and more trouble breaking the algorithm. Nowadays, the underlying devices that are used can be location enabled

R. Pal · M. Sen (✉)
Techno India University, Kolkata, West Bengal, India
e-mail: mainaksen.1988@gmail.com

R. Pal
e-mail: ridam101@gmail.com

O. Bhartia
Dayananda Sagar College of Engineering, Bengaluru, Karnataka, India
e-mail: onam.bhartia@gmail.com

V. E. Balas et al. (eds.), *Data Management, Analytics and Innovation*, Advances in Intelligent Systems and Computing 808,
https://doi.org/10.1007/978-981-13-1402-5_28

like mobile phone, GPS data login devices, etc., and many day-to-day needs of people are being solved by the use of location-based services. Location-based service or the technology related to location-based service utilizes the location of an entity to provide a value-added solution to the user [1]. For example, different navigation applications available on smartphone enable people to board Ola and Uber cabs by defining their location, and people can also find nearby restaurants, shopping malls, and also the road that has less traffic toward their destination. Different apps also support a security alarm to track the near ones of traveler. Remote health care is also an area where these location-based services are used to track a person who needs medical attention.

Since these services are nowadays available at a very lower cost, the location of one side can be sent to the other side very easily through a public channel. Beyond conventional cryptography, people are in need of more powerful crypto algorithm like geo-encryption. The purpose of geo-encryption is to provide security to the transmission of information. In this type of cryptosystem, one could cryptographically bind or attach a location and to some extend time to the ciphertext in such a way that the encrypted text or file can be decrypted only when the location, time etc. constraints are satisfied at the receiver side.

Kealy stated about the mutual relationship between position and location in order to enhance these features. The limitation of current location-based system such as its efficiency and other such problems can be addressed through development in the fields of position technologies, human spatial cognition, and qualitative spatial reasoning. It states about integrating sensor fusion knowledge with spatial fusion knowledge through feedback cycle, which would help in the development of location-based system [2]. Barni mentioned various protocols used in the encryption of data in his paper. It has stated about the homomorphic encryption, oblivious transfer, hybrid protocols, and biometric recognition protocols which were explained in detailed fashion [3]. Biometric feature extraction such as fingerprint, face recognition, etc., can also be deployed for security purpose [4, 5]. Fingerprint of human is one of the most basic features, which can be used for security-related applications [6, 7].

The most basic formation in a fingerprint is a ridge, which forms a certain pattern of black lines. The core of the fingerprint is the innermost recurve that occurs at the center of the pattern. Delta is the feature where three lines converge to form a triangle type structure. Occasionally, ridges split into two ridges known as bifurcation. When a ridge starts or stops at a short distance it is called dot.

The three basic patterns in fingerprint are arch, loop, and whorl. The ridges in the arch pattern start at one side of the finger, rise at the center, and exit from the opposite side. There are no cores or delta in an arch. The ridges in a loop pattern start at one side of the finger, loop in the middle, and exit from the same side from where it started. There are one core and one delta in a loop pattern. The loop can face either left or right. The ridges in a whorl pattern are generally round shaped. There are usually two cores one facing up and one facing down. There are two deltas one in left and other in right. In this paper, we have deployed feature extraction of fingerprint and location-based service to generate a key for encryption. The extracted feature of fingerprint was converted to text, which along with the location-based service, has helped in generation of the final key. The public key generated by the biometric system can be prone to various attacks during the time of communication [8].

The organization of the paper is as follows. After this brief introduction, the necessary background of location encryption is described in Sect. 2. Section 3 states the concept of pattern recognition. Section 4 describes the process for fingerprint identification. Next, the proposed solution has been described in Sect. 5 supported by the experimental results that are shown in Sect. 6. Finally, Sect. 7 concludes the discussion.

2 Location Encryption

Geo-encryption is a type of encryption in which one could encrypt a data for a specific place or geographic area and it would also include the time specification for creation of ciphertext [1]. Thus, the ciphertext generated can be decrypted only within the specific geographic area and time constraint. A plaintext is converted to its corresponding ciphertext using the location that is latitude and longitude of the receiver and the ciphertext can only be decrypted if the location of the receiver is known.

The data is encrypted with the help of location provided by GPS-enabled devices. Initially, a key is generated based on its location and an encryption algorithm is used to form the ciphertext. A key synchronization mechanism must be maintained so that both the sides use the same key for encryption and decryption and in this case, both sides should agree that they will be using this location-based concept at some early stage so that we can also avoid the key distribution step [2, 9]. Once the synchronization is maintained, the data can be encrypted and sent over to the mobile location where it requires to be transferred. After the data is received, it can again be decrypted using the same key to retain the original data.

3 Pattern Recognition

Regularities in data shape or data structure can be defined as pattern. A circular arc can also be termed as pattern. In order to say that one pattern matches with the other pattern, the saved pattern must match with the given pattern in certain criteria and aspects. For example, in case of circular arc, a pattern can be recognized by radius and location of center. Perpendiculars will be drawn on the perimeter of the circle. The point of intersection will be the center and the distance of point on perimeter to the center will be its radius. If radii of both circles are almost equal then the assumption is made that circular arc is similar to the other. Else, it is not similar to the other arc.

Pattern recognition is nothing but a matching problem, which focuses on matching between patterns. When the similarity between two patterns is very high or the dissimilarity is very low, where one of the patterns is stated as model or reference on the knowledge based, we can state whether the second pattern is recognized by the first pattern. Whenever there is a need to recognize a pattern, certain features of the pattern are extracted and compared with the feature domain to conclude whether the patterns are same or different. The features extracted from the circular arc were the radius and the center. On basis of that, it was determined whether the arcs are same or different.

To generate the known pattern, there are two techniques called supervised and unsupervised learning. In case of supervised learning, there are set of known patterns. The features of such similar patterns are collected and generated for all other patterns. Using the stored features, a model is generated which is a feature vector. The feature vector generated is the representative of the same class of patterns. This representative feature vector is stored in the knowledge-based system. Thus, when a new pattern comes such features will be generated and compared with feature vector stored in the knowledge based using the distance between the two feature vectors. If the distance between those two feature vectors is very small, then the unknown pattern is considered to be same as the pattern in the knowledge-based system, otherwise, it is not in the knowledge-based system.

In case of unsupervised learning, there is no known pattern. There is a mixture of patterns from which mixture of feature vectors is generated, without having any information about its source. The feature vector needs to be processed in order to partition the set of feature vectors into a number of subsets, where the feature vector in one set are generated from similar pattern and feature vector from another subset are generated from another pattern. This partitioning of feature vector into subset has to be done by data processing. Thus, from each classified set, a representative feature vector for each set is generated.

4 Process Used for Recognition of Fingerprint

In this section, the proposed model for feature extraction of fingerprint has been stated. Each process has been described in detail stating how it functions.

4.1 Image Acquisition

In image acquisition, the scanned image is acquired in the form of input image. The image acquired should be in specific format such as JPEG, PNG, etc. The image is acquired from scanner or any other suitable device.

4.2 Preprocessing

In preprocessing, a set of operations are performed on the image to render its quality before it is set up for image segmentation. The image is optimized so that the image segmentation can be done using certain nite number of steps. It includes noise removing, edge removing, dilation and filling, and other operations.

4.3 Image Segmentation

Image segmentation is the task where an image is split up into individual objects containing within the image. In image segmentation, an image is stored in the form of matrix where the number within the matrix represents each pixel in the image. There are three matrixes stacked on top of each other, where each matrix represents one color from RBG (Red/Blue/Green). The value of each element in the matrix can range between 0 and 255 (0 indicates dark and 255 indicates bright). In Image segmentation, the image needs to be processed to transform it into different matrixes. The other matrix is known as the object matrix. The object matrix has the same number of rows and columns as the image matrix but it does not have three matrixes stacked on top of each other. The value of each cell in the matrix is an object label number where all pixels belong to a single object having the same labeled number. This kind of output will allow recognizing that there are three distinct objects in an image along with its border, which will help to do different kinds of operations on each of these objects. Object labeling helps in distinguishing each pixel as the part of a connected object. Its basic goal is to identify an object. It basically comprises of four steps in order to apply object labeling algorithm

1. The color image gets converted into grayscale image.
2. The grayscale image is then converted into black and white or binary image.
3. Transformation might be applied if necessary. Some transformations that are available are opening and closing, which is already predefined in various languages or software.
4. Object labeling algorithm—It is an algorithm which changes black and white image to an object matrix. The black and white matrix and the object matrix will have the same dimension.

4.4 Feature Extraction

A feature vector is an *n*-dimensional vector of numerical features representing an object [10, 11]. The vector space associated with these vectors is called feature space. The number of features in a feature vector is referred to as its dimensionality. The similarity of the two objects can be obtained by computing the dot product of the corresponding feature vector. Assigning appropriate values to each feature and computing the dot product of the corresponding feature vector help in mathematically pin downing the similarity among them. The feature vector to pattern matching mapping is one to many, that is, it is not unique—so many patterns can match to the same feature vector. But given a pattern, if the procedure for generation of feature vector is fixed, it will always generate the same feature vector.

There are two types of features:

1. **Shape feature**: Tells about the shape of the object.
2. **Region feature**: Tells about the property of the region enclosed within the object. In gray level, it might be the intensity value but in colored case, it might be the color value or color intensity.

Thus, a single feature does not give a proper description. So, we need multiple features where every feature helps to capture certain property of the pattern and all those features are considered in form of a vector. The position of element within the vector is very important. So, the need was to process these features in particular order and throughout the procedure, it was needful to maintain that same order. So, all these features taken together put in the form of a feature vector, to some extent, pinpoint to a pattern or class of patterns where domain of the class is minuscule.

5 Proposed Methodology

Initially, the image of the fingerprint is obtained. Next, the preprocessing of the fingerprint image was done. For image optimization, the image is cropped in such a way that it forms the minimum-sized square which encloses the whole fingerprint. As a result, for a given fingerprint, the secondary key generated will be unique and same for every given time no matter what the image size is. This image is then worked upon to provide the secondary key for the whole process of this cryptosystem. The entire process of this secondary key generation is divided into two parts—the first part being the image segmentation part and second part being feature extraction.

In the process of image segmentation, the image of the fingerprint was resized into a 60 * 90 pixel image and then it was used for further manipulations. Then, the resized image was divided into squares of 10 * 10 pixels. As a result, 54 blocks of such identical squares were formed with six squares in each row and nine altogether columns.

After this, feature extraction of the segmented image takes place. Now, the number of black pixels in all these squares are calculated respectively. This was done by forming diagonals in the squares. The number of black pixels in all these diagonals is calculated and added to get the total number of black pixels in each square. This number of black pixels is stored in a specific order so that they can correspond with their respective squares.

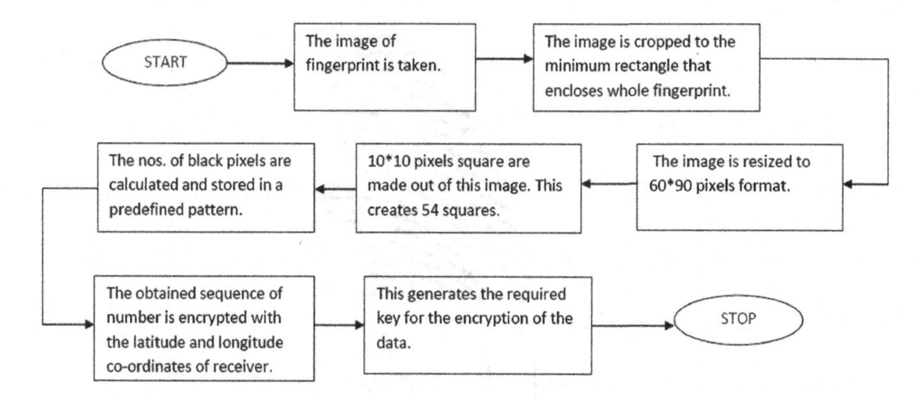

Fig. 1 Operational flow diagram of the proposed method

This sequence of numbers that we get from the above processes is the secondary key. This key is again encrypted with an algorithm using the longitude and latitude of the receiver to form the final key for this cryptographic process. The location coordinates being the shared key in this two-user end-to-end encryption, the secondary key gets decrypted by using the decryption algorithm on it along with the received key. Figure 1 explains the schematic diagram of this proposed algorithm.

6 Results and Discussions

Initially, the fingerprint was scanned through a scanner from Bob and was extracted in the form of input image. Later, the feature of this fingerprint was extracted and was then converted into text (in the form of numbers) as shown in Fig. 2.

Then, this result was encrypted using an encryption algorithm along with the location of the receiver obtained from location-based services shown in Fig. 3. The location of sender(Alice) and receiver(Bob) both were taken by open-source applications based on Android operating system.

The final key was formed using an encryption algorithm in between the stored location of the receiver and biometrics of the sender as shown in Fig. 4. Now the data is converted to ciphertext using symmetric key, where the key will be same for both encryption as well as decryption.

Using this key, sender can encrypt a plaintext to ciphertext and on the receiver side, the receiver can decrypt the ciphertext to its original format as shown in Figs. 5 and 6, respectively.

Then, a second experiment was done to check the algorithm, and in this case Alice was same but Bob was another person altogether whose fingerprint was also scanned at the early stage and the same procedure was followed. Figures 7, 8, 9, 10, and 11 explain the procedure.

Fig. 2 Image conversion to text [12]

Lat/Long: 22.6050889, 88.4290119
Address: BD/241, Gauranganagar, Keshtopur, West Bengal 700101, India

Lat/Long: 12.9051039, 77.5572736
Address: 494, 50th Feet Main Road, 1st Stage, Kumaraswamy Layout, Bengaluru, Karnataka 560078, India

Fig. 3 Location of Alice and Bob

```
The Final Key is:
18 18 18 18 18 18 18 18 18
18 18 18 18 18 18 18 18 24
18 18 18 18 30 51 18 18 18
18 51 35 18 18 18 23 59 58
18 18 18 03 63 50 18 18 18
13 63 12 18 18 17 55 35 48
```

Fig. 4 Formation of final key

```
Enter the plain text:
We are the students of Engineering. CSE is our specialization.


At the sender side, the plain text is:
We are the students of Engineering. CSE is our specialization.
```

Fig. 5 Encryption on sender side

```
The input text to the Decryption Block at the reciever side is:
�(9,�;/,�:;<+,5;:�6-�
                         5.05,,905.��

 �0:�6<9�:7,*0(30A(;065�


The Decrypted text is:
We are the students of Engineering. CSE is our specialization.
```

Fig. 6 Decryption on receiver side

Fig. 7 Image conversion to text [12]

Lat/Long: 22.6050889,88.4290119
Address: BD/241, Gauranganagar,
Keshtopur, Kolkata,
West Bengal 700101, India

Lat/Long: 22.5799586,88.465008
Address: Biswa Bangla Sarani,
Action Area I, New Town
West Bengal 700156, India

Fig. 8 Location of Alice and Bob

```
The Final Key is:
78 78 78 78 78 78 78 78 78
78 78 78 78 78 78 78 78 76
78 78 78 78 75 84 78 78 78
78 66 10 06 78 78 78 77 87
83 78 78 78 66 104 101 78 78
111 99 126 78 78 73 100 103 97
```

Fig. 9 Formation of final key

```
Enter the plain text:
We are the students of Engineering.CSE is our specialization.


At the sender side, the plain text is:
We are the students of Engineering.CSE is our specialization.
```

Fig. 10 Encryption on sender side

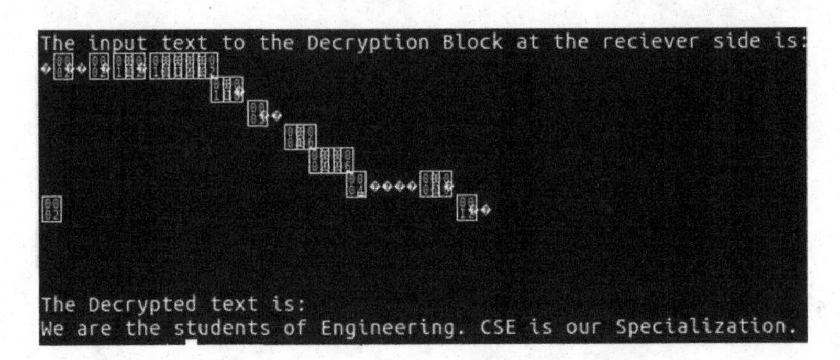

Fig. 11 Decryption on receiver side

7 Conclusion

In order to make a system more secure, researchers are trying to introduce new layers to current existing ones. One such is proposed here and we have seen that without standing at proper location, that is, latitude or longitude no one can decrypt as the key itself needs the value of latitude and longitude and by this it increases one level of security. The inclusion of biometric features of the receiver creates another level so that not only at proper location but receiver with proper identification can only decrypt the ciphertext.

This approach still presents a problem as the encrypted file includes the receiver's address and so people can physically harm an intended receiver. The location obtained by any application is not full proof and can be broken by attacker like if the latitude–longitude pair is encoded by 1 cm and the attacker tries his search within 1 km radius, then a brute force attack might even find the exact latitude–longitude pair with today's high-end processors. Thus, further studies can be done in these fields to make the system more secure.

References

1. Raper, J., Gartner, G., Karimi, H., & Rizos, C. (2007). Applications of location based services: A selected review. Journal of Location Based Services. https://doi.org/10.1080/17489720701862184.
2. Kealy, A., Winter, S., & Retscher, G. (2007). Intelligent location models for next generation location-based services. *Journal of Location Based Services, 1*(4). https://doi.org/10.1080/17489720801905313.
3. Barni, M., Droandi, G., & Lazzeretti, R. (2015, September). Privacy protection in biometric-based recognition systems. *IEEE Signal Processing Magazine*, 6676. https://doi.org/10.1109/MSP.2015.2438131.
4. Chiou, S. Y. (2013). Secure method for biometric-based recognition with integrated cryptographic functions. BioMed Research International. https://doi.org/10.1155/2013/623815.

5. Chang, Y., Zhang, W., & Chen, T. (2004). Biometrics-based cryptographic key generation. In *International Conference on Multimedia and Expo (ICME)*, 22032206. https://doi.org/10.1109/ICME.2004.1394707.
6. Jagadeesan, A., Duraiswamy, K. (2010). Secured cryptographic key generation from multimodal biometrics: Feature level fusion of fingerprint and Iris. *(IJCSIS) International Journal of Computer Science and Information Security, 7*(2), 028037. http://arxiv.org/abs/1003.1458.
7. Goh, A., & Ngo, D. C. L. (2003). Computation of cryptographic keys from face bio-metrics. *Communications and Multimedia Security, 2828*, 113. https://doi.org/10.1007/978-3-540-45184-61.
8. Raghavendra, R., & Busch, C. (2014). Presentation attack detection algorithm for face and iris biometrics. In *European Signal Processing Conference*, EUSIPCO, pp. 1387–1391.
9. Ellul, C., Gupta, S., Haklay, M. M., & Bryson, K. (2013). A platform for location based app development for citizen science and community mapping. In *Progress in Location-Based Services*. Springer, Heidelberg, pp. 71–90. https://doi.org/10.1007/978-3-642-34203-55.
10. Pradeep, J., Srinivasan, E., & Himavathi, S. (2011). Diagonal based feature extraction for handwritten alphabets recognition system using neural network. *International Journal of Computer Science and Information Technology (IJCSIT), 3*(1), 2738. https://doi.org/10.5121/ijcsit.2011.3103.
11. Shrivastava, A., & Srivastava, D. K. (2014). Fingerprint identification using feature extraction: A survey. In *Proceedings of 2014 International Conference on Contemporary Computing and Informatics, IC3I 2014*. https://doi.org/10.1109/IC3I.2014.7019653.
12. https://pixabay.com/en/fingerprint-world-map-swirls-2750393/.

Secure and Energy Aware Shortest Path Routing Framework for WSN

Nikitha Kukunuru

Abstract Wireless Sensor Network is a network in which the sensor nodes are operated in a distributed and self-organizing fashion. Due to this nature the sensor nodes in this network are vulnerable to various malicious attacks. The sensor nodes are resource constrained, i.e., they have limited resources such as bandwidth, energy and memory. Along with Security and Energy Consumption, Delay also needs to be considered during the routing protocol design. This proposes a Trust and Energy Aware Routing Framework which provides resilience to various malicious attacks. The proposed framework helps in finding trusted nodes with maximum residual energy and routes the data through shorter paths. The experimental analysis with varying network parameters as varying the malicious nodes and varying packet size reveal the robustness of proposed approach. Total Throughput, Energy Consumption and End-to-End Delay are considered for experimental evaluation.

Keywords Malicious nodes · Residual energy · Throughput · Trust
Wireless sensor network

1 Introduction

Wireless Sensor Networks have gained an increased research interest in different aspects of the human lives. Routing approaches [1, 2] have been developed for secure WSN. These approaches use authentication and cryptography [3] entities that can't afford to provide the security over various node's misbehavior attacks. Because all these approaches assume that the nodes of WSN are cooperative in nature and also trust worthy. But this hypothesis is not practical for misbehavior attacks.

N. Kukunuru (✉)
Information Technology Department, M.V.S.R Engineering College,
Hyderabad, Telangana, India
e-mail: nikithakukunuru@gmail.com

V. E. Balas et al. (eds.), *Data Management, Analytics and Innovation*,
Advances in Intelligent Systems and Computing 808,
https://doi.org/10.1007/978-981-13-1402-5_29

Recently, the trust based security has become most promising security providing approach to ensure the security for WSN. This is a new technique which can provide security without any use of cryptography techniques. Simply the trust can be defined as a "degree of reliability of other nodes which are helping to source node" In trust based approaches, node's further behavior can be predicted based on its past observations. Several trust based schemes are proposed in earlier to provide security for the WSN. However the limitations of all those approaches are outlined as; most of the approaches didn't focus on the energy consumption [4, 5] during trust evaluation. A new trust evaluation is developed by considering both direct and indirect trusts. However it didn't given much importance for energy evaluation and delay considerations, by which the network life time decreases drastically. To overcome these issues, a new trust, energy and delay routing protocol is given in this paper. This paper considers the trust model proposed of the earlier approach [6]. Along with this trust model, this paper proposes a new energy consumption model and path selection model. These two models try to achieve their objectives such as minimum energy consumption and minimum delay. Based on these three models, a final routing metric is formulated to perform, called as Combined Routing Metric (CRM). Simulation based evaluation proposed routing protocol performs well with respect to energy consumption, network life time and throughput when compared to the state of art. The remaining paper is detailed as: Sect. 2 explains the literature survey. Section 3 gives the details of proposed approach. Simulation results are described in Sect. 4 and finally the conclusions are given in Sect. 5.

2 Literaure Survey

This section presents the details of earlier routing protocols developed to ensure the security in WSN. Trust based secure routing protocols [6–16] were proposed to counter node misbehavior attacks. A Trust Aware Secure Routing Framework (TSRF) [8] counters the node misbehavior attack. In [9], a Trust Aware Routing Framework (TARF) is developed by Zhan et al. to defend against wormhole attack. An Ambient Trust Sensor Routing (ATSR) is proposed by Zahariadis et al. [7] to defend against misbehaving nodes. A novel function which adaptively weights location, trust and energy information drives the routing decisions, allowing for shifting emphasis from security to path optimality.

Secure and Energy Aware Routing Protocol (ETARP) designed by Gong et al. [10] proposed energy efficient and secured routing protocol for wireless sensor networks (WSNs). This protocol detects and finds routes based on the efficient utility with incurring cost and is compared with common AODV (Ad Hoc On Demand Distance Vector) [17] routing protocol.

A trust-based on-demand multipath routing (AOTDV) [11] and light-weight trust-based routing protocol (LTB-AODV) [12] are the two new secure routing protocols proposed based on standard AODV. Reputation system used in AOTDV or LTB-AODV watches for a single specific behavior only, unlike the Bayesian network adopted in ETARP. ETARP monitors multiple node behaviors and makes comprehensive judgments on node status. Furthermore, AOTDV or LTB-AODV focuses only on security issues.

3 Proposed Approach

A new trust and energy aware routing approach is proposed. This lets the sensor node to select an optimal path based on the characteristics of neighbor nodes trust value, energy and distance from the base station. Complete details about the trust evaluation model, energy consumption model and the path selection model are given in the following sections;

A. Trust Evaluation Model

In this model, the sensor nodes are processed for trust evaluation. Every sensor nodes tries to find the trustworthiness of its neighboring nodes. In common the trust is in two forms, self-trust and Recommended Trust (RT). Self-Trust is the trust existing between two directly communicating nodes. Recommended trust is the trust recommended by others. Self-Trust is evaluated by the sensor node itself, called self-evaluated trust (SET) and the recommended trust is obtained through neighboring nodes. SET is the trust between nodes i and j when node i is communicating with node j. RT is the recommended trust by the other bode k to node i about node j. Thus the total trust can be considered as a combination of SET and RT. Let $\mathrm{SET}_{i,j}$ is the self-evaluated trust by node i over node j and $\mathrm{RT}_{i,j}^{k}$ is the recommended trust by node k to node i regarding the trustworthiness of node j. then the total trust can be formulated as

$$\mathrm{TT}_{i,j} = \omega_1 * \mathrm{SET}_{i,j} + \omega_2 * \frac{\sum_{k=1}^{K} \mathrm{RT}_{i,j}^{k}}{G_j} \tag{1}$$

where ω_1 and ω_2 are two arbitrary constants and G_j is the number of neighboring nodes of node j.

In the Fig. 1, node i and node j are the two nodes between which the communication needs to establish. Thus the node i evaluates the trust worthiness of node j directly through $\mathrm{SET}_{i,j}$. G_1, G_2, G_3 and G_4 are the neighboring nodes which are common to both of node i and node j. Here every G_j recommends one trust value about the node j to the node i. Hence the total recommended trust is the sum of all recommended trust values divided by total number of neighboring nodes. RT is formulated according as

$$\sum_{k \in G_j, k \neq i} \mathrm{RT}_{i,j}^{k} = \sum_{k \in G_j, k \neq i} \mathrm{SET}_{i,k} \times \mathrm{SET}_{k,j} \tag{2}$$

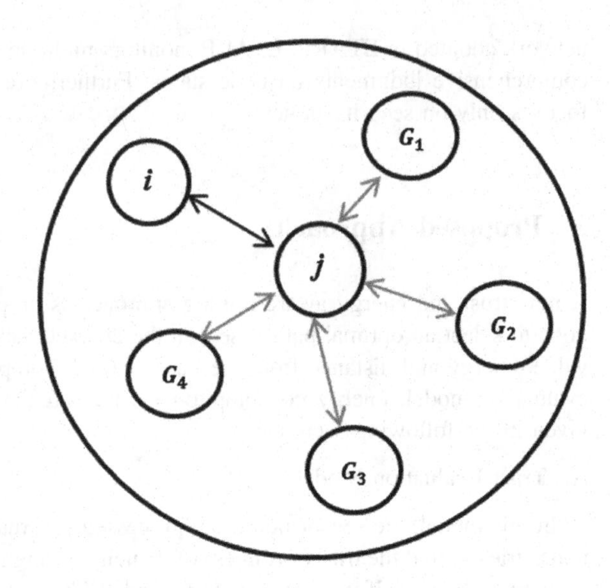

Fig. 1 Trust evaluation diagram

where $\mathrm{SET}_{k,j}$ is the self-evaluated trust by node *k* for node *j*. Note that the node *k* must be a common neighbor to both node *i* and node *j*. $\mathrm{SET}_{i,k}$ denotes the self-evaluated trust of node *i* over node *k*. This entity declares the trustworthiness of recommending node *k* for node *i*. The main advantage with Recommended Trust is the provision of more information about the neighboring nodes of node *i*. Because the node *i* cannot monitor all of its neighboring nodes due to the scarce energy resources. Then speed-up the convergence of trust evaluation, due to this a node can find the misbehaving node earliest.

Finally the evaluated trust in combination with route discovery technique helps in the selection of trusted nodes and in the isolation of misbehaving nodes.

B. Energy Consumption Model

The energy consumed by a node directly depends on the distance by which it is transmitted. In the WSN, distance (l) between two nodes determines the communication model thorough which they are communicating. The communication model is determined through a predetermined threshold, l_0. If the l is less than or equal to l_0, then the amount of energy required is directly proportional to the square of the distance. In this case, communication model termed as free space model. If l is greater than l_0, then the amount of energy required is directly proportional to the fourth power of the distance. In this case communication model is termed as multipath fading model.

$$l_0 = \sqrt{\frac{E_{\mathrm{FS}}}{E_{\mathrm{MP}}}} \tag{3}$$

where E_{FS} is the energy consumed per one bit in the free space model and the E_{MP} is the energy consumed per one bit in the multipath fading model.

The energy that a node consumes for the transmission $(E_{\mathrm{Tx}}(k,l))$ of a message of k bits over a distance l is the sum of energy consumed at transmission circuitry (E_{Tc}) and the energy consumed at amplifier of transmitter $(E_{\mathrm{T\_amp}})$.

$$E_{\mathrm{Tx}}(k,l) = E_{\mathrm{Tc}}(k) + E_{\mathrm{T_{amp}}}(k,l) \tag{4}$$

Form the above expression, the amount of energy consumed for free space mode is written as

$$E_{\mathrm{Tx\_FS}}(k,l) = k \times E_{\mathrm{Tc}} + k \times E_{\mathrm{FS}} \times l^2 \tag{5}$$

Similarly the energy consumed for multipath fading model is written as

$$E_{\mathrm{Tx\_MP}}(k,l) = k \times E_{\mathrm{Tc}} + k \times E_{\mathrm{MP}} \times l^4 \tag{6}$$

One important note is that if a sensor node is behaving as intermediate node between some source node and base station, then it receives information along with transmission. In such case, the node consumes energy for receiving also and the total energy consumption is the summation of the total energy consumed for transmission and the total energy consumed for reception. For a sensor node of receiving k information bits can be determined as

$$E_{\mathrm{Rx}}(k) = k \times E_{\mathrm{Rc}} \tag{7}$$

Finally the total energy (E_{T}) is formulated as

$$E_{\mathrm{T}} = E_{\mathrm{Tx}} + E_{\mathrm{Rx}} \tag{8}$$

where E_{Rx} is the amount of energy consumed by receiving one bit at by the receiver.

C. Path Selection Model

Based on the above trust evaluation model and energy consumption model, this paper proposes a new metric to perform next node selection. The new metric is formulated by combining the trust model energy mode, called as combined routing metric (CRM) and mathematically formulated as

$$\mathrm{CRM} = \alpha \times TT_{i,j} + \beta \times E_{\mathrm{T}} + \gamma \times \mathrm{path\_length} \tag{9}$$

Here α, β and γ are the three arbitrary constants. α gives the significance of trust, β signifies the residual energy and γ signifies the path length. The α, β and γ has to be chosen in such a way that $\alpha + \beta + \gamma = 1$. In the expression (9), the path_length denotes the length of selected path. Here the path can be selected based on two aspects, one is based on the hop count and the other is based on the length of path.

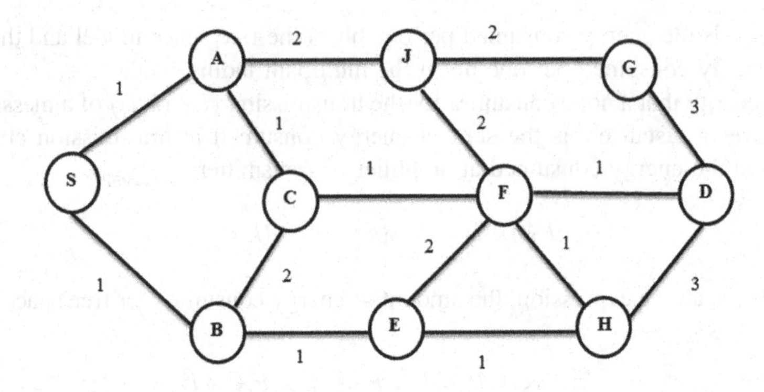

Fig. 2 Example network

For a given source and base station pair, there is possibility of occurrence of multiple paths. From the all possible paths, one path needs to be selected as an optimal path. If focused on delay, the path with minimum hop length and/or minimum hop count has less delay. If the path is selected based on the hop count, then there is a possibility of longer hops. In such case the delay is high even though the hop count is less. In other case, i.e., path selected based on the hop length there is a possibility of path with minimum hop count. So during the path selection both hop length and hop count needs to be considered. An example is shown in Fig. 2.

The available paths between source and destination are shown in Table 1.

From the all possible paths, one path selected based on the Hop_Factor (HF). The path having minimum HF is selected as an optimal path. The expression for HF is given as,

$$\mathrm{HF} = \frac{\text{Path Length}}{\text{Hop Count}} \tag{10}$$

In the case of above example (Fig. 3), the minimum HF is observed for two paths, Path 2 (HF = 1) and Path 6 (HF = 1). In such a case the path is chosen using the number of intermediate nodes (NI). The numbers of intermediate nodes are

Table 1 Possible paths and their metrics

No.	Path	Hop count	Path length	Hop_Factor (HF)
1.	S–A–J–G–D	4	8	2
2.	S–A–C–F–D	4	4	1
3.	S–B–C–F–D	4	5	1.25
4.	S–B–E–H–D	4	6	1.5
5.	S–A–J–F–D	4	6	1.5
6.	S–B–E–H–F–D	5	5	1
7.	S–B–E–F–D	4	5	1.25

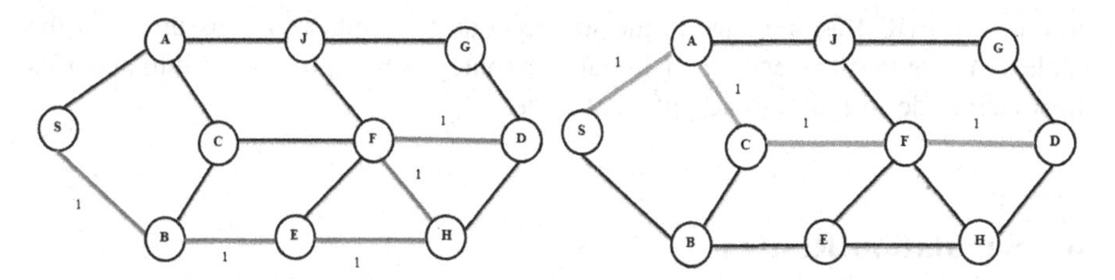

Fig. 3 Possible paths with Minimum HF

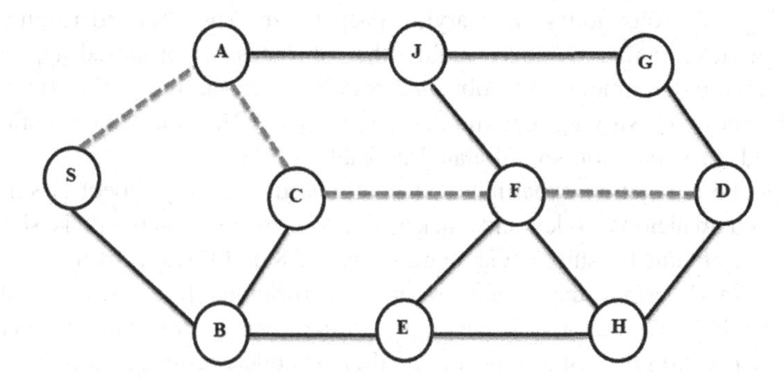

Fig. 4 Final selected path

evaluated as ***NI = Hop Count-1***. For the path 2, the NI is 2 and for path 6, the NI is 3. Thus Path 2 is selected as final optimal path (Fig. 4).

D. Route Recovery

The proposed approach also performs effectively in the case of route Failure. Consider the example shown in Fig. 5.

In the above Fig. 5, the node *A* finds that the total trust if its next node is less than the threshold, then it considers the node *C* as malicious or faulty and forwards

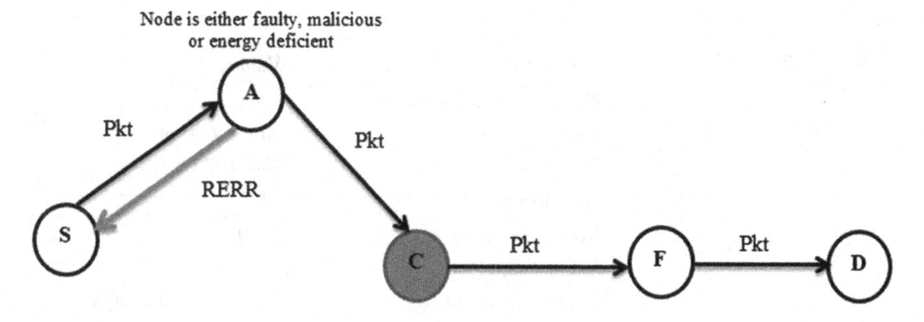

Fig. 5 Route recovery

a route error (RERR) message to the source node *S*. In other case, assume that the node *C* is having less energy; it initiates a route resolve process to initiate to its upstream node to find another efficient node.

4 Simulation Results

The performance is evaluated with respect to throughput, energy consumption and End-to-End Delay. The proposed approach tests over various simulation parameters as varying malicious nodes and varying packet size. The obtained results under varying number malicious nodes realize the robustness of proposed approach in varying trust environments. The obtained results under varying packet size realize the robustness in varying communication models. The simulation parameters considered for simulation are tabulated in Table 2.

Under the simulation experiments, the proposed routing protocol is simulated with varied malicious nodes and packet size. As example network is shown in Fig. 6. The obtained results are shown in Figs. 7, 8 and 9 respectively.

Figure 7a describes the details of energy consumed for different malicious nodes. As the malicious nodes increases, the total energy consumption increases. Since the detection of misbehavior of a malicious node consumes extra energy, the increase in the count of malicious nodes also increases the total energy consumption. From Fig. 7, the energy consumption of proposed routing approach is observed to be less compared to LTB-AODV and TERP. Even though there is a linear increment in the energy consumption, the proposed approach has less energy consumption. The obtained energy consumption results for varying packet size in shown in Fig. 7b. As the packet size increases, the amount of energy required for every sensor to transmit also increases. But due to the proposed energy consumption model, the proposed approach is observed to have less energy consumption.

The total number of packets delivered at destination decreases with the increase in number of malicious nodes. These nodes try to drop the packets by which the throughput of the network decreases. Figure 8a describes the details of throughput analysis for different malicious nodes. Compared with LTB-AODV and TERP, the

Table 2 Simulation parameters

Parameter	Value
Number of sensor nodes	30
Area	600×600 m^2
Packet size	200–1000 bytes
Data format	CBR
Malicious nodes	1–10
Initial energy	50 J
Energy threshold (ε_{thresh})	20% of initial energy
Trust threshold (δ)	0.6

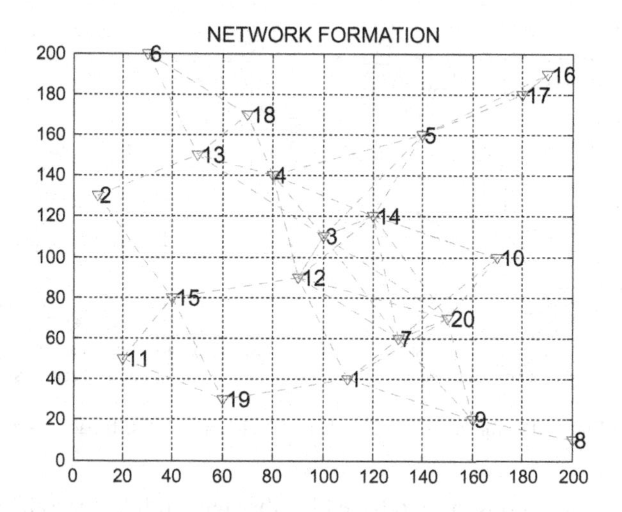

Fig. 6 Example network

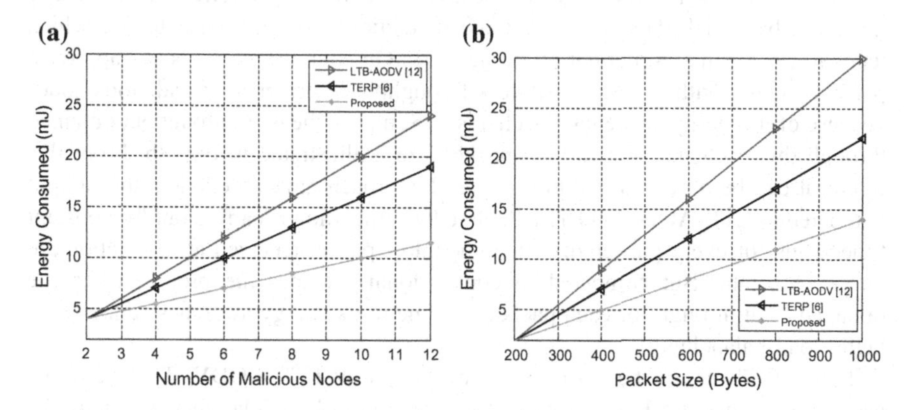

Fig. 7 Energy consumption for varying **a** Number malicious nodes, **b** packet size

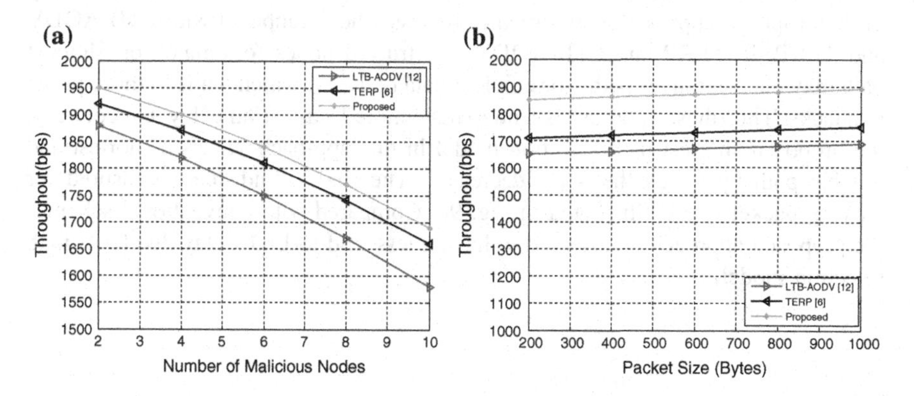

Fig. 8 Throughput for varying **a** Number malicious nodes, **b** packet size

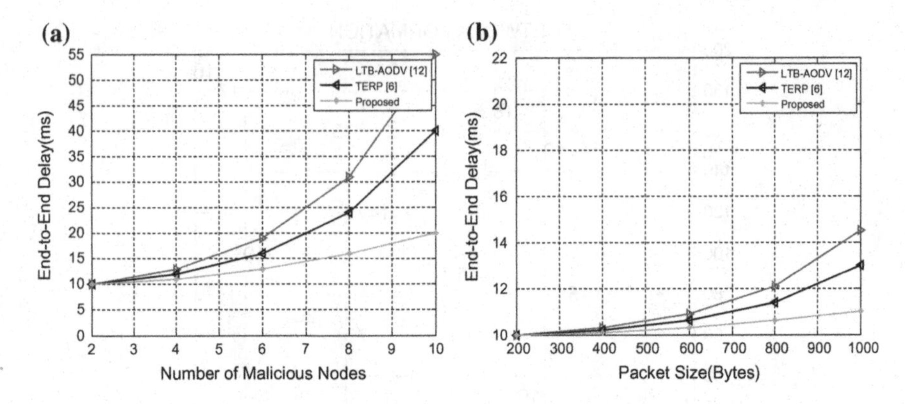

Fig. 9 End-to-End Delay for varying **a** Number of malicious nodes, **b** packet size

proposed approach is observed to have high throughput. In order estimate the trust, LTB-AODV relies on the Intrusion Detection System and thus performs poorly when the network is exposed to misbehavior. TERP simply relies on the energy threshold by which there is no accurate prediction about the misbehavior which tends to reduce the throughput of the network. On contrary the proposed approach performs better with respect to network throughput in presence of malicious nodes because of its energy awareness mechanism. Similarly the total throughput obtained through three schemes for varying packet size is illustrated in Fig. 8b. From this figure, it can be observed that the proposed approach have an efficient throughput compared to LTB-AODV and TERP. Here the simulation is carried out for constant time. Thus for every packet size, the amount of packet received at destination are almost constant. But compared to conventional schemes the proposed scheme obtained high through for each packet size due to its energy awareness that allows better load balancing.

Figure 9 illustrates the performance evaluation of LTB-AODV, TERP and the proposed schemes for End-to-End Delay. The End-to-End delay increases with the increases with the increase in number of malicious nodes due to the more number of disconnections. From the above figure, it can be observed that the End-to-End delay of the proposed approach is observed to be less when compared with LTB-AODV and TERP. In TERP and LTB-AODV, the trusted nodes formulated on shortest path may choose longer paths for efficient data delivery, which tends to more prone to attacks. Thus these approaches tend to have more End-to-End delay. On contrary, the proposed shortest path selection model in this approach selects a shortest and reliable path by which the delay decreases. The End-to-End delay measured for varying packet size is illustrated in Fig. 9b. Compared with conventional schemes the proposed approach is observed to have a less End-to-End Delay due to its path selection model.

5 Conclusion

This paper proposed a new security and energy aware framework to enhance the security and lifetime of a WSN in the presence of various malicious attacks. The proposed framework focuses mainly on the three important factors such as trust, energy and network lifetime. By combining all these aspects a new routing metric called combined routing metric is derived through which the path is selected such that all the aspects are satisfied. Simply focusing only on the trust reduces the network lifetime due to the energy deficiency and only energy makes the network more prone to malicious attacks. Along with these factors, delay is also considered in the proposed framework by which the QoS increases. The simulation results discuss the enhanced performance of proposed approach in comparison with the existing approaches.

References

1. Chen, S., Zhang, Y., Liu, Q., & Feng, J. (2012). Dealing with dishonest recommendation: The trials in reputation management court. *Ad Hoc Networks, 10*(8), 1603–1618.
2. Cho, J., Swami, A., & Chen, I. (2011). A survey on trust management for mobile ad hoc networks. *IEEE Communications Surveys and Tutorials, 13*(4), 562–583.
3. Cordasco, J., & Wetzel, S. (2008). Cryptographic versus trust based methods for MANET routing security. *Electronis Notes Theoretical Computer Science, 197*(2), 131–140.
4. Chang, J.-H., & Tassiulas, L. (2000). Energy conserving routing in wireless ad-hoc networks. In *Proceedings of the 19th Annual Joint Conference of the IEEE Computer and Communications Societies (INFOCOM '00)*, Vol. 1, IEEE, March 2000, pp. 22–31.
5. Ye, F., Chen, A., Lu, S., & Zhang, L. (2010). A scalable solution to minimum cost forwarding in large sensor networks. In *Proceedings of the 10th International Conference on Computer Communications and Network*, Scottsdale, Ariz, USA, 2010, pp. 304–309.
6. Ahmed, A., & Bakar, K. A. (2015). TERP: A trust and energy aware routing protocol for wireless sensor network. *IEEE Transactions on Networking*.
7. Zahariadis, T., Trakadas, P., Leligou, H. C., Maniatis, S., & Karkazis, P. (2012). A novel trust-aware geographical routing scheme for wireless sensor networks. *Wireless Personal Communications, 69*(2), 805–826.
8. Duan, J., Yang, D., Zhu, H., Zhang, S., & Zhao, J. (2014). TSRF: A trust-aware secure routing framework in wireless sensor networks. *International Journal of Distributed Sensor Networks, 2014*(Article ID 209436), 1–14.
9. Zhan, G., Shi, W., & Deng, J. (2012). Design and implementation of TARF: A trust-aware routing framework for WSNs. *IEEE Transactions Dependable and Secure Computing, 9*(2), 184–197.
10. Gong, P. (2015). ETARP: An energy efficient trust-aware routing protocol for wireless sensor networks. *Journal of Sensors*.
11. Li, X., Jia, Z., Zhang, P., Zhang, R., & Wang, H. (2010). Trust-based on-demand multipath routing in mobile ad hoc networks. *IET Information Security, 4*(4), 212–232.
12. Marchang, N., & Datta, R. (2012). Light-weight trust-based routing protocol for mobile ad hoc networks. *IET Information Security, 6*(2), 77–83.
13. Perrig, A., Szewczyk, R., Tygar, J. D., Wen, V., & Culler, D. E. (2002). SPINS: Security protocols for sensor networks. *Wireless Networks, 8*(5), 521–534.

14. Ferng, H.-W,. & Rachmarini, D. (2012). A secure routing protocol for wireless sensor networks with consideration of energy efficiency. In *Proceedings of the IEEE Network Operations and Management Symposium (NOMS '12)*, Maui, Hawaii, USA, April 2012, pp. 105–112.
15. Sadek, A. K., Yu, W., & Liu, K. J. R. (2009). On the energy efficiency of cooperative communications in wireless sensor networks. *ACM Transactions on Sensor Networks, 6*(1).
16. Cho, J., Swami, A., & Chen, I. (2011). A survey on trust management for mobile ad hoc networks. *IEEE Communications Surveys and Tutorials, 13*(4), 562–583.
17. Perkins, C., Belding-Royer, E., & Das, S. (2003). Ad hoc on-demand distance vector (AODV) routing. *RFC*, 3561.

Air Pollution Monitoring System Using Wireless Sensor Network (WSN)

Deepika Patil, T. C. Thanuja and B. C. Melinamath

Abstract Rapid industrialization and urbanization cause the continuous decline in the environmental quality parameters. Today, the world is facing a challenge like global warming which occurs when carbon dioxide (CO_2) and other greenhouse gases accumulate in the atmosphere and absorb sunlight and solar emission that have bounced off the earth's surface. Its impacts are sea level rise, changes in the seasonal patterns, rising temperature, more frequent droughts, and extreme rainfalls. Air pollutions' serious impact on human health and environment requires worldwide awareness and understanding. Existing systems gives real-time air pollution data to pollution monitoring authorities and, these systems are having fixed infrastructure with maintenance, reconfiguration, and reduced sensing issues. Conventional measurements are costly methods and spatially restricted. Therefore, air pollution monitoring becomes a challenging task. Here, we propose Wireless Sensor Network (WSN) based system with low-cost sensors, which collects air pollutant information in real time from different locations. Sensor data is transferred to cloud (ThingSpeak an open source API) for future analysis and calculate Air Quality Index (AQI) Android application can be used to visualize real-time air quality of the location.

Keywords Air pollution · Wireless Sensor Network (WSN) · ThingSpeak AQI · ZigBee

D. Patil (✉) · B. C. Melinamath
Department of Computer Science and Engineering,
Sanjay Ghodawat Institutes, Kolhapur, Maharashtra, India
e-mail: patil.dd@sginstitute.in

T. C. Thanuja
Department VLSI and Embedded System, VTU, Belgaum, Karnataka, India

V. E. Balas et al. (eds.), *Data Management, Analytics and Innovation*,
Advances in Intelligent Systems and Computing 808,
https://doi.org/10.1007/978-981-13-1402-5_30

1 Introduction

Air pollution monitoring requires global attention due to its harmful effect on the environment and human health [1, 2]. Population increase and fast industrialization cause major environmental deprivation like air, water, and land pollution. Air pollution is main environmental health hazard according to World Health Organization (WHO). In 2012, WHO report about 5 million people died because of air pollutants. Nitrogen Dioxide (NO_2), Sulfur Dioxide (SO_2), Carbon Monoxide (CO), Carbon Dioxide (CO_2) Particulate Matter ($PM_{2.5}$ and PM_{10}),and Ground-level Ozone (O_3) are the main causes of air pollution. Traditional methods of estimating air pollution concentration are more expensive. Measuring devices were deployed in remote locations without network connectivity therefore data transmission and data gathering become key challenges. Further traditional monitoring systems do not provide low spatiotemporal resolutions and personal exposures to air pollution information to the controlling authorities [3–5]. Real-time air pollution data collection and personal exposure to air pollutant warnings messages, and notifications can be made possible using advanced sensor and wireless technologies [1] by fixed sensors node, wearable sensors, or sensor mounted on moving vehicles in the monitoring area. In the proposed air pollution monitoring system establish WSNs with low-cost calibrated gas sensors and wireless module.

The objectives of the proposed system are

- Design low-cost Wireless Sensor Network with reconfiguration and adaptability capabilities.
- Designing hardware/software Universal Sensor Interface for pollutant data collection.
- Energy efficiency data transmission using wireless technologies.
- Sensor node evaluation with diverse plug-and-play sensor modules.

1.1 Importance of Air Pollution Monitoring

Today air pollution is crises all around the world but developing countries are face special challenge [6]. Major cities of China, as well India, have among the worst air pollution problems, these two developing economies top the list of the highly air polluted regions of Asia.

Figure 1 shows that statistics of PM2.5 of the world's major cities. In comparison with the other cities of the world, Indian cities have significantly higher level of pollution set by standards of WHO, and USEPA. In India, the major sources pollution are fuelwood and biomass burning industrial and vehicle emission, and crop residue burning. Table 1 shows pollutant level in different cities of India. Sulfur dioxide (SO_2) level is in the normal range but medium and high

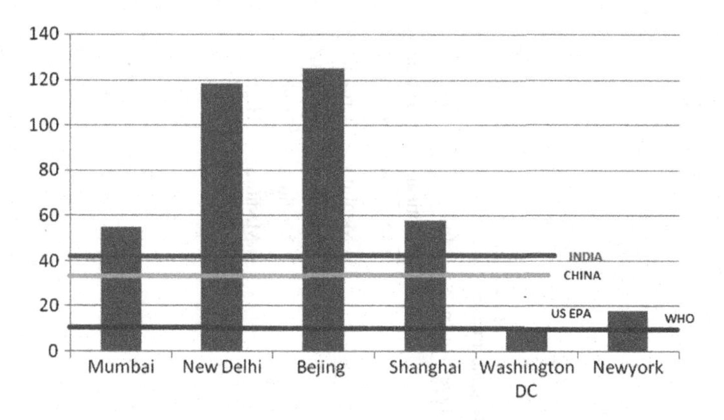

Fig. 1 PM $_{2.5}$ level in the selected Cities of the word in 2014

concentration for Nitrogen Oxide (NO_2) is observed. Particulate Matter (PM_{10}) concentration levels are high in all cities but critically high in Delhi and Kolkata. Central Pollution Control Board (CPCB) has set up a nationwide Programme called as National Air Quality Monitoring Programme (NAMP) which shows critical levels of PM_{10}, CPCB classify the cities based on level of pollutants if it is 1.5 times the standards, then it is 'Critical' and called 'Moderate' when the level crosses fifty percent of the standards. The need of the hour is to recognize the critically polluted cities identify types and sources of air pollution and develop control strategies to defeat this menace.

1.1.1 Sources of Air Pollution Its Impact and Control Strategies

Figure 2 explains different approaches of pollutant measurement and assessment, standards setting and how to decrease emission, this can be accomplished by

- Conducting Environmental Education Awareness and Training programme for the citizen to build up capabilities and skills to improve and protect the environment.
- Concerning experts from all sectors, i.e., Governments regulatory authorities, researchers, academician, industries, and NGOs for setting standards and reducing emissions.
- Using advanced technology and replacement of conventional equipment.

Reducing air pollution is a complex task until we reduce the usage of fossil fuels. But we depend on fossil fuels to power everything from cars to lights in the home.

Table 1 Level of air quality in major Indian cities

Major cities	Sulfur dioxide			Nitrogen dioxide			Particulate matter		
	Average Mg (m^3)	Ind	Res	Average Mg (m^3)	Ind	Res	Average Mg (m^3)	Ind	Res
New Delhi	5	Low	Low	57	Medium	Medium	222	Critical	Critical
Mumbai	5	Low	Low	35	Low	Medium	119	Medium	High
Kolkata	13	Low	Low	66	Medium	High	115	Medium	Critical
Chennai	12	Low	Low	19	Low	Low	65	Medium	Medium
Bangalore	16	Low	Low	29	Medium	Medium	94	Medium	High
Pune	32	Low	Low	58	Low	Low	113	Medium	High

Source Central Pollution Control Board (CPCB). *Ind* Industrial, *Res* Residential

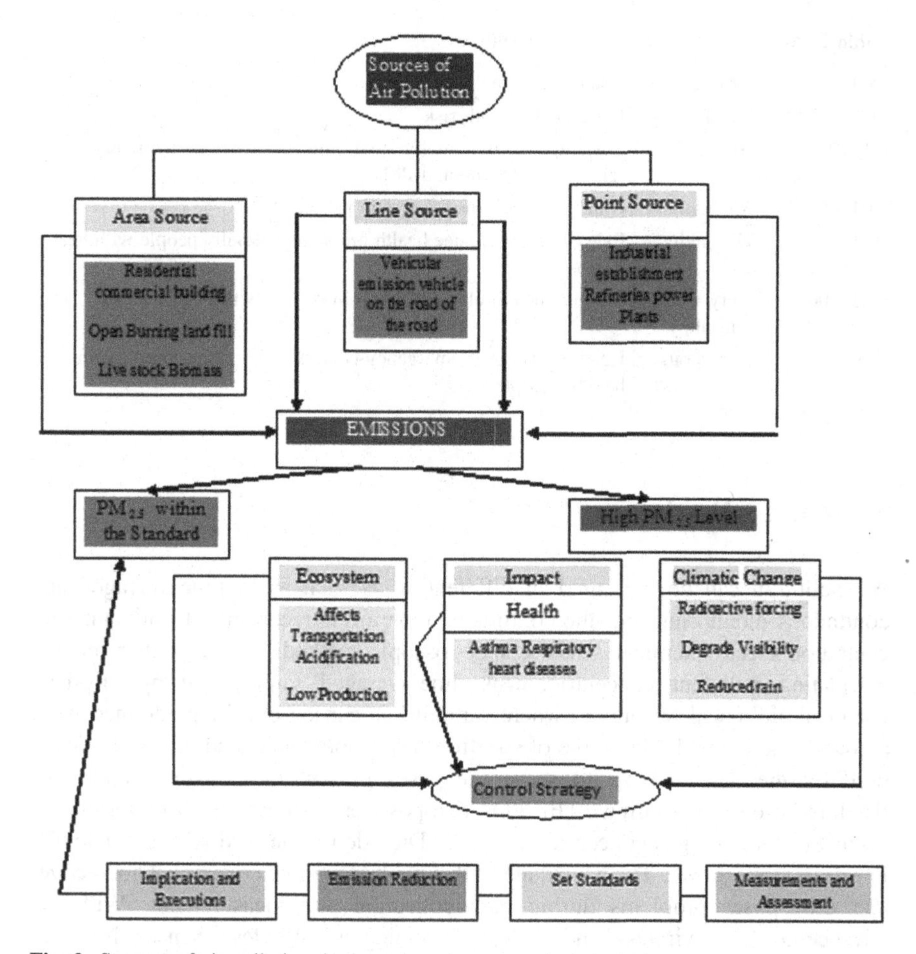

Fig. 2 Sources of air pollution, its impact, and some control strategies

1.1.2 AQI Index and Health Impact

The AQI is an index which reports air quality and tells us that we breathing clean or unhealthy air. AQI is calculated for four major pollutants ground level Ozone (O_3), Particulate Matter ($PM_{2.5}$), Carbon Monoxide (CO), and Sulfur Dioxide (SO_2). Table 2 shows AQI values and its related health impact. Continuous exposure to Ozone (O_3) will increase respiratory symptoms like acute asthma [3]. Air pollution also has major health-related impacts on people with heart or lung disease, elderly people, and children.

Table 2 AQI values and its impact on health

AQI values	Description	Health concern
0–50	Satisfactory	Little or no health risk
51–100	Acceptable	Respiratory problems for small number of people who are sensitive to Ozone and $PM._{2.5}$
101–150	Medium	Unhealthy for sensitive group people
151–200	Unhealthy	Every one experience health problems specially people with heart decease
201–300	Very unhealthy	Triggers health alerts problem in people with heart decease
300–500	Hazardous	Heath warning of emergency condition, breathing problem in healthy people

2 Related Work

Air pollution can be measured in different ways such as in time-averaged and continuous monitoring. For the continuous monitoring/recording of pollutant concentration electrochemical sensors, spectroscopic method is used. Other method calculating pollutant concentration is time-averaged sampling using physical, chemical/biological technique, where sampling is carried out for predefined fixed amount time intervals. Principle of spectroscopy, photometry, and diffusion tube is used by the gas analyzer to sense the gases present in a given sample [7]. Satellite-based monitoring TEMPO (Troposphere Emission Monitoring of Pollutant) measure gas concentration Sulfur Dioxide Ozone, and Nitrogen Dioxide in earth's troposphere. The advantage of this method is more coverage. Cloud cover and dust poses problems during data collection and measurement. With the advancement of wireless and sensor technologies, Wireless Sensor Network (WSN) can be constructed using low-cost sensors. Power consumption and coverage area are the limitation of this technique as it is very hard to transmit the information from source node to destination node within a single hop. To transfer the data to central node, it requires forwarding nodes [8]. An IoT (Internet of Things) based alerting system sends an alert message if Carbon Monoxide and emissions exceed threshold value. Reference [9] proposed an energy efficient wireless sensor network that uses data compression technique for reduce power consumption during data transmission to increase network lifetime [10]. Uses Zigbee module to transmit pollutant data to server and provides low power consumption with increased node availability [11]. Propose vehicular air pollution monitoring by using RFID (Radio Frequency Identification) tags which helps to track vehicles causing pollution. To reduce the power consumption at sensor node and for energy efficient data transmission [12] implemented hierarchical routing protocols where sensor nodes switches between active to sleep mode.

Proposed System

The proposed system measures various air pollutants like CO, CO2, PM10, $PM_{2.5}$ O_3, SO_2, and NO_2. All sensors sense the data and each sensor node transmit the sensed information for the storage at central node. The system maintains following features:

- Network connectivity throughout the operation.
- Fault tolerant mechanism to achieve reliability, accuracy and reduce energy consumption to maximize network lifetime.
- Details of personal exposure to air pollutants using Android application.

2.1 System Design

Designing and testing of real-time air pollution monitoring system in a variety of environmental conditions difficult task. For the superior performance of sensor node, it is vital to include all the parameters. Here, we develop a WSN consisting of sensor nodes with microcontroller and wireless communication modules and Fig. 3 shows sensor node architecture. A Universal Sensor Interface (USI) connects different gas sensor to the microcontroller and provide accurate, reliable interfacing for sensors. Power management is carried out by the microcontroller. For the location tracking, Global Positioning System (GPS) modules are used. The data can be recorded using data logger.

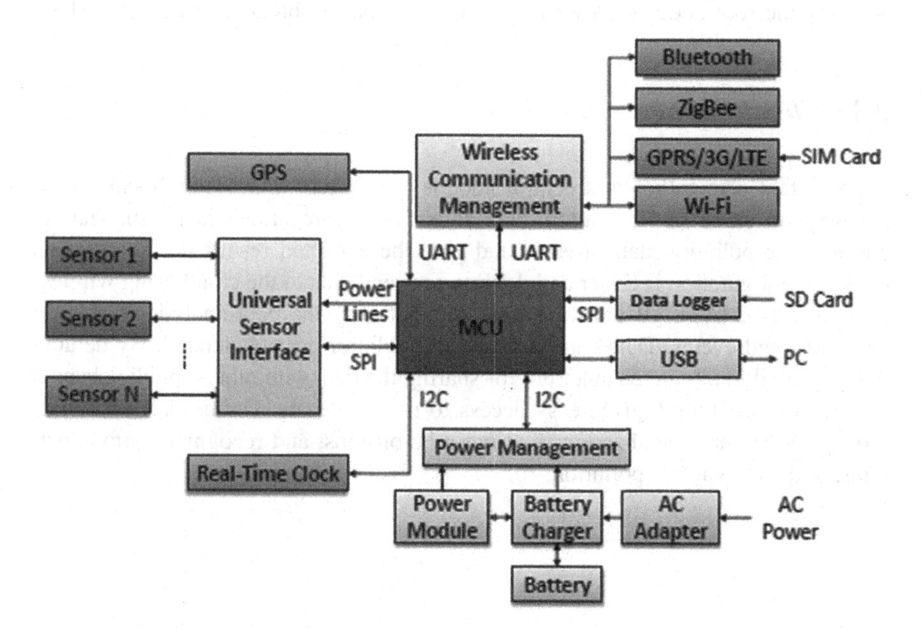

Fig. 3 Sensor node architecture

2.2 Architecture Details

The proposed system architecture is divided into following modules:

1. **Main Body**: It consists of Microcontroller Unit (MCU) Universal Sensor Interfaces to connect different sensor modules, GPS modules. Maximum 16 sensor modules can be connected to Main Body. The can be collected and recorded using Micro-SD card based data logger.
2. **Sensor-Module**: Air quality sensors for detecting various gases are used in this module and the sensor should be small in size more robust when used in outdoor environment, the B4 series sensor is preferred because of fast response and long-term detection and the ability to sense gases with ppb-level resolution with minimum power consumption.
3. **Wireless communication module**: ZigBee wireless communication module is preferred here, but it is possible to connect different wireless modules such as Bluetooth WI-Fi, GPRS/3G/LTE so that during the failure, other module can be utilized.

Data gathering protocols in Wireless Sensor Network (WSN) should implement the following:

- Every node should transmit data in the fixed single time slot.
- Synchronization of all sensor nodes on receiving message from the root node.
- To keep track of data packet transmitted, MAC address should be attached to each packet.
- Only the root node is allowed, store the data packet broadcasted other nodes.

2.3 Cloud Storage

All pollutant data gathered at central node is transferred to the ThingSpeak cloud. ThingSpeak cloud services allow data collection, aggregation, data visualization, and analyze pollution data streams and give the expected results. In ThingSpeak device, Configuration is easier and devices can send data to the cloud using wireless technologies. MATLAB software runs automatically the data analytics based on schedules and events. Data can be stored in public or private channel. By default, data is stored in private channel but for sharing the data with others, public channel is used. Cloud storage gives easy access to pollutant data. Online data analytical tools provide data visualization, find out relationships, and recognize various patterns, and trends in air pollution.

3 Results

In this section, we compare the pollutant data collected from the proposed system and reference data. Temperature and humidity sensor are also used along with gas sensors. Readings obtained from sensors and another reference values are compared to identify the behavior of sensors in different environmental conditions. Experiments were repeated in different locations with varying temperature and humidity. Sensor data is transferred to the sink node using ZigBee protocol with different MAC address. Synchronization and sleep scheduling are performed for increasing the lifetime of the network. On ThingSpeak, cloud pollutant data is stored, which can be analyzed and visualized.

4 Conclusion

The proposed system uses Wireless Sensor Network for monitoring air pollution level in atmosphere. Every sensor node has a collection of sensors for sensing all types of air pollutants and ZigBee wireless module in the sensor system architecture transfers the pollutant data to central or sink node. Using the WSN, monitoring of air pollution at low concentration levels with high energy efficiency can be made possible. Sensed data includes the composition of pollutant in air and is pushed to the ThingSpeak cloud to store, analyze and visualize. AQI is calculated using all major air pollutants and Android application gives real-time personal exposure to air pollution.

References

1. World Health Organization. (2014, March). *7 million premature deaths annually linked to air pollution*. air-pollution/en/.
2. Romley, J. A., Hackbarth, A., & Goldman, D. P. (2010). *The impact of air quality on hospital spending*. RAND Corporation.
3. BBC. (2016). *Polluted air affects 92% of global population, says WHO*. In BBC Health, BBC News. [Online]. Available: http://www.bbe.eom/news/health-37483616. Retrieved on 13, 2016.
4. Dobre, A., Arnold, S. J., Smalley, R. J., Boddy, J. W. D., Barlow, J. F., Tomlin, A. S., et al. (2005). Flow field measurements in the proximity of an urban intersection in London, UK. *Atmospheric Environment, 39*(26), 4647–4657.
5. Yu, O., Sheppard, L., Lumley, T., Koenig, J. Q., & Shapiro, G. G. (2000). Effects of ambient air pollution on symptoms of asthma in Seattle-area children enrolled in the CAMP study. *Environmental Health Perspectives, 108*(12), 1209–1214.
6. Ahuja, D., & Tatustan, M. (2009). Sustainable energy for developing countries. *Surveys Integrating Environment and Society*. https://sapiens.revues.org/823.
7. Working principle of gas analyzers used in NAAQM. Available: www.environnement-sa.fr/wp-content/uploads/2012/09/3VAQMS2012ENs.pdf. Retrieved 09/05/2016.

8. Kandekar, A. B., Kala, N. A., Jain, N. P., Tatiya, N. V., & Pawar, D. P. (2016, April). Monitoring and controlling system for industrial pollution using IoT. *International Journal of Modern Trends in Engineering and Research, 3*(4), 87–90.
9. Roopashree, J., Raghunath, C. R., & Ravikumar. (2015, June). Low power EMC optimized wireless sensor network for air pollution monitoring system. *International Journal on Recent and Innovation Trends in Computing and Communication, 3*(6), 3532–3537.
10. Khodve, S. R., & Kulkarni, A. N. (2016, March). Web based air pollution monitoring system (Air pollution monitoring using smart phone). *International Journal of Science and Research (IJSR), 5*(3), 266–269.
11. Mishra, S. A., Tijare, D. S., & Asutkar, G. M., Dr. (2011, May). Design of energy aware air pollution monitoring system using WSN. *International Journal of Advances in Engineering & Technology, I*(2), 107–116.
12. Deshmukh, S., Jagtap, A., Inamdar, S., & Mahadik, G. (2016). Real time traffie management and air quality monitoring system using IoT. *International Journal of innovative Research in Computer and Communication Engineering, 4*(4), 7026–7033.

Dynamic Range Implementation in Wi-Fi Access Point Through Range Adaptation Algorithm

K. B. Jagan, S. Jayaganesh and R. Neelaveni

Abstract A normal Wi-Fi access point (AP) works with a particular modulation and coding scheme (MCS). Each MCS value corresponds to a particular throughput and range. Because of this, a Wi-Fi Access point supports only a particular range at any given time. When the client is moving away from the access point, it loses the connectivity and if the Wi-Fi AP is working with the highest MCS index value, then the range supported by the AP becomes very small. Hence through dynamically changing the MCS value as the client is moving away or towards the AP, the range supported by the Wi-Fi access point can be made to change dynamically. This is achieved through the implementation of a Range adaptation algorithm. The algorithm works by knowing the position of a client can be deduced from the SNR value of its received signal. After knowing the position the highest Vht-Mcs value with a range that can support the client's position is chosen. In this paper, the relation between SNR and MCS is studied and then an algorithm based on this is developed for changing range dynamically.

Keywords IEEE 802.11ac · Modulation and coding scheme (MCS)
Signal-to-noise ratio · Very high throughput (VHT) · Wi-Fi access point (AP)
BPSK · QPSK · QAM · Multicast · MIMO

K. B. Jagan (✉) · R. Neelaveni
Department of Electronics and Communication, MNM Jain Engineering College,
Chennai, Tamil Nadu, India
e-mail: jagan.kb38@gmail.com

R. Neelaveni
e-mail: veniganesh5@gmail.com

S. Jayaganesh
Department of Electrical and Electronics, MNM Jain Engineering College,
Chennai, Tamil Nadu, India
e-mail: sjaynesh17@gmail.com

V. E. Balas et al. (eds.), *Data Management, Analytics and Innovation*,
Advances in Intelligent Systems and Computing 808,
https://doi.org/10.1007/978-981-13-1402-5_31

1 Introduction

The expansion for Wi-Fi is wireless fidelity and this is used as a certification mechanism for devices that work with the IEEE standard 802.11. This standard incorporated the physical as well as the data link layer of Wi-Fi. In 1997 Wi-Fi protocol was officially standardized and many of the gadgets like smartphones, TVs, laptops, game consoles depend on this technology. Even within much advancement to Wi-Fi, it still suffers from problems like obstacles to its speed, range, the reliability of data being transmitted, and security. Compared to other connections, the speed of Wi-Fi protocol is around 1–54 Mbps. This is very slow to physical line based connections which have a speed from 100 Mbps to several Gbps. When it comes to range, a typical Wi-Fi access point can support a range of few tens of meters working with IEEE 802.11n standard. Though this is very insufficient for a large space, for a small house, this is more than enough. This disability of Wi-Fi makes it impossible to use for large structures. In order to get additional range, many opt for repeaters or additional modems, but this is not very economical. Security also possesses a great risk here, its encryption technique is not very secure. Early method of protection for Wi-Fi connectivity was wired equivalent protection or WEP which later led to development of other standards like WPA, etc. Since Wi-Fi's physical layer uses RF waves for communication, it is prone to problems like interference from other sources and complex pattern of propagation which is not under the control of the administrator. However, this paper will mainly focus on the range issue. For any Wi-Fi access point, range always seems to be a problem because it simply cannot operate with large radius. This problem may due to many reasons like interference, absorption of signal, or operating environment being very noisy. There are different modulation and coding schemes at which the Wi-Fi access point operates to increase signal-to-noise ratio (SNR). If the SNR value is good then it means that the Wi-Fi access point operates well even in a very noisy environment.

1.1 Modulation and Coding Scheme

Modulation and coding is used to alter the carrier wave according to the message so that the message is transmitted in a reliable manner through an unfriendly channel. Here, the useful information is mapped on to the carrier wave and at the receiver, proper decoding and demodulation techniques can be used to retrieve the original message signal back. AWGN or additive white Gaussian noise was used as a model to develop the modulation and coding index. With the new standard IEEE 802.11n in 2007, MCS index was also introduced. White Gaussian noise has a flat spectral density for all the frequency spectrum. Modulation and coding schemes are based on parameters like the type of coding method, channel width, modulation

Table 1 MCS index for 802.11n and 802.11ac

			Theoretical throughput for single spatial stream (in Mbit/s)							
MCS index	Modulation type	Coding rate	20 MHz channels		40 MHz channels		80 MHz channels		160 MHz channels	
			800 ns GI	400 ns GI	800 ns GI	400 ns GI	800 ns GI	400 ns GI	800 ns GI	400 ns GI
0	BPSK	1/2	6.5	7.2	13.5	15	29.3	32.5	58.5	65
1	QPSK	1/2	13	14.4	27	30	58.5	65	117	130
2	QPSK	3/4	19.5	21.7	40.5	45	87.8	97.5	175.5	195
3	16-QAM	1/2	26	28.9	54	60	117	130	234	260
4	16-QAM	3/4	39	43.3	81	90	175.5	195	351	390
5	64-QAM	2/3	52	57.8	108	120	234	260	468	520
6	64-QAM	3/4	58.5	65	121.5	135	263.3	292.5	526.5	585
7	64-QAM	5/6	65	72.2	135	150	292.5	325	585	650
8	256-QAM	3/4	78	86.7	162	180	351	390	702	780
9	256-QAM	5/6	N/A	N/A	180	200	390	433.3	780	866.7

techniques, number of spatial streams, etc. Combination of all these parameters corresponds to a MCS value. The modulation techniques used here are binary phase shift keying (BPSK), quadrature phase shift keying (QPSK), and quadrature amplitude modulation (QAM) and all the Vht-Mcs values are derivatives from these three modulation schemes. For channel widths 20 and 40 MHz, the number of MCS values for them is 77 from which eight are the fundamental MCS values for 20 MHz and thus form the basic data rates. IEEE 802.11n uses High throughput MCS (HtMcs) whereas IEEE 802.11ac uses Very high throughput MCS (Vht-Mcs). As the name suggests in 802.11ac the throughput is increased through the addition of two more channel widths, they are 80 and 160 MHz. The 80 MHz doubles the spectral width of 802.11n in addition to that there is another 160 MHz channel width. IEEE 802.11ac uses four digit apart channel numbers like its predecessors (Table 1).

Within a wide channel, there is one frequency which is defined as the primary and others as secondary. The Wi-Fi AP usually sends the beacon denoting its presence only in the primary channel and not in the secondary channel. If a part of 80 MHz channel is used then the primary is 44 whereas the secondary ones are 36, 40 and 48 (Fig. 1).

IEEE 802.11ac has a much simpler type of coding compared to IEEE 802.11n. The 802.11n standard has more than 70 options but 802.11ac has just 10. Out of the 10, 7 of the MCS values are mandatory and many manufacturers are trying to implement 256 QAM thus making all the 9 MCS values commercially possible. 802.11ac supports 256 QAM which corresponds to Vht-Mcs 8 and 9. This constellation has 16 amplitude levels and 16 phase shifts. Compared to 8 phase shifts and 8 amplitude levels in 64 QAM constellation supported by 802.11n. This higher constellation points in 256 QAM packs more data thus making 802.11ac faster compared to its predecessors. However, MCS index 9 is not applicable for channel width of 20 MHz. There is about a 10% increase in throughput using the shorter

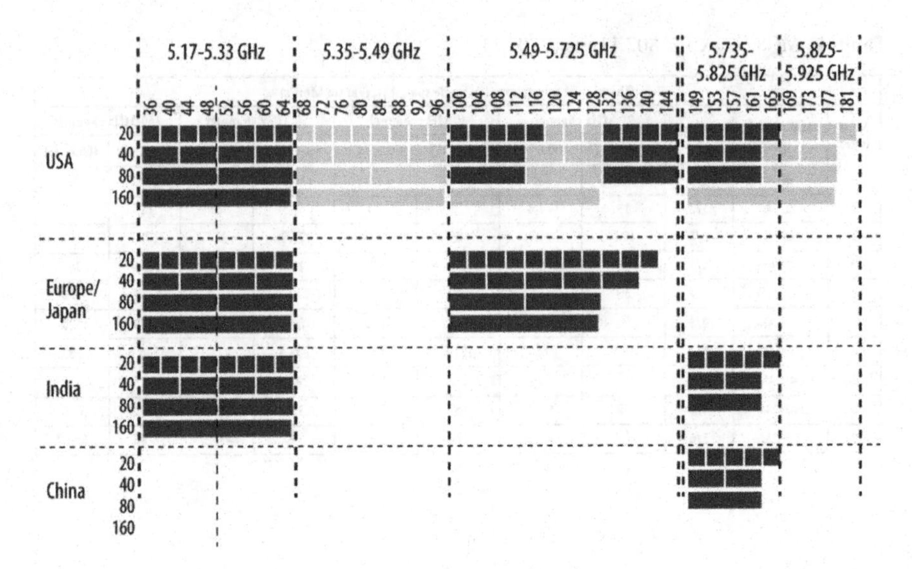

Fig. 1 Channel map 802.11ac

guard interval. Note the increase in throughput as more spatial streams are brought into play. As a means of forward error correction, convolutional coding is usually used. If '*n*' represents the total number of bits and '*k*' represents the number of useful bits of information, then k/n is given as the code rate and the number of redundant bits is given as $n - k$. Here values in the coding scheme such as 1/2, 3/4, or 5/6 denote that there is one redundant bit added after every single, third, and fifth bits, respectively. The received data is then decoded by the means of certain specialized algorithms at the receiver. The motive of this method is to make the received data more reliable that is sent by the transmitter. This procedure also removes any need for retransmitting the data since errors are corrected by the receiver itself.

1.2 Signal-to-Noise Ratio

The ratio of intended signal power to the background noise power is given as SNR or signal-to-noise ratio. The intended signal here is the Wi-Fi signal. Devices that work in the ISM (Industrial scientific and medical) band, fluorescent lamps, Game console joysticks, wireless camcorders, and others may act as a source of noise for Wi-Fi. However, noise due to the co-channel interference is not included here. An SNR ratio of above 41 dB (decibels) is considered to be an excellent connection, while SNR between 22 and 40 dB is a good connection. Between 16 and 20 dB corresponds to poor connection and a ratio of 11–15 dB is the minimum value for an unreliable connection. The MCS index table links the client's SNR values to the MCS indexes. This mapping is done to determine the best data rate that a client can achieve based on the connection between itself and the Wi-Fi access point. The signal-to-noise ratio is also very much related to the RSSI value. RSSI stands for relative received signal strength.

$$\mathrm{SNR} = \frac{P_{\text{signal}}}{P_{\text{noise}}} = \frac{V_{\text{rms}}^2}{V_{\text{qn}}^2} \tag{1}$$

$$\mathrm{SNR_{dB}} = 10\log_{10}\left(\frac{P_{\text{signal}}}{P_{\text{noise}}}\right) = 20\log_{10}\left(\frac{V_{\text{rms}}^2}{V_{\text{qn}}^2}\right) \tag{2}$$

The two values of RSSI are of great importance; these are minimum receiver sensitivity (MRS) and the expected receiver sensitivity (ERS). The MRS values for IEEE 802.11 corresponds to the minimum relative received signal sensitivity values such that a radio receiver can effectively be able to decode the received signal's modulation type involved and coding scheme (Vht-Mcs index) with packet error rate (PER) less than 10%. Many modern radios with 802.11 standards usually provide good receiver sensitivity than the minimum required value. The ERS gives the typical receiver sensitivity of the clients such that they can achieve any given MCS index at lower RSSI than the required to pass testing. For example, the MRS for IEEE 802.11ac 20 MHz PPDU at MCS 8 is −57 dBm. But most IEEE 802.11ac radio receivers can decode this PPDU at even lower RSSI value such as −62 dBm.

2 Bit Error Rate

Eb/N0 is the energy per bit to the noise power density ratio; it is also called a BER or bit error rate. It is a very important parameter in digital communication. It gives us the SNR value per bit. When comparing BER, SNR measurement is very useful without considering about bandwidth of the digital signals. In description Eb implies signal energy associated with user data bit, in other words, Eb is the ratio between signal energy to the message signal bit rate. When the power of the power of the signal is taken in watts and message bit rate is bits per seconds then the unit for energy per bit is given in joules. The noise spectral density is given as N0 and the power of noise in a 1 Hz bandwidth is measured as W/Hz or J. Since Eb and N0 have the same units their ratio is dimensionless. IT is however represented as db or decibels. It tells the overall efficiency of the data transmission without going in deep into the modulation or the coding scheme used or bandwidth. Hence, it helps to avoid confusion as which definition of bandwidth is to apply to the signal. In interference-limited channels, the Eb/N0 must be used with care because of the additive white noises might be taken into the account in interference.

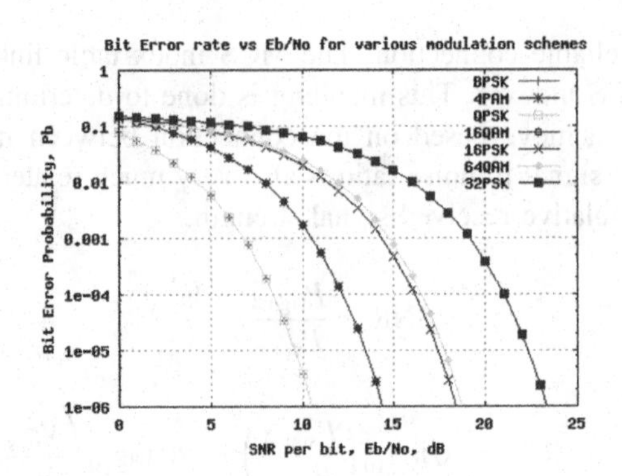

3 Related Works

A commercially available Wi-Fi access point works with a particular Modulation and coding scheme based on the link adaptation algorithm. Though there are several algorithms proposed for efficiently choosing the best transmission rate, these algorithms concentrate more on the reliability of the connection. The earliest link adaptation was the ARF or Auto rate Fall-back by Kamerman and Monteban. Here, the Access point would initially work on the highest transmission rate (MCS). When an ACK is missed after a successful transmission it would retransmit the same data with the same transmission rate. If the ACK is missed again the transmission rate is decreased and the timer is started. When either the timer expires or the next received ACK reaches a threshold *N*, the transmission rate increases. A probe packet is sent at the new rate if there is a miss then it rolls back to its original value. After this method, there are many other rate adaptation algorithms developed which are used to select the best transmission rate. These algorithms cannot perform well in environments where the signal strength changes very quickly. The problem with these approaches is that they mainly concentrate on improving the transmission rate and not the range. Hence as the transmission rate increases the range of the Wi-Fi access point decreases. There is another rate adaptation scheme which uses different MCS values to protect the connection from noise. Like the previous algorithm, this method also favors noise protection more than Wi-Fi range. In a user's perspective, a Wi-Fi access point should have good throughput as well as range. If the throughput is very high and the range of the Wi-Fi is only few meters then the complete ideology of wireless transmission is lost. Another way is to use Cantenna (Fig. 2).

A Cantenna is nothing a simple half cut cylindrical reflector or a full can type reflector which is used to reflect the signal from the main antenna and thereby increasing the range. The method is only valid if the Wi-Fi Access point is placed close to any nonreflective surface like wall or cupboard, etc. Since the reflector makes the antenna highly directive, if the access point is to be placed in large open areas then the antenna cannot function as an isotropic antenna. Cantenna also

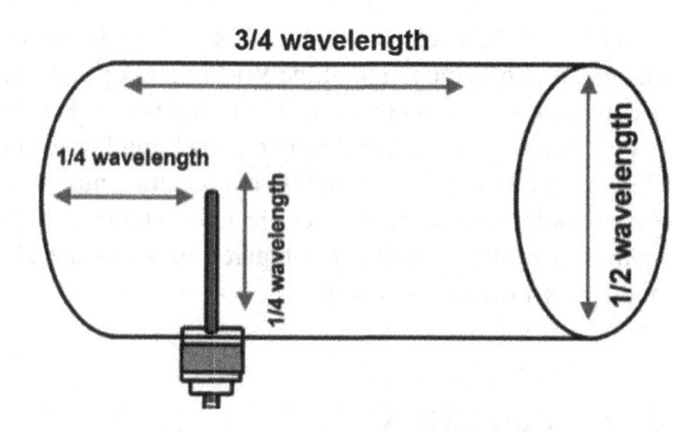

Fig. 2 Cantenna

requires a lot of hardware modification to the existing Wi-Fi access points. Cantenna is not the best solution for a centralized Wi-Fi access point. Compared to the above two methods, a most often used way to increase the range of an access point is to employ Wi-Fi repeaters. These work as any other conventional repeater. They simply work by receiving Wi-Fi signal. Repeaters also increase the range of Wi-Fi by 2 times. Sometimes, repeaters cost as much as Wi-Fi Access point. Wireless distribution service is supported by many routers nowadays, this involves in setting up many number of routers connected wirelessly with each other. This will be like setting up a single and a large ranged network. This can be done in two modes, first one involves in something called as wireless bridging where routers are used to create a point-to-point link. They do not support connections from clients. This is particularly useful in office spaces where one might want to join separate buildings under a single network. Other one involves wireless routing techniques where clients can join to the bridging routers. However, setting up a WDS system is very difficult, particularly between non-similar routers.

Speed can be an issue with these types of products, as there are two wireless connections. Latency is important as it dictates how responsive things feel. Latency refers to the delay between sending a request and receiving the right reply for the request. Latency can also increase as the number of repeaters in a network increases.

For example, a high latency wireless network is one which takes time to send a request and receive a reply when a link is clicked on a website, thus making it appears to take some time in loading the next page. For example, a wireless network with high latency means clicking on a link on a website and it takes a while to send the request and receive the reply, making it appear to take a while to load the page. Though this extender seems to do a good job in increasing the range, they are not economical.

Installing WiFi extenders for the increase in range is very costly. The other way to extend a range of a Wi-Fi access point is to choose the channel number which has the least number of Wi-Fi access points working with it. Channel number corresponds to the Wi-Fi access point's and client's operating band in the 2.4 or 5 GHz range. Since interference increases, the background noise power level increases which causes a decrease in the SNR value. So by choosing the right operating channel number, the range of the Wi-Fi access point can be increased greatly which is achieved with an increase in the throughput to. But the problem with this method is that there are several operating channel number for both the access point and the clients. Different clients have different channel number. Even when the client's and Access point's channel number are not matching, they do work but with a significant loss in the throughput compared to when the client and Wi-Fi access point both work in the same band.

4 SNR and MCS

Each MCS value is supported only for a particular range of SNR after which the throughput drastically reduces. Since the adaptation mechanism is for IEEE 802.11ac standard, the MCS of it is represented as very high throughput MCS or Vht-Mcs. Vht-Mcs 0 supports the maximum range with a throughput of 7.2 Mbps theoretical speed and practically around 4.4 Mbps are achieved. Vht-Mcs 0 has BPSK 1/2 coding scheme. All the intermediate MCS, i.e., MCS-1, MCS-2, MCS-3, MCS-4, MCS-5, MCS-6, and MCS-7, have QPSK 1/2, QPSK 3/4, 16-QAM 1/2, 16-QAM 3/4, 64-QAM 2/3, 64-QAM 3/4, 64-QAM 5/6, 256-QAM 3/4, 256-QAM 5/6, respectively (Table 2).

Higher the QAM value more the number of points in the constellations hence higher is the data rate. Similarly, the range supported by Vht-Mcs 8 is 76 m and by Vht-Mcs 0 is 1186 m. Since 20 MHz bandwidth only supports only till Vht-Mcs 8, the graph is plotted till Vht-Mcs 8 86.7 Mbps theoretical and 59 Mbps practically. The throughput curve is described for each Vht-Mcs value is given in the graph. The throughput curve remains almost constant till a particular SNR value. After a certain SNR is reached, the throughput for the particular Vht-Mcs decreases rapidly. This SNR value after which the throughput decreases is called as the threshold value for the Vht-Mcs (Fig. 3).

Table 2 MCS index and their supported range

Vht-Mcs	SNR (in dB)	MAX supported range (in m)
0	2.82096	1186
1	6.70927	768
2	8.53929	624
3	12.2181	412
4	15.2977	296
5	20.7808	160
6	22.0238	140
7	23.7965	116
8	27.9118	76

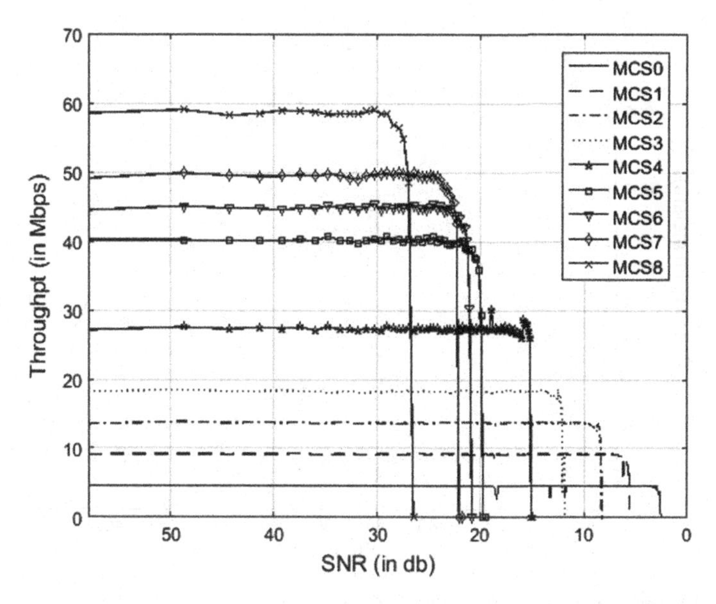

Fig. 3 Plot showing the relation between different MCS and their supported SNR

Transmit power is very important parameter in wireless communication. Lower Vht-Mcs values correspond to higher ranges and higher the transmit power higher will be the range of AP. According to the above statement, the range of an AP depends on two factors one is Vht-Mcs value and the other one is TxPower of AP. If the AP is working in the highest Vht-Mcs value say 8, then it needs lower TxPower like 11 dB since range supported by Vht-Mcs 8 value is around the range supported by 11 dB. If the AP is operating in TxPower higher than 11 dB, then it would be of no use as the range is limited by the Vht-Mcs 8. Similarly, the lowest MCS value 0 needs a TxPower of 20 dB so that the right range for Vht-Mcs 0 is achieved. If the AP is working power values less than 20 dB then the full range of Vht-Mcs 0 is not achieved since range gets limited due to the TxPower.

5 SNR and Distance

Signal-to-noise ratio can be used as an indicator of a client's position form the access point. When the client is very near to the access point's antenna the radio energy received by the client is very large hence the signal power is very large. If we assume the environment to have a constant additive white Gaussian noise, then the signal-to-noise ratio value is very high as the numerator (signal power is high.) Now the client is little away from the AP, hence there is drop in the signal power value so the SNR value dips. If the client keeps moving away from the AP, then the signal power decreases exponentially with respect to the distance between the AP

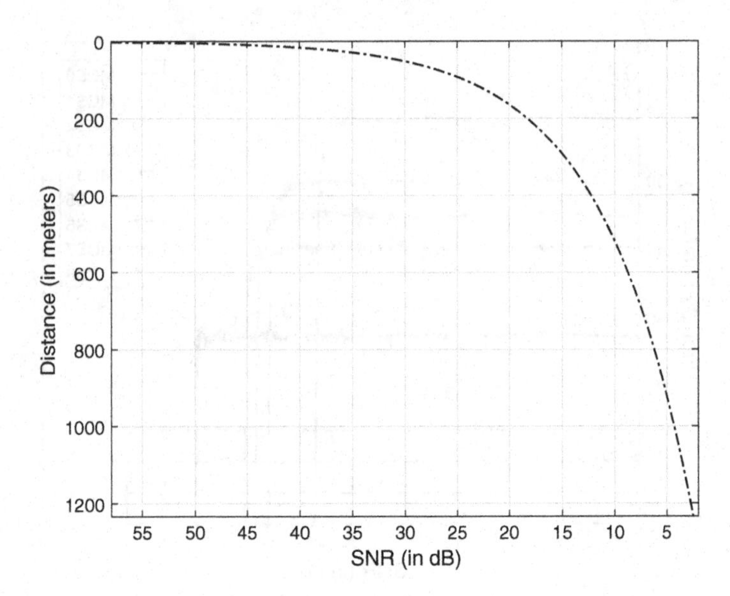

Fig. 4 SNR decreases with distance from AP

and client, hence there is an exponential decrease in the SNR value. Because SNR value decreases as the distance increase, SNR feedback is one of the ways to notify the AP of the client's position. In order for SNR feedback to work perfectly, the noise throughout the AP's maximum range of support must be constant otherwise it would lead the AP to assume wrong client's position (Fig. 4).

6 Range Adaptation Algorithm

Based on the above details, the range adaptation algorithm (RAA) is developed for the dynamic range implementation. Note all these processes happen in the access point.

(1) Start with highest Vht-Mcs for the given channel width.
(2) Create a table based on the MCS index and their corresponding supported range of SNR.
(3) Count the number of clients connected to the AP.
(4) If only one client is connected to AP then proceed to Step 5.
(5) Client calculates the SNR from the packets sent by the AP.
(6) As the client moves away from AP SNR value keeps decreasing.
(7) When the client moves to such a distance where the SNR has reduced to the lowest threshold value for the particular Vht-Mcs, at which the AP is operating, then reduces to the next Vht-Mcs value.

(8) If the SNR keeps decreasing below the lowest threshold value for the current Vht-Mcs then switch to the next lower Vht-Mcs value.
(9) Keep repeating Step 8 if SNR keeps decreasing till Vht-Mcs 5. Else if the SNR increases as the client now is moving towards the AP then move to Step 10.
(10) As the SNR is increasing switch to the next highest Vht-Mcs value if the SNR value has reached the lowest threshold value of the next highest Vht-Mcs value.
(11) Keep repeating Step 10 if the SNR value keeps increasing till the highest Vht-Mcs.
(12) If there are more than two clients then follow then limit the lowest Vht-Mcs to 6.
(13) Implement multicast priority scheme if more than two clients are present.

The following algorithm is used in each spatial streams of 802.11ac.

7 Range Adaptation

Through changing the Vht-Mcs index value, the range of the Wi-Fi Access point can be increased. The SNR can be used as the parameter of reference and the Vht-Mcs can be decremented or incremented based on that. The SNR value is present in the packets transmitted by the access point and client. The client obtains the SNR tag from the packet. Then the client notifies the SNR at its position through the frame in the ACK packet sent by it. Each Vht-Mcs value is mapped to a particular range of SNR, Where the throughput for the particular Vht-Mcs is high. The initial Vht-Mcs of operation is Vht-Mcs 8 and the value decrements from this as the client moves away from the AP as the SNR value decreases and increments as the client moves towards the AP since the SNR value now increases. Once the client crosses, the effective SNR range for a particular VHT-MCS, the next lower value (if the client moves away from AP) or higher value (if the client moves towards the AP) Vht-Mcs is adopted by the AP (Fig. 5).

For each range of the decrement is not done below Vht-Mcs 5 as the throughput for the corresponding Vht-Mcs below 5 is very low and the SNR supported by them is low which leads to a poor connection. Since for a good connection the SNR should be above 22 dB Vht-Mcs 5, the lowest supported SNR is up to 20 dB. By applying the algorithm based on the above description, the range of SNR supported by the Access point working under constant Vht-Mcs 8 is decreased by 7 dB and the range is increased from 76 to 170 m. There is however a small decrease in the throughput but that is traded off with the increase in Range; still, the overall throughput will not decrease beyond 30 Mbps which is considered to be a good throughput level for WLAN. If two clients are connected to the access point, then the Vht-Mcs is decreased till Vht-Mcs 6 so that the second client, if it is stationary and is near to the AP then its throughput would not change that much but the range for the person who is moving away will see a great increase in the coverage of the

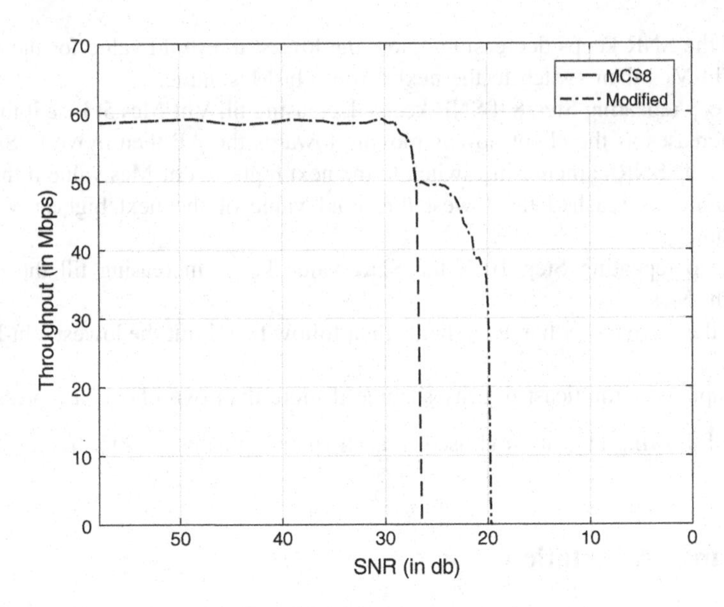

Fig. 5 Comparison Between MCS8 and RAA

AP. In case of multiple clients connected the AP, then priority is given by the AP based on whether the majority of the clients are near or close to the AP and how far they are. If the majority of the clients connected are away from the AP, i.e., beyond the range of Vht-Mcs 6, then the AP applies the algorithm till Vht-Mcs 5. Another priority scheme is if majority clients are far but randomly positioned then the access point makes the best effort to choose the right Vht-Mcs by computing the mean SNR of all the clients (Fig. 6).

When comparing range of Vht-Mcs 8 and range adaptation algorithm, it is absolutely clear that the range with range adaptation algorithm if more. There is almost a 100 m extra range available. Both the graphs under this section are plotted for the case, A client is connected to an access point and it is moving away from the AP. Readings are noted at every 4 m intervals.

Consider a case where there is a single client connected to a Wi-Fi access point. The client is moving away from the access point at 1 m/s at time $t = 0$. Since the client is initially deployed close to the access point the SNR value of the signal received by the client has a very high value. The Wi-Fi access point on start works at the maximum possible Vht-Mcs possible for the operating channel width (here Vht-Mcs is 8 since the operating channel width is 20 MHz). As the client is moving away from the access point, there is a decrease in the SNR. This is where "Range Adaptation Algorithm" kicks in. Since SNR value is degrading, the access point understands that the client is moving away from it. Once the SNR crosses the threshold SNR for a particular Vht-Mcs, the access point decreases its Vht-Mcs value to the next lower one. When the client crosses the range of the particular Vht-Mcs, there is a decrement in the Vht-Mcs value. Here the MCS decreases from

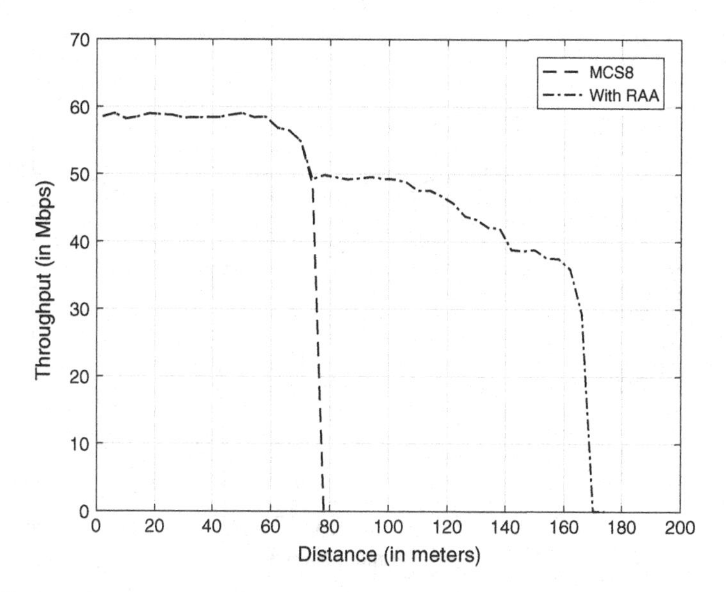

Fig. 6 Supported range comparison bet. MCS8 and RAA

8 to 7, this happens at t = 60 s. As the client keeps moving further, the SNR decreases and when it goes beyond the threshold of Vht-Mcs 7, there is another decrement in the MCS value. The process continues till Vht-Mcs 5. At t = 68, the client has crossed the range of Vht-Mcs 5 and the algorithm does not allow further decrease in the Vht-Mcs value since they all correspond to lower SNR values. There is no connectivity between the access point and client since the client has gone to the region beyond the supported range of Vht-Mcs 5. This is shown by zero throughput value between the client and access point. At t = 184, the client starts moving back towards the access point. As the client is moving towards the access point, the SNR value improves. Once the client reaches, the coverage range of Vht-Mcs 5 connection is once again established between the client and access point. The SNR value improves as the client moves towards the access point. The access point starts with Vht-Mcs 5 and this value keeps incrementing as the client crosses the threshold value of Vht-Mcs 5 the net MCS is 6 and the algorithm works such that the process continues till the maximum Vht-Mcs for the client's position. The simulation here is done for 300 s. Hence, the access point does not reach the highest Vht-Mcs value since the client does not reach that range before t = 300. From this example, we can clearly see that the "Range Adaptation Algorithm" dynamically changes the Vht-Mcs value so that the range changes dynamically (Fig. 7).

The main advantage of using RAA is that it provides dynamic range and also tries its best effort to provide high throughput at the same time. One may argue that it would be far easier and wise to operate an access point at the Vht-Mcs which supports the largest range. But the problem is that with higher range there are fewer throughputs, i.e., Vht-Mcs 5 supports 170 m range but supports only 40 Mbps connection. On the other hand, the highest Vht-Mcs 8 has a throughput of 60 Mbps

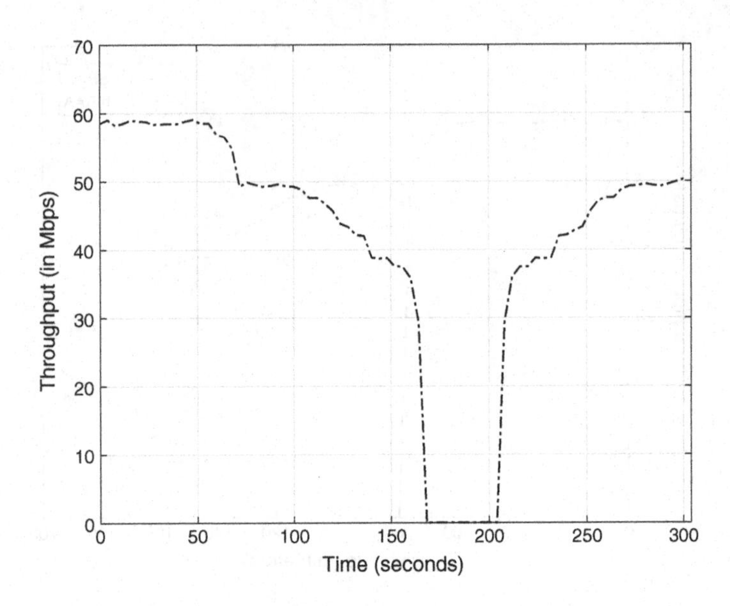

Fig. 7 Client moving away and toward AP

but supports only a range of 70 m effectively. The RAA algorithm makes the access point to have higher throughput when the client is present near it and lowers throughput systematically as the client crosses each threshold range. If the access point were to work at constant Vht-Mcs 5 then throughput will be 40 Mbps in all cases and in case of Vht-Mcs 8 then the range will be very limited. RAA takes the advantage of both these and tries to improve the throughput and range of the access point. Hence by switching between each MCS value, the overall network efficiency improves by a great deal, as the client gets the highest throughput available for its position. Vht-MCS 5 provides the highest range, i.e., 160 m of all the other Vht-Mcs values used in the algorithm but it also has the lowest throughput of 40 Mbps, but Vht-Mcs 8 has the highest throughput of 58 Mbps but the range supported by it is only 75 m, so the range adaptation algorithm takes the advantages of both the MCS values. Since the drop in the throughput between Vht-Mcs 5 & Vht-Mcs 8 has a very high step change, so in order to rectify this the algorithm proceeds through Vht-Mcs 6 and Vht-Mcs 7. Since they provide a smoother transition.

7.1 Multicast Connection

Since the algorithm works best for a single Wi-Fi access point and client connection, multicast connection always seems to be a problem. Hence, developing a multicast algorithm is also very difficult. The easiest way to proceed is to assume the appropriate Vht-Mcs value set by furthest client. This will cause the Wi-Fi access point to work with the lowest set Vht-Mcs value. Example assume that there are only 2 clients connected to the access point if one of the client is in motion and other client is in complete rest then the Vht-Mcs value would decrease since one of the client is in motion. This will cause a decrease in throughput for the second client even when this client is not moving. So to address this situation, the Vht-Mcs would decrease till 6 and not 5. If there are more than two clients connected to access point, then different decisions have to be made by the access point based on the client's positions. If majority of the clients are located in a particular Vht-Mcs range then based on the SNR value of the majority clients, that particular Vht-Mcs would be chosen. If all of them are randomly arranged then the average of all the SNR values of all the clients connected to the Wi-Fi access point would be calculated and the Vht-Mcs for the average SNR would be chosen. The abovementioned technique might be useful for older standards of Wi-Fi. All the standards before 802.11ac had single-user MIMO. They used a primitive beamforming method which involves increasing the signal power over a selected region of the access point's territory to increase the data rate. For the newer standard like 802.11ac, the SU-MMIO has been developed to Multiuser MIMO. However, this is yet to be proven and implemented in commercial Wi-Fi products. The main idea here is that if there are multiple clients connected to the access point such that they are located in sufficiently different directions then a beamformed transmission may be sent to each of the clients at the same time (Fig. 8).

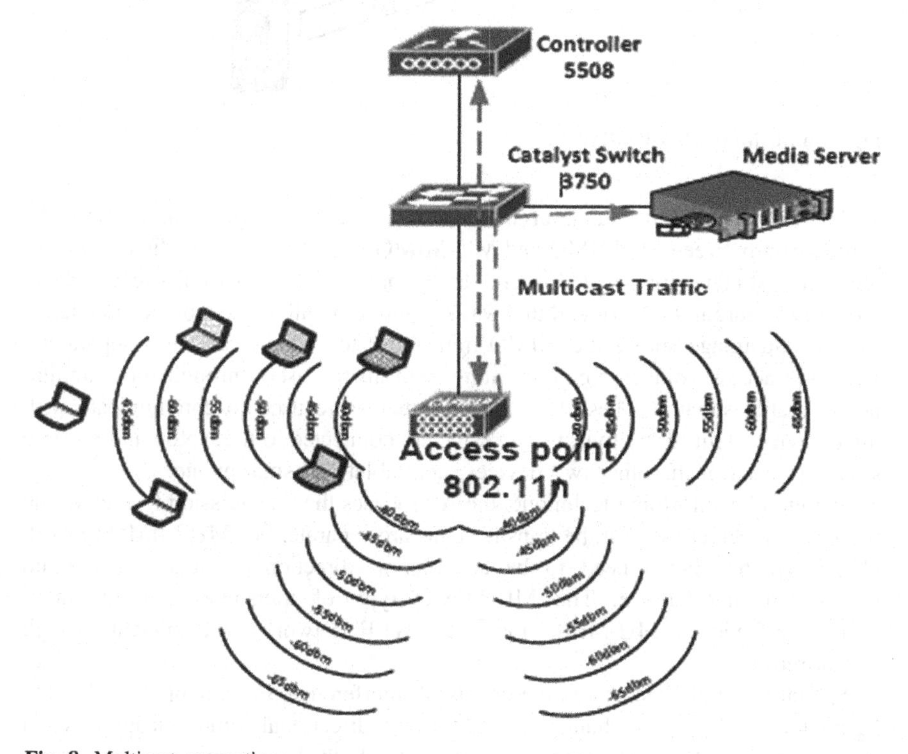

Fig. 8 Multicast connection

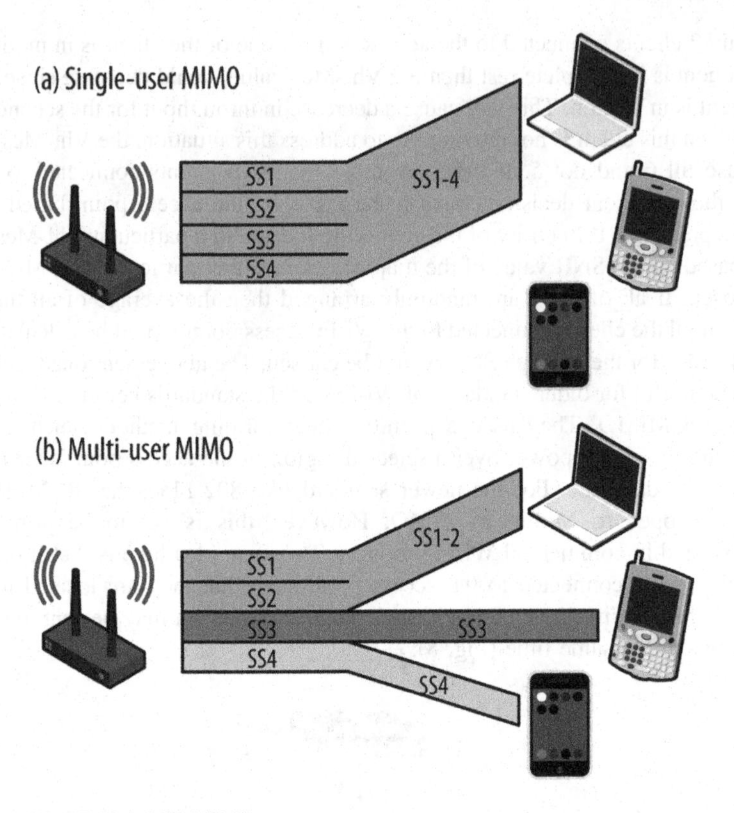

Fig. 9 **a** SU-MIMO **b** MU-MIMO

To understand better look at the following images. Figure 9a, b show the comparison between SU-MIMO and MU-MIMO in 802.11ac. In the first figure, all the spatial streams are directed at one device only. Multiple spatial streams was a common feature in 802.11n and this was supported by all access points and clients. The second image shows the MIMO transmitter to address each user separately. Here, the access point is communicating with all the users through four simultaneous spatial streams. These four spatial streams are used to communicate with three devices. Out of the four, two streams are connected to the laptop and a single stream to each of the other two devices, i.e., tablet and smartphone.

Through beamforming technique, the AP focuses the transmission to each of the respective receivers so that the transmissions are separate. For MU-MIMO to work effectively the clients need to be at different directions so that there is no inter-stream interference. The MU-MIMO type of transmission needs more up-to-date feedback. MU-MIMO makes the Wi-Fi network more efficient through spatial reuse.

Spatial reuse also decreases the effects of interference. For example, in the first figure, the wireless radio channel is used for omnidirectional communication. When the AP sends a data, the energy is received by both the smartphone and laptop, and the channel can support only one transmission at a time, i.e., the AP can communicate with any one of the device at a time. High-density networks have small coverage area so that same channel can be used multiple times. MU-MIMO was built on this small cell approach, thus enabling even more tightly packed networks. In the second figure, MU-MIMO is in use. As a result, the AP can send data independently and simultaneously to each device. Thus, the antenna radiates power only in the direction of the connected devices. So the AP with MU-MIMO does a better job in spatial reuse and there is less interference too.

MU-MIMO works through multiple antennas. These antennas are spaced several wavelengths apart so that there is less interference between them. The multiple antennas help in providing multiple spatial streams. There are around eight spatial streams available in 802.11ac. Each spatial stream works independently and can connect to one client (or more but for now assume that they are connected to one client only). Hence there can eight independent AP—client connections. Since each spatial stream works independently and can switch between each Vht-Mcs value independently, this makes the access point to operate at multiple Vht-Mcs value for each of the clients since they are connected through different streams.

To understand better, assume that there are two clients (client 1 and 2) connected to the AP. They are connected to the AP through 2 independent spatial streams. Initially the two clients are deployed at a same distance from the AP but at different directions such that beamforming happens. Now one of the clients moves away from the AP say client 1 and other client is stationary say "client 2". Since client 1 is in motion for some time and then stops, because of this, the range adaptation algorithm kicks in and decides the access point's operating Vht-Mcs for that spatial stream. Say the client is now in the range of Vht-Mcs 6 then the AP switches to Vht-Mcs 6. But for the stationary client 2, the operating Vht-Mcs is still the initial one, i.e., Vht-Mcs 8 since the two clients are connected through two different spatial streams. This scheme can work till eight clients provided they their distance between them is large so that beamforming can take place. This method is more efficient compared to all other schemes discussed above (Fig. 10).

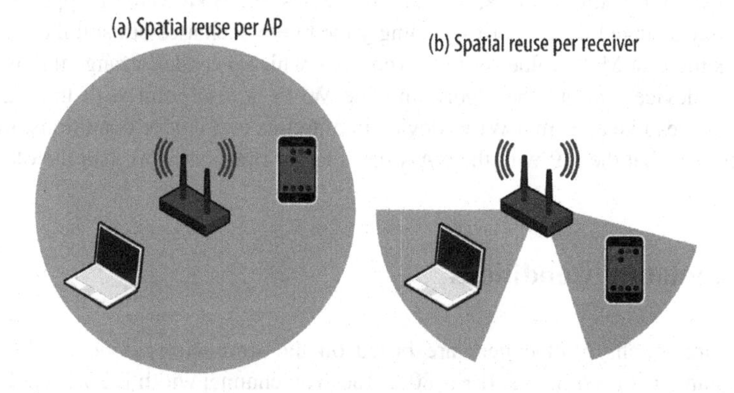

Fig. 10 Improved spatial reuse through MU-MIMO

It is to be noted that even though a single spatial stream there can be many clients connected to access point. So in order to decide the best operating Vht-Mcs value, the range adaptation algorithm incorporates both the cases. Range adaptation works fine for multiple spatial streams with each stream connected to one client. In case of single stream multiple clients, the above-discussed methods need to be used. So to make the algorithm more effective, use the Vht-Mcs selection for single spatial stream in all the spatial streams.

7.2 Outdoor Versus Indoor

Performances of a Wi-Fi device in outdoor conditions and indoor conditions vary since indoors have a lot of walls and other obstacles. Because of this factor, the measured SNR value will also be different. SNR measured outdoor will be less and the Wi-Fi device supports a larger range. The SNR range maps a higher range of connectivity (in terms of meters) in the case of outdoors compared to inorders. This principle is true for a WiFi access point without the Range adaptation algorithm too. But still if the WiFi Access point is equipped with RAA then it's supported range will be higher than Wi-Fi Access points without RAA even in indoor cases. Always indoor supported range of any Wi-Fi Access point will be less than outdoor cases. This is true even for a normal Wi-Fi access point which does not have the range adaptation algorithm, but the range supported when the device has range adaptation algorithm is very large. When considering indoor conditions, the range supported is comparatively less. A normal Wi-Fi access point supports usually half the outdoor range, i.e., say if a device supports 90 m of outdoor range then it has an indoor range of 45 m. The proportion is not always constant since the real world has a lot of variables, hence if the indoor has many Wi-Fi access points at close vicinity then the noise levels increases and then the indoor range will further decrease. The advantage with the range adaptation algorithm is that even when the indoor environment has many noise sources, it still provides a throughput and range support for the particular SNR value. When there are many noise sources, the SNR value decreases, the SNR value mapped for each range may change but and correspondingly the range decreases but still the algorithm chooses the best MCS value for the given SNR which provides a range that is higher than the device without the algorithm. The Wi-Fi access point with the algorithm performances like a normal Wi-Fi device in outdoors and indoor conditions, the only difference is that the AP with the RAA has a larger range support than the other one.

8 Simulation Conditions

The entire graphs plotted here are based on the simulations done in NS-3. The Wi-Fi standards used here is IEEE 802.11ac. The channel width is 20 MHz because of this MCS value till 8 is used. The channel number is 36. Short guard enabled for the simulation. The transmit power is 20 dB. The propagation loss models used for the simulation are “Constant Speed Propagation Delay Model” and “Friis Propagation Loss Model”.

9 Conclusion

In this paper, a way for increasing the range of an access point is implemented. The increase in range causes a decrease in throughput, which is acceptable as the range gets increased by 100 m. There is a lot of hardware methods present to increase the range of the Wi-Fi access point, but through the implementation of a range adaptation algorithm, cost increase for implementation is less compared to other techniques. SNR based selection of range is one of the easiest ways and it is sent along with packet by the access point. By getting this value from the tag, the client can send its SNR in the ACK packet. The access point’s transmit power is also an important parameter to consider. Even if an access point is working with a particular high range supporting Vht-Mcs, if the transmit power is not higher or equal to the minimum support power for the particular MCS value, then range gets restricted due to the transmit power. Therefore, the access-point should work in appropriate transmit power for a given Vht-Mcs. As IEEE 802.11ac supports Multiuser MIMO each client connected to the AP can work at the best Vht-Mcs for their position as Range Adaptation Algorithm is used in each spatial streams.

10 Future Work

This algorithm is rather difficult to implement with traditional techniques since we need separate circuits for the implementation of different modulation techniques. But with the advancement of cognitive radio technology, it will be easy to implement this technique practically. There needs to certain improvement with respect to the fact that more than one client is connected to the access point. Rather than selecting the appropriate Vht-Mcs based on the mean SNR value of the majority clients who are very far other improved algorithms for the priority needs to be developed. As SNR feedback method will perform poorly when there are many localized noise sources present in the AP’s working range. Here the AP is made to works at constant TxPower of 20 dB. Since AP needs to work with the appropriate transmit power for the corresponding Vht-Mcs, the range adaptation algorithm can also include spatial reuse if both the transmit power and MCS are optimized. With future developments in the IEEE 802.11 standards and MCS index, both the range and throughput can also be increased. The following algorithm uses SNR feedback technique with constant noise assumption. Algorithms with other standardized parameter for consideration would make the range adaptation more effective.

References

1. Matthew Gast 802.11ac: A Survival Guide. Chapter 1, *Introduction to 802.11ac, Beamforming and Multi-User MIMO (MU-MIMO)*, p. 7.
2. Perahia, E., & Stacey, R., Next Generation Wireless Standards 802.11n 802.11ac. Chapter 1, *Environments and applications for 802.11n*, p. 12.
3. Perahia, E., & Stacey, R., Next Generation Wireless Standards 802.11n 802.11ac. Chapter 7, *Very High Throughput PHY*, p. 182.
4. David Forney, G., Jr., Fellow, IEEE, Modulation and coding for Linear Gaussian Channels. *EEE Transactions on Information Theory*, Vol. 44, No. 6, October 1998.
5. CISCO 802.11ac: The Fifth Generation of Wi-Fi., Technical White Paper, aironet-3600-series.
6. IEEE 802.11 Working Group Project Timelines, 2012-11-03.
7. Kadu, V. V., Manually Designed Wi-Fi Cantenna and its Testing in Real-Time Environment. *International Journal of Engineering Research and Development*. e-ISSN:2278-067X, p-ISSN:2278-800X, https://www.ijerd.com. Vol. 3, Issue 2 (August 2012), pp. 01–06.
8. Holland, G., Vaidya, N., & Bahl, P. (2001). A rate-adaptive MAC protocol for multi-hop wireless networks. In *Proceedings of ACM MobiCom* 01 (pp. 236–251), July 2001.
9. Shami, A., Maier, M., & Assi, C. (2010, January 23). *Broadband access networks: Technologies and deployments* (p. 100). Springer Science Business Media.
10. IEEE Standard 802.11-2007, p. 531. (Revision of IEEE Std 802.1 1-1999) ("Official IEEE 802.11 working group project timelines". January 26, 2017. Retrieved 2017-02-12.)
11. Gudipati, A., & Katti, S., Automatic Rate Adaptation. *SIGCOMM'11*, August 15–19, 2011.
12. Sun, W., Choi, M., & Choi, S. (2013, July). IEEE 802.11ah: A Long Range 802.11 WLAN at Sub 1 GHz. *Journal of ICT Standardization, 1*, 83–108. https://doi.org/10.13052/jicts2245-800X.125.
13. Kassner, M. (2013, June 18). Cheat sheet: What you need to know about 802.11ac. In: TechRepublic. 18. Juni 2013, abgerufen am 6. Dezember 2013.
14. Kelly, V. (2014-01-07). New IEEE 802.11ac™ Specification Driven by Evolving Market Need for Higher, Multi-User Through- put in Wireless LANs. IEEE. Archived from the original on 2014-01-12. Retrieved 2014-01-11.
15. Yang, J. C.-N. (2001, October 10). What is OFDM and COFDM? Shoufeng, Hualien 974, Taiwan: Department of Computer Science and Information Engineering National Dong Hwa University.

Air Quality Parameter Measurements System Using MQTT Protocol for IoT Communication Over GSM/GPRS Technology

Anil Thosar and Rohan Nathi

Abstract Increasing populations and industrial activity are threatening our environment. Air pollutants are emitted at a very high rate. With the technological innovation and advancement, smart cities have come into existence where Information Technology is used for better optimization and resources planning hence promoting the sustainable growth and development. Internet of Things and Data Analytics together can lead an organization to better cost management, avoiding equipment failures and improving business operations. We propose Air Quality Measurement System which can effectively keep a track over Air quality in atmosphere. Internet is used for end to end connectivity. The parameters sensed by the system are sent to the server through a Cellular Network System. Communication between device and server is deployed over lightweight Message Queuing Telemetry Transportation (MQTT) protocol.

Keywords Embedded systems · Internet of things (IoT) · Internet protocol MQTT protocol

1 Introduction

In recent times, air pollution is one of the major threats. Air quality has been affected due to enormous amount of toxic emissions from industrial, vehicles, and day-to-day human activities. Air quality has an impact on many aspects of life; energy efficiency and work productivity are affected by poor air quality. Air quality control and monitoring have therefore gained momentum in recent times. Need to

---

A. Thosar (✉) · R. Nathi
Department of Electronics Engineering, K.J. Somaiya College of Engineering, Mumbai, Maharashtra, India
e-mail: anil.thosar@somaiya.edu

R. Nathi
e-mail: rohan.an@somaiya.edu

V. E. Balas et al. (eds.), *Data Management, Analytics and Innovation*, Advances in Intelligent Systems and Computing 808,
https://doi.org/10.1007/978-981-13-1402-5_32

control air pollution has become an absolutely necessary factor to provide a safer future for the next generation.

Data collected at various parts of city, industrial and other areas will help us to understand the pattern of the air pollution, which will further help in identifying the places that are prone to health hazards. It will also aid the government authorities to plan the measures to reduce air pollution.

The motive of the designing and implementing the device prototype is to detect certain harmful air pollutants and log the amount at which they are present over a server with minimum usage of network resources. The device shows variations in measured values in a predefined range when exposed to certain pollutants which will be discussed further in more details.

Internet of things (IoT) has gained attention in the recent times for ease in remote monitoring, data analysis, surveillance applications, etc. Internet Technology has changed from human-centric usage nature to device-centric usage. The protocols existing in the today's Internet technology are heavy weighted in terms of network resources utilization and exchanges a lot of data to establish reliable communication.

The IoT network essentially consists of embedded devices connected to Central Monitoring Centre, and these devices have limited processing capabilities and hence are called constrained resources, as these devices have in less battery capacity, low memory, and bandwidth.

The IoT device deployed in fields generally uses those Internet protocols which do not take into the consideration these constrained devices characteristics. Hence, newer protocols such as MQTT, MQTT-SN, and CoAP, etc. are being developed to ideally suit the requirements of these constrained devices and make them smart [1].

Wireless access to the Internet is provided by GSM/GPRS technology (location where there is a network signal). This makes it possible for IoT devices to access the internet from any remote location.

2 Literature Survey

Objective of this research [2] article is to present a system model which can facilitate the assessment of health impacts caused due to indoor air pollutant as well as outdoor pollutant, here a sensing network based microcontroller equipped with gas sensors, optical dust particle sensor, humidity, and temperature sensor has been used for air quality monitoring. The design included various units mainly: sensing unit, processing unit, power unit, display unit, and communication unit. This work will apply the techniques of electrical engineering with the knowledge of environmental engineering by using sensor networks to measure Air Quality Parameters.

Paper [3] uses a monitoring framework that uses open software and hardware which is based on Libelium's gas sensing capable motes. The sensed data by these motes will be sent to a computing platform, which is dynamically configurable that supports the scaling of both real-time incident management and longer term strategic planning decisions. Hence, the proposed monitoring framework in [3] provides timely distributed measurements of various air quality metrics which would aid in evaluating the impact of emissions of toxic components in air.

Paper [4] uses an implementation of Message Queuing Telemetry Transportation (MQTT) using a CC3200 node by Texas Instruments. Wi-Fi network establishes the wireless Internet access to the node. The system is incorporated with TI-RTOS. The system proposed by the authors consists of a sensing unit and an actuation control based on CC3200. Onboard sensors, push buttons, and LEDs form the perception layer in the proposed prototype. The data is sent over Wi-Fi. CC3200 is powered by USB cable connected to host PC. This data is published to IBMs MQTT broker. Push buttons on CC3200 also act as publishers, publishing their state to the broker.

3 Proposed System

Proposed system consists of Sensor Module which comprises of DHT11 and MQ135 Sensor. Open hardware Arduino Uno collects data from sensor modules DHT11 and MQ135, DHT11 MQ135 and Arduino Uno forms a sensing platform. All the ADC conversions and calibration are done by Arduino Uno and transferred to Open CPU GSM Module with the help of serial converter. This sensing module periodically sends data to Open CPU GSM module.

Once data is received by OCGM, it is sent over Internet network through GSM/GPRS network. The system periodically publishes data to MQTT-IoT Eclipse message broker over Topic "AirMonitor". Client (can be Web client or android App) can subscribe to the same topic ("AirMonitor") and start receiving message [5].

Figure 1 shows a block diagram for Air Quality Parameter Measurements System Sensor Module with OCGM that can be installed at monitoring site, while clients can be at any corner of world. MQTT broker software can be deployed over server of the service provider.

Entire end to end system implementation is shown in Fig. 2 where the web client utility and mobile app client utility has subscribed to the same topic over which sensing device publishes the data. Hence all the subscribed clients receive the data.

4 System Modeling

4.1 *Hardware and Software Used*

In order to carry out this project, we made use of the following set of hardware:

- Arduino Uno open source hardware (controller ATmega328).
- Open CPU GSM Module.

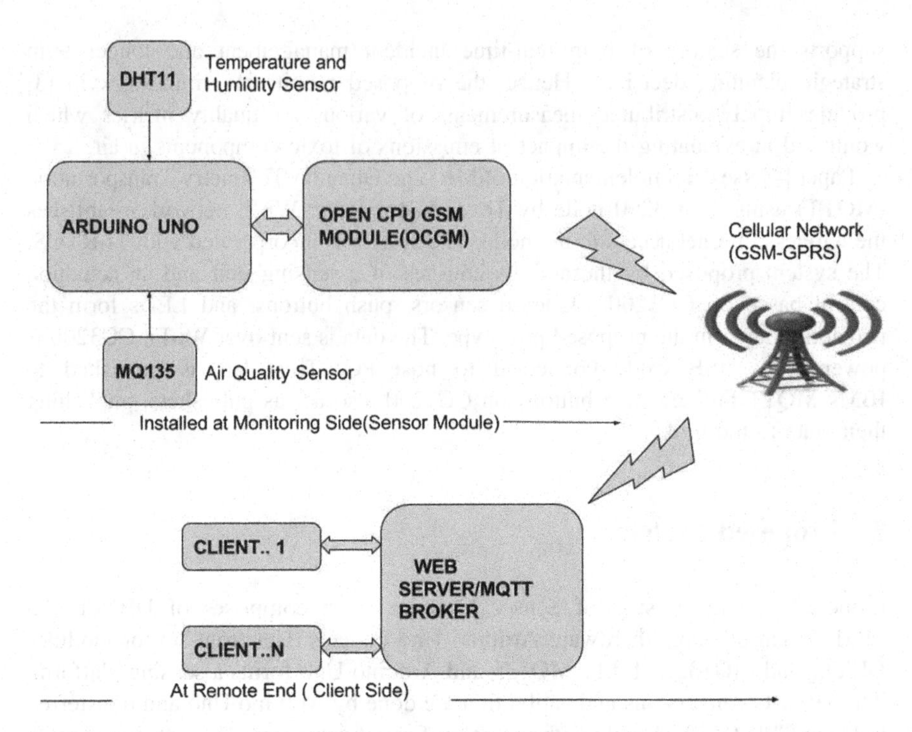

Fig. 1 Block diagram of Air Quality Parameter Measurements System

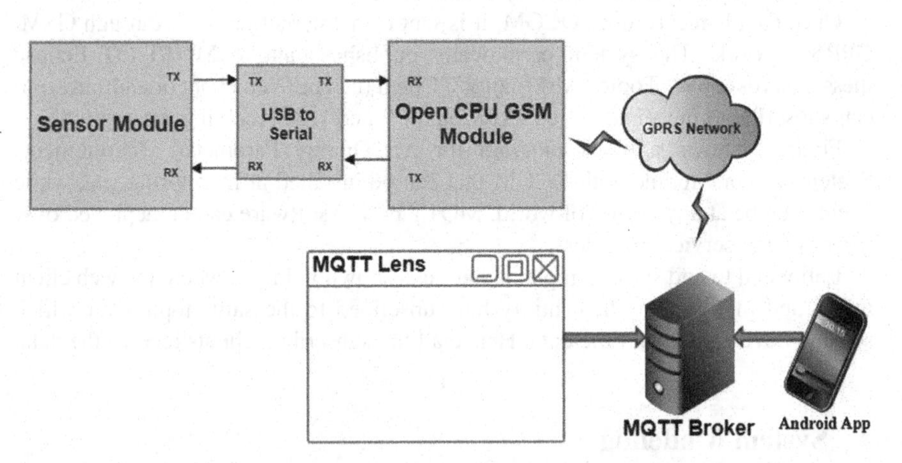

Fig. 2 End to end system implementations where MQTT lens and MQTT client (an Android App) receives the data from sensing device. Sensor Module and Open CPU GSM Module communicates with each other over serial converter

- DHT11 Temperature and Humidity Sensor Module.
- MQ135 CO_2 Gas Sensor.
- Serial to USB converter.

The software comprises of:

- Arm compiler.
- C editor.
- Q-Flash tool.
- Docklight (COM port monitor).
- MQTT lens utility (Web Client).
- MQTT client (Android App).

4.2 Sensor Module

Sensor Module consists of sensors for measuring CO_2, temperature, and humidity content in Air and Arduino Uno. MQ135 and DHT11 are deployed for that measuring purpose. Arduino Uno is used for collecting these measured parameters.

1. **MQ135**: MQ135 gas sensor has a sensitive chemical material (SnO_2), which has lower conductivity in clean air. A simple electrical circuit can convert change in conductivity into proper electrical signals that corresponds to output of fifteen signal of gas concentration. It is a low-cost sensor and suitable for different application in air monitoring [6]. To evaluate the sensor for its response to CO_2 gas, it was exposed to smoke for a span of one minute, which is shown in Fig. 3 [7, 8].
2. **DHT11**: DHT11 is a composite digital humidity and temperature sensor. It has a calibrated digital signal output for humidity and temperature. The sensor

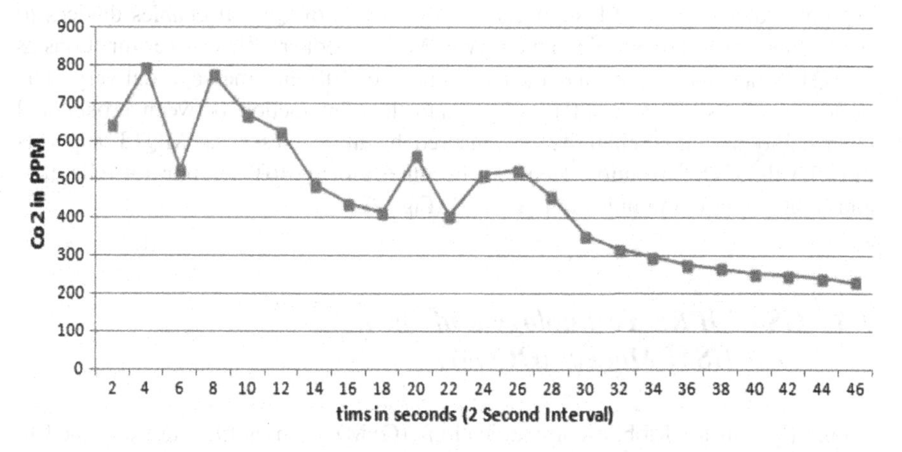

Fig. 3 Variation in CO_2 level for smoke captured for a time span of two seconds interval

comprises of NTC temperature measurement component and a resistive sensitive wet component for humidity measurement, both the components are connected with a 8-bit microcontroller [9, 10].

4.3 MQTT Lens (Web Client)

MQTT lens is client's side utility provided by Google Chrome. MQTT Lens connects to an MQTT broker and is able to subscribe and publish to MQTT topics. The connection is established over Port Number 1883.

4.4 MQTT Client (Android App)

Android Smartphone supports this app after the installation from Google Play Store. This app can subscribe/unsubscribe topics; topic subscriptions are made as per the client requirements.

This app gives the notification for a new message received on several topics subscriptions. Port Number 1883 is used by default for communication.

4.5 MQTT

The MQTT is the Message Queuing Telemetry Transport protocol that is based on providing subscribe publish architecture; it enables the two entities to communicate with each other, independently without request–response approach over a network. It is lightweight protocol making which provides reliability and some levels of assurance delivery. MQTT supports different levels of QoS. It enables devices to send data (publish) of a topic to the server (MQTT Broker). This server functions as a MQTT message broker and manages entire end to end message delivery. The protocol defines 13 types of messages and their interaction between broker and clients. Payload and reliability are ensured by these messages. MQTT supports three levels of QoS solution. Hence, it becomes a better IoT solution for reliability and lightweight communication [11, 12] (Fig. 4).

4.6 GSM/GPRS Technology and Open CPU GSM Modem (OCGM)

Global System for Mobile communications (GSM) is an architecture designed to serve cellular calls over long distance. General Packet Radio Service (GPRS) is a packet-oriented mobile data service deployed on 2G and 3G cellular

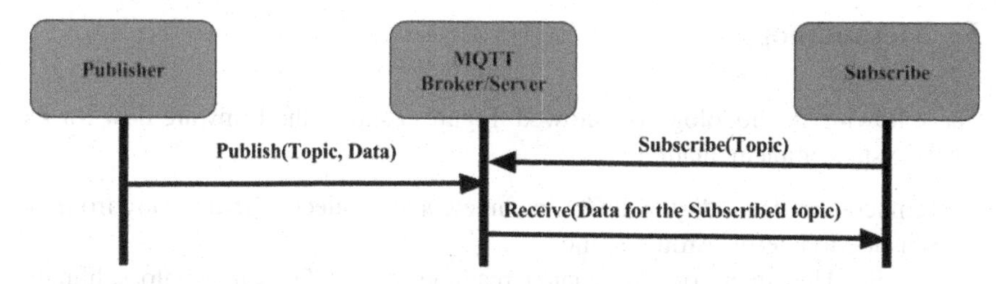

Fig. 4 Publish Subscribe Architecture for MQTT protocol, publisher can subscribe to a topic where it has to publish (send) data, while subscriber can subscribe to the same topic where publisher is publishing and starts receiving the messages

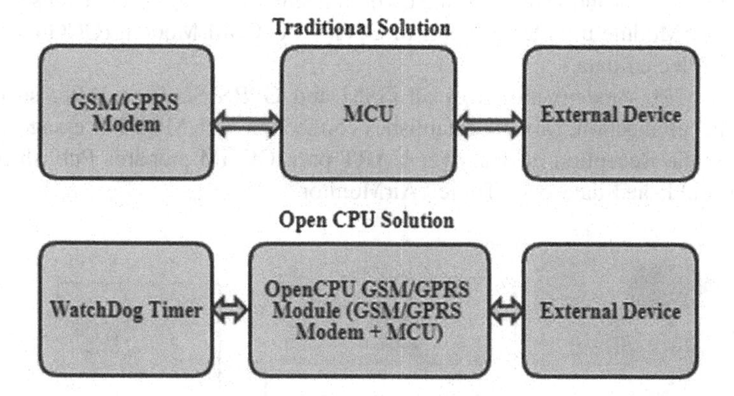

Fig. 5 Open CPU solution that represents embedded controller and GSM modem

communication system which is an extension of GSM technology that supports higher data rate. An Open CPU GSM/GPRS modem is a class of wireless 2G modem, which is designed for communication over the GSM and GPRS network. Open CPU is an embedded development solution for the M2M field. Open CPU concept combines microcontroller sending AT commands to GSM modem on a single entity (silicon chip). Embedded applications can be conveniently designed based on it. In the OpenCPU solution, GSM/GPRS module acts as the main processor. It enables the customer to create innovative application and download it directly into Quectel module to run. SIM (Subscriber Identity Module) card is required to activate communication with the network. So, GSM/GPRS module with OpenCPU solution facilitates customer's product design and accelerates the application development [13] (Fig. 5).

5 Methodology

The following methodology is followed. Figure 6 shows the firmware flow for the entire system implementation.

- Temperature, Humidity, and CO_2 parameters are collected periodically from the Sensor Module by Arduino Uno.
- Arduino Uno then converts analog readings from CO_2 sensor into calibrated CO_2 level (in PPM).
- DHT11 is an integrated type of Sensor which has inbuilt microcontroller that is responsible for producing the digital output which is equivalent to temperature and humidity sensed in environment.
- Arduino UNO and Open CPU are communicating with each other over serial port.
- Sensor Module uses UART port of Open CPU GSM Modem (OCGM) to send the collected data.
- Open CPU modem initializes all GSM and GPRS Services. After successful GPRS connection, Modem establishes connection with MQTT Message Broker.
- After the Reception of data over UART port, OCGM prepares Publish packets and published data over Topic "AirMonitor".

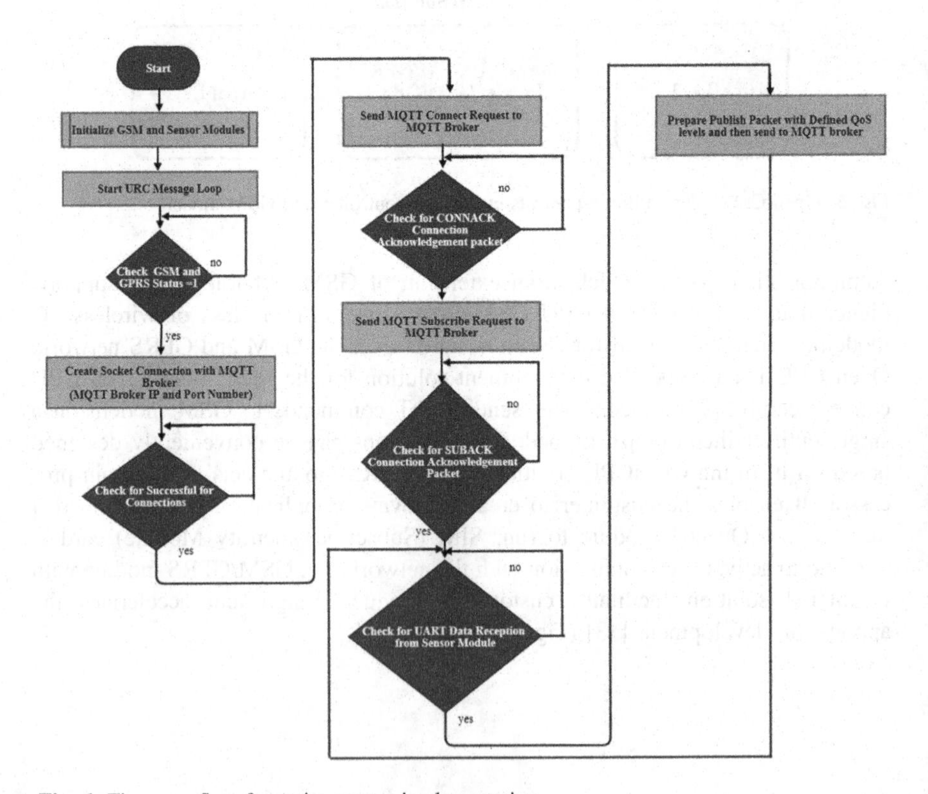

Fig. 6 Firmware flow for entire system implementation

- MQTT Lens (Web Client) and MQTT Client (Android App Client) establishes a connection with MQTT broker, after the successful establishment, it makes a subscription to topic "AirMonitor". Every time device publishes data is received by these clients.

6 Implementations and Results

The methodology discussed above has been successfully implemented; IoT Eclipse MQTT message broker has been configured to perform all message transaction. The device sends data to this configured broker.

All the debug messages are captured at device end to monitor the status of device while connecting to MQTT broker and sending data to the server. Debug messages are captured over Docklight Software (which is COM port) and it is shown in Figs. 6 and 7. Table 1 summarizes all MQTT message size captured for Air Quality

```
Communication
ASCII | HEX | Decimal | Binary
****<---- ROHAN FINAL YEAR MTECH PROJECT: MQTT QOS: 0 V0.15 ---->****

<-- RIL is ready -->

<-- GSM Network Status:2 -->
<-- Signal strength:99, BER:99 -->
 Automatic Time At Zone executed Successfully
<-- GPRS network status:0 -->

<-- CFUN Status:1 -->

<-- SIM Card Status:1 -->
<-- GPRS network status:2 -->
<-- Sys Init Status 2 -->

<-- GSM Network Status:1 -->
<-- Signal strength:25, BER:0 -->
 Automatic Time At Zone executed Successfully
<-- Module has registered to GPRS network -->
<-- Register GPRS callback function -->
<-- Configure PDP context -->
<-- Activating GPRS... -->
<-- Waiting for GPRS ACTIVED. -->

<-- Sys Init Status 3 -->

<-- SMS module is ready -->
<-- BEFORE m_SocketConnState = 0. -->
making connection in GPRS<-- CallBack: active GPRS successfully. -->
<-- Register socket callback function -->
<-- enter socket
<-- Create socket successfully, socket id=0. -->
<-- Convert Ip Address successfully,m_ipaddress=198,41,30,241 -->
```

Fig. 7 Debug messages captured on Docklight Software

Table 1 Summary of all the MQTT Message Packet Size captured for Air Quality Parameter Measurement Application

MQTT message type	Message size (bytes)
Connect	18
Connect ack	2
Subscribe	17
Subscribe ack	2
Message size (received from sensor modules)	67
Publish packet	125

```
Colors&Fonts Mode   COM18   115200,
Communication
ASCII  HEX  Decimal  Binary
making connection in GPRS<-- CallBack: active GPRS successfully. -->
<-- Register socket callback function -->
<-- enter socket
<-- Create socket successfully, socket id=0. -->
<-- Convert Ip Address successfully,m_ipaddress=198,41,30,241 -->
<-- Connecting to server(IP:198.41.30.241, port:1883)... -->
<-- Waiting to SOC Connect : WOULD_BLOCK -->
<-- Callback: socket connect successfully. -->

Sending MQTT Request Connection to MQTT Broker

Connection Request Sent to MQTT Broker

 <-- Callback socket Event.-->

 CONNACK:: Connection Acknowledgement:: Recieved from MQTT broker.
 CONNACK Packet Size:2 bytes
                                                                                10/2/2017
12:07:19.055 [RX] - Recieved Data from Sensor Module: HUMIDITY||33.00||Percent||TEMPERATURE||32.00||C||Co2||
261.75||ppm||
 Length of Data Recieved from Sensor Module: 67 bytes

 Publishing to Topic: AirMonitor  QoS level:0
 Publish Packet Size : 125 bytes
                                                                                10/2/2017
12:07:26.832 [RX] - Recieved Data from Sensor Module: HUMIDITY||33.00||Percent||TEMPERATURE||32.00||C||Co2||
261.75||ppm||
 Length of Data Recieved from Sensor Module: 67 bytes

 Publishing to Topic: AirMonitor  QoS level:0
 Publish Packet Size : 125 bytes
```

Fig. 8 Debug messages captured on Docklight Software

Parameter measurement application (Fig. 8). MQTT Client is Android Utility also receives messages sent by device which is shown in Fig. 9.

Status of Open CPU GSM module was seen in every phase while connecting to GSM network, after successfully connection activating GPRS services (GSM status network: 1 indicates that device has successfully registered to GSM Network and can avail the GPRS Services). Once GSM/GPRS Services are activated device attempts to connect to the MQTT message broker, after successful connection Publish message is sent. MQTT lens (Web Client Utility) receives the messages published by device and simultaneously Android App also receives the same Publish message (Fig. 10).

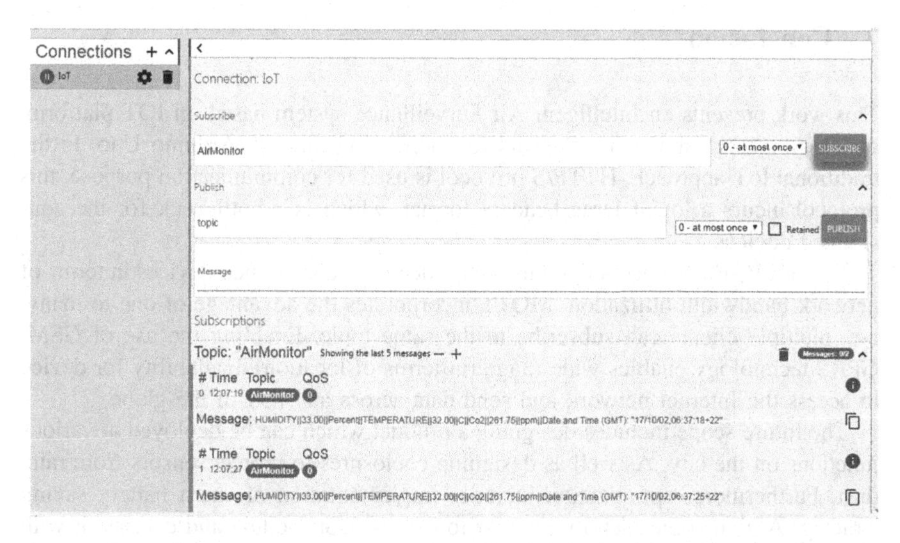

Fig. 9 Client has made subscription to topic "AirMonitor" over MQTT broker, after successful subscription acknowledgement by the broker, it starts receiving message over MQTT lens (Web Client Utility)

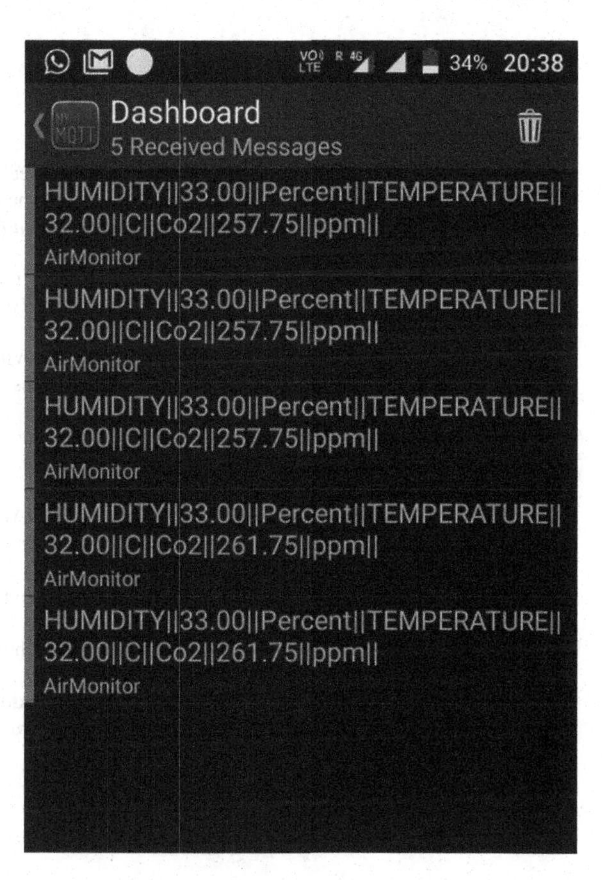

Fig. 10 Client subscribed to topic "AirMonitor" receiving message over MQTT lens (Mobile App Utility)

7 Conclusion

This work presents an Intelligent Air Surveillance system based on IOT platform, which connects Air Quality Sensors for monitoring through Arduino Uno. In the traditional IoT approach, HTTP/S protocol is used for communication purpose, this protocol incurs a lot of large headers format, which is a bottleneck for the constrained devices.

Use of MQTT protocols lightens the burden over constrained devices in terms of network bandwidth utilization. MQTT incorporates the advantage of one to many, i.e., multiple clients can subscribe to the same topic. Enabling the use of GSM/GPRS technology enables wide range (in terms of location) availability for device to access the Internet network and send data across any part of the globe.

The future scope includes designing a product which can be deployed at various junctions on the city. As well as designing enclosures to protect sensors from rain, dust. Furthermore, optimization includes implementation of smart battery saving strategy. Assign some identity number to each sensor module and corelate it with junction name, so that user can search for data at some specific junction and some specific area by conventional names. All these would be challenges for real-life product.

References

1. Kraijak, S., & Tuwanut, P. (2015). A survey on internet of things architecture, protocols, possible applications, security, privacy, real-world implementation and future trends. In *IEEE 16th International Conference on Communication Technology (ICCT)*, 2015 (pp. 26–31). https://doi.org/10.1109/icct.2015.7399787.
2. Jangid, S., et al. (2016). An embedded system model for air quality monitoring. In *3rd International Conference on Computing for Sustainable Global Development (INDIA Com)*, 2016.
3. Mansour, S., Nasser, N., Karim, L., & Ali, A. (2014). Wireless Sensor Network-based Air Quality Monitoring System. In *International Conference on Computing, Networking and Communications (ICNC)*, 2014.
4. Kodali, R. K. (2016). An implementation of MQTT using CC3200. In *2016 International in Conference on Control, Instrumentation, Communication and Computational Technologies (ICCICCT)*, 2016.
5. Mansour, S., Nasser, N., Karim, L., & Ali, A. (2014). Wireless sensor network-based air quality monitoring system. In *International Conference on Computing, Networking and Communications (ICNC)*, 2014.
6. MQ135 Datasheets retrieved from URL http://china-total.com/Product/meter/gas-sensor/MQ135.pdf.
7. MQ135 library for Arduino Uno retrieved from https://playground.arduino.cc/Main/MQGasSensors.
8. Petty, S., Summary of Ashrae's Position on Carbon Dioxide (CO_2) Levels.
9. DHT11 Datasheet retrieved from URL https://www.adafruit.com/product/386.

10. DHT11 library for Arduino Uno https://playground.arduino.cc/Main/DHT11Lib in Spaces. Retrieved from URL https://www.ashrae.org/File%20Library/docLib/Technology/…/TC-04-03-FAQ-35.pdf.
11. Publish Subscribe Architecture. Available https://msdn.microsoft.com/en-us/library/ff649664.aspx.
12. Oasis MQTT version 3.1.1. Available: http://docs.oasis-open.org/mqtt/mqtt/v3.1.1/os/mqtt-v3.1.1-os.pdf.
13. Quectel Wireless Solutions "GSM/GPRSM66". Retrieved from http://www.quectel.com/product/prodetail.aspx?id=73. Retrieved 2015-05-04.
14. Arduino Software and Hardware Uno Tutorials January, 2012 retrieved from URL http://arduino.cc/en/Tutorial/.

Comprehensive Analysis of Routing Protocols Surrounding Underwater Sensor Networks (UWSNs)

Anand Nayyar, Vikram Puri and Dac-Nhuong Le

Abstract In recent times, Underwater Sensor Networks (UWSNs) has become highly important for performing all sorts of underwater operations. It is somewhat cumbersome to implement terrestrial sensor networks routing protocols in UWSN due to high propagation delay, packet delay, and energy efficiency. So, lots of efforts are going on by researchers dedicated towards proposing efficient routing protocols for UWSN. As UWSN have specific characteristics in terms of rapid dynamic topology change, limited bandwidth, high energy consumption, high latency, packet delay issues, and security problems, designing a routing protocol to overcome all the aforesaid mentioned issues is quite a daunting task. In this research paper, we primarily focus on surveying various routing protocols available till date for data routing in UWSNs. In addition, comparison of protocols is also mentioned on the basis of various characteristics like routing technique, packet delivery ratio, energy efficiency, packet delay and localization to give a clear picture of the benefits and shortcomings of each and every enlisted protocol for UWSN.

Keywords Underwater sensor networks · Routing protocols · Acoustic networks Sensor communications · Routing · AUV (Autonomous Underwater Vehicles) UUV (Undermanned Underwater Vehicles) · NS-2

A. Nayyar (✉)
Graduate School, Duy Tan University, Da Nang, Vietnam
e-mail: anandnayyar@duytan.edu.vn

V. Puri
GNDU Regional Campus, Jalandhar, Punjab, India
e-mail: vikrampuri@acm.org

D.-N. Le
Department of Educational Testing and Quality Assurance,
Haiphong University, Haiphong, Vietnam
e-mail: nhuongld@dhhp.edu.vn

V. E. Balas et al. (eds.), *Data Management, Analytics and Innovation*,
Advances in Intelligent Systems and Computing 808,
https://doi.org/10.1007/978-981-13-1402-5_33

1 Introduction

The ocean is a huge dump of water that has attracted lots of people across the nook and corner of this world to take a deep dive and solve precious mysteries under its lap. Almost 75% of the earth's planet surface is covered by water. The deep oceans till today remained a harsh challenge for human beings to carry out search operations. Current technologies in the form of sensors do not meet the technical specifications for installation and deployment of low-cost and power-efficient equipment.

Nowadays, there is an emergent requirement of underwater monitoring [1] in terms of searching for underwater sea resources, capturing underwater scientific data in terms of marine resources, detection of all sorts of underwater incidents like cutting or disruption of optical fiber cables or pollutions via chemicals spill or issue of oil spilling, but the existing technologies are not up to the mark in terms of meeting up accurate technical requirements. Underwater sensor networks have specific requirements in terms of algorithms and protocols for monitoring and routing of data. There are tons of protocols proposed for Terrestrial Wireless Sensor Networks (TWSNs) [2, 3] but they are not fully suitable and reliable to be implemented in Underwater Sensor Networks (UWSNs) due to various shortcomings in terms of: Lack of quality bandwidth, dynamic change in network topology, speedy energy consumption, and high latency in terms of transmission.

UWSNs are connected networks and face frequent and long term delays in transmission due to harsh underwater environments. UWSNs are considered highly important for all sorts of military operations. Various UUVs (Undermanned Underwater Vehicles) and AUVs (Autonomous Underwater Vehicles) are developed by organizations fully equipped with underwater sensors for performing various tasks of underwater sea exploration operations, collecting data and monitoring. But all these sensors face lots of challenges and difficulties considering different scenarios (Fig. 1).

Various UWSN routing protocols are developed by researchers for enhancing the reliability and efficiency for carrying out undersea operations, but every protocol has its own pros and cons.

The main key properties cum issues for routing efficiency in UWSNs are [4, 5]:

- High Propagation Delays: In underwater scenarios, the transmission of data via radio signals don't work in an efficient manner and this marks the usage of acoustic communication. The main issues surrounding acoustic communication are less bandwidth and high propagation delays. The speed of propagation of acoustic signals in water is 1.5×10^3 m/s which is near to about five orders of magnitude and is lower than radio propagation speed which is 3×10^8 m/s.
- Node Mobility: Underwater environments are exposed to water currents, which in turn produces fluctuations in network topology. Nodes, if not efficiently anchored can move from one position to another, thereby making UWSN a dynamic topology.

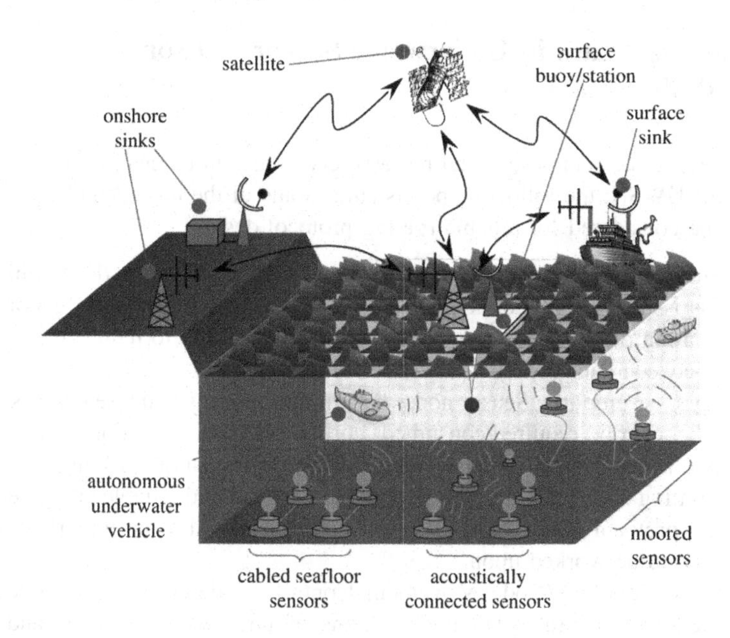

Fig. 1 Underwater sensor network: complete scenario

- Error Prone and Limited Bandwidth: Considering the scenario of underwater communication. The channel has limited bandwidth capacity in terms of frequency and range of transmission (measured in terms of KHz). It suffers from other issues like high bit error rates due to multipath, doppler effect and noise.
- High Energy Consumption: Considering the various types of sensor networks like industrial, agriculture, military, terrestrial UWSN, most of the sensors deployed are battery powered with limited energy capacity. Sensor nodes deployed underwater performs different sorts of operations like routing, mobility, and topology control which drains the battery soon.
- Maintenance Issues: UWSNs design and development incur high costs and limited organizations manufacture distribute in the market. The cost of acquiring underwater sensors is very high and maintenance requires lots of efforts and costs to keep the network up and running.
- Other Issues: UWSNs are prone to corrosion leading to node failures. Underwater channels face transmission issues due to multipath and fading. UWSNs are deployed in harsh environmental conditions which makes routing a serious challenge.

Considering all the shortcomings of UWSN, researchers are doing research at all layers of UWSN with ultimate objective to propose an energy efficient routing protocol for UWSN to perform hustle free underwater operations.

2 Routing Issues in Underwater Sensor Networks (UWSNs)

There are a lot of open issues and research challenges in efficient routing protocol design for UWSN, the following points enlist some of the issues/challenges which need to be considered for reliable routing protocol design:

- Propagation Delay: As UWSN faces lots of high propagation delay which is a serious issue, so a model is desired for calculating accurate propagation delay and routing protocol needs to reduce propagation delay to reduce transmission time between data packet from sender to receiver.
- Energy Consumption: Sensor nodes have limited energy and research has shown that the energy required/consumed during network operation especially in UWSN as compared to other sensor networks is almost 10 times more in transmitting and receiving. So, till date, no accurate and reliable energy efficient routing protocol is there for UWSN which is regarded as an important area of research to be worked upon.
- Dynamic Topology/Node Movement Control: Underwater sensor operations face a lot of natural calamities in terms of underwater habitants and water currents which make the nodes move from one place to another. Terrestrial UWSN has lots of mobility models which are not suitable to be deployed for UWSN. So, appropriate mobility model is need of the hour for UWSN for carrying out all sorts of underwater operations.
- Secure Routing: Security requires consistent research to combat all sorts of threats coming from outside world to keep operations of UWSN error free. Various cryptographic techniques when integrated with routing protocols are helpful for end-to-end routing and prevents disruptions from all sorts of attacks like Man-in-the-middle attack, DoS attack, DDoS attack etc.
- Efficiency and Reliability: Considering underwater acoustic channels, various issues arise in terms of quality transmission like limited bandwidth, link quality, high bit error rates, and RF channels. Routing protocol should be proposed in a such a manner that improves the efficiency and reliability of communication and performs routing with high rate of reliability and fault tolerance.
- Utilization of Intelligent Algorithms [6]: Some of the protocols of UWSN makes use of Optimization and Intelligent Algorithms for routing. In terms of efficient routing protocol, various Intelligent Algorithms should be deployed making use of modern techniques like Genetic Algorithms, Swarm Intelligence, Fuzzy Logics etc.

3 Comparison of Underwater Sensor Network with Terrestrial Sensor Networks

The following points enlist the differences:

1. Real-World Deployment: The sensor nodes are deployed densely in terrestrial networks, whereas the deployment is sparse as in case of UWSNs.
2. Incurring Cost: Terrestrial sensor networks are relatively cheap to deploy as compared to UWSN because UWSN demands sophisticated hardware and specialized sensors which few companies manufacture and the maintenance cost is also relatively high.
3. Energy Consumption: The energy required for transmission in nodes is high in UWSN as compared to terrestrial networks due to complex signal processing.
4. Memory Requirements: Sensor nodes in terrestrial networks require less memory capacity due to limited data gathered, whereas in UWSN, data can be classified as Ordinary, Intermediate and Emergent and this puts more storage requirements on sensors in UWSN.
5. Intelligent Algorithms/Protocols: Terrestrial Sensor Networks do not demand high-quality intelligent algorithms and protocols for operations, whereas UWSN being a complex network and always dynamic in nature puts a lot of stress on sensors to operate in harsh environments which in turn makes requirement of sophisticated routing protocols making use of Intelligent Algorithms/Protocols derived from Fuzzy, Swarm, and Genetic Based Optimization techniques for successful operation (Table 1).

Table 1 Basic differences between underwater environment and terrestrial environment

Basis of difference	Underwater environment (acoustic communication)	Terrestrial environment (RF technology)
Energy consumption	Very high	Somewhat low
Propagation delay	High	Low
Bandwidth	Low	High
Rate of data transmission	Low	High
Noise and environmental interferences	High	Low
Dynamic topology operation	High	Low
Efficiency	Low	High
Speed of propagation	Low (between 1200 and 1400 m/s)	High (3×10^8 m/s)

4 Routing Protocols for Underwater Sensor Network (UWSN)

Underwater sensor networks (UWSNs) have tons of routing protocols for ensuring accurate packet delivery among sensor nodes operating underwater [6–9].

1. **Vector-Based Forwarding Protocol (VBF)**

Vector-Based Forwarding Protocol (VBF) [10, 11] is a location-based routing protocol designed especially for UWSN for improvising delay and transmission rates. Apart from handling energy efficiency, the VBF routing protocol handles the mobility in a reliable way. Vector-Based Forwarding protocol is also termed as "Routing Pipe" in which path is established between sender and receiver nodes and transmission is done via routing pipe.

Protocol Operation: Every packet transmitted in the network contains diverse information like: Sensor node position, destination node position (Target), forwarder and ranging field.

On receiving a data packet, a sensor node calculates its relative position to the forwarder making use of distance to the forwarder and Angle of Arrival (AoA) of the signal. Every node operating in UWSN is integrated with specific hardware required to calculate distance and AoA of the signal. If the node determines that it is quite close enough to the routing vector as per predefined distance threshold value, it puts its own computed position in the packet and packet forwarding is performed; otherwise, the packet is discarded. In this way, all the packet forwarders in network formulate a "Routing Pipe": all the nodes in the pipe are only eligible for packet forwarding and those nodes which are not close to the routing vector are not used for any sort of packet forwarding.

Tests and Results: VBF protocol is tested via simulation and results state that the protocol is highly efficient for UWSN in terms of energy efficiency, delay and transmission rate because of utilization of routing pipe technique. VBF protocol plays an important role in congestion reduction in UWSN operations.

2. **Hop-By-Hop Vector-Based Forwarding (HH-VBF)**

Hop-By-Hop Vector-Based Forwarding (HH-VBF) [12] is another location-based routing protocol for UWSN and has functionality similar to Vector-Based Forwarding Protocol [10, 11] and this protocol also makes of Routing Pipe in a much efficient manner. Rather than using a unique pipe from source to destination node, HH-VBF creates a "Routing Pipe" for every forwarder node in the network. Via this approach, the main problems surrounding VBF protocol in terms of low data delivery in sparse networks and sensitivity to routing pipe radius are solved.

Protocol Operation: In HH-VBF, if a sensor node receives a packet either from source or forwarder, it calculates the vector from source node to destination node. Via this, the routing pipe changes each hop in the network. On receiving the packet, the receiver computes the vector from sender to sink and calculates the distance. If the distance is smaller than threshold value, it forwards the data packet and act as "Candidate Forwarder". This node maintains a Self-Adaption time.

Forwarding Mechanism: On receiving a packet, a node holds the packet for some time. When the waiting time expires, the sensor node with smallest desirableness factor forwards the packet first. HH-VBF protocol supports overhearing meaning that node calculates its distance to the different vectors from packet forwarders to the sink node. Nodes compare predefined minimum distance threshold with distance values and take the final decision to forward the packet or not.

Tests and Results: Researchers have tested HH-VBF protocol using NS-2 simulator via parameters similar to LinkQuest UWM1000 scenario, i.e., 1000 m × 1000 m × 500 m to check performance in terms of node density, node mobility, and energy efficiency. Results state that HH-VBF is efficient in determining better data delivery paths as compared to VBF in sparse networks. HH-VBF outperforms VBF in terms of mobility and energy efficiency.

3. **Depth-Based Routing Protocol for UWSN (DBR)**

Depth-Based Routing Protocol (DBT) [13] for UWSN is based on Greedy Algorithm.

Protocol Operation: In this protocol, every sensor node in the network acts individually, based on its depth and depth of the forwarding nodes, to make the final decision to forward the packet or not.

On receiving the data packet, the sensor node first gathers the information of the packet previous hop and then compares the depth of the previous hop with its own depth. If the node is highly closer to the water surface, it will forward the packet, otherwise, the packet will be dropped as the packet comes from the next best node which is highly closer to the water surface.

DBR protocol is highly efficient in terms of congestion control and energy efficiency.

Tests and Results: DBR protocol is tested on NS-2 Simulator along with Aqua-Sim (NS-2 Extension Package for Simulating UWSN) on 500 m x 500 m x 500 m 3D UWSN area. The performance of protocol is measured on parameters like: packet delivery ratio, packet delay and energy consumption.

In DBR protocol, every node is equipped with Depth Sensor to reduce the energy level during transmission. The shortcoming of DSR is Broadcasting, which in turn makes routing somewhat complex as node candidates would be increased for packet forwarding. DSR is not suitable for dynamic topology based scenarios.

4. **Hop-By-Hop Dynamic Addressing Based (H2-DAB) Routing Protocol for UWSN**

H2-DAB (Hop-By-Hop Dynamic Addressing Based) [14] routing protocol was designed primarily for improvising packet delivery ratio, latency, and energy consumption without adding any additional network hardware in sensor nodes.

Protocol Operation: H2-DAB is highly robust, scalable and energy efficient routing protocol making use of multi-sink architecture. Every sensor node in the network is assigned routing address, composed of two parts: (1) Node ID: a unique

ID for floating nodes, to locate the nodes for forwarding data packets; (2) Hop ID: used by floating nodes to receive HELLO packets.

The routing in H2-DAB works in two main phases: During the first phase, routes are created by making use of Dynamic HopID to every floating node in the network. In the next phase, the data is delivered to the nodes by using HopID.

The default HopID on the node remains unchanged until a HELLO packet is not received. After receiving the HELLO packet, the node updates the HopID and decreases the value of maximum hop count by 1. The routing process ends when HELLO packet reaches either anchored node or when hop count becomes 0. The sensor node with smallest backup link wins the overall network competition.

Tests and Results: H2-DAB protocol was tested using NS-2 simulator on 300 sensors nodes in the 3-D UWSN area of 1500 m × 1500 m × 1500 m with floating nodes forming layers at distance of 250 m from surface to bottom and communication range of sensor nodes is 500 m. The H2-DAB protocol was tested on two parameters: packet delivery ratio and packet delay. The results state that H2-DAB protocol is nearly 90% efficient in packet delivery ratio in both dense and sparse networks, in addition to this, H2-DAB is highly energy efficient routing protocol for UWSN.

5. **Focused Beam Routing Protocol for Underwater Sensor Networks (FBR)**

Focused Beam Routing Protocol (FBR) [15] is a cross-layer scalable routing protocol for UWSNs. It is highly preferred protocol for both static and mobile-based UWSN network topologies for routing without any need of clock synchronization.

Protocol Operation: FBR protocol was primarily designed to combat the issue of flooding and reducing energy at constant intervals of time in sensor nodes. The candidate forwarders are located on the basis of angle of cone which is constructed from source to destination.

The node which wants to transmit the data forwards the RTS message with first power level to the geographical area and the nodes in the corresponding area reply back with CTS message. If the source node does not receive any sort of reply, it increases the power level to next and sends a new RTS message. This procedure works iteratively until the sender nodes receive the CTS message. If the maximum power level of the node reaches by broadcasting RTS message in the area and nonreceipt of CTS message, the source node will shift the cone and searches for new candidate forwarding nodes in the left and right side of the cone.

Tests and Results: FBR protocol was tested using Discrete Event Underwater Acoustic Network Simulator on 200 km^2 area and the performance is compared with Dijkstra's shortest path finding algorithm. The results and performance state that FBR protocol is efficient as compared to Dijkstra Algorithm and can discover routes where minimum energy can be consumed with minimal knowledge of network topology. FBR protocol performs better in terms of energy efficiency and packet delay.

6. **Path Unaware Layered Routing Protocol (PULRP)**

Path Unaware Layered Routing Protocol (PULRP) [16] is an efficient routing protocol especially designed for dense well-connected 3-D UWSN.

Protocol Operation: PULRP protocol operates in two main phases: Layering Phase and Communication Phase. In the layering phase, concentric layers of spheres are created around sink node with every node belonging to any and only one of the spherical layers. The sphere's radius is determined on the basis of the probability of efficient forwarding of packets and overall latency in terms of packet delivery from source to destination.

In the communication phase, the routing path is created dynamically, from source node to destination node across different concentric layers.

Tests and Results: PULRP routing protocol is tested on varied simulation test beds primarily for two parameters: sucess rate and packet delay. The simulation is carried on 3-D UWSN network area of 100 m × 100 m × 100 m and packet generation is done at an average of 10 seconds and simulation time is 20 min. The PULRP protocol is compared with Dijkstra's Shortest Path Algorithm and Underwater Diffusion Algorithm (UWD) and the simulation results state that PULRP protocol achieves high sucess rate in terms of packet delivery and packet delay and does not require localization, time synchronization, and even do not require any sort of routing table maintenance.

7. **Adaptive Routing**

Adaptive Routing [17] was mainly proposed to accomplish multiple objectives/tasks with different application requirements in terms of packet delay, packet delivery ratio, and consumption of energy.

Protocol Operation: In UWSNs, sensor nodes are deployed randomly forming dynamic topology and give scientific data in terms of water quality back to sink node at regular intervals of time. Data packets can be categorized as Ordinary, Intermediate or Emergency packet depending on the data of water quality reporting. If the quality of the water is good and there is nothing important to report, the packets can be treated as Ordinary or Intermediate and can be delivered back to sink node without considering the delay and energy consumption. But in case of water pollution or any water current occurrence, the packet is termed as emergent and requires quick delivery to sink node with less delay and moderate energy utilization.

In the discovery of neighboring node, each node broadcasts a HELLO Packet. In order to reduce packet redundancy in the network, ACKs for successful packet delivery at sink are broadcasted via Epidemic Routing Approach. Any node in the network, when receives ACKs will delete the next packets and broadcasts the ACKs to the existing network.

In order to calculate Priority, it is done on the basis of information vector, which contains packet emergency level, the age of packet, node spatial-temporal density, and energy level of the node. The packet priority will only be calculated when a node discovers a new neighbor. After packet priority calculation, the node will take the appropriate routing decision to forward the packet to the destination node.

Tests and Results: The simulation-based results demonstrate that Adaptive Routing outshines other routing schemes like Epidemic Routing and Single-Copy Routing in terms of packet delivery ratio, packet delay, and energy efficiency in the overall network.

8. **GPS-Free Routing Protocol for Deep Water (DUCS)**

Distributed Underwater Clustering Scheme [18–20] is regarded as novel GPS-free clustering scheme especially designed for UWSNs.

Protocol Operation: DUCS is a self-organizing routing protocol in which sensor nodes in the network are divided into clusters and every cluster is designated a cluster head in which every single node is connected via single hop. The cluster heads take the data gathered by sensor nodes and perform data aggregation tasks on the data received. After data aggregation tasks, it is the duty of cluster heads to send the data back to the sink node. In order to keep the energy level optimized in the overall network, DUCS routing protocol can designate any sensor node as "Cluster Head" at regular intervals of time in dynamic fashion.

DUCS routing protocol operation is divided into specific stages: During the setup phase, clusters are created in sensor network. At steady-state phase, data transfer happens between the sensor nodes and these two phases are iterated at regular intervals of time.

Tests and Results: DUCS protocol is tested on NS-2 simulator on the basis of performance parameters like routing overhead, packet delivery ratio, and number of alive nodes per data sent. The simulation was performed with 50–200 nodes on area of 75 m × 75 m × 2000 m with random walk mobility model at speed of 1.5 m/s and simulation ran for 200 s. Simulation results state that DUCS protocol outperforms LEACH protocol in terms of routing overhead and is almost four times better as compared to LEACH protocol.

9. **A Low Propagation Delay Multipath Routing (MPR)**

Multipath Routing Protocol (MPR) [21] is highly energy efficient routing protocol proposed for UWSN in order to improvise packet delay.

Protocol Operation: MPR protocol constructs several multiple sub-paths during path formation from source node to destination node. Multiple sub-paths are formed at 2-hop distance using relay node. Multiple sub-paths lay a strong foundation of collision avoidance as receivers receive data packets from different relay nodes at different intervals of time.

When a data packet is fetched by the relay node, it checks the transmission schedule in order to take a decision whether a collision can occur or not. The relay node creates appropriate time slots in case of collision, otherwise, packets are forwarded to sink node.

Multipath Routing Protocol is advantageous in various scenarios like: (1) Less propagation delay during routing; (2) every single node in the network is required to take 2-hop information in order to stay connected in dynamically changing topology; (3) reduces packet delay from source to destination as the packet is sent via multiple paths and collision is avoided in the network.

MPR protocol operates in three main phases. During the first phase, the routing path is calculated via propagation delay so that source node needs to attain only 2-hop neighboring node information. During the second phase, intermediate candidate node is selected. During the third phase, nodes use propagation delay information to decide which node will act as Intermediate Node.

Tests and Results: MPR protocol is tested on NS-2 simulator using 3-D UWSN of region area 2000 m × 2000 m × 500 m, transmission range of 100 m, Time Slot Length 200 ms and parameters are similar to Link Quest UWM 1000 at Bit rate of 10 kbps and performance is measured on the basis of packet delay, packet delivery ratio, throughput and routing overhead. MPR protocol is compared with VBF and HH-VBF protocol. Simulation results state that MPR outshines in packet delay, packet delivery ratio, and routing overhead as compared to VBF and HH-VBF. But MPR protocol makes use of various matrix operations which put stress on nodes in terms of energy consumption which is higher as compared to VBF and HH-VBF protocols.

10. **HydroCast: Pressure Routing for UWSN (HydroCast)**

Pressure Routing Protocol for UWSN makes use of depth information in order to find routes for packet forwarding from source nodes to sink node.

Protocol Operation: It makes use of the dead end recovery technique making this protocol highly efficient along with nodes clustering. HydroCast [22] is efficient in performing localization on data without any requirement of expensive distributed localization algorithms. The way of selecting a forwarding set like that of cluster is based on nodes progress which is determined using parameters like packet delivery ratio and distance between source to destination. The process of forwarding is performed along with maximum progress node and side by side cluster is chosen.

Tests and Results: HydroCast protocol is tested on QualNet Simulator using 100–450 nodes on UWSN area of 1000 m × 1000 m × 1000 m, simulation ran for 3600 seconds and is tested on various parameters like packet delivery ratio, packet delay, and routing overhead. The protocol is compared with DBR Protocol and simulation results state that HydroCast is highly efficient in terms of packet delivery and delay because it makes use of adaptive timer setting at each hop.

11. **Multipath Power Control Transmission (MPT)**

Multipath Power Control (MPT) [23] routing protocol was designed for UWSN for improvising packet delay and enhancing energy efficiency in sensor nodes. MPT is novel and intelligent routing protocol that combines efficient power utilization with multipath routing and packet combination is performed at the destination node.

Protocol Operation: MPT protocol operates in three phases:

In the initial phase, the source node starts the initialization of the route request packet. All the intermediate nodes transmit route request packet and route reply packet is replied by destination node. As the route request packet is broadcasted, the source node can receive multiple replies from multiple paths with varied hop counts. Now, it is the duty of the source node to select the optimum path among available paths considering the energy level of the nodes. The energy level determination is done on the basis of information collected during path establishment.

The packets are then transmitted via the selected path. The destination node checks the packet integrity on receipt of packet and verifies for any error if there. If no error is reported, the packet automatically gets routed to application layer otherwise, packet buffering is performed.

Tests and Results: MPT protocol is tested on NS-2 simulator using 3-D UWSN network comprising of 512 nodes on area of 4000 m × 4000 m × 2000 m and the data rate is 10 kbps. Packet generation is done every 10 seconds and packet size is 200 bytes. Simulation is performed for 10000 seconds and repeated almost 100 times and performance is measured on basis of energy consumption and average packet delay. Simulation results state that MPT performs well in packet delay and maintains overall energy efficiency in sensor network.

12. **Minimum Cost Clustering Protocol (MCCP)**

Minimum Cost Clustering Routing Protocol [24] is cluster-based routing protocol for UWSNs and is formed on the basis of cost metric considering three significant parameters: (1) overall energy utilization of nodes being part of cluster for data transmission to cluster head; (2) Residual energy of cluster head along with nodes being member of the cluster; (3) Location between sink node and cluster head in network.

Protocol Operation: In MCCP protocol, the clusters are created in a distributed manner. During initialization of network, every node in the network is treated as cluster head and automatically becomes part of a cluster. The cluster head node makes neighboring nodes part of the cluster. After the cluster is created, the cost is determined on the basis of above mentioned three parameters and the cost is broadcasted to all nodes operating at 2-hops. In case the sending node has better cluster incurring cost, the receiving node will extract head ID from packet and transmits back JOIN message. But if the cost is not good, then the node broadcasts INVITE message. In this manner, cluster head nodes as well as member nodes of cluster are created.

In order to balance traffic load and energy level of nodes, MCCP protocol performs re-clustering to reselect cluster heads and cluster members at regular intervals of time.

Tests and Results: MCCP protocol was tested on NS-2 simulator [25] on 100 nodes in a UWSN region of 100 m × 100 m using dynamic topology with initial battery power of 2 J of nodes. The MCCP protocol's performance is measured on the basis of clustering performance, packet delivery ratio, and overall energy efficiency and compared with HEED protocol. Simulation results state that MCCP protocol outperforms HEED in every parameter and network performance is enhanced when MCCP protocol is utilized.

13. **Information Carrying Based Routing Protocol (ICRP)**

Information Carrying Based Routing Protocol (ICRP) [26] is highly reactive and nonlocalized routing protocol designed especially for UWSN in order to enhance scalable routing and energy efficiency in the network. It comprises of two parts: Information Carrying Mechanism, in this data packet carries the control packets which leads to the development of routing path from source to destination. The delay of control packets is eliminated, which overall enhances packet delay of the protocol.

Protocol Operation: ICRP protocol operates in three phases: (1) Discovery of Route; (2) Maintenance of Route; (3) Route Retraction.

The route discovery process is started by the source node. When a source node wants to transmit any packet, it broadcasts route discovery packet in the network. On receiving route discovery packet, other nodes in the network transmits the same packet and records the reverse path. When the sink node receives the packet, the sink node also records the reverse path. On receiving the data packet, acknowledgement message is sent by sink node to source node with the same reverse path, otherwise, it is transmitted through reverse route.

During the second phase, a reverse route is created. Each destination node will use one path, and every path has a time property which denotes the time that route path is not used for transmission and is known as route lifetime. The longer the route lifetime, the longer the time the path is not utilized. When the value of lifetime exceeds the threshold value, the path is termed as "INVALID" and rediscovery starts.

In phase third, when the lifetime of route in routing table exceeds the threshold value, the route becomes invalid and gets canceled. If any data packets are still pending to be transmitted to the destination, new routing path is discovered and routing table gets updated.

Tests and Results: ICRP protocol is simulated using 100 sensors in 3D-UWSN region of 100 m x 100 m x 100 m with communication speed of 600 bps, bandwidth at 5 kbps and initial energy of sensor nodes is set to 1000 J. The protocol is tested on two parameters: energy consumption and packet delay. Simulation results state that ICRP protocol is highly efficient in maintaining energy efficiency and packey delay in UWSN.

5 Performance Comparison of Routing Protocol for Underwater Sensor Network (UWSN)

See Table 2.

Table 2 Enlists the Performance Comparison of UWSN routing protocols

Protocol name	Parameters				
	Routing technique	Packet delivery ratio	Energy efficiency	End-to-end delay	Localization requirement
VBF	Location-based routing	Low	High	High	Yes
HH-VBF	Vector-based flooding	High	Medium	Medium	Yes
DBR	Depth-based flooding	High	Medium	High	Partially
H2-DAB	Addressing-based flooding	Medium	High	High	No
FBR	Vector-based flooding	Medium	Medium	High	Yes
PULRP	Layered based	High	High	High	No
Adaptive routing	Priority-based	Flexible	Flexible	Flexible	Yes
DUCS	Distributed clustering-based	Medium	High	High	No
MPR	Multipath	High	High	High	No
HydroCast	Source-based clustering	High	High	High	No
MPT	Path-based	Medium	Medium	High	No
MCCP	Distributed clustering-based	Low	High	High	Yes
ICRP	Path-based	Medium	Low	High	No

6 Conclusion and Future Scope

Efficient Routing in UWSN is a new area of research, having lots of potential to propose novel routing protocols in this area as limited set of research results are proposed till date. The routing protocols in every network have common objectives with regard to optimal energy efficiency, best packet delivery ratio, less packet delay, low routing overhead with no congestion and best latency. In this research paper, routing protocols for UWSN are presented. The performance comparison of protocols is also discussed in terms of various network parameters. Although most of the routing protocols are doing exceptionally well in terms of performance of

UWSN, still lots of challenges need to be solved, for example, efficient utilization of resources, scalability, stability in dynamic topology, security and many more.

Future Scope

Considering the challenges which need to be addressed for efficient working of UWSN, the goal would be to propose a robust, highly scalable, energy efficient, and above all secure routing protocol for underwater sensors with intelligent localization schemes to operate exceptionally well in harsh underwater conditions of all sorts.

References

1. Akyildiz, I. F., Pompili, D., & Melodia, T. (2005). Underwater acoustic sensor networks: Research challenges. *Ad Hoc Networks, 3*(3), 257–279.
2. Heidemann, J., Stojanovic, M., & Zorzi, M. (2012). Underwater sensor networks: Applications, advances and challenges. *Philosophical Transactions of the Royal Society A, 370*(1958), 158–175.
3. Lanbo, L., Shengli, Z., & Jun-Hong, C. (2008). Prospects and problems of wireless communication for underwater sensor networks. *Wireless Communications and Mobile Computing, 8*(8), 977–994.
4. Yingzhuang, L. (2007). Underwater sensor networks. Ship. *Electronic Engineering, 6,* 049.
5. Domingo, M. C., & Prior, R. (2008). Energy analysis of routing protocols for underwater wireless sensor networks. *Computer Communications, 31*(6), 1227–1238.
6. Li, N., Martínez, J. F., Meneses Chaus, J. M., & Eckert, M. (2016). A survey on underwater acoustic sensor network routing protocols. *Sensors, 16*(3), 414.
7. Sharma, A., & Gaffar, A. H. (2012) A survey on routing protocols for underwater sensor networks. *International Journal of Computer Science & Communication Networks, 2*(1), 74–82.
8. Ayaz, M., Baig, I., Abdullah, A., & Faye, I. (2011). A survey on routing techniques in underwater wireless sensor networks. *Journal of Network and Computer Applications, 34*(6), 1908–1927.
9. Gomathi, R. M., Manickam, J. M. L., & Sivasangari, A. (2016). A comparative study on routing strategies for underwater acoustic wireless sensor network. *Contemporary Engineering Sciences, 9*(2), 71–80.
10. Xie, P., Cui, J. H., & Lao, L. (2006, May). VBF: Vector-based forwarding protocol for underwater sensor networks. In *Networking* (Vol. 3976, pp. 1216–1221).
11. Su, C., Liu, X., & Shang, F. (2010, November). Vector-based low-delay forwarding protocol for underwater wireless sensor networks. In *2010 International Conference on Multimedia Information Networking and Security (MINES)* (pp. 178–181). New York: IEEE.
12. Nicolaou, N., See, A., Xie, P., Cui, J. H., & Maggiorini, D. (2007, June). Improving the robustness of location-based routing for underwater sensor networks. In *Oceans 2007-Europe* (pp. 1–6). New York: IEEE.
13. Yan, H., Shi, Z., & Cui, J. H. (2008). DBR: Depth-based routing for underwater sensor networks. In *NETWORKING 2008 Ad Hoc and Sensor Networks, Wireless Networks, Next Generation Internet* (pp. 72–86).
14. Ayaz, M., & Abdullah, A. (2009, December). Hop-by-hop dynamic addressing based (H2-DAB) routing protocol for underwater wireless sensor networks. In *ICIMT'09. International Conference on Information and Multimedia Technology, 2009* (pp. 436–441). New York: IEEE.

15. Jornet, J. M., Stojanovic, M., & Zorzi, M. (2008, September). Focused beam routing protocol for underwater acoustic networks. In *Proceedings of the third ACM international workshop on Underwater Networks* (pp. 75–82). New York: ACM.
16. Gopi, S., Kannan, G., Chander, D., Desai, U. B., & Merchant, S. N. (2008, May). Pulrp: Path unaware layered routing protocol for underwater sensor networks. In *IEEE International Conference on Communications, 2008. ICC'08* (pp. 3141–3145). New York: IEEE.
17. Guo, Z., Peng, Z., Wang, B., Cui, J. H., & Wu, J. (2011, August). Adaptive routing in underwater delay tolerant sensor networks. In *2011 6th International ICST Conference on Communications and Networking in China (CHINACOM)* (pp. 1044–1051). New York: IEEE.
18. Domingo, M. C., & Prior, R. (2007, October). Design and analysis of a GPS-free routing protocol for underwater wireless sensor networks in deep water. In *International Conference on Sensor Technologies and Applications, 2007. SensorComm 2007* (pp. 215–220). New York: IEEE.
19. Domingo, M. C. (2011). Securing underwater wireless communication networks. *IEEE Wireless Communications, 18*(1).
20. Gkikopouli, A., Nikolakopoulos, G., & Manesis, S. (2012, July). A survey on underwater wireless sensor networks and applications. In 2012 *20th Mediterranean Conference on Control & Automation (MED)* (pp. 1147–1154). New York: IEEE.
21. Chen, Y. S., Juang, T. Y., Lin, Y. W., & Tsai, I. C. (2010). A low propagation delay multi-path routing protocol for underwater sensor networks. 網際網路技術學刊, *11*(2), 153–165.
22. Noh, Y., Lee, U., Lee, S., Wang, P., Vieira, L., Cui, J., et al. (2014). Pressure routing for underwater sensor networks. *IEEE Transactions on Mobile Computing, 12*(16), 1–9.
23. Zhou, Z., & Cui, J. H. (2008, May). Energy efficient multi-path communication for time-critical applications in underwater sensor networks. In *Proceedings of the 9th ACM International Symposium on Mobile Ad Hoc Networking and Computing* (pp. 221–230). New York: ACM.
24. Wang, P., Li, C., & Zheng, J. (2007, June). Distributed minimum-cost clustering protocol for underwater sensor networks (UWSNs). In *IEEE International Conference on Communications, 2007. ICC'07* (pp. 3510–3515). New York: IEEE.
25. Nayyar, A., & Singh, R. (2015). A comprehensive review of simulation tools for wireless sensor networks (WSNs). *Journal of Wireless Networking and Communications, 5*(1), 19–47.
26. Liang, W., Yu, H., Liu, L., Li, B., & Che, C. (2007, August). Information-carrying based routing protocol for underwater acoustic sensor network. In *International Conference on Mechatronics and Automation, 2007. ICMA 2007* (pp. 729–734). New York: IEEE.

Network Intrusion Detection in an Enterprise: Unsupervised Analytical Methodology

Garima Makkar, Malini Jayaraman and Sonam Sharma

Abstract Be it an individual, or an organization or any government institution, cyber-attack has no boundaries. Cyber-attacks in the form of Malware, Phishing and Intrusion into an enterprise network have become more prevalent these days. With advancement in technology, the number of connected devices has increased vastly leading to storage of very sensitive data belonging to different entities. Cybercriminals attempt to access this data as it is very lucrative for them to monetize this information. Due to the sophistication in technology used by cybercriminals, these attacks have become more difficult to detect and handle, making it a major challenge for governments and various enterprises to protect their sensitive data. Traditional detection methods such as antivirus and firewalls are limited only to known attacks, i.e., the attacks which have occurred in the past. Nowadays the growing advancement in the field of technology has led to unique and different types of attacks for which the traditional detection methods fail. In this paper, we will propose our methodology of Intrusion detection which will be able to handle such threats in near real time.

Keywords Anomaly detection · Cyber security · Intrusion detection
K-means

G. Makkar (✉) · M. Jayaraman · S. Sharma
Tata Consultancy Services, Bengaluru, Karnataka, India
e-mail: garima.makkar@tcs.com

M. Jayaraman
e-mail: malini.jayaraman@tcs.com

S. Sharma
e-mail: sonam.sharma2@tcs.com

V. E. Balas et al. (eds.), *Data Management, Analytics and Innovation*,
Advances in Intelligent Systems and Computing 808,
https://doi.org/10.1007/978-981-13-1402-5_34

1 Introduction

With Internet and connected devices becoming a necessity in life, the amount of personal information collected and stored by the systems have increased drastically. As digital footprints increase, sensitive details of the customers such as bank details and passwords are being collected and stored by the companies for making financial transactions easier. Enterprise also has other company-related sensitive information stored in their network. All these sensitive information collected and stored by an enterprise attracts a lot of hackers. The ability to monetize these sensitive information have tempted cybercriminals to attack more frequently and more sophisticatedly. According to 2017 Data Investigation Report provided by Verizon, there have been 42,068 attempts of comprising confidentiality and integrity of accounts in the surveyed companies, out of which 1935 attempts were successful in accessing internal data. These figures show the seriousness and scale of these cyber threats. As the scale of the network activity increases, it becomes very difficult to manually check and identify attacks in an enterprise. There are various security mechanisms available these days to secure computer systems against duplication, destruction, unauthorized use, virus attacks, and alterations. But no perfect resolution to arrest these attacks exists. Due to the high rate of network activity, the time taken to detect and contain a data breach is very high. With higher time to contain, the cost of data breach also increases. Hence it has become critical to identify the attacks at the very first time of occurrences and act accordingly to constraint any possible damage. Cyber security has now shifted from being an additional protection to a mandatory requirement. All these have led to a growing importance of cyber security in both government and private organizations.

Cyber threats have many forms such as Cyber-espionage, Insider and privilege misuse, Physical theft and loss, etc. Out of these, "Intrusion" is becoming more common method of attack. In this type of attack, a cybercriminal enters a network and gains access to internal information. Within a network, one way of detecting malicious activities is by using an Intrusion detection system (IDS). An IDS is a software application that examines a system or network for any malicious activity or violation in the policy of an enterprise. The basic job of this system is to inspect the log file of the host and network to identify suspicious activity. Typically, the techniques in Intrusion Detection can be categorized into Signature/Misuse Detection technique and Anomaly Detection. In Signature detection technique, the network/host activity is compared to known attack pattern to identify attacks. This method, however, cannot be used to detect unknown attacks. The anomaly detection method detects unusual patterns based on its understanding of the data and triggers an alert. With the advancement of technology and the enormous growth in the network traffic, it has been noticed that both these method perform sub-optimally. With sophistication in the cyber-attacks, the patterns of intrusion vary for every attack making signature matching method inefficient. As the network traffic increases, there are many subgroups of system having patterns different from a majority, leading to a very high false positive in anomaly detection method.

This has led to an active research of using machine learning techniques (both supervised and unsupervised) to distinguish between normal instances and attacks.

This paper is divided into following sections: Sect. 2 discusses about literature review and critical research gaps highlighted in this paper. In Sect. 3, the problem statement for our analysis is defined. The proposed methodology is stated in Sect. 4. The results and conclusions are explained in Sects. 5 and 6.

2 Literature Review

In this section a detailed study about various techniques detecting intrusions is provided. The two main approaches to find intrusions are rule based expert system and statistical approach. The rule based expert system helps to detect known attacks in high rate. Ratha et al. [1], detected attacks pertaining to a biometric process and clustered them into six groups. But the problem with this system lies in the flexibility of the rules, that is, a little variation in the sequence of attack can make an instance from being intrusive to non-intrusive. In the statistical approach of intrusion detection, supervised and unsupervised techniques are being explored. Few research papers related to these approaches for intrusion detection have been discussed below.

2.1 Supervised Algorithms on Intrusion Detection

Various researchers have worked on various different techniques to give prominent output for IDS. Intrusion Detection using SVM has been explored by many. For example, Goyal and Kumar [2], have implemented genetic algorithm for identifying harmful attacks in the network. The algorithm focuses on features like connection status, network service to destination and protocol type to generate a set of rules identifying an attack type. This approach creates different set of rules identifying types of attacks. Their approach proved to be efficient with 100% accuracy for the classification of intrusions and detections. Khan et al. [3], used DGSOT (i.e., Dynamically Growing Self-Organizing Tree) as it is considered as an improvement over traditional clustering algorithms like, hierarchical partition and agglomerative clustering. Between two classes, analysis on clustering discovered the boundary points, which were then used to coach support vector machine (SVM). For gain in training time and loss in precision, they compare their approach to Rocchio Bundling algorithm and casual choice. In the same year, Ganapathy et al. [4] have implemented a multi-layer classification technique for anomaly detection in cell phone ad hoc networks. Their system also used an amalgamation of a tree classifier and multi-class Support Vector machine algorithm. Also due to fast adaptation and high detection rate, the Artificial Neural Networks are considered as one of the famous supervised machine learning techniques for detecting Intrusions in real time. Muna et al. [5]; Norouzian et al. [6] etc. are some of the IDS work being done using ANN technique.

2.2 Unsupervised Algorithms on Intrusion Detection

The classification techniques require labelled data for training which is quite difficult to generate. While dealing with big network data, it becomes difficult to flag or classify each record manually. This demands an approach that classifies the instances when dataset is not labelled. Unsupervised learning algorithms like Clustering have gained lot of importance in this respect. In the case of unlabeled data set, unsupervised anomaly detection algorithms are used to find the anomalies assuming that maximum records in the data are normal and hence do not involve training data. These techniques are based on the implicit belief that normal instances are more recurrent than aberrations in the test data. If this presumption fails then such techniques lead to high false alarm rate. Such divergence assumes that the testing data includes less anomalies and the development model is vigorous to these less anomalies.

Most of the work on intrusion detection based on unsupervised techniques is done using Clustering. Portnoy et al. [7], implemented unsupervised technique to detect attacks present in unlabeled data. They used incremental K-means clustering algorithm to group the entire log data containing both normal and abnormal instances. And based on the number of instances each cluster is then labeled accordingly. It was seen that clusters with small number of instances came out as attacks and then these clusters are used to spot intrusions in the test data. Suseela et al. [8], have presented a paper on hierarchical Kohonen Net (K-Map) for detecting intrusions. This approach does not involve costly point-to-point computation in forming the clusters. Another advantage of this approach was the reduced network size. And for implementation of K-map, the Subsets were selected aimlessly that included both normal and attack records from the KDD-Cup'99 dataset. Another technique for enhancing IDSs, called Biclustering, was presented by Lappas and Pelechrinis [9]. Their approach proved to be useful in extracting important knowledge about the association between features and processes. Though this method has its own advantages, it cannot be used to detect abnormal intrusions. Also, Qing et al. [10] developed a system related to data mining for finding out the intrusions in their dataset. Considering that the traffic is huge and there are large number of attacks present they decided to choose Fuzzy C-means technique and hence were able to improve the performance in terms of both accuracy and detection rate.

2.3 Hybrid Techniques on Intrusion Detection

In general, the supervised techniques give better accuracy when the attacks are known. While the presence of unknown attacks in test data deteriorates its accuracy significantly. And if we look at unsupervised algorithms, then the presence of known or unknown attacks hardly make any difference in their performance. This resulted some of the researchers to use a combination of both the techniques.

For example, a research related to ANN was done by Zhi-song pan et al. [11], who implemented a hybrid neural network and decision tree algorithm for detecting known attacks in an organization. They tested classification abilities of C4.5 algorithm and neural networks for intrusion detection and then concluded which algorithm is better to detect a particular class of an attack. Similarly, Shekhar [12], distinguished between normal and abnormal instances in a computer network based on a hybrid of ID3 decision tree and k-means grouping methods. Some authors like Wang et al. [13], Gaikwad et al. [14], etc., have defined IDS based on a hybrid of Artificial Neural Network (ANN) and fuzzy clustering. Their method proved to be useful for overcoming the problem of low precision detection and weak stability detection. Primarily many researcher rely on conventional methods like K-nearest neighbor, Self-Organizing Map; Neural Network, Naïve Bayes, etc. Some studies are Bahrololum et al. [15], Muda et al. [16], Om et al. [17] where they have leverage these methodologies for intrusion detection, detection rate, accuracy, false alarm rate normal, and abnormal instances.

Thus all in all it can be stated that the main objective is to find out the system with best accuracy utilizing the different methods for finding attacks in the dataset. All the papers that have been presented in respect to IDS are based on various different techniques with few implementing too many techniques. While our analysis is based on just one technique where feature extraction is done in a unique way so as to get better output as compared to others.

3 Problem Statement

The objective of this paper is to propose a methodology to detect an intrusion in an enterprise network by classifying whether the host is affected or not. We also try to identify the abnormal activities of the detected hosts to understand the attacks. For this study, we have taken unflagged datasets of a network. Only the activities of a user after the connection is made are captured. Under such circumstances it becomes essential to understand each and every aspect of the data.

The current techniques in intrusion detection classify the intrusion as either host or instance based. The aim of Host-based Intrusion detection system is to collect and analyze the information about the activities going on a specific single system, also called as host. Basically, these systems can spot the suspicious activities related to a particular User Agent. The methodology that is being applied in this paper does not detect host-based intrusions. The proposed methodology implements an IP based intrusion detection. Important features from the dataset are extracted which are then used to carry out the "IP" based intrusion detection. As the quantity of available traffic data is generally large, manual labelling of each record is both time consuming and ineffective. Hence, we propose a system of grouping similar instances and labelling the groups. Going by this methodology, we apply K-means clustering on the logs generated to determine whether a new record is intrusive or not. Also this will give us the information about abnormal activity that took place in an enterprise during the time period for which the log data was collected.

4 Analytical Methodology to Detect Intrusion

How does an individual or an enterprise or an organization decide how to use and manage their data? What is the best possible way to maintain the balance between sharing of data and its privacy on computer networks? Since the beginning, Data is considered as a form of valuable asset making organizations naturally concerned about moving it. The future of the cyber security for most traditional enterprises is still unclear since most of these traditional organizations are still wondering the best strategy to secure their personal information and infrastructure.

Keeping this in mind, we now explain the procedure that is followed for our experimentation. The first subsection gives the detailed description about the data used in the analysis, followed by feature extraction and experiment analysis.

4.1 Data

For our analysis, we have used data from MACCDC [18]. This data set has been captured by Bro software. Bro is an open-source Unix which at times is compared to Network based Intrusion Detection System, i.e., collecting information from the entire network itself as data travels across the network system. This software helps in capturing different kinds of log files of an enterprise. For our purpose, we are considering only three log files namely, Connection, HTTP and DNS. Below is the brief description of each of them:-

a. ***CONN.LOG***: This log summarizes details of each UDP and TCP connection in a single line containing around two million records. And due to the presence of such detailed information, conn_log can be used to extract plenty of useful statistics. For example: The "duration" column tells for how long a connection is lasted. Another column called "resp_bytes" records how many bytes are sent by an IP address to a particular client. Thus, understanding each column can help us to find which columns are important from our analysis point of view.
b. ***HTTP LOG***: This file contains all HTTP requests and responses summary sent over a network. Each record starts with a ts (timestamp), a UID (Unique Connection Identifier) and a 4-Tuple connection (i.e., Originator Host/port and respondent Host/port). And the remaining columns explain the activity that is occurring for each particular UID. For example, the "host" field records the list of hosts an IP is pinging to. Similarly to check if the status of HTTP request made is Ok or denied, we can refer to "status_code" field in this log.
c. ***DNS LOG***: Containing 366,170 records, this log records all the DNS queries along with their responses. Using information present in this log we can find about the type of queries made by a client, for example, whether that query uses TCP or whether it is referring to some spoofed source address, etc. All such questions can be answered using the content present in the DNS log.

The entire dataset contains information about each connection made and its activities in the interval of 8 hours in an organization. And with the number of records we can easily infer the amount of network traffic. These logs depict the activities of every host in an enterprise transparently. For every connection, an entry is made in the logs and then activities for that particular connection is recorded. But whether a connection to destination will be made or not depends upon what that particular host has asked for. For example, the host may ask for a destination that is listed in the list of malicious domains or a destination that does not have a good reputation, etc. And all such type of necessary information about both users and hosts is being contained in these logs.

4.2 Feature Extraction

The second step we did is to extract features from all the above mentioned logs so as to distinguish outbound communications from the organization. The selection of feature is purely done on the basis of known intrusion behavior and in accordance with the algorithm used in our experiment. For each host in the network, a feature vector is generated which includes 11 features listed in Table 1. These features can be categorized into four groups: features describing company's policy, features based on activities of the host, features related to volume of network traffic, and destination based features. We describe this extraction from each of the log files in detail below.

Conn.log file is considered as a backbone of the dataset since it is a mass of merged data types used to find the state of a connection throughout its lifetime. The first feature that we extracted denotes the number of times a particular connection is made. Using this we can check the system activities accordingly. Malware can cause sudden spikes in the volume of host's network traffic and such interesting

Table 1 Clustering features (Modified and appended from Yen et al. [19])

Types of features	Explanation
Features describing company's policy	Blocked connections
	Third-level connections
	Blocked domains
	Unpopular IP's
Features based on host	"New" user agent
Features related to volume of traffic	Number of logins
	Sudden connection spikes
	Domain spikes
Destination based features	Unknown destinations
	New external domains
	Destinations contacted without HTTP referrer

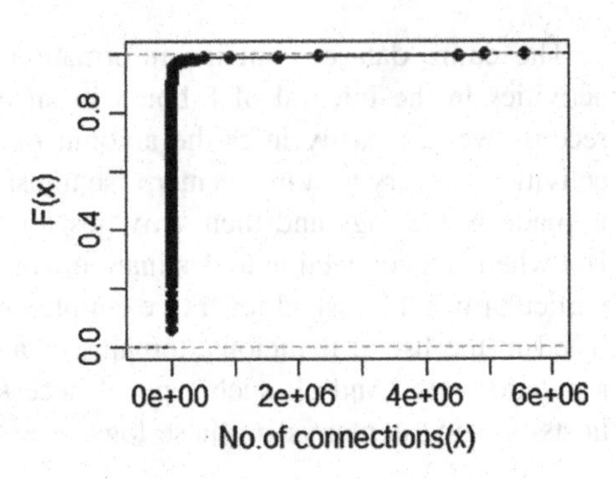

Fig. 1 Cumulative distribution function for number of connections in conn_log

activities can easily be captured using amount of duration when an IP is generating large volume of traffic abnormally. To find out the suitable threshold for large volumes of traffic, we count the duration for which each connection lasted. Figure 1 shows cumulative distribution frequency for the number of connections made.

HTTP contains major information about host activities and hence is considered as one of the most important log. In an enterprise, hosts are more uniform in their software arrangements as compared to their academic networks. Thus the cases in which user agents install new software is of interest to us. Though there is lack of visibility about host's machine, these logs generated by Bro captures a variable named "User_agent" which can be used to find the host's software configurations. A "User_Agent" string contains the type of application, software version, and the operating system. Count of new user agent strings from the host can be an important feature for identifying attack. An important thing to note is that in our dataset there are many user_agents which are closely related to each other, for example: "Mozilla/5.0 (X11; Linux i686; rv: 2.0.1) Gecko/20100101 Firefox/4.0.1", and "Mozilla/5.0 (X11; Linux i686; rv:10.0.2) Gecko/20100101 Firefox/10.0.2". In such cases, taking count of user agents would be misleading as count function will consider these two as separate entity. To avoid this, levenshtein distance is used as the counting function. Levenshtein distance measures the similarity between the source string and the target string. Using levenshtein distance, strings that are closely related to each other will be grouped into one category. A user agent is considered new entity only if the levenshtein distance is greater than a threshold. And this is how new user_agent strings can be assimilated and only the count of different "user_agent" will be calculated.

We are keen to identify the hosts which communicate with new or unknown destinations that are rarely or never contacted within the organization. Here, we assume that popular destinations (or websites) have less probability of being compromised as compared to obscure destinations which may be an indicator of abnormal (or suspicious) behavior.

Each enterprise has its own unique network policies and whenever some IP pings another IP, the success of these connections will depend upon these policies. For example, a connection to some external destination can be denied if it has a low reputation or is forbidden for employees. The blocked connections (or domains) are hence an important indicator of host misconduct. Also for the sites that has not been categorized yet, the client must agree to the enterprise's policies before proceeding further. The connections which require this type of acknowledgement are referred as *third-level connections*. This information is extracted by counting the number of connections that are blocked or third-level.

Another important feature is to find out the "new external" domains contacted by a host. Firstly, we record the history of all external destinations contacted by each host over a period of time. And after that a connection is taken as new if it has not been contacted by any host during our period of observation. We are also interested to find the count of new domains contacted by hosts without a HTTP referrer. This usually happens when a user contacts new sites or is directed to advertisements by search engine. And hosts visiting new sites without referrer are regarded more suspicious. Also one of the indications of suspicious activity is when a host contacts unpopular IPs. Though it is normal when communication is occasional but is considered as suspicious when frequent. In the next section, we will discuss how these features are combined so as to find anomalies in our dataset.

4.3 K-Means Clustering as an Unsupervised Intrusion Detection Approach

With limited truth about which hosts are anonymous and behaving suspiciously, we solve the problem of detecting intrusions using an unsupervised learning technique called Clustering. Many employees are there in each department of an organization who are performing particular job functions so we would be able to find out the groups of users performing alike behaviors, while infected hosts with distinct behaviors as "Outliers".

Clustering is a technique of grouping abstract objects into buckets of similar objects. In the present analysis this technique is applied so as to find patterns in a dataset. A good clustering technique results into low inter-cluster similarity and high intra-cluster similarity, i.e., the objects in the same cluster are more alike to each other as compared to those in other clusters.

4.4 Intrusion Detection with K-Means Clustering

In the present experiment K-means clustering technique has been applied to differentiate between normal incidents and attacks. The K-means is an iterative clustering technique aiming to find out the local maxima in each iteration. This is one of the most widely used algorithm for the purpose of clustering because of the belief that it is easy to implement. The principal reason behind implementing K-means is to differentiate between normal and abnormal data points which behave in a similar manner into separate partitions known as K-th cluster centroids. The following steps explain how k-means algorithm works in detecting intrusions:

i. Mention the "K" desired number of clusters C-1, C-2,..., C-K.
ii. Randomly assign each point in the data set to a cluster.
iii. Compute centroids for each cluster.
iv. After computing centroids, reassign the data points to the closest cluster centroid.
v. Compute centroids again for each of these clusters.
vi. Repeat the last two steps until no further improvement is possible, i.e. no more switching of the data points between the clusters is occurring.

In following section we have leveraged aforesaid methodology to showcase some intrusion detection scenarios. There is high potential to extend this initiative to many more what if scenarios, intrusions type and technical approaches.

5 Results and Experimentation

Given the description of feature extraction in Sect. 4.2, each host is represented by a multi-dimensional vector explaining these features and further the analysis of detecting intrusions has been performed on this dataset. The parameters used for the execution of k-means clustering algorithm are discussed below:-

i. ***Distance Function***: To create clusters, Euclidean distance is used for measuring inter and intra-cluster distances. For n-dimensional space, the Euclidean distance can be calculated as

$$d(p,q) = \sqrt{\sum_{i=1}^{n} (p_i - q_i)^2}$$

where d is the distance between vectors p and q.

ii. ***Number of clusters***: K-means algorithm requires the number of clusters to create be defined in advance. To extract optimal number of clusters, we plot total "within groups sum of squares (WSS)" with the number of clusters. And a bend in the plot gives us the optimal number of clusters to be specified in the algorithm. The following figure shows the WSS for our dataset.

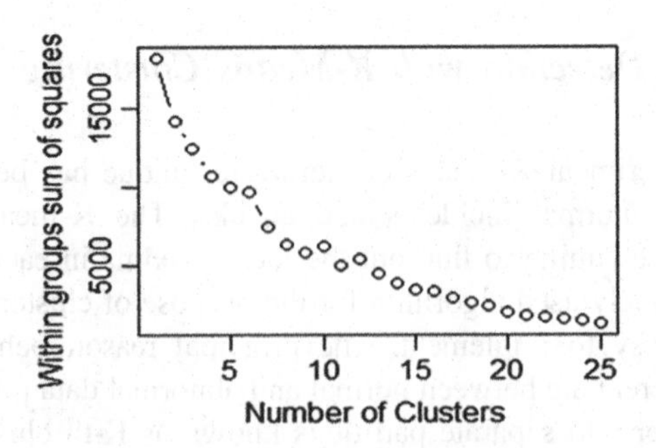

Fig. 2 WSS plot (varying number of clusters to classify intrusion correctly)

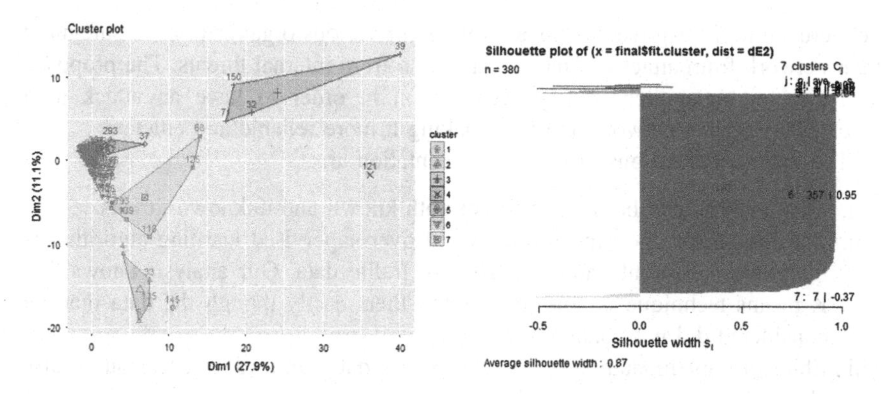

Fig. 3 Representation of clusters

As discussed above, this paper aims to look for the IP based intrusions and for which we applied K-means algorithm. The most important requirement for its application is to determine the number of clusters, which we have determined using WSS plot shown in Fig. 2. From this plot we chose $k = 7$ for our analysis purpose. K-Means algorithm was applied with 7 as the cluster count.

Once the clusters were obtained, the cluster composition was studied to identify the fit. Both the cluster plot and silhouette plot in Fig. 3 indicates the cluster fit has been good for this dataset for $k = 7$.

Classification of Clusters: The important job is to classify extracted clusters into classes of normal and intrusive. For that we assume that majority of records in the data set indicates the clusters with most number of instances as normal while remaining as attacks. And the remaining are being labelled as anomalous clusters.

After getting clusters, it was seen that 90% of the instances lied in cluster 6 while the remaining 10% in other six clusters. Also out of these six clusters, three contained only 1% of the instances. Going by our assumption of clusters with majority of the instances as normal, we can now easily classify Cluster no. 6 as normal while all other clusters as Intrusive.

On checking the composition of the clusters, it was identified that factors like the number of times a connection is made by a particular IP, the sum of bytes sent, number of without referrer connections made, etc., were very high compared to cluster 6. Thus, by using this approach, we were able to identify Hosts having anomalous activities.

6 Conclusion

The practicality of intrusion detection using unsupervised machine learning algorithms is being investigated in this paper. The Intrusion Detection System is considered as one of the powerful defense technology these days. Since it has proved to be a crucial tool for governments as well as for various organizations to safeguard their network from attacks of hackers and also from internal threats. The proposed analysis can be applied in any organization in order to have an attack alert mechanism against network and thus making it more reliable and efficient.

The major implications of this experimentation are:-

i. This analysis can be used to detect both known and unknown threats.
ii. The reason for choosing unsupervised over supervised learning algorithm is the extreme vibrant nature of network traffic data. Our analysis shows that K-means technique is able to detect outliers easily though the data that we considered did not contain any flags.
iii. Efficient implementation of K-means in big data can lead to reduction in false alarm rate for unknown attacks.
iv. Most of the work done in past is instance based. But this work works well for both instance based as well as IP-based intrusion detection.

References

1. Ratha, N. K., Connell, J. H., & Bolle, R. M. (2001). Enhancing security and privacy in biometrics-based authentication systems. *IBM Systems Journal, 40*(3), 614–634.
2. Goyal, A., & Kumar, C. (2008). GA-NIDS: A genetic algorithm based network intrusion detection system. Northwestern University.
3. Khan, L., Awad, M., & Thuraisingham, B. (2007). A new intrusion detection system using support vector machines and hierarchical clustering. *The VLDB Journal—The International Journal on Very Large Data Bases, 16*(4), 507–521.
4. Ganapathy, S., Yogesh, P., & Kannan, A. (2011). An intelligent intrusion detection system for mobile ad-hoc networks using classification techniques. *Communications in Computer and Information Science, 148,* 117–122.
5. Jawhar, M. M. T., & Mehrotra, M. (2010). Anomaly intrusion detection system using hamming network approach. *International Journal of Computer Science & Communication, 1* (1), 165–169.
6. Norouzian, M. R., & Merati, S. (2011). Classifying attacks in a network intrusion detection system based on artificial neural networks. In *2011 13th International Conference on Advanced Communication Technology (ICACT)*. IEEE.
7. Portnoy, L., Eskin, E, & Stolfo, S. (2001). Intrusion detection with unlabeled data using clustering. In *Proceedings of ACM CSS Workshop on Data Mining Applied to Security (DMSA-2001.*
8. Sarasamma, S. T., Zhu, Q. A., & Huff, J. (2005). Hierarchical Kohonenen net for anomaly detection in network security. *IEEE Transactions on Systems, Man, and Cybernetics, Part B (Cybernetics), 35*(2), 302–312.
9. Lappas, T., & Pelechrinis, K. (2007). Data mining techniques for (network) intrusion detection systems. *Department of Computer Science and Engineering UC Riverside, Riverside CA, 92521.*
10. Qing, Y., Xiaoping, W., & Gaofeng, H. (2010). An intrusion detection approach based on data mining. In *2010 2nd International Conference on Future Computer and Communication (ICFCC)*, Vol. 1. IEEE.
11. Pan, Z.-S. et al. (2003). Hybrid neural network and C4. 5 for misuse detection. In *2003 International Conference on Machine Learning and Cybernetics*, Vol. 4. IEEE.

12. Gaddam, S. R., Phoha, V. V., & Balagani, K. S. (2007). K-Means + ID3: A novel method for supervised anomaly detection by cascading K-Means clustering and ID3 decision tree learning methods. *IEEE Transactions on Knowledge and Data Engineering, 19*(3), 345–354.
13. Wang, G. et al. (2010). A new approach to intrusion detection using Artificial Neural Networks and fuzzy clustering. *Expert Systems with Applications, 37*(9), 6225–6232.
14. Gaikwad, D. P. et al. (2012). Anomaly based intrusion detection system using artificial neural network and fuzzy clustering. *International Journal of Engineering, 1*(9).
15. Bahrololum, M., Salahi, E., & Khaleghi, M. (2009). Anomaly intrusion detection design using hybrid of unsupervised and supervised neural network. *International Journal of Computer Networks & Communications (IJCNC), 1*(2), 26–33.
16. Muda, Z. et al. (2011). Intrusion detection based on K-Means clustering and Naïve Bayes classification. In *2011 7th International Conference on Information Technology in Asia (CITA 11)*. IEEE.
17. Om, H., & Kundu, A. (2012). A hybrid system for reducing the false alarm rate of anomaly intrusion detection system. In *2012 1st International Conference on Recent Advances in Information Technology (RAIT)*. IEEE.
18. NETRESEC. (2012). *U.S. National CyberWatch Mid-Atlantic Collegiate Cyber Defense Competition (MACCDC) Netresec*. http://www.netresec.com/?page=MACCDC.
19. Yen, T.-F. et al. (2013). Beehive: Large-scale log analysis for detecting suspicious activity in enterprise networks. In *Proceedings of the 29th Annual Computer Security Applications Conference*. ACM.
20. Denning, D. E. (1987). An intrusion-detection model. *IEEE Transactions on Software Engineering, 2*, 222–232.

Multi-stage Greenfield and Brownfield Network Optimization with Improved Meta Heuristics

Avneet Saxena and Dharmender Yadav

Abstract Facility Location decisions are part of the company's strategy which are important due to significant investment and decision is usually irreversible. Facility locations are important because (a) it requires large investment that cannot be recovered, (b) decisions affect the competitiveness of the company, and (c) decisions affect not only costs but the company's income (d) Customer satisfaction and trade off of decision based on service level. In a supply chain network design, facility locations (for example, warehouse, distributor, manufacturing, cross dock, or retailers locations) play an important role in driving efficient distribution planning and satisfying customer service level. The problem is to identify optimal set of facilities that can serve all the customer's demand with given service level at minimal cost. This problem becomes complex when possible locations are not known (Green field problem) and cost is the major driver for selection of facilities. This paper addresses this problem in two parts (a) first, improving the existing methodology (clustering) to achieve optimum clustering solution. With improved clustering outcome, we get better set of facility locations to be installed for Greenfield scenario. (b) Second, extend solution towards mathematical optimization based on cost, demand, service level, priority of facilities and business-based constraints (Brown filed Problem). This solution will help in selecting the best set of available facility location respecting business constraint so that customer demand is met in cost. Clustering algorithm, improved Meta Heuristic and GLPK (Open Source package to solve LP/MILP Problems) based mathematical optimization has been developed in Java. Currently we are pursuing further research in this direction to improve network and facility decisions.

Keywords Facility location · Network optimization · Meta heuristics
Decision scenario's

A. Saxena (✉) · D. Yadav
TCS, Think Campus, Electronic City, Phase-2, Bengaluru, Karnataka, India
e-mail: avneet.saxena@tcs.com

D. Yadav
e-mail: dharmender.yadav@tcs.com

V. E. Balas et al. (eds.), *Data Management, Analytics and Innovation*,
Advances in Intelligent Systems and Computing 808,
https://doi.org/10.1007/978-981-13-1402-5_35

1 Introduction

Creating a network of facilities requires a huge investment and have significant impact on the overall cost of a supply chain. This challenge lead to identify optimal set of facilities that can serve all the customer's demand with given service level at minimal cost. This problem becomes complex when possible locations are not known (Green field problem) or optimal location are not known in existing network (Brown filed problem) under operational constraints. Facility location decisions are critical and difficult due to following reasons:

a. They require a large investment that cannot be recovered
b. These decisions affect competitiveness of a company
c. They affect not only the cost structure but also revenues
d. They have implications on customer satisfaction and trade off of decisions based on service level (for instance, trade of service versus cost benefits).

Using an appropriate method and system (for example, general purpose decision tool) for deciding facility location can reduce the cost of delivery without compromising on the services provided to customers. To decide upon a facility location and customer allocation, customers are grouped into optimally sized clusters based on transportation cost, facility fixed cost, handling cost at facility, inventory cost and business constraints in order to minimize cost of each service delivery to each cluster. We can use this approach to evaluate multiple scenarios. Figure 1 explains possible factors involved in deciding the network optimization and facility location decisions scenarios. This includes both green filed and brown filed decisions.

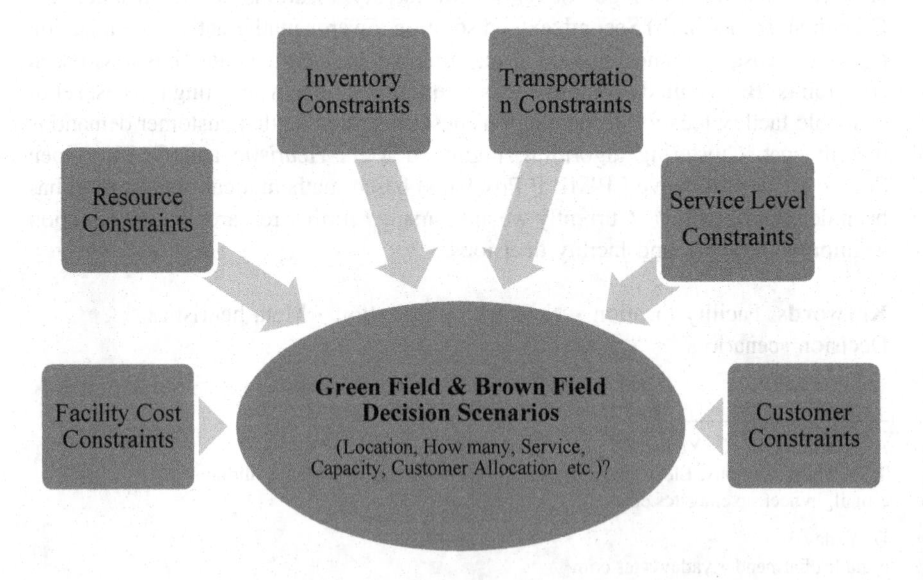

Fig. 1 Factors to consider in Facility Location Decisions

For green filed decision we have majorly consider the potential customer location and demand whereas for brownfield decision cost optimization has been done under all operational constraints and decisions what-if scenarios.

Improved hybrid meta heuristic-based Clustering technique is applied to identify the cluster of customers (i.e., indicative facilities) to minimize distance in supplying goods or products to retailer or customers. K-means algorithm is one such available algorithm usually applied in aforesaid scenario. However, solution obtained from k-means algorithm is sensitive to initial seed (selection of cluster seeds at staring of clustering). This usually led to local or sub optimal solution. In this paper, we have addressed this problem in two stages.

(a) In first stage, we have worked on a green field scenario, i.e., we do not know the exact location of any facility. Hybrid Meta Heuristic (K-means, Genetic Algorithm, and simulated annealing) approach has been used to overcome the limitation of traditional solution and improve overall Greenfield Solution.
(b) In second stage, we have worked on brownfield scenario, i.e., network and facility location optimization under various operational constraints and business decisions context. This lead to mathematical optimization based on cost, demand, service level, priority of facilities and business based what-if scenarios. Mixed Integer Linear Programming (MILP) based solution suing GLPK (Open source optimizer) has been developed and integrated with JAVA interface.

Overall we propose multi-stage (Greenfield, Brown filed) solution methodology using hybrid Meta heuristics (combination of K-means, genetic algorithm and simulated annealing) and Mixed Integer Linear Programming for cost based optimality. Details of aforesaid approach is highlighted in subsequent sections of this paper.

2 Summary of Literature Review

There has been an increased focus on facility related decisions and customer service allocation due to the dynamic nature demand, need for higher level of customer service level, increased cost pressures, limited capacity, introduction of new trends and products, higher levels of inventory and rising logistics cost. Some of these problems have been highlighted in literature by previous researchers, e.g., retail site selection [1, 4]; logistics and transportation planning [6, 7, 10]; marketing [12]; Lin and Chang [3], Wang and general research in facility location and allocation [11]; Huh and Lim [8]; Prabha and Saranya [2]; Boudahri1 [5], Nikola [9]. As mentioned, last two decades, researcher's has explored and worked in many different direction of network optimization and facility location and with their research they have addressed certain problem and open new avenues for practitioners. In this paper we have cater some of the research gaps and developed innovative solution methodology from practitioner point of view. In this paper we have improved clustering based facility decision by using hybrid Meta heuristics and then develop various practical business decision scenarios from cost optimization perspective. This paper is an effort to create a method and system to solve these pain areas in dynamic environment across all business units and industries.

3 Problem Description

Traditionally, facility locations are identified using clustering techniques with distance optimization. But those locations may not be optimized with respect to cost as well as no practical constraints are not considered like targeted lead time for transportation, service level, etc., since installment of facilities includes various costs like fixed cost(land cost), variable cost like labor, transportation costs varies with location. To overcome the limitation of facility decision and allocation we have employed the heuristic k-means along with Meta Heuristics (GA + SA) and MILP based mathematical model. Developed methodology is also translated into the system or decision tool which facilitate in variety of decisions.

Hence, the objective of this problem is to identify optimal set of facilities that should serve all the customer‘s demand considering given service level and business constraints so that total cost including transportation, facilities fixed cost, inventory holding cost, penalty cost for unsatisfying demand is minimum.

4 Two Stage Approach: Greenfield and Brownfield Decisions

Our analytical solution consist of two stages. Stages one we are solving the Greenfield scenario by optimizing the k-means clustering results using genetic algorithm and simulated annealing and the outcome of stage one is passed to stage two. In stage two, we have described the MILP based mathematical model to solve the brownfield scenario along with business constraints. Figure 2 explains the complete flow of the two stages solution methodology along with input, approach and outcome.

4.1 Stage-1: Improved Clustering for Greenfield Facility Location Decisions Using Hybrid Meta Heuristics

Following is a steps outline for improved hybrid Meta heuristics which includes K-means, Genetic algorithm, and simulate annealing.

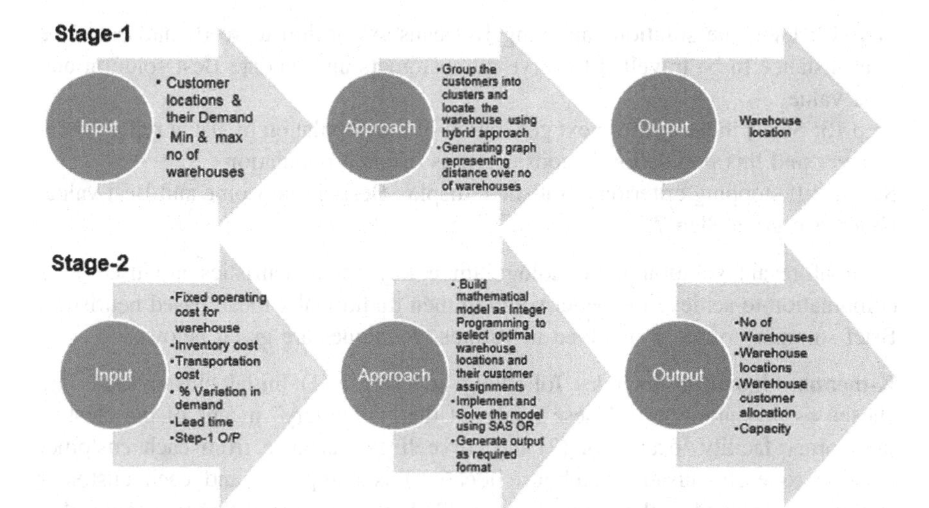
Stage-1
Input
• Customer locations & their Demand
• Min & max no of warehouses
Approach
•Group the customers into clusters and locate the warehouse using hybrid approach
•Generating graph representing distance over no of warehouses
Output
Warehouse location
Stage-2
Input
•Fixed operating cost for warehouse
•Inventory cost
•Transportation cost
• % Variation in demand
•Lead time
•Step-1 O/P
Approach
• Build mathematical model as Integer Programming to select optimal warehouse locations and their customer assignments
•Implement and Solve the model using SAS OR
•Generate output as required format
Output
•No of Warehouses
•Warehouse locations
•Warehouse customer allocation
•Capacity

Fig. 2 Multistage Greenfield and Brownfield Facility Location Methodology

Outline of hybrid algorithm:

Step 0: Initialize population size, crossover rate, mutation rate and stopping criterion using GA
Step 1: Generate population of the initial seeds
Step 2: Generate solutions using K-means Algorithm and calculate total distance to be travelled to serve all customers
Step 3: Assign seed with minimum distance as Best solution and its distance as Best value
Step 4: Apply the simulated annealing for each seed in the population
Step 5: Generate solutions using K-means Algorithm and calculate total distance to be travelled to serve all customers and this population is considered as original population
Step 6: Update Best solution and Best value
Step 7: Select the seeds for generating offspring seeds. These seeds are considered as intermediate population
Step 8: Generate the offspring seeds by applying crossover operator to the Intermediate population,
Step 9: Generate the offspring seeds by applying mutation operator to the Original population
Step 10: Generate solutions applying K-means Algorithm to offspring seeds and calculate total distance to be travelled to serve all customers and, update Best solution and Best value
Step 11: Apply the simulated annealing for each offspring seed in the population

Step 12: Generate solutions applying K-means Algorithm to seeds and calculate total distance to be travelled to serve all customers and, update Best solution and Best value
Step 13: Select the seeds for next generation of the population by using replacement strategy and this population is considered as original population
Step 15: If stopping criterion is satisfied, display Best chromosome and Best value. Otherwise go to Step 7.

In aforesaid solution methodology three key Meta heuristics are used in a combination to achieve the better solution then traditional k mean based heuristics. Brief summary of steps involved in various techniques are given below:

K-means Algorithm includes following key steps: (1) Initial cluster seeds are chosen as per initial seed. These represent the "temporary" means of the clusters (temporary facility locations); (2) the curve linear distance from each customer location to each cluster (warehouse location) is computed, and each customer location is assigned to the closest cluster. (3) For each cluster, the new centroid is computed—and each seed value is now replaced by the respective cluster centroid; (4) The curve linear distance from a customer to each cluster centroid is computed, and the customer location is assigned to the cluster centroid with the smallest distance; (5) The cluster centroids are recalculated based on the new membership assignment and (6) Steps 4 and 5 are repeated until no customer location moves to other clusters or stopping criterion satisfied.

Genetic Algorithm includes following key steps (Nikola [9]: (1) Generate random population of n chromosomes (suitable solutions for the problem); (2) Evaluate the fitness $f(x)$ of each chromosome x in the population; (3) Create a new population by repeating following steps until the new population is complete; (4) Select two parent chromosomes from a population according to their fitness (the better fitness, the bigger chance to be selected); (5) 1 With a crossover probability cross over the parents to form new offspring (children). If no crossover was performed, offspring is the exact copy of parents; (6) with a mutation probability mutate new offspring at each locus (position in chromosome); (7) Place new offspring in the new population; (8) Use new generated population for a further run of the algorithm. If the end condition is satisfied, stop, and return the best solution in current population. Run this solution in loop until stop condition met.

Simulated Annealing includes following key steps: (1): Read temperature (T), temperature reduction parameter, stopping criterion1 and stopping criterion2; (2): Take initial seed x; (3): Assign initial seed to current seed *xcurrent* and as best seed x *be*. Assign its value to (x) and as best value (*best*). (4) Generate neighbor seed x' of *xcurrent* and apply k-means algorithm and assign its value to (x'); (5) If (x') < (x) assign x' to *xcurrent* and $f(x')$ to $f(x)$ and go to step 6 otherwise go to step 7; (6) If (x') < (*best*) assign x' to *xbest* and $f(x')$ to $f(best)$ and go to step 8; (7) Calculate probability for accepting x' by using $e^{\frac{f(x')-f(x)}{T}}$ Generate random number $\in (0,1)$ for x'.

If $e^{\frac{f(x')-f(x)}{T}} >$ random number, assign x' to *xcurrent* and (x') to (x); (8) Repeat the process from 4 to 7 until stopping criterion 1 is satisfied; (9) Reduce the T by using temperature reduction factor; (10) Repeat steps from 4 to 9 until stopping criterion 2 is satisfied.

4.2 Mathematical Model to Solve the Brownfield Business Scenario

The key steps to developing a facility location problem (when facilities are available):

1. Collect complete information about customer demand, available location, its cost parameters and business rules
2. Modify cost parameters related to facility locations generated
3. Implement mathematical model using MILP
4. Run scenarios (from step 2 to 6) to identify the right number of facility locations
5. Allocate customers to the right facility suggested by the model output
6. Implement the solution
7. Revisit the solution when customer location and demand changes.

Brown field facilities locations is solved using mathematical modeling with MILP to identify facilities to be served to customers to minimize the cost considering business constraints like maximum distance with in which facility can serve customer, percentage of demand satisfied with in maximum distance, etc. However, the results from these systems are sometimes impractical. Due to that we have to create multiple scenarios. The solution provides the best results possible from distance perspective, cost perspective and considers business constraints. We are using a detailed scenario analysis to demonstrate the cost and service benefits of this system over existing techniques. This system can help organizations make facility related decisions easily in the long as well as the short run from the perspective of cost, service, inventory, etc.

Mathematical formulation of mathematical model for Facility Location and allocation as follows:

Objective Function:
Minimize

$$Z = \sum_{j=1}^{W} F_j * O_j + \sum_{i=1}^{I}\sum_{j=1}^{W} H_j * L_i * AC_{ij} + \sum_{j=1}^{W} SS_j * H_j * O_j + \sum_{i=1}^{I}\sum_{j=1}^{W} TC_{ij} * L_i * D_{ij} * AC_{ij}$$

Subject to

$$\sum_{j=1}^{W} AC_{ij} = 1 \forall i$$

$$\sum_{i=1}^{I} AC_{ij} \leq 10000 * O_j \forall j$$

$$\sum_{i=1}^{I} AC_{ij} \geq O_j \forall j,$$

$$L_i * \sum_{j=1}^{W} AC_{ij} \geq L_i \forall i$$

where

$j = 1,\ldots W$	Warehouses
$i = 1,\ldots I$	Customers
$k = 1,\ldots,W1$	Number of warehouses fixed
Li	Demand Qty of Customer i
Dij	Distance b/w the Customer i and warehouse j
Fj	Fixed Cost of Warehouse j
Hj	Handling Cost of Warehouse j
Oj	assume value 1 if the Warehouse is open; 0 otherwise
$ACij$	assume value 1 if the customer i is assigned to warehouse j; 0 otherwise
$TCij$	Transportation cost b/w the customer i and warehouse j.

5 Results and Discussions

In this paper we have propose this multi-stage green filed and brownfield solution using hybrid meta heuristics based clustering and MILP based cost optimization. Extended MILP based mathematical model is developed by exploiting limitation of earlier models. This mathematical model is translated into the GLPK based MILP module which can facilitate us to run various scenarios in dynamic network and facility location environment. This improved approach enhances industry vision and helps in reducing their supply chain cost without impacting services to customers. Figure 3 shows the model interface to support multistage facility location and customer allocation solution. The major technical advantages of this are

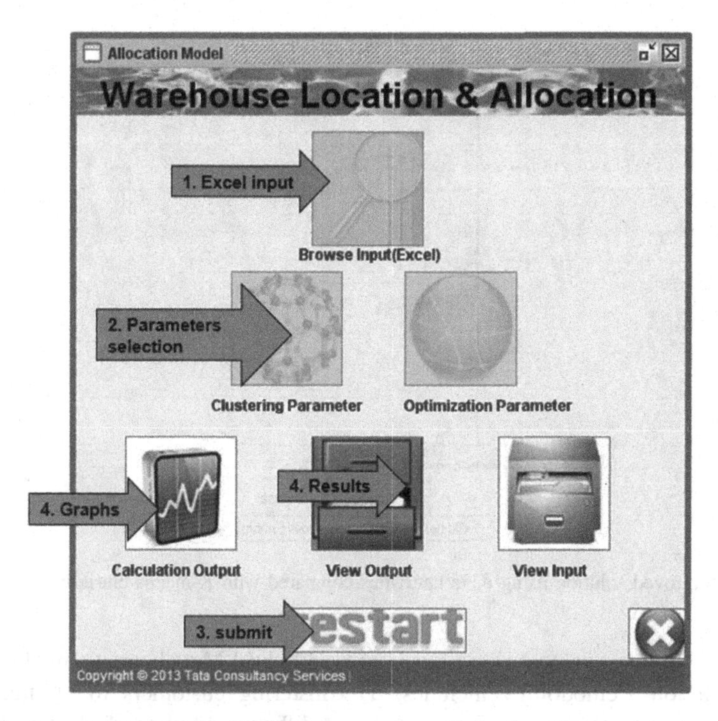

Fig. 3 Integrated model developed in Java & GLPK

- GA will explore search space by generating initial seeds for clustering
  - SA will exploit search space to obtain local optimal solution.
  - Mathematical model will consider all practical constraints and not limited to distance optimization.
- By combing these methods with clustering it will improve probability to obtain global optimal solution.

Some demonstrative examples with results are highlighted in following section.

5.1 Case Example Stage 1 with Result (Greenfield Clustering Solution)

Let us assume that we have 100 customer locations (latitude and longitude) and demand. In this examples, let's use traditional k-means clustering approach and hybrid meta heuristics based clustering approach to identify how many number of facilities in network to open, where to open and which particular customer's to serve. This is traditional Greenfield facility location decision scenarios where we

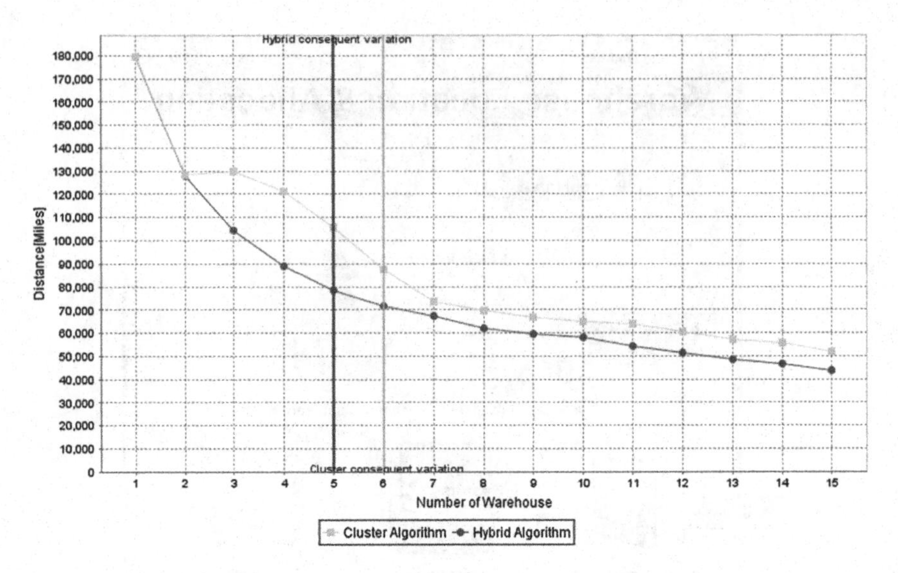

Fig. 4 Improved solution using meta heuristics compared with K-means clustering

have shown improvement using our improved hybrid Meta heuristic methodology. Our solution methodology includes (1) Clustering customers to optimize total distance from customers to warehouses for different number of clusters (2) The solution also take care of overlapping of infeasible locations like sea, out of cities, etc.; (3) User needs to take judgmental decision regarding how many warehouses needs to operated. Indicative Output snapshot as follows:

The graph shows the impact of number of warehouses on the total distance traveled by the product. E.g. in Fig. 4, for two warehouses the total traveled distance is 130,000 miles whereas for 5 warehouses it is only 800,000 miles with Improved Meta Heuristics. It is good notice that for same 5 facility location if we obtained traditional solution then we have to cover 110,000 miles. This clearly indicates 30,000 miles saving from our suggested improved clustering methodology. Table 1 shows the possible locations of warehouses (Name of city and state) e.g. new York city in PA state. The selection of number of facility using clustering is the user's judgmental decision but it provide an insight to select the appropriate number of warehouses on the basis of traveled distance, for example, after seven warehouse distance has not been reduced significantly.

Table 1 Outcome of clustering with facility locations

City	State	Lattitude	Longitude
Fairbanks	Ak	64.837	−147.71
York	PA	39.96	−76.72

5.2 Case Example Stage 2 with Results (Brownfield Decision Optimization)

Brown field facilities locations is solved using mathematical modeling with MILP to identify facilities to be served to customers to minimize the cost considering business constraints like maximum distance with in which facility can serve customer, percentage of demand satisfied with in maximum distance, etc. However, the results from these systems are sometimes impractical. Due to that we have to create multiple scenarios. The solution provides the best results possible from distance perspective, cost perspective and considers business constraints. We are using a detailed scenario analysis to demonstrate the cost and service benefits of this system over existing techniques.

Example: To optimize the number of warehouse locations, operating cost and also decide the warehouse capacity and allocate the customers to warehouse in optimized manner. (From clustering analysis, we have seven possible warehouse locations and 100 customer locations). Major inputs are facility locations, customer locations and their demand per year, fixed operating cost for warehouse, Handling cost per unit and Transportation cost. The selected warehouse locations are only 5 among 7, for ex. $w1$, $w2$ locations with capacities has shown in the Table 2.

The capacity of selected warehouses is decided, e.g., Warehouse W1 has 100,000 capacity. Table 3 is the allocation results of warehouse to customers that is yearly amount of quantity to be transferred from warehouse to customer.

Figure. 5 shows the impact of the number of warehouses on total cost. Selecting number of warehouses to be served is user option. The graph also shows comparison of cost-based solution obtained from phase 1 and phase 2. This depicts that how many number of warehouses needs to be opened from number of warehouses available.

We suggest that researchers and practitioners can use the proposed algorithm perform a ‘what-if’ scenario analysis. This can help managers understand and visualize the facility decision suggested, which is extremely important as these decisions cannot be revisited frequently. It is envisaged that such facility location

Table 2 Snapshot of warehouse capacity

Warehouse	Capacity
W1	100,000
W2	200,000

Table 3 Snapshot of warehouse to customer allocation

Warehouse	Customer	Demand
W1	Customer6	1000
W1	Customer2	3000
W3	Customer3	2000

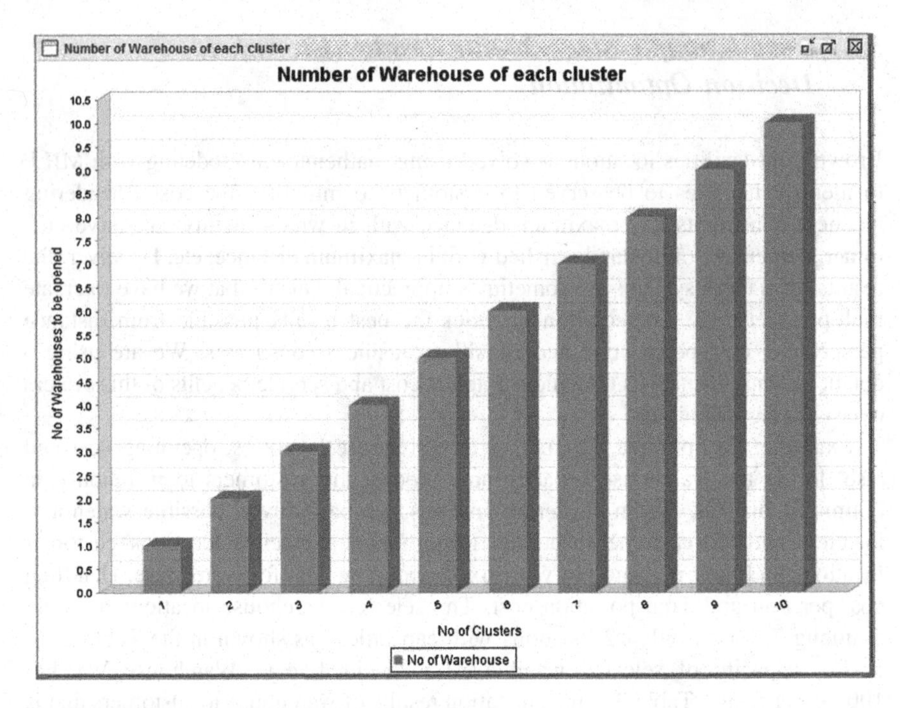

Fig. 5 Number of warehouses and their impact based on cost based optimization

and allocation decision support system when applied at various levels in a complex environment, can help achieve real-life business benefits and optimization of facility decisions.

6 Conclusions

In this paper we have proposed hybrid (combination of K-means, Genetic algorithm and simulated annealing) multi-stage (hybrid heuristic and Mixed Integer Liner Programing) solution methodology for solving Greenfield and Brownfield facility location problem. Significant research has been conducted before conceptualization of two stage solution and its deployment. It is realized that there is plenty of scope of improvement available in this area due to dynamic supply chain orientation. Primarily, this is due to lack of accurate business perspectives, research assumptions and other practical implications. Previous researcher work has highlighted as summary to identify the gaps and improvement directions.

We have demonstrated how our suggested methodology can help organizations to achieve the best improve solution as well allow user to run various business scenarios. We have improved Greenfield facility location decision using our improved Meta heuristics and then subsequently performed cost optimization addressing all operational constrains with various what-if scenarios. Suggested hybrid multi-stage solution methodology rather helps us to achieve optimal solution and leave no space for sub optimality. The input of improved clustering is used for MILP optimization based on aforesaid constraint with cost minimization as objective. This helps to identify right number of facilities, facility locations, capacity at facility and customer allocation to facility. Depending upon business requirement and constraints, various scenarios can be generated. There is enormous potential for hybrid Meta heuristics and two phase approach from a business and research perspective. Our effort continues in this direction for real life implementation of the proposed method to achieve increased business benefits.

References

1. Rana, K. (1991). Order-picking in narrow-aisle warehouse. *International Journal of Physical Distribution & Logistics Management, 20*(2), 9–15.
2. Arun Prabha, K., & Saranya, R. (2011). Refinement of K-Means clustering using genetic algorithm. *Journal of Computer Applications (JCA), IV*(2).
3. Lin, C.-Y., & Chang, C.-C. (2005). A new density-based scheme for clustering based on genetic algorithm. *Fundamenta Informaticae, 68,* 315–331.
4. Cormier, G., & Gunn, E. A. (1992). A review of warehouse models. *European Journal of Operational Research, 58*(1), 3–13.
5. Boudahri, F., Bennekrouf, M., Belkaid, F., & Sari, Z. (2012). Application of a capacitated centered clustering problem for design of agri-food supply chain network. *IJCSI International Journal of Computer Science Issues, 9*(4), No 1.
6. Grabmeier, J., & Rudolph, A. (2002). Techniques of cluster algorithms in datamining. *DataMining and Knowledge Discovery, 6,* 303–360.
7. Maulik, U., & Bandyopadhyay, S. (2002). Performance evaluation of some clustering algorithms and validity indices. *IEEE Transactions on Pattern Analysis and Machine Intelligence, 24*(12), 1650–1654.
8. Huh, M.-H., & Lim, Y. B. (2009). Weighting variables in K-means clustering. *Journal of Applied Statistics, 36*(1), 67–78.
9. Marković, N., Ryzhov, I. O., & Schonfeld, P. (2017). Evasive flow capture: A multi-period stochastic facility location problem with independent demand. *European Journal of Operational Research, 257*(2), 353–704.
10. Roodbergen, K. J., & de Koster, R. (2001). Routing methods for warehouses with multiple cross aisles. *International Journal of Production Research, 39*(9), 1865–1883.
11. Tsai, C. Y., Liou, J. J., & Huang, T. M. (2008). Using a multiple-GA method to solve the batch picking problem: considering travel distance and order due time. *International Journal of Production Research, 46*(22), 6533–6555.
12. Zhang, G. Q., Xue, J., & Lai, K. K. (2002). A class of genetic algorithms for multiple-level warehouse layout Problems. *International Journal of Production Research, 40*(3), 731–744.

Implementing Signature Recognition System as SaaS on Microsoft Azure Cloud

Joel Philip and Dhvani Shah

Abstract The use of information technology in varied applications is growing exponentially which also makes the security of data a vital part of it. Authentication plays an imperative role in the field of information security. In this study, biometrics is used for authentication purpose and also describes the combinational power of biometrics and cloud computing technologies that exhibit the outstanding properties of flexibility, scalability, and reduced overhead costs, in order to reduce the cost of the biometric system requirements. The massive computational power and unlimited storage provided by cloud vendors make the system fast. The purpose of this research is to precisely design a biometric-based cloud architecture for online signature recognition on Windows Tablet PC, which will make the signature recognition system (SRS) more scalable, pluggable, and faster, thereby categorizing it under "Bring Your Own Device" category. For extracting the features of the signature to uniquely identify the user, Webber local descriptor (WLD) process is used. The real-time implementation of this feature extraction process as well as the execution of the classifier for the verification process is deployed on Microsoft Azure public cloud. For performance evaluation, total acceptance ratio (TAR) and total rejection ratio (TTR) are used. The proposed online signature system gives 78.10% PI (performance index) and 0.16 SPI (security performance index).

Keywords Biometrics · Signature recognition · Cloud computing
Microsoft Azure · Web and worker role · REST API · Webber local descriptor

---

J. Philip (✉)
Universal College of Engineering, Mumbai, Maharashtra, India
e-mail: tjoelphilip@gmail.com

D. Shah
St. John College of Engineering and Management, Palghar, Maharashtra, India
e-mail: shahdhvani08@gmail.com

V. E. Balas et al. (eds.), *Data Management, Analytics and Innovation*,
Advances in Intelligent Systems and Computing 808,
https://doi.org/10.1007/978-981-13-1402-5_36

1 Introduction

In this digitally centered world, advances in various technologies make life tranquil by the provision of various sophisticated knowledge hubs through the innovation of diverse devices. However, each hi-tech innovation has high probability of concealed intimidations to its operators. The advancements in digitization have brightened the concerns regarding the privacy and theft of users' information. With increase in the use of Internet-enabled applications, the amount of digital data has increased exponentially and has become very important, and users try to secure their data with strong scrambled passwords. However, the mismanagement of these security measures and broken security technologies are rising at an alarming rate. This results in cards being redone or forged and being misused. These cumulative efforts to fight against cyber security have given rise to the development and use of secure biometric-based system. One of the highest primacies in the domain of information security is validating that whether the user who is trying to gain access to the information has the authorization or not. Such access is usually accomplished by a person attesting their identity by submitting the proof for his identification as a part of the authentication process. The basic three methods for authentication are what we know—passwords or other personal information, the other is what we have—smart cards or tokens, and the final method is what we are—biometrics technology.

Biometric-based solutions are capable of securing private financial dealings and confidential data privacy. The necessity built upon biometric-based solutions can be brought into being on various spheres such as government-based institutions, the defense ministry, and commercial applications. Network security setups focused on enterprise-wide network, government ID cards, secure electrical banking, investments and retailing networks, and law enforcement. Some of the sectors already benefiting from these technologies are health and social services [1].

In this study, 10 signatures of each user captured from Tablet PC are stored on the Azure Blob storage. From the client side, after capturing the signatures, a call is made to the RESTful API, which is deployed as an Azure App Service (Web Role) on the Azure cloud platform making client architecture thin. RESTful API acts as the communication layer between the client side and the server side, i.e., the cloud platform. The Webber local descriptor (WLD) process is implemented as a worker role service. The WLD algorithm is the heart of this research since it uniquely identifies each user and thus encapsulates security into various applications and specifically banking applications.

The unbounded storage capacity and high processing power of cloud architecture make this SRS a highly scalable, flexible, faster, and low-cost biometric system. This research can be published as Software as a Service (SaaS) to the clients. Thus, this SaaS can be used for all the systems where authentication is required, e.g., bank, attendance systems, etc. The only concern is regarding the security of the features stored as text files on cloud storage as they are stored on cloud. But this can be resolved by implementing the cryptographic and security-related algorithms in the future. Internet connectivity is a must on client devices to exploit this service. Hence, this research is a low-cost, portable, scalable, and flexible biometric-based SRS.

2 Related Work

The authors [2] have discussed biometrics in five different parts. They have described how these systems work, their strengths, and weaknesses, where they can be effectively deployed. The paper [3] describes the major advantages and disadvantages of each technology with a definite indication of some appropriate technologies and where they can be applied [3]. H. B. Kekre, V. A. Bharadi [4] propose a segmentation technique for fingerprint and palmprint where a dynamic threshold is generated for each of the inputs. In [5], the authors used digitizer to capture handwritten signatures and they the concept of neural networks for the classification and recognition of the signatures. To detect the line strokes from signature image, Kaewkongka, Chamnongthai, and Thipakom [6] proposed Hough transform, which extracts the parameterized Hough space from the signature skeleton as a unique characteristic feature of signatures. Armand, Blumenstein, and Muthukkumarasamy [7] have applied grouping of the modified direction feature to train and test two neural network-based classifiers. In [8], the authors Doroz and Wrobel present a new technique of identifying handwritten signatures, whose main cornerstone is on the mean differences, which has been improved appropriately. An another approach is also discussed, where evaluating the variation is signature pixels by calculating their locations [9]. Online signature recognition is referred as dynamic signature recognition. Rhee and Cho [10] perform online signature recognition using segmentation methodologies. In [11], the researchers have used Gabor filters to extract the feature vectors of the dynamic signature. In [12], the authors illustrate how biometric authentication is accomplished on a mobile device by migrating the biometric recognition process to the cloud servers. In [13], authors Bharadi and Philip have projected a robust architecture for employing online SRS on a public cloud like Windows Azure. In [14, 15], Shah and Bharadi have proposed an architecture that uses an embedded system and Azure platform to develop a portable authentication system. Also, the use of the encryption algorithm makes the system viable for secure communication providing end-to-end security.

3 Banking Applications

Signature recognition is one of the most important research areas in the field of identity recognition based on biometrics which can be effectively used in banking applications. The technology is also regarded as a front subject in many fields such as pattern recognition and signal processing. Generally, there are two main types of signature recognition, namely, off-line and online signature recognition.

Off-line signature recognition deals with the analysis of the signature image alone. The major drawback of this type of signature recognition is that the signature image alone constitutes a limited database for analysis, difficult to make an effective determination of the validity of the signature.

However, this research work aims to promote online signature recognition in banking applications, which consists of digitizing the signature as it is being produced. With this method, the information obtained will contain not only the signature image in numerical values, i.e., text files, but also time domain information, such as signing speed and acceleration apart from pressure points, strokes, and acceleration as well as the coordinates, i.e., X, Y. All this information can be combined to determine the validity of a signature much more effectively than what an off-line recognition system is capable of, which leads to higher level of accuracy for easy detection of fraudulence or forgery in bank cheques, as dynamic characteristics are very difficult to duplicate. For this research, online signature recognition for banking applications requires a Windows Tablet PC and Active Stylus—used to scan signature dynamically.

4 Webber Local Descriptor (WLD)

An experimental psychologist, Ernst Weber, during the nineteenth century, observed that the ratio of the increment threshold to the background intensity is a constant [16]. This relationship, known since as Weber's law, can be expressed as follows:

$$\Delta I = k, \tag{1}$$

where ΔI signifies the increment threshold (just noticeable difference for discrimination), I denotes the initial stimulus intensity, and k signifies that the proportion on the left side of the equation remains constant despite variations in the I term. The fraction $\Delta I/I$ is known as the Weber fraction. Weber's law, more merely stated, says that the size of a just noticeable difference (i.e., ΔI) is a constant fraction of the original stimulus value. So, for example, in a noisy environment one must shout to be heard while a whisper works in a quiet room [17].

The proposed descriptor, WLD, is centered on Weber's law. It has numerous advantages, such as detecting edges smartly, robustness to noise and illumination change, and its influential representation ability. WLD is based on a physiological law. It extracts features from an image by simulating a human sensing his/her surroundings.

The core of this research is the extraction of features for each user so as to identify them uniquely. In this study, a feature vector extraction method called WLD is used and executed by the Azure worker role.

5 System Methodology

The proposed architecture is a blend of two technologies: Biometrics and cloud computing. Biometrics helps to authenticate a user uniquely, whereas cloud technology provides the storage of signatures and gives the computing platform to extract the biometric features from the signatures using WLD algorithm (Fig. 1).

5.1 Implementation Steps

Take signature trials from users through a dynamic interface on Windows Tablet, which distinguishes user input from Active Stylus storing its *X*, *Y* coordinates and pressure information as input values. Enrollment operation involves taking the signatures obtained through the Windows Tablet PC and uploading them onto the blob storage of the Azure cloud through Web Role. In the background, feature

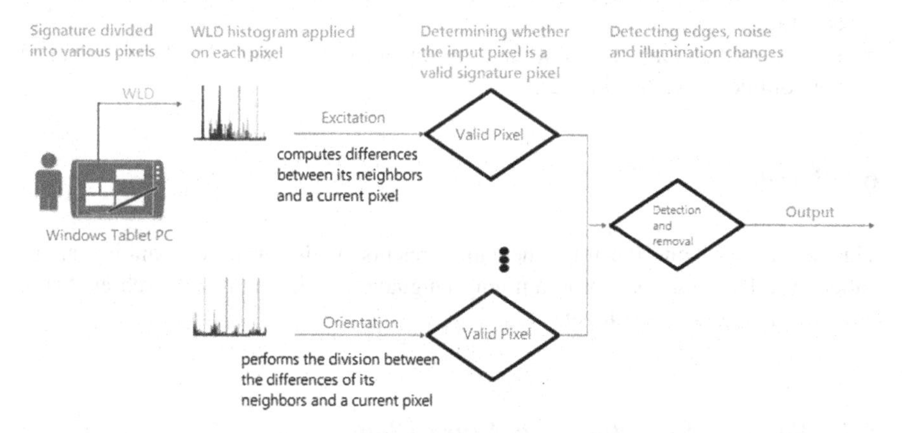

Fig. 1 *WLD process* used for feature extraction process

extraction process of each user is published as a worker role on the uploaded signatures on the Azure platform to automate the extraction process. After feature extraction process, the classifier parameters of each user are stored on back to blob storages as text files for the verification stage. The enrollment operation step-by-step execution is as follows:

It primarily captures the 10 signatures of a particular user. The signatures are captured using Windows Tablet PC (using Microsoft Ink Library). Each user is given a unique user id which helps us to identify his 10 signatures. After the signatures are captured, POST method published as Web Role on Azure is called. These 10 signatures along with the user id are given to the API. The API receives two things from the client; first is the user id of a user and second, the 10 signatures of that user. Now, API does the following jobs:

- The API makes a connection with the Azure Blob storage using the connection string. On the cloud side, storage account is created first which provides connection information to be used within the connection string.
- Also, the user id is sent to the service bus queue which automatically activates the cloud service responsible to execute the WLD algorithm published as worker role on Azure.
- The WLD process is executed on the captured signatures and the unique features of each are stored on the cloud storage for the verification process. The features are saved in a text file named "Classifier.txt" within the container of each user.
- During the verification process, the real-time signature of the user is sent to the blob storage and the WLD process is initiated on the cloud, which identifies the unique features of the signature and these classifier parameters are compared with the already stored parameters on blob storage during the training process.
- The confidence value is generated, and the user is verified/authenticated accordingly.
- For the verification process, we use the signatures 8, 9, and 10 and then calculate the confidence value (Fig. 2).

6 Results

This section explains the implementation results of the proposed banking application. We have used C# programming language for the entire research as Azure provides a great support for it.

6.1 Sending Signatures on Azure Blob

The client-side application is responsible for capturing signature from Tablet PC, 10 signatures per user, storing them locally until they have successfully stored on Azure Blob as shown in Fig. 3. The client is not directly connected to the blob

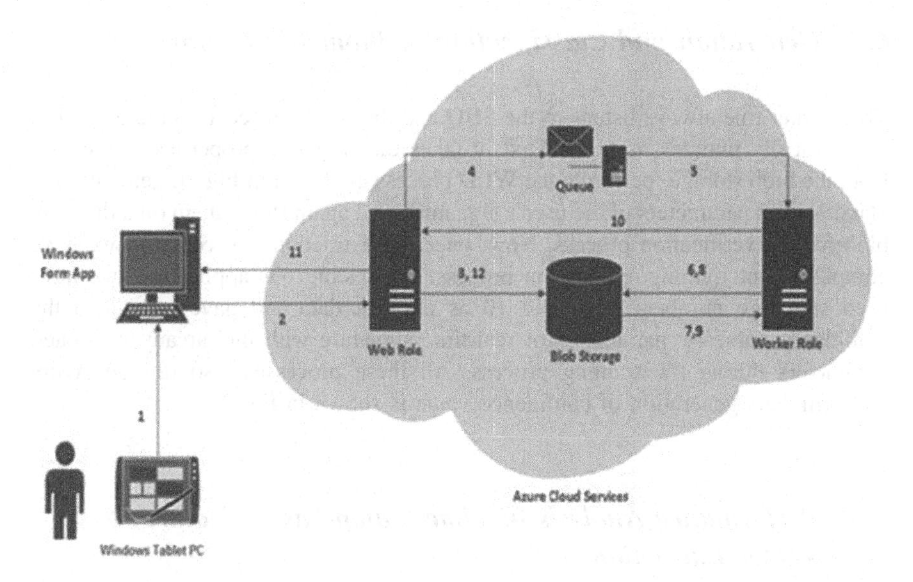

Fig. 2 *Proposed architecture*

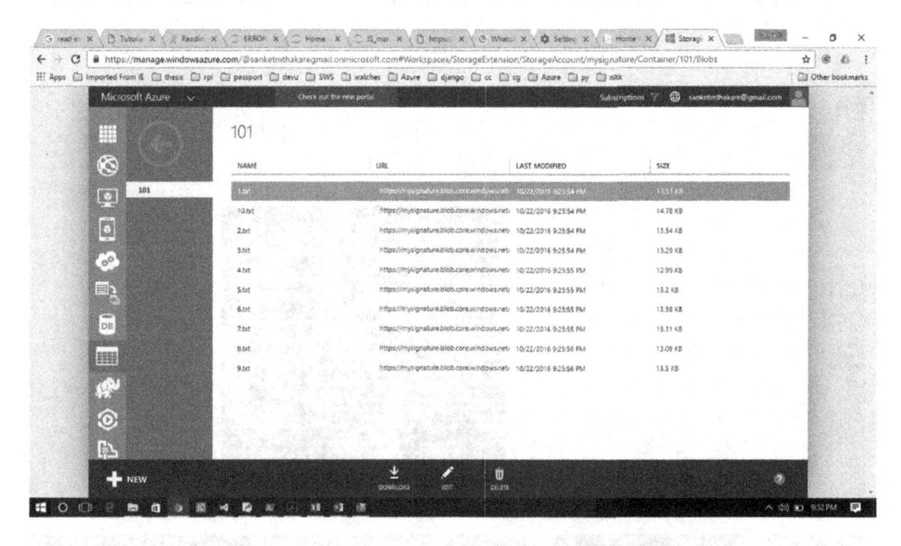

Fig. 3 *Blob storage* on Azure platform holding the training signatures of users

storage. Web Role is the entry point for all HTTP requests on the cloud, which forwards the text file to blob and also sends the unique user id to the service bus queue.

6.2 Generation and Classification of Biometric Features

The worker role always listens to the SBQ and the instant it receives the user id, it automatically initiates its tasks. First, it takes the signature properties of the user from the blob storage, performs the WLD process as shown in Fig. 4, generates the classification parameters of the user's signature, and again stores them on a different blob for the verification process. Now, once the extraction is successful, we have completed our training/registration process. For testing our application, we have used signature numbers 8, 9, and 10 as our test data and have calculated the confidence value by parameters of real-time signature with the already generated parameters during the training process. All these processes also run on Azure platform. The generation of confidence value is shown in Fig. 5.

6.3 Performance Analysis of Time Stamp-based Feature Vector Extraction

In the performance analysis of the time stamp-based feature vector extraction, the feature vector's performances are compared with different performance parameters

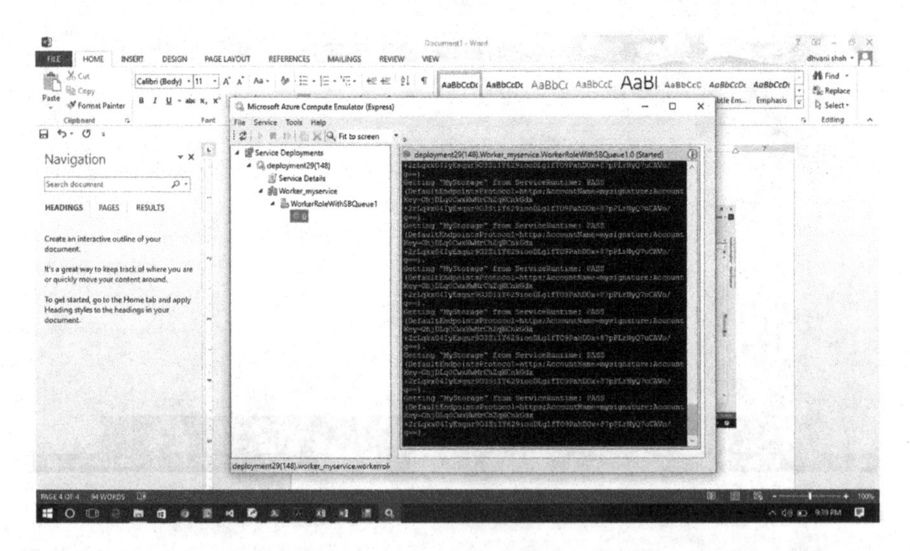

Fig. 4 *WLD extraction as worker role* on Azure platform. As one cannot view the ongoing of the process inside a worker role on Azure, there is an emulator for the same

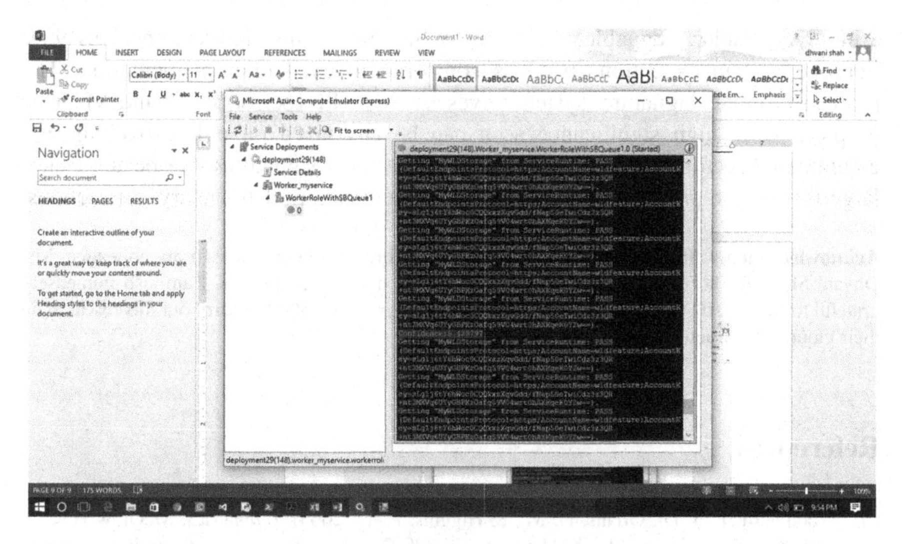

Fig. 5 *Matching process* on Azure as a worker role

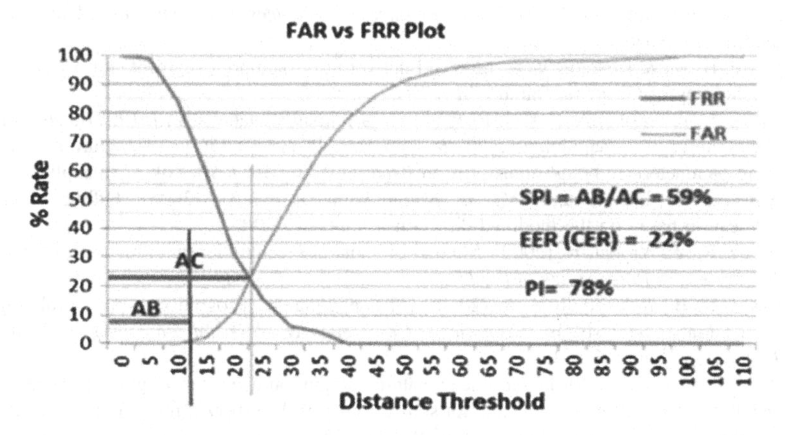

Fig. 6 *FAR versus FRR plot* showing crossover (ROC curve)

or metrics as shown in Fig. 6. Performance index value of 78.10 and security performance index as 0.16 is achieved, which is desirable.

7 Conclusion

This research addresses the issues faced by traditional online SRS and how they are resolved by implementing the online SRS on cloud. The proposed architecture has the public cloud architecture. In this research, the scalable architecture is discussed in depth and successfully employed on Microsoft Azure cloud which aims to devise a highly scalable, pluggable, and faster online SRS. This architecture gives 90% improvement in execution speed as compared to existing implementation. The proposed online signature system gives 78.10% PI (performance index). This proposal is of high significance as it can be envisaged on online banking and e-commerce sectors, where dynamically handwritten signatures can be used on a large for a more precise and well-defined authentication of monetary transactions.

Acknowledgments It is my great pleasure to express my sincere gratitude to my colleague, Ms. Dhvani Shah, for her unwavering support to carry out this entire research. I am also immensely grateful to Microsoft Azure for awarding me a research grant of $5000 to carry out this research on their cloud infrastructure.

References

1. Woodward Jr., J. D., Orlans, N. M., & Higgins, P. T. (2003). *Biometrics*. McGraw-Hill.
2. Nanavati, S., Thieme, M., & Nanavati, R. (2002). *Biometrics: Identify verification in a networked world*. Wiley Computer Publication.
3. Khushk, K. P., & Iqbal, A. A. (2005). An overview of leading biometrics technologies used for human identity. In *Proceeding of Engineering Sciences and Technology*, University of Sindh, Hyderabad.
4. Kekre, H. B., & Bharadi, V. A. (2009). Fingerprint & palmprint segmentation by automatic thresholding of Gabor magnitude. In *ICETET*.
5. Pacut, A., & Czajka, A. (2001). Recognition of human signatures. In *IEEE Transactions*.
6. Kaewkongka, T., Chamnongthai, K., & Thipakom, B. (1999). Off-line signature recognition using parameterized hough transform. In *Proceedings of 5th ISSP*, vol. 1, Australia.
7. Armand, S., Blumenstein, M., & Muthukkumarasamy, V. (2006). Offline signature verification based on the modified direction feature. In *ICPR*.
8. Doroz, R., & Wrobel, K. (2009). Method of signature recognition with the use of the mean differences. In *Proceedings of the ITI*.
9. Kekre, H. B., & Bharadi, V. A. (2010). Off-line signature recognition using morphological pixel variance analysis. In *International Conference & Workshop on Emerging Trends in Technology*, Mumbai, India.
10. Rhee, T., & Cho, S. (2001). On line signature recognition using model guided segmentation and discriminative feature selection for skilled forgeries. In *Proceedings of Sixth International Conference on Document Analysis and Recognition*.
11. Kekre, H. B., & Bharadi, V. A. (2010). Gabor filter based feature vector for dynamic signature recognition. *International Journal of Computer Applications, 2*.
12. Bommagani, A. S., Valenti, M. C., & Ross, A. (2014). A framework for secure cloud-empowered mobile biometrics. In *Proceedings of MILCOM*.
13. Bharadi, V. A., & Philip, J. (2016). Signature verification SaaS implementation on Microsoft Azure cloud. In *ICCCV*.
14. Shah, D., & Bharadi, V. (2016). IoT based biometrics implementation on Raspberry Pi. *Procedia Computer Science, 79*.
15. Shah, D. K., Bharadi, V. A., Kaul, V. J., & Amrutia, S. (2016). End-to-end encryption based biometric SaaS: Using Raspberry Pi as a remote authentication node. In *ICCUBEA*.
16. ArticSoft: Biometrics—Problem or solution, Whitepaper.
17. Chen, J., Shan, S., He, C., Zhao, G., Pietikainen, M., Chen, X., & Gao, W. (2009). WLD: A robust local image descriptor. In *IEEE Transactions on Pattern Analysis and Machine Intelligence*. Accessed August 17, 2017.

Target-Controlled Packet Forecast and Communication in Wireless Multimedia Sensor Networks

S. Ambareesh and A. Neela Madheswari

Abstract The two main factors which are vital in present multimedia applications are Target-controlled packet forecast and communication. The degradation of Quality of Service (QoS) is because the packets miss their targets and become useless and are often dropped. As the consumption of real-time hypermedia applications and Internet of Things (IoT) has grown into more, multimedia data communication is a key cause to endorse the QoS of citizens. To accomplish the QoS prerequisite in Wireless Multimedia Sensor Networks (WMSNs) the mixture of multiple communication methods is stimulated for packet sending, counting Conventional Network Coding (CNC), Analog Network Coding (ANC), Plain Routing (PR) and Direct Broadcast (i.e., No-Relaying, NR). The combination and integration of communication methods lowers packet falling probability, but complicates the packet transferring and forecast process instead. Hence, an exhaustive search scheme is introduced to get the optimal forecast sequence and equivalent communication method for target constrained multimedia broadcasts in WMSNs. With respect to promote computing proficiency for the formulated problem, two heuristic methods based on Markov chain approximation and dynamic graph is proposed.

Keywords Quality of service · Target constrained packet scheduling and transmission · Wireless multimedia sensor networks

---

S. Ambareesh (✉)
CSE, Anna University, Chennai, Tamil Nadu, India
e-mail: ambihce@gmail.com

A. Neela Madheswari
CSE, Mahendra Engineering College, Mallasamudram, Tamil Nadu, India
e-mail: neela.madheswari@gmail.com

V. E. Balas et al. (eds.), *Data Management, Analytics and Innovation*,
Advances in Intelligent Systems and Computing 808,
https://doi.org/10.1007/978-981-13-1402-5_37

1 Introduction

Wireless Multimedia Sensor Networks (WMSNs) are the networks of wirelessly connected sensor nodes which include multimedia devices, for example cameras, microphones and they are skilful to obtain video and audio streams, motionless images, as well as scalar sensor data. Real-time multimedia are the applications in which audio-visual aid information has to be carried and passed in real time. Multimedia is a word that represent several forms of material, containing audio, visuals, animatronics, images, text, etc. The greatest illustrations are nonstop or connected mass media such as animatronics, audial and video that are built on time, i.e., each audio clip or video casing has a time duration linked with it, signifying the performance. Multimedia records should be accessible always in continuous manner, in authoritative in which they are related with their time stamp. E.g., consider a video should process 30 frames per second so that the user can view the output video continuously without any interruption, if the network fails to render 30 frames per second or if the network renders 25 frames per second, then there will surely an interruption occurring in the output video. Therefore, real-time hypermedia presentations naturally have restriction on time, i.e., the records has to be conveyed in factual period. Figures 1 and 2 are the examples of real-time applications.

There are numerous examples for multimedia applications like online education, Internet Gaming, Live Streaming, etc., all these real-time multimedia applications have strict deadline constraint. For example, if we take a you tube application consider the user is watching a video in you tube, if video is not played back continuously the user may lose interest to watch the video, hence all real-time application have strict deadline constraint. In these real-time applications, packets nothing but the data or information consume strict target restrictions and must reach at their termini before their limits. If they are reached on time, they turn into unacceptable and are fallen, which shrink and reduce the Quality of Service (QoS). It is the complete representation of a computer system, majorly it is the enactment visualized by the users of the system. Mainly to mark the service quality, numerous linked features of the network package are frequently measured, like fault rates, bit amount, output, broadcast delay, accessibility, jitter, etc. This paper is allocated into 8 sections, in which, Sect. 1 provides the introduction of WSN and WMSN.

Fig. 1 Video conferencing of real-time application

Fig. 2 Social networking sites of real-time application

Section 2 provides the detailed Literature Survey of latest peer reviewed papers related to WMSN. Section 3 provides the discussion on the Existing System. Section 4 provides the Problem Statement with respect to packet dropping node in WMSN. Section 5 provides the details of Proposed System with the approaches to minimize packet dropping in WMSN. Section 6 provides the detail System Architecture, Data Flow Diagram (DFD) and Class Diagram. Section 7 concludes research work with Future Enhancement. Section 8 provides References used for research work.

2 Literature Survey

In [1], the latent effect on technical investigation as well as abundant presentations on WMSNs have drawn more attention. To govern steady as well as source effective path and to offer variable stages of QoS warranty for hypermedia, the broadcast of multiple forms of information depends on a routing protocol. As there are numerous problems like inadequate network resources, complex procedures of multiple media presentations and dynamic fluctuations of network situation is an exciting task in WMSNs. The tests as well as necessities, a complete review on routing of presentation necessities and key procedures are illustrated in this paper. The proposed directing resolutions in this paper deals with the five key types based on their architecture along with optimization qualities, provision on QoS, multiple media responsiveness, energy effectiveness, bottle neck prevention, optimizing. At last, the open investigation topics in steering metrics are depicted with some effective investigation zones concerning routing in embryonic WMSNs presentation states are discussed. The objective of this survey is to deal with the ventures and current tendencies in routing in WMSNs. This paper deals with the experiments in the design of transmitting packets in WMSNs, and then surveys on recent research progress in area of WMSNs. More importantly, future enhancement is focused on research areas of WMSN/IoMT systems.

In [2], stability and reliability in wireless communication is one of the major fact for connecting people using the smart devices in the cities. In this paper, a societal leaning smartphone-constructed adaptive broadcast device is proposed to progress the system connectivity and quantity in Internet of Things (IoT) for smart cities. To make the network connectivity strong, a societal leaning double mart grounded relay medley scheme is explored to encourage the relay smartphones to transmit the packets for others. For gaining maximal throughput in IoT based on smartphones, and also the relay scheme selection is dogged by combining numerous kinds of broadcast methods in an optimum manner to make maximum use of wireless scale resource. A method based on a firefly procedure is determined to answer the high computational complexity. The proposed mode has two steps. First, by the friendship the social features of smartphones are modeled and a relay selection method has been offered. Secondly, to educate the interaction amongst NC and spatial recycle by synchronously triggering associations in an ideal way a variety spatial recycle aware relay system selection process is proposed. To deal with the computational complexity and to achieve the optimal network performance, a firefly procedure based empirical approach is obtained.

In [3], the IoT relates to the real life scenario where most of the belongings, objects or human being in everyday life should interconnect with added systems and deliver facilities on Internet. Substances identify, sense, interacting as well as process the abilities to mark the IoT model a real time. IoT defines IEEE 802.15.4 standard as the major inter-connection procedures. The IEEE 802.15.4 standard gives Guaranteed Time Slot (GTS) apparatus which complements QoS for the on-time documents communication. Even there are many QoS structures in IEEE 802.15.4 standard, even though major difficulty of endways delay resides. For overcoming this end-to-end delay problem, a supportive Medium Access Control (MAC) protocol for on-time information broadcast is proposed. The presentation of the proposed scheme is illustrated through the simulations. The proposed scheme improves network performance which is demonstrated. The proposed method overcomes the difficulty of GTS tradition on less duty sequence along with the straight broadcast amid finale systems. The planned method similarly lowers the delays produced by PAN controller relays meanwhile systems using the planned method can straightly transfer the on-time information not checking a PAN controller. Since the proposed method selects the pathway with the improved linkage feature, which reduces the energy intake by re-broadcast, and increases the system performance. The energy intake of the proposed method is higher to both the IEEE 802.15.4 standard and ESS system.

In [4], deterministic delay constraints are difficult to guarantee due to the fundamentally stochastic environment of wireless vanishing stations. Using the thought of operative size, the proposed system provides statistical delay guarantees. Considering a large amount of user setup where different kind of users have different delay QoS restrictions. The resource distribution is derived strategy which exploits the sum video feature and spread over to any quality metric with hollow rate-quality plotting. The resource distribution policy is extended to imprison the video quality based adaptive user subcarrier transfer in wideband networks as well as imprison the impact of adaptive variation and coding. Another difficulty of fairness driven resource distribution is solved whereby the concept is to improve the lowest video quality through users.

Finally, user presence and forecast strategies are derived which enable selection of a large number of user subcategory such that all nominated users can meet their geometric delay condition. The cinematic users with differentiated QoS [5] necessities can attain similar video quality with massively diverse resource necessities. The concept of effective capacity is used in this paper to provide a framework for statistical delay provisioning for multiple users sharing a wireless network. The resource distribution policies were prolonged to capture video quality based adaptive user subcarrier project in wideband networks as well as the effect of adaptive modulation and coding. This paper [6] focuses on the video quality driven resource sharing which are referenced where the comparable perceptual feature optimizations is proposed in a client based environment without contact to the reference. This has the main advantages that the user dignified video quality includes the effect of channel misrepresentations along with the source alterations which is opposite to the server dignified video quality which only imprisons source distortions. The quality assessment to the user is enabled which shrinks the server capacity and ignores upholding a large session state for each user at the server.

3 Existing System

For lowering of packet communication interrupt in WMSNs, some work has been made on unlike facets, such as QoS-based routing mechanisms, target concerned stand in line approaches [7] in relays, well-organized access approaches in Media Access Control (MAC) layer [8], and cross-layer optimization procedures that not only reflect the transmission rate in physical layer systematically [9], but also relaying method in network layer. Even though some QoS methods have been specified in IEEE 802.15.4 standard [10] for WSNs, the difficulty of endwise interval quiet resides. The supportive MAC protocol aimed at on-time data broadcast was discovered in [11], which mainly concentrates on the inspiration aided method in star net topology and can be seen as a thoughtful of complement of IEEE 802.15.4 regular standard. A multi-user system for users with interruption QoS limitations has stayed measured in [12], and the supply division rule has also been extracted for sum video worth improvement.

It is broadly addresses that Predictable non-physical-layer Network Coding (NC) prominently improves the grid material when packet target restrictions are not measured [13]. The broadcast nature of wireless channel is advantageous in CNC. The transmission time can be reduced by authorizing a relay to encrypt at minimum two packages, these are acknowledged disjointedly from dissimilar foundation nodes, into single package as well as transmit it to termini causing decryption of the projected packages. Therefore, NC is likewise one of the efficient solutions to hand the problem of packet broadcast with limited restrictions with supportive communications. But, the decrypting interval in CNC may be higher if the terminus nodes cannot obtain enough amount of packets for deciphering [14].

Hence, CNC ought to be cautiously applied in target-controlled packet communications. Compared with CNC, Analog Network Coding (ANC), as a kind of Physical layer Network Coding (PNC), can be added lower diffusion time by permitting two signals to be communicated instantaneously on or after the spring nodules and depend one upon another at the relay nodule [15]. However, ANC has further severe restraints on network topologies (star, mesh, bus, etc.,) and channel conditions. Hence, it is stimulated to combine the unlike program approaches to fullfill their benefits.

4 Problem Statement

This paper is intended to solve following problems,

1. Reduce the packet drop.
2. Ensure more packets are reached in deadline.

The problem is to be solved by integration of scheduling methods and proper scheduling of packets according to their deadline. The main neutral is to find a superlative system adaptively to reduce the amount of packets mislaid their limits rendering to dissimilar network. Packet Dropping Probability can be calculated as the proportion of the amount of packages omitted by their targets to the entire amount of packages that has to be transferred.

5 Proposed System

In direction to agreement with present programs in multi-rate WMSNs, an adaptive incorporation system for unlike communication approaches is proposed, counting CNC, ANC, PR, and NR. Two methods one based on Markov Chain and other based on Dynamic Chart method are proposed, to top quality the optimum transmission approaches and series of packets.

5.1 Markov Chain Approximation Approach

Markov guesstimate has been used to answer the Extreme Biased Configuration delinquent [16, 17], such as, scheming the Carrier Sense Multiple Access (CSMA) device to attain the optimum throughput, selecting path in wire line networks, or deciphering the channel assignment problem in wireless LANs. Many real and key complications can be communicated in the practice of Extreme Biased Configuration delinquent, and so is the optimization problem in this paper. Markov guesstimate method obtains the procedures which are very adjacent to the finest outcomes, when the amount of formal conversions is huge ample. As the extent of the states, i.e., the likely arrangement and broadcast systems, growths hastily by total number of packets, and the parallel density is still very great which is illustrated. Furthermore, a method with less computational complexity is extracted. A transition matrix is defined. In which the number of vertex is the number of source + number of destination. The transition matrix value is 1 when the packet is transferred at time t_i from node m to node n, else it is 0.

The transition matrix is updated in every time interval to check if more packets meet their deadline. According to this, an element in transition is set 1 or 0. The transition is made according to constraint of maximum possible transition possible at that time. Due to this updation of transition matrix at every time interval, a near optimum schedule would be achieved in later time period.

5.2 Dynamic Chart Method

The Dynamic grid centered methodology is introduced and delinquent is solved by building a chart for the collection of suitable packet(s) of each broadcast. Dynamic graph approach is based on following statements,

- Statement 1: Amongst the total optimum well-ordered set barriers, the unintentional that additional packages are supportively transferred through CNC is infrequent.
- Statement 2: In PNC-based optimal ordered set barriers, once a packet in one subset misses its target, the packet(s) in its subsequent subset(s) will also miss its targets, if any.

Hence, based on the overhead statements, three directions on packet forecast and broadcast are defined

- Direction 1: The maximum number of packets in one subset should not be greater than two.
- Direction 2: The packet(s) has to be broadcasted successfully which are scheduled in subset.
- Direction 3: The broadcast of packets should be done sequentially one after the other subset if not the packets of the subsets fails to meet their deadlines.

So, considering all the above directions and statements, the efficiency of one subset broadcast is observed as the change among the number of packets in the division and the number of fallen packets affected by this broadcast. The subset with the maximum broadcast efficiency should be organized with the maximum importance.

6 System Design

6.1 System Architecture

System architecture also called as system planning is the intangible strategy that defines the anatomy and performance of an organization. It describes the organization apparatuses or building wedges and offers a plan using which merchandises can be obtained, and organizations established, which will toil together to appliance the complete organization.

Configuration Based on configuration details like number of nodes, area of simulation, deadlines for packet and transmission range, configuration module generates the TCL script and invokes on the simulator.

Simulator Node communicate through simulator module.

Node Node module has three applications,

- Packet Generator: Packet Generator application generates and sends the multimedia packet, i.e., source node at configured rate.
- Packet Reception: Packet Reception application receives the packet and compare the deadline of packet, with the receive time. If the packet is received at time it is accepted, if the packet misses the deadline the packet is dropped.
- Relay Application: Relay application can be either implement Markov Chain approximation Scheduling approach or can implement Dynamic Graph Scheduling approach.

Measurements Collects the statistics, i.e., the information from simulator, for example how many are sent or how many packets are dropped.

System planning is publicized in Fig. 3 as follows.

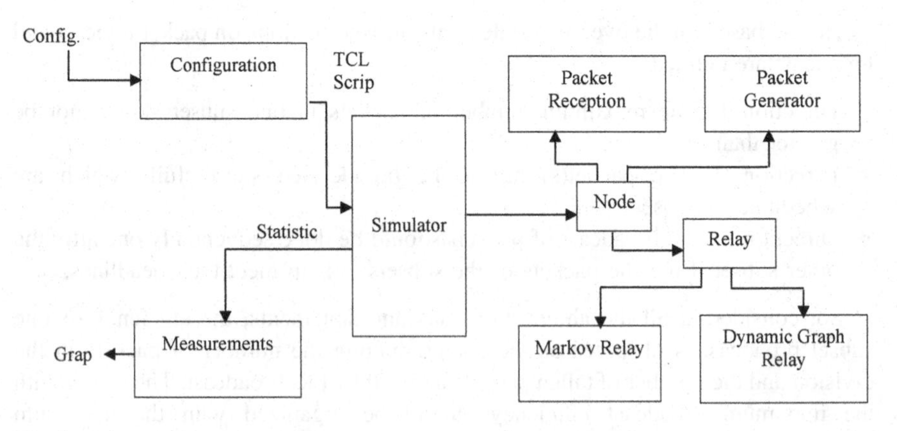

Fig. 3 System architecture

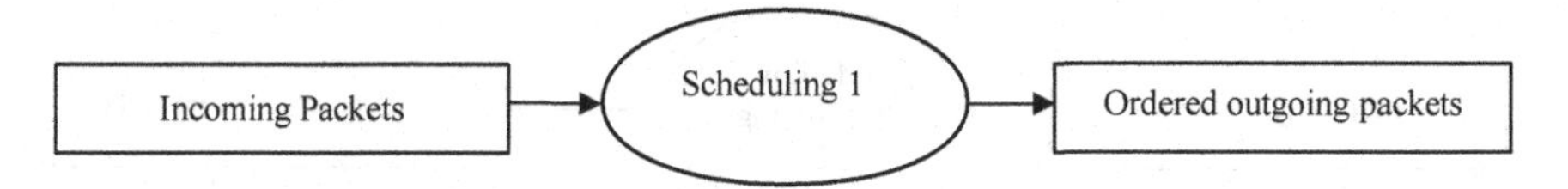

Fig. 4 Level 0 DFD

6.2 *Data Flow Diagram*

A Data Flow Diagram (DFD) is a pictorial demonstration of "stream" of data over a statistics organization. DFDs are being recycled for the conception of information handling (organized project). On a DFD, information objects moves from an exterior data basis or an interior data basis to an inner data basis or an exterior data basis, through an inner procedure.

6.3 *Level 0 DFD*

A perspective level can also be called Level 0 DFD illustrates the interface amid the organization as well as exterior mediators that enact as documents bases and documents descends. Another context illustration is the system's interfaces with the external world are modeled virginally in terms of data flows across the system edge. The context Fig. 4 illustrates the entire system as a distinct process, and gives no evidences as to its interior group.

Scheduling is the main process in the project where we need input as packets and give an output in ordering of packets. Here the input is Incoming Packets which undergo Scheduling process and the output is generated, i.e., ordered outgoing packets.

6.4 *Level 1 DFD*

Level 1 DFD illustrates the reason for division of the structure into sub-structures (procedures), everyone that contracts with one or the other of the data streams from an outward mediator either to an outward mediator, it also deliver total functions of the structure as an entire. It also recognizes interior data basis which should be existing in accordance to complete the task, and displays the stream of facts amid the numerous portions of the structure.

The incoming packets can undergo either Markov chain approximation scheduling method as showed in Fig. 5 or dynamic graph method as showed in Fig. 6 and generate the ordered outgoing packets.

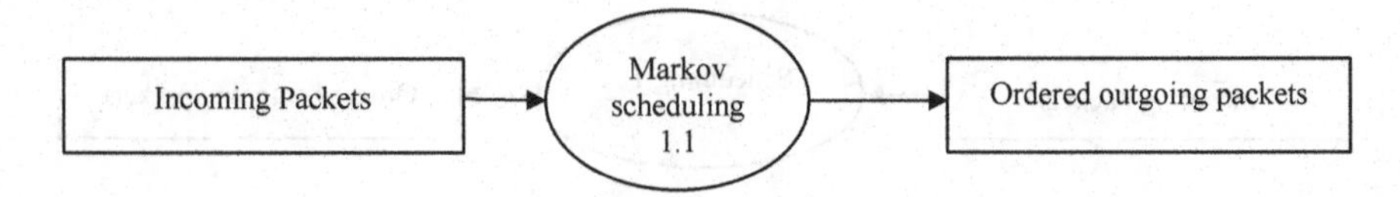

Fig. 5 Level 1 DFD depicting Markov scheduling

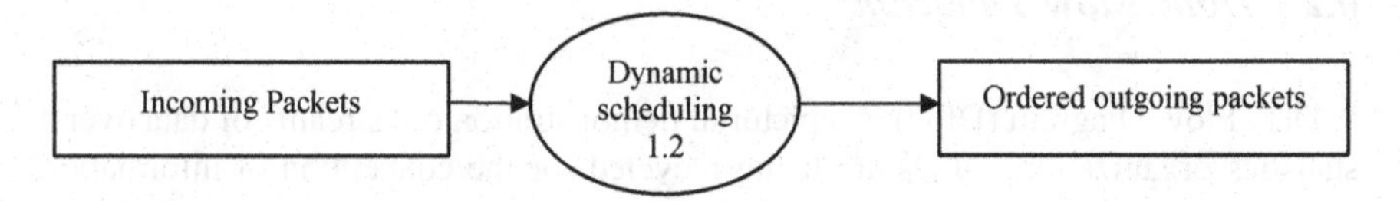

Fig. 6 Level 1 DFD depicting dynamic graph scheduling

6.5 Level 2 DFD

Level 2 DFD illustrates the division of sub structure into substitute procedures, every one of that treaties with one or the other of the records streams from an exterior agent, as well it provides total of the functions of the structure as an entire. Along with that it recognizes interior data basis which should be existing in demand for the structure to perform the task, and illustrates the stream of information amid the numerous fragments of the structure.

Figure. 7 shows the Level 2 DFD is the sub-process depicting the Markov chain approximation scheduling method, where firstly a flow matrix needs to initialized (1.1.1), further the initialized flow matrix needs to optimized (1.1.2) and later packet is scheduled (1.1.3).

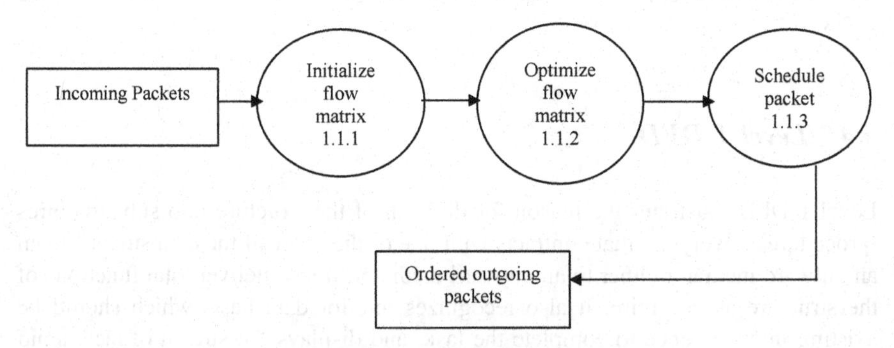

Fig. 7 Level 2 DFD depicting Markov scheduling in detail

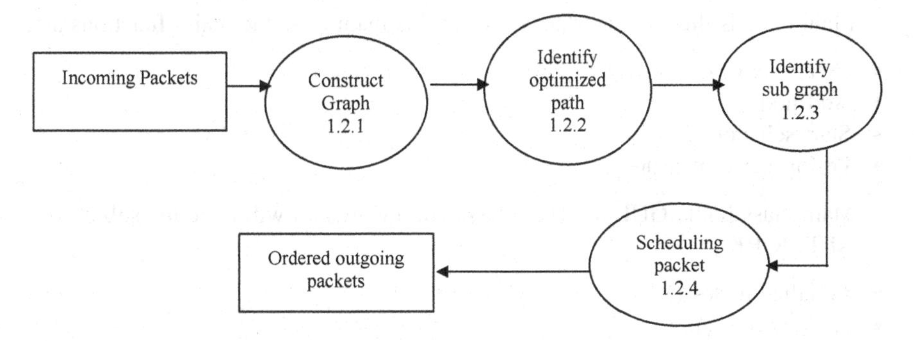

Fig. 8 Level 1 DFD depicting Markov scheduling in detail

Figure. 8 illustrates Level 2 DFD is the sub-process depicting the Dynamic Graph scheduling method, where firstly a graph is constructed (1.2.1), further the optimized path is identified (1.2.2) then the sub graph (1.2.3) is identified and later packet is scheduled (1.2.4).

6.6 Class Diagram

A class figure in the Modeling Language (ML) is a type of stationary organization figure which defines complete construction of a scheme which presents the attributes classes and their relation.

Class figure is illustrated in Fig. 9 as follows,

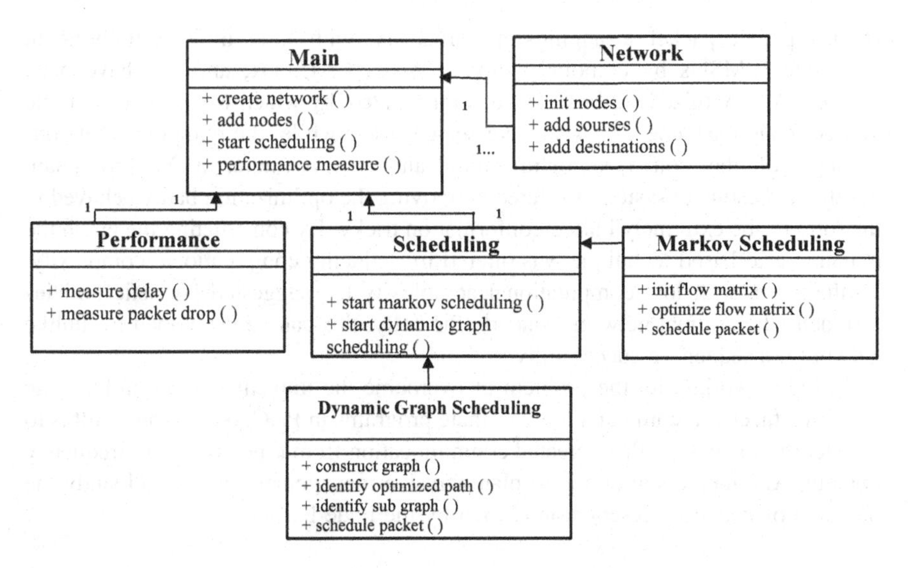

Fig. 9 Class diagram

Main class is the user interface class, in this main class the major functions are,

- Create network
- Add nodes
- Start scheduling
- Performance measure

Main class is the GUI interface class. The main class will have the sub-class.
NETWORK

- Initialize nodes ()
- Add sources ()
- Add destinations ()

SCHEDULING
Scheduling has two sub-process called Markov Scheduling and Dynamic Graph Scheduling.

- Start Markov Scheduling ()
- Start Dynamic Graph Scheduling ()

PERFORMANCE

- Measure delay ()
- Measure packet drop ()

7 Conclusion

In this paper, packet arranging and broadcast with time limit restrictions in multi-rate WMSNs by conjointly linking ANC, CNC, PR, and NR have been focused. An optimized method has been formulated, by which we can excellent the ideal diffusion scheme via exhaustive search. Meanwhile, the computational convolution of the conveyed optimization anomaly is more, a Markov chain constructed estimate system is offered by moving the optimization badly behaved in the form of the extremely biased conformation tricky. By constructing the graph the dynamic grid based technique was offered for reducing computational complexity. Furthermore, with low computational complexity the suggested heuristic policies can tactic the optimal network concert efficiently, this can be developed for unlike network consequences successfully.

In future work, with the problem of overcome the implementation problematic from two facets. In command to coordinate programs in PNC, one of the result is to transfer the forecast of PNC-created communication from time domain to frequency domain. Alternative solution is to plan abrasive management policy and study the outcome of network presentation carried by the node barrier.

References

1. Shen, H., & Bai, G. (2016). Routing in wireless multimedia sensor networks: A survey and challenges ahead. *Journal of Network and Computer Applications*.
2. Ning, Z., Xia, F., Hu, X., Chen, Z., & Obaidat, M. (2016). Social-oriented adaptive transmission in opportunistic internet of smartphones, accepted. In *IEEE Transactions on Industrial Informatics*.
3. Kim, J., Barrado, J., & Jeon, D. (2015). An energy-efficient transmission scheme for real-time data in wireless sensor networks. *Sensors*.
4. Khalek, A., Caramanis, C., & Heath, R. (2015). Delay-constrained video transmission: Quality driven resource allocation and scheduling. In *IEEE Journal of Selected Topics in Signal Processing*.
5. Fortino, G., & Trunfio, P. (2014). *Internet of things based on smart objects: Technology, middleware and applications*. Springer.
6. Chen, H., Chan, H., Chan, C., & Leung, V. (2013). QoS-based cross-layer scheduling for wireless multimedia transmissions with adaptive modulation and coding. In *IEEE Transactions on Communications*.
7. Xiao, Y., Thulasiraman, K., Fang, X., Yang, D., & Xue, G. (2012). Computing a most probable delay constrained path: NP-hardness and approximation schemes. In *IEEE Transactions on Computers*, May 2012.
8. Mao, Z., Koksal, C., & Shroff, N. (2013). Online packet scheduling with hard deadlines in multi-hop communication networks. In *Proceedings IEEE*, April 2013.
9. Anbagi, I. S., Erol-Kantarci, M., & Mouftah, H. T. (2014). Delay-aware medium access schemes for WSN-based partial discharge measurement. In *IEEE Transactions on Instrumentation and Measurement*.
10. Ning, Z., Song, Q., Guo, L., Chen, Z., & Jamalipour, A. (2016). Integration of scheduling and network coding in multi-rate wireless mesh networks: Optimization models and algorithms. *Ad Hoc Networks*.
11. Mali, G., & Misra, S. (2016). TRAST: Trust-based distributed topology management for wireless multimedia sensor networks. In *IEEE Transactions on Computers*.
12. Ning, Z., Liu, L., & Xia, F. (2016). CAIS: A copy adjustable incentive scheme in community-based socially-aware networking. In *IEEE Transactions on Vehicular Technology*.
13. Li, X., Wang, C.-C., & Lin, X. (2011). On the capacity of immediately decodable coding schemes for wireless stored-video broadcast with hard deadline constraints. In *IEEE Journal on Selected Areas in Communications*, May 2011.
14. Yeow, W.-L., Hoang, A. T., & Tham, C.-K. (2009). Minimizing delay for multicast-streaming in wireless networks with network coding. In *INFOCOM*, IEEE, April 2009.
15. Katti, S., Gollakota, S., & Katabi, D. (2007, October). Embracing wireless interference: Analog network coding. *ACM SIGCOMM Computer Communication Review*.
16. Zhan, C., & Xu, Y. (2010). Broadcast scheduling based on network coding in time critical wireless networks. In *IEEE International Symposium on Network Coding (NetCod)*, June 2010.
17. Wang, X., Yuen, C., & Xu, Y. (2012). Joint rate selection and wireless network coding for time critical applications. In *Wireless Communications and Networking Conference (WCNC)*, IEEE, April 2012.

Author Index

V. E. Balas et al. (eds.), *Data Management, Analytics and Innovation*,
Advances in Intelligent Systems and Computing 808,
https://doi.org/10.1007/978-981-13-1402-5

Printed in the United States
By Bookmasters